A FIRST COURSE IN

Calculus

Including Analytic Geometry

Third Edition

By

RANSOM V. LYNCH

The Phillips Exeter Academy

EXETER, NEW HAMPSHIRE

GINN AND COMPANY

Home Office: Lexington, Massachusetts 02173

0–663–41967–0

PICTURE SOURCES

Frontispiece: General Dynamics Corporation
Page 59: Commonwealth of Massachusetts/Image Specialists
Page 128: NASA/Judy Poe
Page 304: Bureau of Reclamation, Department of the Interior/Image Specialists
Page 362: Goddard Space Flight Center/Judy Poe
Page 408: (top) Rutherford Platt
(bottom) Rideout from Black Star

Preface

A year of calculus is perhaps the ideal culmination of a high-school mathematics program. As a summing-up, it calls on all the skills and information acquired in previous years and applies them to problems that are inherently interesting. Though these problems are diverse, the subject itself, in contrast to many twelfth-grade courses, has a natural coherence and unity. For the student who does not plan to continue the study of mathematics in college, a calculus course provides a survey of one of the greatest achievements of the human mind and can be an important element in a liberal education. For the student who is undecided, it provides an excellent sample of the principal concern of collegiate mathematics. For the student who is intent on mathematical or scientific studies, it provides a running start, opens many doors normally closed to college freshmen, and can significantly change his or her entire undergraduate career.

This book is based on extensive classroom experience. A first preliminary version grew out of my own ten years' experience teaching calculus in schools and my consequent dissatisfaction with existing texts. Three years' experience with that text led to a revision, a revision that, in its turn, had three years of use in a variety of classes. The present text reflects not only the many suggestions and criticisms that came out of this extensive classroom use but also a critical analysis by mathematicians. It is a flexible book, adaptable to a variety of programs and classes.

Although this book looks forward to the further study of calculus and to its applications, it also looks back over the high-school years. The course for which the original version was written is the last in an integrated four-year sequence. Accordingly, the book has been planned in such a way that it uses and reinforces the content of previous courses; it shows the student new and interesting applications of the mathematics already learned. Thus, although the text includes a brief, self-contained treatment of analytic geometry (the straight line in Chapter 1, conic sections and other curves in Chapter 8), much of the material in Chapters 2–7 not only assumes that the student already has had some experience with graphs and a slight acquaintance with the conics but is specifically designed to review and extend that experience, in anticipation of the more formal development in Chapter 8. The treatment of the circular functions, in §§ 2.2–2.4, is similarly self-contained, but it is highly condensed and thus makes the pedagogical if not the mathematical assumption that the student already has some familiarity with these functions; indeed, they are used in Chapter 1. Other prerequisites are standard courses in elementary and intermediate algebra and in geometry, including some work in three dimensions.

The first four chapters provide an intuitive introduction to the methods of calculus and to the sorts of problems with which elementary calculus deals. These chapters could be used, in a class that does not have time for a full course, as a substantial unit of what is commonly called "polynomial" calculus. Chapter 1 introduces some of the basic terminology and notation of sets, not with any intention of teaching set theory, but merely to take advantage of the clarity and simplicity that the language of sets provides. In addition, Chapter 1 reviews the basic analytic geometry of the straight line. Chapter 2 introduces the concept of function and some of the algebra of functions, reviews the properties of the circular functions, and introduces the idea of the derivative. Many years of trying to teach the function concept have convinced me that the ordered-pair definition, used in Chapter 2, seems to students the most concrete and is therefore the most easily understood; hence the concept of functions as mappings, which is actually more widely used in calculus, is introduced only secondarily. Chapter 3 illustrates some of the principal applications of the derivative, and Chapter 4 is an intuitive introduction to the nature and use of the indefinite integral. Teachers who wish to have the definite integral available at this point will find it entirely possible to combine with Chapter 4 an informal treatment of §§ 9.1–9.3, but the proofs of Chapter 9 are best postponed until after Chapter 5 has been studied.

A major issue in the teaching of elementary calculus is how much rigor and abstraction are appropriate at what levels. It is particularly true in the high school that students come to this subject at many levels of intellectual maturity and with varying amounts of mathematical motivation. The most able high-school students have tough, vigorous minds that will not forever be satisfied with the postponement of proofs to more advanced courses. On the other hand, many students of only moderate ability can profitably study calculus in high school, and too much rigor too soon may have the effect of scaring many of these out of further mathematical studies. Chapter 5 is designed to serve the needs of both groups. For the more sophisticated and demanding, it provides an introduction to the concepts and methods of proof that are used, in more advanced courses, to develop the mathematical foundations of analysis. The chapter can, however, be presented informally, with little attention to rigorous detail, as a second exposure to the crucial concept of limit and as an intuitive description of continuity; the rest of the book is so arranged that such a presentation should not handicap the student.

Chapters 6 through 12 present the traditional topics of first-year calculus. Chapter 9 is a long one; some teachers may find it desirable to break the chapter by introducing Chapter 10 immediately after § 9.3. Also, §§ 9.4 and 9.7–9.9 may be omitted without affecting the work of subsequent chapters.

Exercises marked [A] and [B] are of parallel content and only moderate difficulty; in some instances, concepts developed in these exercises are used in later work. Where additional practice may be desired, sets of Miscellaneous Exercises have been provided; these cover a considerable range of difficulty. A few particularly demanding exercises are marked [C]. Answers to most of the exercises are in the back of the book.

I am indebted to Professor Angus E. Taylor for many valuable suggestions, and to Professor Edward J. Cogan, whose careful scrutiny of the manuscript has saved me from many blunders; they have not, however, seen the final manuscript, and any blunders that may remain are entirely my responsibility. I am also grateful to my colleagues at Exeter and to our students; for six years they have taught and studied successive approximations to the following pages, and if these approximations have converged to a teachable result, much of the credit is theirs.

R. V. L.

Preface to the Third Edition

In preparing this third edition I have again tried to adapt the text to current practice in the teaching of beginning calculus, insofar as a consensus can be determined. The changes made since the first edition include the earlier introduction, in Chapter 4, of the definite integral and the fundamental theorems of calculus. Chapter 5 has been completely rewritten, with a view to improving its organization and decreasing its difficulty, and also to make possible the introduction, in § 5.5, of L'Hôpital's Rule for quotient forms. Material on rational approximations to integrals and to functions has been added in Chapters 9 and 10; the teacher who is so inclined will find these chapters a source of interesting problems for the computer or the programmable calculator. Nonquotient applications of L'Hôpital's Rule have also been added to Chapter 10.

My debt to the students and teachers who have used this book and taken the trouble to offer suggestions for its improvement has multiplied over the years. I welcome this opportunity to acknowledge that debt and express my thanks. To a considerable extent, this is their book.

R. V. L.

To the Student

This book has been written for you to read. You must recognize, however—perhaps you have already learned this—that reading mathematics is very different from ordinary reading. Read with pencil and paper at hand, writing out any steps that have been left for you to supply. Read a proof through several times, if necessary; be sure, before you are through, that you understand every step. At various points you will find references to previous results; if you don't remember them, *look them up.* The most important results are numbered in such a way as to make this easy: within a given section, key equations are numbered consecutively in parentheses, at the right, and they are referred to thereafter by these numbers; in later sections, they are referred to by these numbers and by the number of the section in which they occur, so that § 1.4, equation (3), is the equation numbered (3) in the fourth section of Chapter 1. Similarly, Theorem 5.3-1 is the first theorem in the third section of Chapter 5.

You will find answers to many of the exercises in the back of the book. It is my conviction that students who are mature enough to study calculus are mature enough to use answers constructively. Don't cheat yourself by looking first at the answer. Work the problem first, then look at the answer, and, if you were wrong, *try again.*

A professor of English once said to me, "No one who is ignorant of calculus can be called educated." I hope you will think of your study of this book not only as a key to science or to more mathematics, but also as an important element in a liberal education, an introduction to one of the noblest creations of the human mind.

R. V. L.

Contents

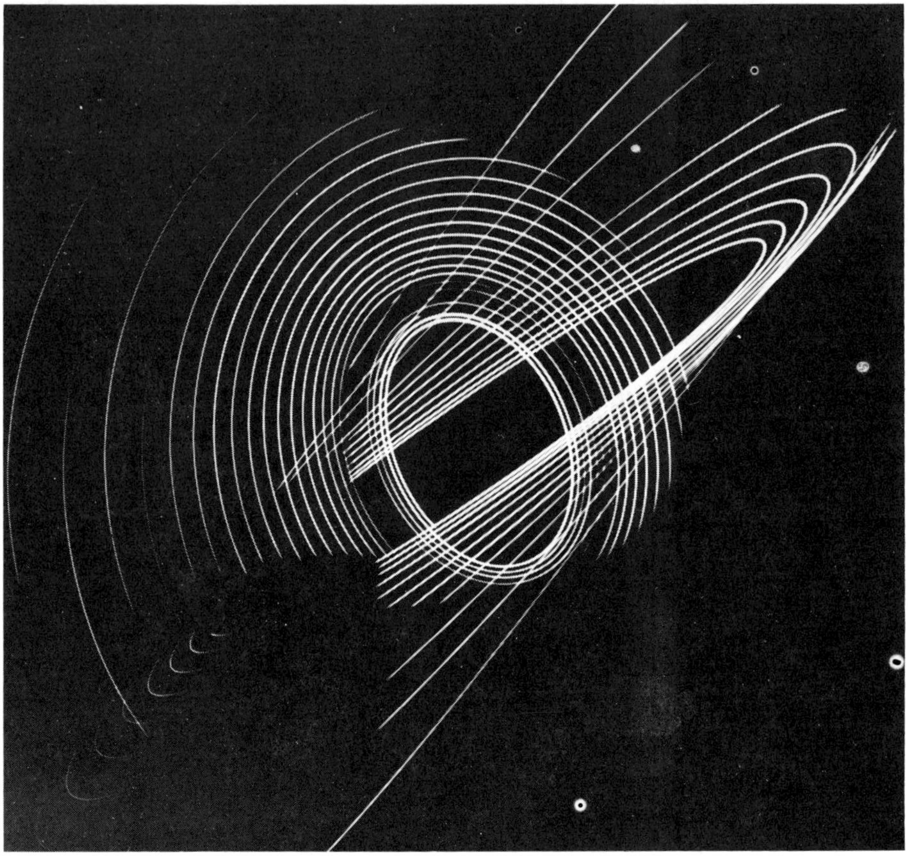

Through the use of calculus, it is possible to determine what the orbit of an earth satellite will be before the satellite is launched. The photograph shows a laboratory simulation of some possible orbits.

1

Introduction and Review

1.1 Foreword

Calculus, like plane geometry or any other branch of mathematics, is an abstract system consisting of certain assumptions, called axioms or postulates, and of some of the logical consequences of those assumptions, called theorems. The assumptions which form the basis for geometry are a set of statements about elements called points and lines, and these elements therefore form the subject matter of plane geometry. The assumptions at the foundations of calculus are a set of statements about elements called real numbers, and these therefore form the subject matter of calculus. These elements are familiar, for they also form the subject matter of arithmetic and of elementary algebra. Calculus, however, uses certain properties of real numbers largely ignored in more elementary work, to develop the idea of a *limit*. This concept, which is the root idea of calculus, leads to techniques of tremendous power, applicable to a great variety of physical and mathematical problems.

It would be possible for us to trace the logical development of calculus from its basic assumptions, proving each theorem in rigorous detail, but this important and interesting task is better postponed to a more advanced course. The development which follows will therefore rely heavily on intuition and proceed with a minimum of proof and of mathematical rigor.

Though calculus is fundamentally concerned only with certain properties of numbers, it was originally developed as a tool for solving problems in physics and geometry, and it still finds in those fields not only a continuing source of problems but also a major aid to understanding. Most of what follows will therefore be based on geometrical arguments. Hence, as soon as we have introduced some basic terms and notation, we shall review some of the fundamentals of analytic geometry and then use them as a point of departure for calculus.

1.2 Some basic terms

Set. Recognition and classification are among the most fundamental of mental activities. These particular activities involve a concept which is basic in mathematics, the concept of a set. Some familiar examples are the set of pupils in your mathematics class, the set of Sundays in October, and the set of prime numbers.

We shall not attempt a formal definition of the term "set," nor of the concept of belonging to a set. Informally and intuitively, we shall agree that a set has been adequately described when we have some way of deciding that a given object does or does not belong to the set. If the set contains only a finite number of elements, they can be listed, and we can determine whether an object does or does not belong to the set by comparing it with the list. Thus, for example, the set A of decimal digits can be described as

$$A = \{0, 1, 2, 3, 4, 5, 6, 7, 8, 9\}.$$

Note the use of a single letter, in this case A, as a name for the set, and also the use of braces, $\{\quad\}$, to enclose the names of the elements of the set; these are conventions in general use. In the same way, we can designate the set of positive prime numbers less than 10 as

$$B = \{2, 3, 5, 7\}.$$

To indicate that an object belongs to a specified set we use the symbol $\in$, which is a modification of the Greek letter ϵ (epsilon). Thus, for example, the notation "$x \in S$" should be read "x is an element of S," or "x belongs to S." With reference to the sets A and B, defined above, we may say $6 \in A$ (6 is an element of A), $3 \in B$ (3 belongs to B), and so on. To indicate that an object does not belong to a specified set, we draw a line through the $\in$ symbol: thus, for example, $4 \notin B$ (4 does not belong to B). Compare the familiar notation $\neq$ for "is not equal to" or "does not equal."

Two sets are the same if and only if they contain exactly the same elements; this is a definition of what we mean by equality of sets. Hence the order of listing elements is immaterial, although one order may be more convenient than another, and repetitions, since they have no significance, are ordinarily eliminated. For example,

$$\{2, 7, 5, 3\} = \{2, 2, 3, 5, 3, 7\} = B,$$

but

$$\{2, 3, 7\} \neq B.$$

To list all the elements of a set is inconvenient if their number is large, impossible if it is infinite. In such cases, a set can be specified by some description or rule. Thus, for example, the set C of all positive integers is specified by the description "the set of all positive integers." A conventional way of writing this description, convenient in practice, is

$$C = \{x: x \text{ is an integer and } x > 0\},$$

read "C is the set of all x such that x is an integer and x is greater than 0." In general, the notation

$$\{\_\_\_: \_\_\_\_\_\},$$

called the set-builder, is read "the set of all . . . such that . . .". * Another example of the use of the set-builder notation is

$$D = \{y: y \text{ is a real number and } -1 \leq y \leq 1\}.$$

We can also use this notation for the sets A and B described above:

$$A = \{z: z \text{ is a decimal digit}\};$$

$$B = \{r: r \text{ is a prime number and } 0 < r < 10\}.$$

There may be many ways to describe the same set; thus

$$\{x: x \text{ is a real number and } x^2 - 4x + 4 = 0\} = \{2\}.$$

Sets can be combined in a variety of ways; indeed, there is a substantial and significant algebra of sets. The only operations of this algebra that we shall need to use, however, are *union* and *intersection*.

Definitions.

If A and B are sets, then the union of A and B, written $A \cup B$, is the set of all elements that belong to at least one of A and B:

$$A \cup B = \{x: x \in A \text{ or } x \in B \text{ (or both)}\}.$$

If A and B are sets, then the intersection of A and B, written $A \cap B$, is the set of all elements that belong to both A and B:

$$A \cap B = \{x: x \in A \text{ and } x \in B\}.$$

Variable. The formulas of mathematics afford many examples of the use of a letter to represent any one of a set of numbers. In the formula for the area of a circle, $A = \pi r^2$, for example, it is generally understood that the radius r can be any positive number. In this case, the letter r is used as a *variable.*

* A common alternative form of the set-builder notation uses a vertical line where we use a colon. In this notation the set C would be described as "$C = \{x | x \text{ is an integer and } x > 0\}$." We shall use the colon, in order to avoid confusion with the vertical-line notation for absolute value.

Definition. A variable is a symbol that stands for, and may be replaced by, any element of a specified set.* This set is called the replacement set, or the domain, of the variable.

In calculus as in algebra, most of the variables used are letters of some alphabet, and most of them have sets of real numbers as their domains. When there is no danger of confusion, we shall therefore not bother, in describing domains, to keep repeating such specifications as "x is a real number." Thus, in describing such a set as D, above, we shall simply say $D = \{y: -1 \leq y \leq 1\}$.

It is important to note that D is a set of numbers. The variable y appears in our description of D, but $y \notin D$; it is a so-called "dummy" variable, and almost any other symbol could replace it. Thus, $\{x: -1 \leq x \leq 1\}$, $\{\Delta: -1 \leq \Delta \leq 1\}$, and even $\{\text{John Doe}: -1 \leq \text{John Doe} \leq 1\}$ are all names of the same set D.

The number line. The idea of numbering the points on a line is familiar to anyone who has used a ruler. The basic principle involved, which we assume, is that it is possible to construct a scale on any line in such a way that there is a unique real number corresponding to each point and a unique point corresponding to each real number. A line with such an associated scale is called a number line, the number associated with each point is the coordinate of the point, and the set of all coordinates on a line is a one-dimensional coordinate system; the point with coordinate zero is the origin of the system. The correspondence thus established between points and numbers is such a natural one that we often find it convenient to use the word "point" to refer to a number.

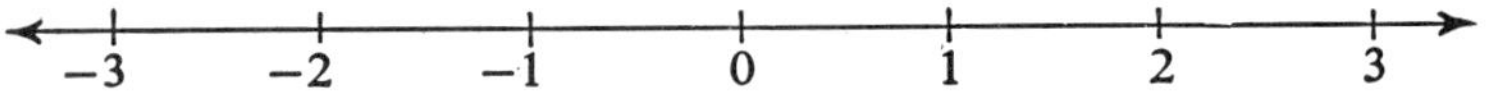

Fig. 1

We shall also assume that, associated with each pair of points on a number line, there is a unique non-negative number, the distance between the points, which we shall define in terms of the difference of their coordinates. Thus, for example, we shall take the distance between the points (numbers) 3 and -2 to be $3 - (-2) = 5$. We wish, however, to consider only the size of a distance, not its direction, and if we should happen to compute this dis-

* Strictly speaking, a variable is replaced by a symbol representing a specific element of its domain, and not by the element itself. We do not write *numbers* in place of variables in a formula, we write *numerals*, which are symbols for numbers. We shall not insist, in this text, on the distinction between things and the symbols which represent them.

tance in the form $(-2) - 3 = -5$, we would suppress the minus sign in the result and again write the distance as 5. With particular, specific numbers we can always perform the subtraction in such a way as to avoid a negative result, but we run into a difficulty if we try to make general statements about the distance between *any* two numbers. Is the distance between x_1 and x_2 the number $x_1 - x_2$ or is it $x_2 - x_1$? Without some way of knowing which of x_1 and x_2 is the larger, we can't tell. What we need is a measure of distance that will be positive if $x_1 \neq x_2$ and zero if $x_1 = x_2$. Such a measure is the absolute value. If x is any real number, then the absolute value of x, written $|x|$, is defined thus:

Definition. $$|x| = \begin{cases} x & \text{if } x \geq 0. \\ -x & \text{if } x < 0. \end{cases}$$

Thus, for example, $|2| = |-2| = 2$, and if $x = -3$, then $|x| = -x = -(-3) = 3$. An equivalent definition is

$$|x| = \sqrt{x^2},$$

since, by definition, the radical sign always indicates the non-negative square root. Thus, for example, $|-2| = \sqrt{(-2)^2} = \sqrt{4} = 2$. Consequences of either definition are that the absolute value of a number is never negative, that the absolute value of each number is the same as the absolute value of its negative, and hence that $|x_1 - x_2| = |x_2 - x_1|$. We may therefore make the following definition:

Definition. On a number line, the distance between the point with coordinate x_1 and the point with coordinate x_2 is $|x_1 - x_2|$ (or $|x_2 - x_1|$).

Since $|x| = |x - 0|$, it follows that $|x|$ may be interpreted as the distance between the point x and the origin.

The student will find it useful, in what follows, to have some facility in working with equations and inequalities which involve absolute values. A summary of some of the rules for manipulating inequalities may be helpful:

1. If $a > b$ and $b > c$, then $a > c$.
2. If $a > b$, then $a + c > b + c$.
3. If $a > b$ and $c > d$, then $a + c > b + d$.
4a. If $a > b$ and $c > 0$, then $ac > bc$.
4b. If $a > b$ and $c < 0$, then $ac < bc$.
5. If $a > b > 0$, then $1/a < 1/b$.

A relationship of fundamental importance in calculus, called the triangle inequality, involves absolute values.

▶ **Triangle Inequality.** For all real numbers x and y, $|x + y| \leq |x| + |y|$.

The student is asked to supply a proof of this inequality in Ex. 8, page 10.

Example 1. For what x is $|x| = 6$?

Solution: A geometrical approach to this problem is easy and intuitive. Since $|x|$ is the distance on the number line between the point x and the origin, and since the only points 6 units from the origin are 6 and -6, we have $x = \pm 6$.

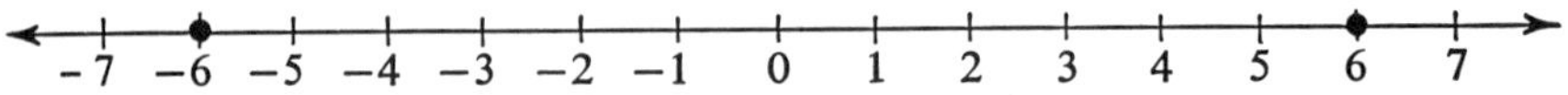

Fig. 2

An algebraic approach is both more formal and more cumbersome: Either $x \geq 0$, in which case $|x| = x$ and $\underline{x = 6}$, or $x < 0$, in which case $|x| = -x$ and $-x = 6$, or $\underline{x = -6}$.

The algebraic approach seems like rather a lot of trouble for this particular problem, but the technique will be found useful in more complicated problems, where geometrical intuition may fail.

Example 2. For what x is $|x| < 6$?

Solution: In terms of distance, the question asks for those points which are less than 6 units from the origin; it should be intuitively clear that these are just the numbers between -6 and 6, and the solution can therefore be concisely indicated as $-6 < x < 6$. (Note the use of *open* dots in the diagram to show that -6 and 6 are excluded.)

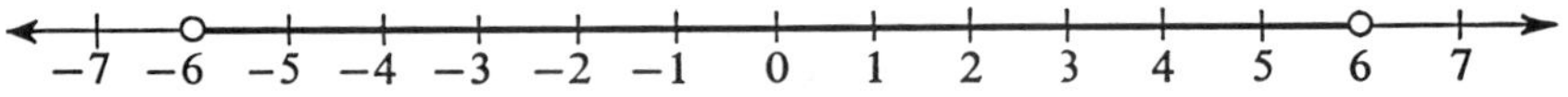

Fig. 3

Algebraically, if $x \geq 0$, then $|x| < 6$ means $x < 6$, and the given inequality will therefore be satisfied if $0 \leq x < 6$. If, alternatively, $x < 0$, we have $|x| = -x$, and the given inequality is equivalent to $-x < 6$ or (multiplying both sides by -1 and applying Rule 4b, above) to $x > -6$; hence the given inequality will also be satisfied if $-6 < x < 0$. Combining these results, we have $-6 < x < 6$, as before.

The set $\{x: -6 < x < 6\}$ is called an *interval*, and the points -6 and 6 are the *endpoints* of the interval. Are the endpoints elements of this interval?

Example 3. For what x is $|x - 3| \leq 5$?

Solution: Since $|x - 3|$ is, by definition, the distance between the point x and the point 3, we are asked for those points whose distances from 3 are not greater than 5. These are the points which lie from 5 units less than 3, or -2, to 5 units more than 3, or 8; that is, $-2 \le x \le 8$.

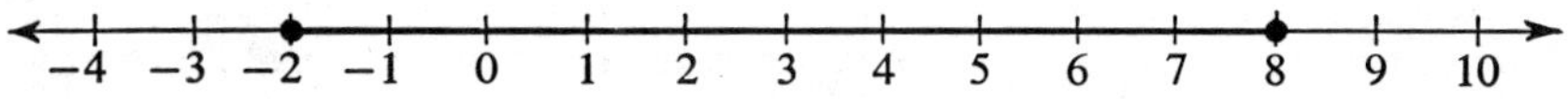

Fig. 4

Alternatively, if $x - 3 \ge 0$, then $|x - 3| = x - 3$ and hence $0 \le x - 3 \le 5$; whereas if $x - 3 < 0$, then $|x - 3| = -(x - 3)$ and hence $-(x - 3) \le 5$, or $-5 \le x - 3 \le 0$. Combining these results, we have

$$-5 \le x - 3 \le 5,$$

or, adding 3,

$$-2 \le x \le 8 \qquad \text{(by Rule 2).}$$

The set $\{x\colon -2 \le x \le 8\}$ is also an interval, with endpoints -2 and 8. Are the endpoints elements of this interval?

Example 4. Solve for x: $|x|^2 - 5|x| - 6 = 0$.

Solution:

$$(|x| - 6)(|x| + 1) = 0;$$

$$|x| = 6 \quad \text{or} \quad |x| = -1.$$

If $|x| = 6$, we have $x = \pm 6$ from Example 1; since the absolute value of a real number is, by definition, a non-negative number, the equation $|x| = -1$ has no real solution.

Intervals and neighborhoods. Not only are most of the domains encountered in calculus sets of real numbers, as noted above; they are often sets of a special sort, called intervals. An interval, roughly speaking, is a piece of the number line, as was suggested in Examples 2 and 3. It may consist of all the points (numbers) of the number line which lie between two given points, called the endpoints of the interval; in this case, it is called a bounded interval. If the given points are a and b, $a < b$, then the interval is denoted $[a, b]$ if the endpoints are to be considered a part of the interval, and (a, b) if they are not. Since it is also possible to include only one endpoint, we shall have occasion to use the notations $[a, b)$, indicating that a is an element of the interval but b is not, and $(a, b]$, indicating that b belongs in the interval but a does not. The set of all points to one side of a given point is an unbounded interval, and the given point is its endpoint. The set of all points to the left of a is denoted $(-\infty, a]$ if a is included in the interval, and $(-\infty, a)$ if it is not, and the set of all points to the right of a is either $[a, \infty)$ or (a, ∞), according as a is or is not an element of the set. Finally, it is convenient to consider the set of all real numbers to be an unbounded interval; it may be denoted $(-\infty, \infty)$. In general, the square bracket, "[" or "]", is used to indicate that a corresponding endpoint is an element of the interval, and the round parenthesis, "(" or ")", to indicate that the endpoint is not included or that there is no endpoint. The symbols $-\infty$ and ∞ (sometimes written $+\infty$) do not represent numbers and are therefore not endpoints of intervals. The preceding discussion is summarized and illustrated by sketches in the following definition:

Definition. If a, b, and x are real numbers, $a < b$, then the following sets are *intervals*:

$(-\infty, a) = \{x\colon x < a\}$

$(-\infty, a] = \{x\colon x \leq a\}$

$[a, b) = \{x\colon a \leq x < b\}$

$[a, b] = \{x\colon a \leq x \leq b\}$

$(a, b] = \{x\colon a < x \leq b\}$

$(a, b) = \{x\colon a < x < b\}$

$[b, \infty) = \{x\colon b \leq x\}$

$(b, \infty) = \{x\colon b < x\}$

$(-\infty, \infty) = \{x\colon x \text{ is a real number}\}$

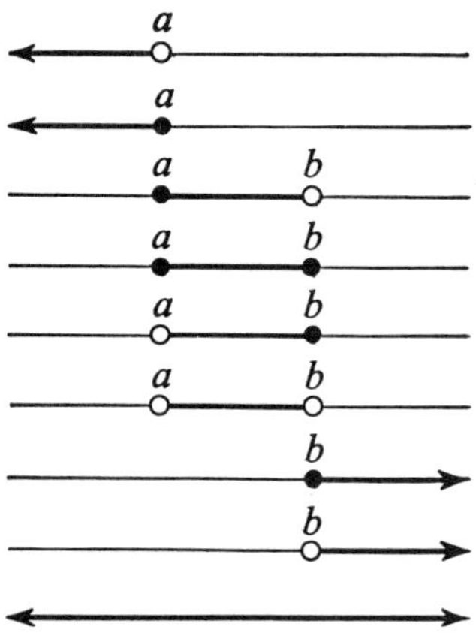

Each of a and b is an *endpoint* of each interval in whose definition it occurs. An interval with two endpoints is said to be *bounded*; all others are *unbounded*. A bounded interval is *closed* if, like $[a, b]$, it contains both of its endpoints; it is *half-open* if, like $[a, b)$ or $(a, b]$, it contains exactly one of its endpoints; it is *open* if, like (a, b), it contains neither endpoint.

We shall have frequent need, in what follows, of a particular type of interval, a *neighborhood.*

Definition. A *neighborhood* of the real number c is an open interval of which c is an element.

In most cases, it will be convenient to consider symmetric, or *circular*, neighborhoods of c, that is, neighborhoods of which c is the midpoint. If $h > 0$, such a neighborhood can conveniently be described as $(c - h, c + h)$, or as $\{x\colon |x - c| < h\}$.

Finally, we shall also need the concept of a neighborhood of c from which c itself has been deleted; such a set is a *deleted*, or *punctured*, neighborhood.

Definition. A <u>punctured neighborhood</u> of c is a neighborhood of c from which c alone has been removed.

If $h > 0$, such a set is conveniently described as $\{x: 0 < |x - c| < h\}$.

Exercises 1.2 [A]

1. Describe each of the following sets by listing its elements:
   (a) $\{x: x \text{ is an even integer and } x \in (20, 30)\}$
   (b) $\{z: z \text{ is an even integer and } z \in [20, 30]\}$
   (c) $\{s: s \text{ is an integer and } |s| < 2\}$
   (d) $\{y: y^3 - y = 0\}$
   (e) $\{x: x$ wrote *Treasure Island*$\}$

2. Use the set-builder notation to describe
   (a) $\{1, -1\}$.
   (b) $\{11, 13, 17, 19\}$.
   (c) $\{0\}$.
   (d) {Maine, New Hampshire, Vermont, Massachusetts, Rhode Island, Connecticut}.

3. Solve for x: (a) $|x| = 4$
   (b) $|x| = -3$
   (c) $|x + 2| = 3$
   (d) $2|x - 1| = 5$

4. (a) If $A = \{1, 3\}, B = \{2, 4\}$, and $C = \{3, 5\}$, describe $A \cup B, B \cup C$, and $A \cup C$.
   (b) Prove or disprove: The union of two intervals is an interval.
   (c) What can you say of sets A and B if $A \cup B = B$?
   (d) Show that a punctured neighborhood is the union of two open intervals.

5. Express as an interval or a union of intervals, and represent by a diagram:
   (a) $\{x: |x - 1| \leq 1\}$
   (b) $\{y: y < 1 \text{ or } y \geq 3\}$
   (c) $\{z: |z - 2| < 3\}$
   (d) $\{t: 1 < |t - 1| < 2\}$
   (e) $\{x: 0 < |x - 1| < 1\}$
   (f) $\{x: 0 \leq |x - 1| < 1\}$

6. Solve each of the following equations for x: (a) $|x|^2 - |x| - 12 = 0$; (b) $|x|^2 + 7|x| + 12 = 0$; (c) $|x|^3 - 4|x| = 0$.

7. By considering all possible combinations of sign, prove

(a) $|x| \cdot |y| = |xy|$ (b) $\dfrac{|x|}{|y|} = \left|\dfrac{x}{y}\right|$

8. Prove the triangle inequality, $|x + y| \leq |x| + |y|$. Consider the following cases:

(a) x and y have the same sign.

(b) $x > 0$, $y < 0$, and $|x| > |y|$.

(c) $x \geq 0$, $y < 0$, and $|x| < |y|$.

Why is it unnecessary to consider in detail $x < 0$, $y \geq 0$?

9. Solve for x: $|x| + |x - 1| = 3$. (*Hint:* You must consider three possibilities: $x < 0$, $0 \leq x < 1$, and $x \geq 1$.)

10. Solve for x: $\dfrac{1}{x} < 2$, expressing your answer as an interval or a union of intervals. (*Hint:* Rules 4a and 4b for inequalities must both be considered.)

Exercises 1.2 [B]

1. Describe each of the following sets by listing its elements:

(a) $\{t: t \text{ is an integer and } t^2 < 10\}$

(b) $\{z: z \text{ is an integer and } |z| < 3\}$

(c) $\{x: x \text{ is an integer and } |x + 2| < 4\}$

(d) $\{r: r^2 - 6r + 6 = 0\}$

(e) $\{y: 2y^3 - 5y^2 + 4y - 1 = 0\}$

2. Express as an interval or a union of intervals, and represent by a diagram:

(a) $\{y: |y + 3| < 1\}$

(b) $\{x: 0 < |x + 1| < 2\}$

(c) $\{x: \sqrt{x^2} = x\}$

(d) $\{z: |2z - 5| < 3\}$

(e) $\{t: t^2 - 4t > 0\}$

(f) $\{x: 0 < |x - 5| < 4\}$

3. Solve: (a) $15|y|^2 + |y| - 6 = 0$

(b) $|x|^2 + 2|x| - 1 = 0$

(c) $|s|^3 - 2|s|^2 - 5|s| + 6 = 0$

4. Prove: $|x| - |y| \leq |x - y| \leq |x| + |y|$.
(*Hint:* Use the triangle inequality.)

5. Solve for x: $|x| + |x - 2| = 6$.

6. (a) Simplify $(-1, 2) \cap [1, 3)$.
   (b) Simplify $[-2, 2] \cap (-3, 5)$.
   (c) What can you say of sets A and B if $A \cap B = A$?
   (d) Is the intersection of two intervals always an interval?

7. Solve for x: $\dfrac{1}{x - 2} > 3$.

8. Solve for x: $|x + 2| + |x| < 5$.

9. Show that an open interval has neither a largest nor a smallest element. (*Hint:* For each $x \in (a, b)$ find a $y \in (a, b)$ such that $y > x$, and a $z \in (a, b)$ such that $z < x$.)

1.3 Coordinates in a plane; the straight line

A two-dimensional coordinate system in a plane is commonly established as follows: A pair of perpendicular number lines, called axes, is chosen in such a way that the scale on each axis has its origin at the intersection of the axes. One of these axes, the axis of abscissas, is often called the x-axis, and the other, the axis of ordinates, is often called the y-axis, though the letters x and y may be replaced by any others. If any pair of real numbers is given, such as $(2, -3)$, this pair, in the given order, defines a unique point which can be found thus: draw a perpendicular to the x-axis at the point with coordinate 2, and a perpendicular to the y-axis at the point with coordinate -3; the intersection of these lines (marked P in Fig. 5) is the required point. The number 2 is called the abscissa, or x-coordinate, of the point, and the number -3 is the ordinate, or y-coordinate, of the point.

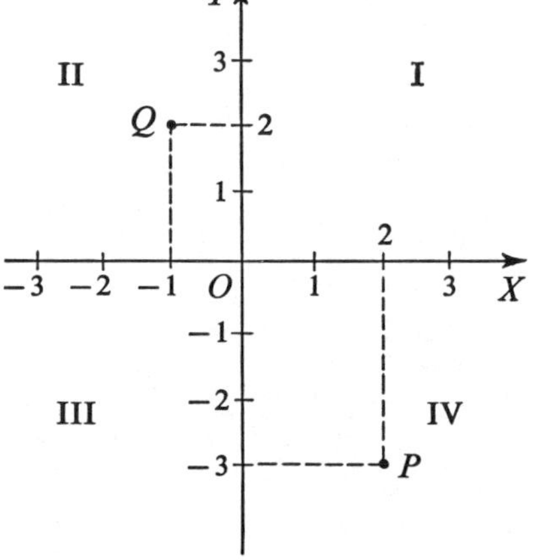

Fig. 5

Conversely, it is possible to determine the coordinates (abscissa and ordinate) of any point Q by drawing through Q lines perpendicular to the axes and reading, from the two scales, the numbers at which these lines meet the axes. In Fig. 5, the point Q has coordinates $(-1, 2)$. Note that coordinates are, by convention, always indicated as a pair of numbers in parentheses, with the abscissa first, so that $(-1, 2)$ means (abscissa -1, ordinate 2) and is not the same as $(2, -1)$ = (abscissa 2, ordinate -1); the order of the numbers is a significant and essential part of the notation.

By virtue of the coordinate system thus established, to every ordered pair of real numbers there corresponds a unique point of the plane, and to every point there corresponds a unique ordered pair of numbers; the numbers are called (rectangular) *Cartesian coordinates* of the point. It is often convenient to identify the ordered pair with the point and speak of "the point (a, b)" rather than "the point with coordinate-pair (a, b)."

There is evidently some risk of confusing the coordinate-pair (a, b) with the open interval (a, b). We must therefore be careful to state explicitly which is meant whenever the context fails to make this point clear.

The axes divide the plane into four regions, called *quadrants*, which are conventionally numbered I, II, III, and IV, as shown in Fig. 5.

Distance in the plane. Consider any two points, $P_1(x_1, y_1)$ (read "P-one, with coordinates x-one, y-one") and $P_2(x_2, y_2)$. Draw through P_1 a line parallel to the y-axis and through P_2 a line parallel to the x-axis. The intersection Q of these lines has coordinates (x_1, y_2). All points of the line P_2Q

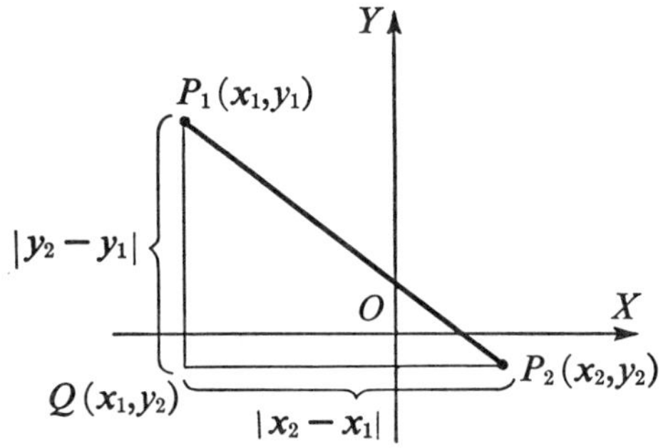

Fig. 6

have the same ordinate, y_2, and their abscissas can be used as a one-dimensional coordinate system on P_2Q. In terms of this coordinate system, the distance P_2Q, as defined in § 1.2, is $|x_2 - x_1|$. Similarly, the distance P_1Q is $|y_2 - y_1|$. The familiar Pythagorean theorem then leads us to the following formula:

▶ The <u>distance</u> between the points $P_1(x_1, y_1)$ and $P_2(x_2, y_2)$ is

$$P_1P_2 = \sqrt{(x_2 - x_1)^2 + (y_2 - y_1)^2}. \tag{1}$$

Midpoint. It is left to the student as an exercise (Ex. 7 of Exercises 1.3 [A]) to verify the following formula:

▶ The <u>midpoint</u> of the segment joining $P_1(x_1, y_1)$ and $P_2(x_2, y_2)$ is

$$P\left(\frac{x_1 + x_2}{2}, \frac{y_1 + y_2}{2}\right). \tag{2}$$

Slope. We define the slope of a line as follows:

▶ **Definition.** If $P_1(x_1, y_1)$ and $P_2(x_2, y_2)$ are any two points on a line L which is not perpendicular to the x-axis, then the <u>slope</u> of L is the ratio

$$m = \frac{y_2 - y_1}{x_2 - x_1} = \frac{y_1 - y_2}{x_1 - x_2}. \tag{3a}$$

A line which is perpendicular to the x-axis does not have a slope.

The quantity $y_2 - y_1$, the difference of ordinates of P_1 and P_2, is commonly designated Δy (read "delta y"). Note that the symbol Δy does not indicate the product of two numbers, Δ times y, as it would in elementary algebra, but the difference between two values of y. Similarly, Δx denotes $x_2 - x_1$, the difference between the two corresponding values of x. In this notation, the formula for slope is

▶ $$m = \frac{\Delta y}{\Delta x}. \tag{3b}$$

If P_1 and P_2 are chosen in such a way that $\Delta x = 1$, then $m = \Delta y$; m, in other words, denotes the difference in y-values (ordinates) which corresponds to a unit difference in x-values (abscissas). The ratio m is sometimes called the <u>rate</u> at which y changes with x.

If a line slants up toward the right, down toward the left, then Δy and Δx will be both positive or both negative, and the slope m will therefore be positive. Similarly, a line which slants down to the right and up to the left has a negative slope. On a line perpendicular to the y-axis (often called a <u>horizontal</u> line), the ordinates of all points are the same; hence $\Delta y = 0$ and $m = 0$. On a line perpendicular to the x-axis (often called a <u>vertical</u> line) the abscissas of all points are the same; hence, for any two of its points, $\Delta x = 0$, and, since division by zero is impossible, our definition of slope fails for such a line. It is important to distinguish between these two cases: a horizontal line has a slope, namely 0, whereas a vertical line has no slope.

The particular choice of the points P_1 and P_2 on the line L does not affect the value of the slope. Thus, in Fig. 7, the choice of P_1 and P_2 gives $\Delta y = NP_2$, $\Delta x = P_1N$, and therefore $m = NP_2/P_1N$. The choice of Q_1 and Q_2, instead, gives $m = MQ_2/Q_1M$, but since the triangles P_1NP_2 and Q_1MQ_2 are similar, these two values of m are the same. This property, that the slope is the same no matter what two points are used to determine it, is a unique and definitive property of the straight line.

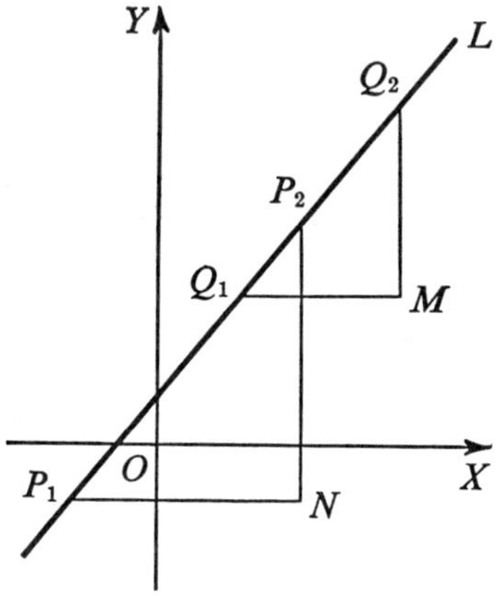

Fig. 7

If the scales on both axes are the same, we have the following definition:

► **Definition.** The inclination of a line which meets the x-axis is the angle α, measured in the positive (counterclockwise) direction, from the x-axis to the line. A horizontal line has inclination 0°.

It follows that $0° \leq \alpha < 180°$.

The relation between slope and inclination is a simple one. As can be seen from Fig. 8, if α is acute, then

$$\tan \alpha = \frac{\Delta y}{\Delta x} = m.$$

If α is obtuse, as in Fig. 9, it follows from trigonometry * that

$$\tan (180° - \alpha) = \frac{|\Delta y|}{|\Delta x|} = |m|.$$

* It is assumed that students using this book have a knowledge of trigonometry. However, the circular functions are discussed in some detail in §§ 2.2, 2.3, and 2.4.

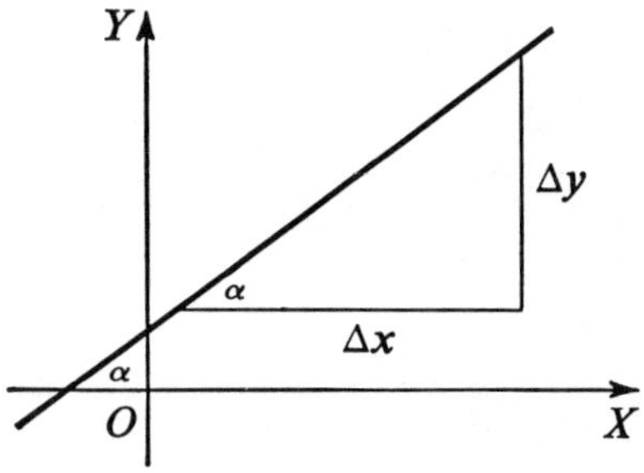

Fig. 8

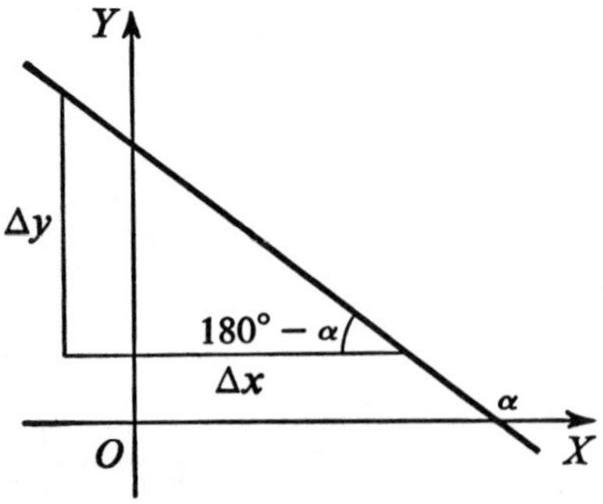

Fig. 9

But $|\tan (180° - \alpha)| = |\tan \alpha|$, and since both m and $\tan \alpha$ are negative when α is obtuse, it follows that

$$\tan \alpha = m \tag{4}$$

in every case.

Note that a vertical line has $\alpha = 90°$, and the fact that $\tan 90°$ is not defined agrees with the fact that such a line has no slope. If a line with acute α is rotated so that α approaches $90°$, then $m = \tan \alpha$ takes larger and larger values, increasing without bound. If α approaches $90°$ through obtuse values, then m decreases without bound.

The use of a scale on one axis which is different from the scale on the other axis introduces various distortions, principally in those geometric concepts that relate to distance; in particular, the foregoing relationship between slope and inclination is invalid if different scales are used. In a geometric problem, therefore, an accurate representation requires the use of the same scale on both axes. It may happen, however, that an accurate representation fails to emphasize the details of principal interest or is merely unreasonably hard to draw; in such cases, we shall choose scales to suit our convenience. In working with problems from physics, moreover, we shall have occasion to consider graphs in which the scales on the two axes represent quantities which have no common measure (distance and time, for example) and in which the concept of using the same scales is therefore meaningless.

Parallel lines. If two lines L_1 and L_2 are parallel, as shown in Fig. 10 at the top of the next page, then $\alpha_1 = \alpha_2$, and therefore, by (4), $m_1 = m_2$ (or, if $\alpha_1 = \alpha_2 = 90°$, neither m_1 nor m_2 exists). Conversely, if $m_1 = m_2$, then $\tan \alpha_1 = \tan \alpha_2$, and, because for a given $\tan \alpha$ there is only one value of α such that $0° \leq \alpha < 180°$, it follows that $\alpha_1 = \alpha_2$ and therefore L_1 is parallel to L_2. It is also true that, if neither m_1 nor m_2 exists, then L_1 is parallel to L_2. These results can be summarized in the statement:

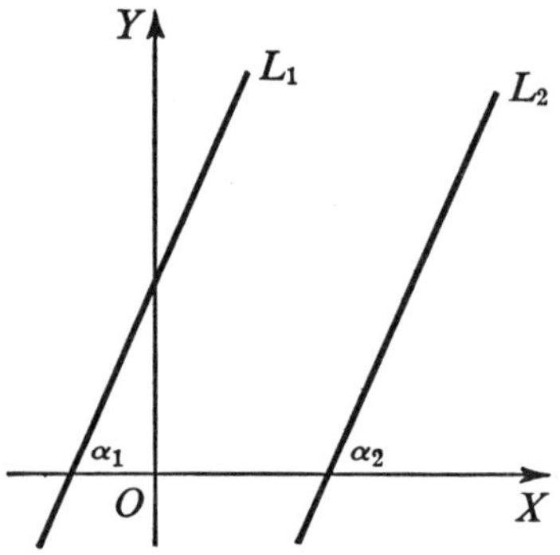

Fig. 10

▶ Parallel lines have the same slope or no slope, and nonvertical lines with the same slope or lines with no slope are parallel.

Perpendicular lines. If L_1 and L_2 are two perpendicular lines not parallel to the coordinate axes (Fig. 11), and if α_2 is the larger inclination, then

$$\alpha_2 = 90^\circ + \alpha_1,$$

and

$$\begin{aligned} m_2 &= \tan \alpha_2 \\ &= \tan (90^\circ + \alpha_1) \\ &= -\cot \alpha_1 = -1/\tan \alpha_1 \\ &= -1/m_1, \end{aligned}$$

and therefore $m_1 m_2 = -1$. Conversely, it can be shown that if $m_1 m_2 = -1$, then $\alpha_2 - \alpha_1 = \pm 90^\circ$ and L_1 is perpendicular to L_2. This demonstration is left to the student as an exercise (Ex. 1 of Miscellaneous Exercises 1.3).

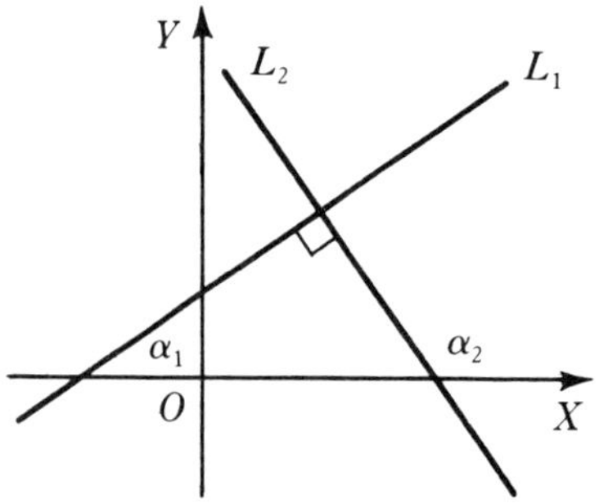

Fig. 11

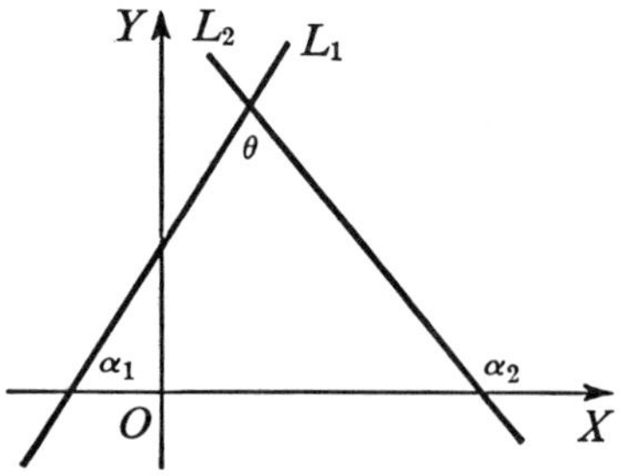

Fig. 12

Angle between two lines. If L_1 and L_2 intersect at an angle $\theta \neq 90^\circ$, as shown in Fig. 12, then

$$\theta = \alpha_2 - \alpha_1,$$

and $$\tan\theta = \tan(\alpha_2 - \alpha_1) = \frac{\tan\alpha_2 - \tan\alpha_1}{1 + \tan\alpha_2 \tan\alpha_1},$$ or

$$\tan\theta = \frac{m_2 - m_1}{1 + m_2 m_1} \tag{5}$$

In a specific application of (5), the slope of either line may be designated m_1. Interchanging m_1 and m_2 merely changes the sign of $\tan\theta$, and since $\tan(180° - \theta) = -\tan\theta$, either order will give us the tangent of one of the angles between L_1 and L_2, the acute angle if $\tan\theta$ is positive, the obtuse angle if it is negative. If it is important to distinguish between these two supplementary angles—if only one of them gives the required information—a rough sketch will almost always show which one should be used.

Since $\tan\alpha_1 = m_1$ and $\tan\alpha_2 = m_2$, it is also possible to find θ, at least approximately, by obtaining approximations to α_1 and α_2 from a calculator or a table of tangents and using $\theta = \alpha_2 - \alpha_1$.

Example 1. Find the slope and the inclination of the line joining $(2, -3)$ and $(5, 1)$.

Solution: From (3), $$m = \frac{1 - (-3)}{5 - 2} = \frac{4}{3},$$

and from (4), $$\tan\alpha = \tfrac{4}{3},$$

whence $$\alpha = 53°8'$$

Example 2. Find the value of k if the line joining $(3, -7)$ and $(-1, 4)$ is (**a**) parallel, and (**b**) perpendicular to the line joining $(2, k)$ and $(k, -3)$.

Solution: (**a**) If two lines are parallel, they have the same slope:

$$-\frac{11}{4} = \frac{k+3}{2-k},$$

$$-22 + 11k = 4k + 12,$$

$$7k = 34,$$

$$k = \tfrac{34}{7}.$$

(**b**) If two lines are perpendicular, the product of their slopes is -1:

$$-\frac{11}{4} \cdot \frac{k+3}{2-k} = -1,$$

$$11k + 33 = 8 - 4k,$$

$$15k = -25,$$

$$k = -\tfrac{5}{3}.$$

Example 3. Draw the line with slope 2 which passes through $(-2, -1)$.

Solution: We are given the slope $m = \dfrac{\Delta y}{\Delta x} = 2$; hence $\Delta y = 2\,\Delta x$, and, along the required line, any change in x is associated with a change in y which is twice as great. Thus, if we take $\Delta x = 1$, then $\Delta y = 2$, and this takes us from the given point $(-2, -1)$ to $(-2 + 1, -1 + 2)$, or $(-1, 1)$. Also, if $\Delta x = 2$, then $\Delta y = 4$, and this takes us from the given point to $(0, 3)$. We then plot these points and draw the line through them (Fig. 13).

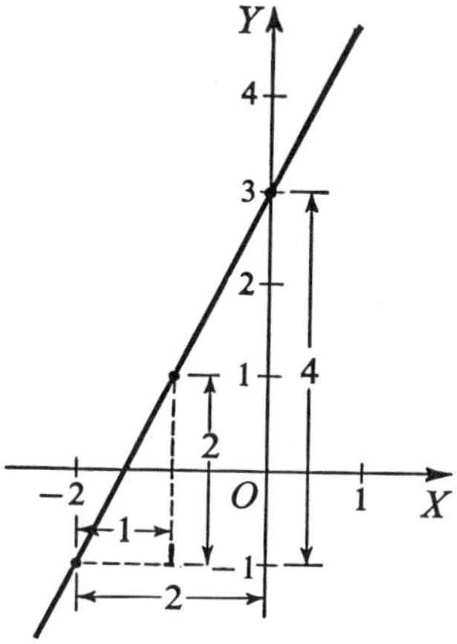

Fig. 13

Two points are, of course, sufficient to determine a line, but the third point affords a convenient check.

Example 4. Show that the points $A(-2, 3)$, $B(0, 2)$, and $C(4, 0)$ are collinear.

Solution: There are many ways of attacking this problem. Two of them are shown here.

METHOD 1. By the distance formula (1), $AB = \sqrt{4 + 1} = \sqrt{5}$, $BC = \sqrt{16 + 4} = 2\sqrt{5}$, and $AC = \sqrt{36 + 9} = 3\sqrt{5}$. Hence $AB + BC = AC$, and the three points are therefore collinear. (Why?)

METHOD 2. By the slope formula (3):

$$\text{slope of } AB = \frac{3 - 2}{-2 - 0} = -\frac{1}{2}, \quad \text{slope of } BC = \frac{2 - 0}{0 - 4} = -\frac{1}{2},$$

and since there is only one line through B with slope $-\frac{1}{2}$ (why?), AB and BC are the same line, or the three points are collinear.

The second method, using slopes, is the easier.

Example 5. The vertices of a triangle are $A(-1, 1)$, $B(1, 4)$, and $C(2, -5)$. Find the angles of the triangle.

Solution: By (3),

$$\text{slope of } AB = \tfrac{3}{2},$$
$$\text{slope of } BC = -9,$$
$$\text{slope of } CA = -2.$$

The angles between these lines can now be found by (5), but there are two angles in each case, and it is necessary to decide whether the acute or the obtuse angle is the required interior angle of the triangle. Figure 14 suggests that A is the only

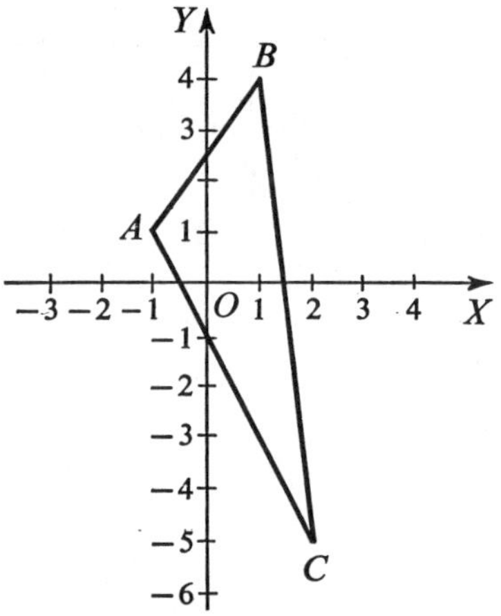

Fig. 14

obtuse angle involved, and a careful consideration of the slopes confirms this. We then find

$$|\tan A| = \left|\frac{-2 - \frac{3}{2}}{1 + (-2)(\frac{3}{2})}\right| = \frac{7}{4},$$

but since A is obtuse, we must take $\tan A = -\frac{7}{4}$, and $A = 119°45'$.

$$|\tan B| = \left|\frac{-9 - \frac{3}{2}}{1 + (-9)(\frac{3}{2})}\right| = \frac{21}{25},$$

and since B is acute, we take $\tan B = \frac{21}{25}$, and $B = 40°2'$.

$$|\tan C| = \left|\frac{-9 + 2}{1 + (9)(2)}\right| = \frac{7}{19},$$

and since C is acute, we take $\tan C = \frac{7}{19}$, and $C = 20°13'$. A convenient approximate check is to add the values found for A, B, and C. Their sum should be close to 180°. An exact check is the trigonometric identity $\tan(A + B) = -\tan C$ or the identity $\tan A + \tan B + \tan C = \tan A \tan B \tan C$.

Example 6. Prove that the medians to the equal sides of an isosceles triangle are equal.

Solution: In many problems of this sort, it is possible to save a good deal of work by a judicious placement of the coordinate axes. In this case, we can take advantage of the symmetry of the triangle, and also give each vertex a zero coordinate, by placing the axes as shown in Fig. 15. If the midpoints of the equal

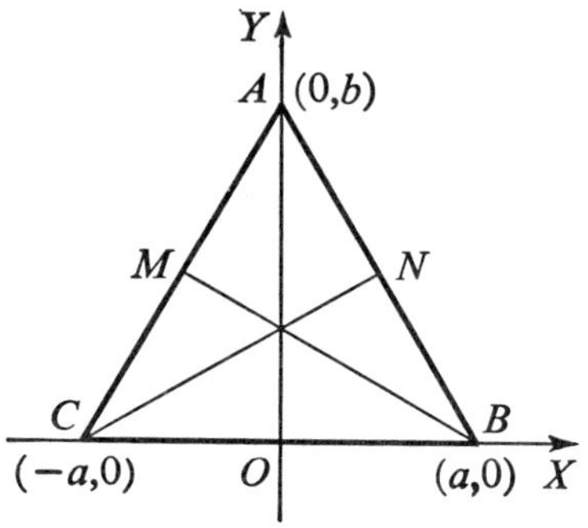

Fig. 15

sides AB and AC are N and M, respectively, then we find by (2) that the coordinates of N are $(a/2, b/2)$ and those of M are $(-a/2, b/2)$. Hence, by (1),

$$CN = \sqrt{(9a^2/4) + (b^2/4)} = BM.$$

Exercises 1.3 [A]

1. Find the distance between the given points:
   (a) $(-2.9, 5.9)$, $(0.3, -0.5)$
   (b) (h, k), $(2h, 2k)$
   (c) $(-2\frac{2}{5}, \frac{5}{7})$, $(\frac{3}{7}, -\frac{4}{5})$

2. Find the slope of the line joining the given points:
   (a) $(-1\frac{2}{3}, 2\frac{1}{2})$, $(\frac{4}{5}, -\frac{7}{10})$
   (b) $(6.3, -2.7)$, $(1.8, 4.5)$
   (c) $(2.03, 3.65)$, $(2.03, -8.47)$
   (d) $\left(\frac{a}{b}, \frac{b}{a}\right)$, $\left(\frac{a+b}{a-b}, \frac{a-b}{a+b}\right)$

3. Show that the distance formula (1) gives $|x_2 - x_1|$ as the distance between (x_1, y_1) and (x_2, y_1), and gives $|y_2 - y_1|$ as the distance between (x_1, y_1) and (x_1, y_2).

4. Are the following sets of three points each collinear?
   (a) $(2, 3)$, $(-5, 2)$, $(9, 4)$
   (b) (a, b), (c, d), $(2a - c, 2b - d)$
   (c) $(3.2, 12.7)$, $(-1.8, 2.2)$, $(-4.8, -8.2)$
5. Find the value of k if the lines joining $(2, k)$ to the origin and to $(10, 0)$ are perpendicular.
6. Find the angles of the triangle with vertices $(-2, 2)$, $(4, 3)$, and $(3, -2)$.
7. Verify formula (2). (*Hint:* Show that $P_1P_2 = 2P_1P = 2PP_2$.)
8. Prove: The diagonals of a parallelogram bisect each other.

Exercises 1.3 [B]

1. Find the point on the x-axis which is equidistant from $(2, 3)$ and $(4, -2)$.
2. A line through $(4, -1)$ has slope $\frac{1}{2}$. At what point does it cross the y-axis?
3. Show that the following points are collinear for any value of t:
   $$(x_1, y_1),\ (x_2, y_2),\ \text{and}\ (x_1 + t(x_2 - x_1),\ y_1 + t(y_2 - y_1)).$$
4. Find the angles of the triangle with vertices
   (a) $(-2, 1)$, $(2, 2)$, $(6, -2)$. (b) $(-9, 3)$, $(3, -2)$, $(8, 10)$.
5. Given $A(1, 5)$ and $B(5, -8)$, find a point P between A and B such that $2AP = PB$.
6. Find the vertices of a triangle if the midpoints of its sides are $(1, 6)$, $(2, 3)$, and $(4, 4)$.
7. Prove: If the diagonals of a parallelogram are equal, the parallelogram is a rectangle.
8. Prove: The diagonals of a rhombus meet at right angles.

Miscellaneous Exercises 1.3

1. Show that if $\tan \alpha_1 \tan \alpha_2 = -1$, then $\cos(\alpha_1 - \alpha_2) = 0$ and consequently $\alpha_1 = \alpha_2 \pm 90°$.
2. Show that $(0, 3)$, $(5, -2)$, $(1, 0)$, and $(8, -1)$ are the vertices of an isosceles trapezoid, and find the angles of the trapezoid.
3. Three vertices of a parallelogram are $(0, 0)$, $(6, 4)$, and $(8, 0)$.
   (a) Find the three possible locations of the fourth vertex.
   (b) Show that all three possible parallelograms have the same area.

4. Find the slopes of the lines that make an angle $\tan^{-1}\frac{1}{2}$ with the line joining (4, 8) and (−1, 3).
5. Find the slopes of the lines that bisect the angles between a line with slope 2 and a line with slope $\frac{3}{4}$.
6. Given $P_1(x_1, y_1)$ and $P_2(x_2, y_2)$, show that the point

$$P\left(\frac{sx_1 + rx_2}{r + s}, \frac{sy_1 + ry_2}{r + s}\right)$$

divides P_1P_2 so that $P_1P/PP_2 = r/s$.
7. Use the result of Ex. 6 to find the centroid of the triangle with vertices (−3, 2), (3, 10), and (5, 2).
8. Prove that the midpoint of the hypotenuse of a right triangle is equidistant from all three vertices.
9. Prove that a triangle with two equal medians is isosceles.
10. Prove that the median of a trapezoid is parallel to the bases and equal to half the sum of the bases.
11. Prove that the diagonals of an isosceles trapezoid are equal.
12. Prove that the midpoints of the sides of any quadrilateral are the vertices of a parallelogram.

1.4 Equation and graph

An equation in two variables, such as $2x + 3y - 6 = 0$ or $x^2 + 4y^2 = 36$ or $\tan x = \log y$, determines a set of pairs of values of the variables that satisfy the equation. Thus, for example, the equation $y = x^2$ is satisfied by $x = 1$, $y = 1$, by $x = -2$, $y = 4$, and by many more number-pairs. It is convenient and conventional to write these number-pairs in the form (x, y); thus, the two examples just given would be written (1, 1) and (−2, 4). This notation naturally suggests that we interpret these number-pairs as coordinates of points in a two-dimensional coordinate system. If we do, we have a *graph* of the equation.

▶ **Definition.** The graph of an equation is the set of points whose coordinates satisfy the equation.

We shall sometimes use the word "curve" as a synonym for "graph," even though the graph may be a straight line. Note that if we focus our attention on the set of *coordinates*, rather than on the set of points that they determine, we have a set each of whose elements is an ordered pair of numbers. Such sets, which are called relations, play an important part in calculus; we shall consider them at some length in the next chapter.

Linear equation and straight line. The equation

$$ax + by + c = 0, \tag{1}$$

in which a and b are not both zero, is called a linear equation because its graph is a straight line. To show this, we consider a particular point of the graph, say $P_1(x_1, y_1)$. Because P_1 belongs to the graph, we have

$$ax_1 + by_1 + c = 0.$$

If, now, $P_2(x_2, y_2)$ is any other point of the graph, we have

$$ax_2 + by_2 + c = 0.$$

Subtracting, we get $\quad a(x_1 - x_2) + b(y_1 - y_2) = 0,$

and if $b \neq 0$, $x_1 \neq x_2$, $\quad \dfrac{y_2 - y_1}{x_2 - x_1} = -\dfrac{a}{b}.$

It follows, by formula (3a) of § 1.3, that P_2, which is *any* point of the graph (other than P_1), lies on the line L which passes through P_1 and has slope $-a/b$. Hence *all* points of the graph lie on L. Conversely, if $P_3(x_3, y_3)$ is any point on L, it can be shown that the coordinates (x_3, y_3) satisfy (1), and therefore P_3 is an element of the graph. This demonstration is left to the student. $\left(\text{Show that } \dfrac{y_3 - y_1}{x_3 - x_1} = -\dfrac{a}{b} \text{ and hence that, for some number } c,\ ax_3 + by_3 + c = 0.\right)$

We have explicitly assumed, in the preceding discussion, that $b \neq 0$. If $b = 0$, then $a \neq 0$ (why?), and we have

$$ax_1 + c = 0, \quad x_1 = -\frac{c}{a},$$

and

$$ax_2 + c = 0, \quad x_2 = -\frac{c}{a};$$

conversely, if $\quad x_3 = -\dfrac{c}{a}, \quad$ then $\quad ax_3 + c = 0.$

Hence, in the case $b = 0$, a point is an element of the graph if and only if its abscissa is $-c/a$, no matter what its ordinate may be, and the graph is therefore the line that is perpendicular to the x-axis at the point $(-c/a, 0)$. Thus, in every case, the graph of (1) is a straight line.

Given any two distinct points, $P_1(x_1, y_1)$ and $P_2(x_2, y_2)$, an equation of the line through them may be found thus: If $x_1 \neq x_2$, and $P(x, y)$ is any point on the line P_1P_2, then the slope of PP_1 is the same as the slope of P_1P_2; that is,

$$\frac{y - y_1}{x - x_1} = \frac{y_2 - y_1}{x_2 - x_1}, \quad \text{or}$$

▶ $$\mathbf{y - y_1 = \frac{y_2 - y_1}{x_2 - x_1}(x - x_1).} \tag{2}$$

This result is called the two-point equation of the straight line. If we clear of fractions, we get

$$(y_1 - y_2)x + (x_2 - x_1)y + (x_1y_2 - x_2y_1) = 0,$$

which is just equation (1) with $a = y_1 - y_2$, $b = x_2 - x_1$, and $c = x_1y_2 - x_2y_1$.

We have assumed $x_1 \neq x_2$; if, on the contrary, $x_1 = x_2 = k$, then the line P_1P_2 is perpendicular to the x-axis at $(k, 0)$, and $P(x, y)$ is on this line if and only if $x = k$, or $x - k = 0$. This also is equation (1) with $a = 1$, $b = 0$, and $c = -k$, and (1) therefore represents an equation of any straight line; it is called the *general* form of that equation.

We note that the fraction which appears on the right-hand side of (2) is the slope m; hence that equation can be rewritten

▶ $$\mathbf{y - y_1 = m(x - x_1),} \tag{3}$$

the point-slope equation of the straight line.

The ordinate of a point where a graph meets the y-axis is called a y-intercept of the graph; similarly, the abscissa of a point where a graph meets the x-axis is an x-intercept. If, then, a line passes through $(0, b)$, its y-intercept is b, and substituting $(0, b)$ for (x_1, y_1) in (3), we have

▶ $$\mathbf{y = mx + b,} \tag{4}$$

the slope-intercept equation of the straight line.

Example 1. Write an equation of the line through $(3, 5)$ and $(-1, 2)$.

Solution:

METHOD 1. Substituting the given coordinates in (2), we find

$$y - 5 = \frac{2 - 5}{-1 - 3}(x - 3), \quad \text{or} \quad 3x - 4y + 11 = 0.$$

METHOD 2. The slope of the required line is $\frac{2 - 5}{-1 - 3} = \frac{3}{4}$, and we can use this in (3):

$$y - 5 = \tfrac{3}{4}(x - 3), \quad \text{or} \quad 3x - 4y + 11 = 0.$$

Example 2. Find the slope of the line $2x + 3y - 6 = 0$.

Solution:

METHOD 1. Perhaps the easiest way to do this is to solve the equation for y in terms of x, thus putting the equation in the slope-intercept form (4):

$$y = -\tfrac{2}{3}x + 2.$$

Comparing this with (4), we see that the coefficient of x, $-\frac{2}{3}$, must be the required slope. Note that we can also read the y-intercept, 2, from this form, and this information is sufficient to permit a reasonably accurate sketch of the line, if one is wanted, without further computation.

METHOD 2. A more basic but mechanically less convenient approach to this problem is to find two points of the line and then use the definition of slope (formula (3a) of § 1.3). Since 0 is usually the easiest number to substitute, it is convenient to try $x = 0$ and $y = 0$ in turn, thus finding that (0, 2) and (3, 0) lie on the line. Hence the slope is

$$m = \frac{0 - 2}{3 - 0} = -\frac{2}{3}.$$

Example 3. Write an equation of the line through $(2, -1)$ perpendicular to $3x - y + 7 = 0$.

Solution: Using either method of Example 2, we find that the slope of the given line is 3. The slope of the required line is then $-\frac{1}{3}$, and (3) therefore becomes

$$y + 1 = -\tfrac{1}{3}(x - 2), \quad \text{or} \quad x + 3y + 1 = 0.$$

Exercises 1.4 [A]

1. Find the slope and the intercepts, and sketch:

(**a**) $x - 5y + 15 = 0$

(**b**) $3x + 2y - 12 = 0$

(**c**) $5x - 3 = 0$

(**d**) $x + y + k = 0$

2. Write an equation of the line

(**a**) through $(-3, 3)$ and $(7, 2)$.

(**b**) through $(3, -5)$ and $(-2, -5)$.

(**c**) through $(a, 0)$ and $(0, b)$. Show that if $ab \neq 0$, then this equation can be put in the form $\dfrac{x}{a} + \dfrac{y}{b} = 1$, known as the <u>intercept form</u> of the equation of the line.

(**d**) through $(-3, 2)$, with slope $\frac{2}{3}$.

(**e**) with slope $\frac{1}{2}$ and y-intercept -3.

(**f**) through $(-3, 0)$, with inclination $30°$.

(**g**) through $(4, 3)$, parallel to $2x - y + 1 = 0$.

(**h**) through the origin, perpendicular to $3x + 2y - 5 = 0$.

3. Three vertices of parallelogram $ABCD$ are $A(0, 0)$, $B(3, 7)$, and $D(8, 1)$. Find equations of BC, DC, and AC.

4. The diagonals of a square lie on the coordinate axes and are each 4 units long. Write equations of the sides of the square (see Ex. 2(c), above).

5. What is the distance between the point (x_1, y_1) and the line $y = k$? Consider the cases $y_1 < k$ and $y_1 \geq k$. Using absolute-value notation, write a single formula which will be valid in all cases.

6. Show that $ax + by + c_1 = 0$ is parallel to $ax + by + c_2 = 0$ if $c_1 \neq c_2$.

7. Find an angle between $2x + y - 5 = 0$ and $x - 3y + 14 = 0$.

8. Show that the altitudes of the triangle with vertices $(-2, 0)$, $(0, 6)$, and $(3, 0)$ are concurrent.

Exercises 1.4 [B]

1. Write an equation of the line

(**a**) through $(\frac{1}{3}, 2\frac{1}{2})$ and $(4\frac{2}{3}, -1)$.

(**b**) through $(0, 4)$ and $(0, -3)$. What is the common name of this line?

(**c**) through $(-2, 4)$ and $(3, -6)$.

(**d**) through $(3, 5)$ and $(7, 5)$.

(**e**) through $(2, 2)$, with inclination $135°$.

(**f**) through $(4, -1)$, with slope -2.

(**g**) through (x_1, y_1), parallel to $ax + by + c = 0$.

(**h**) through (x_1, y_1), perpendicular to $ax + by + c = 0$.

2. Two vertices of equilateral triangle ABC are $A(-4, -1)$ and $B(4, -1)$, and the origin is inside the triangle. Find C.

3. The vertices of triangle ABC are $A(-2, -2)$, $B(4, 1)$, and $C(1, 7)$.

(**a**) Find equations of the medians of ABC.

(**b**) Show that the medians are concurrent.

4. Show that $bx - ay + c_2 = 0$ is perpendicular to $ax + by + c_1 = 0$.

5. For what values of k is $2x + ky + 3 = 0$ parallel to $kx + 4y + 7 = 0$?

6. The sides of a triangle are $2x - 7y + 22 = 0$, $7x + 8y + 12 = 0$, and $9x + y - 31 = 0$. Find (**a**) the vertices, and (**b**) the midpoints of the sides.

7. (**a**) Find an angle between the lines $a_1x + b_1y + c_1 = 0$ and $a_2x + b_2y + c_2 = 0$.

(**b**) Under what conditions will the lines described in part (a) be (**i**) parallel? (**ii**) perpendicular?

8. Find the slope and the distance from the origin of the line $x \cos \theta + y \sin \theta - p = 0$. (This equation is called the normal form of the equation of the straight line.) (*Hint:* Draw a figure, show that θ ($\neq 0$) can be taken as an angle between the line and the y-axis, and use elementary trigonometry.)

Miscellaneous Exercises 1.4

1. Sketch:

(**a**) $y = |x|$

(**b**) $y = |x + 1|$

(**c**) $y = |x| + 1$

(**d**) $y = \dfrac{|x|}{x}$

(**e**) $|x| + |y| = 1$

(**f**) $y = |x| + |x - 2|$

2. For any number $k \neq -7$, the line $3x - y = k$ is parallel to the line $3x - y + 7 = 0$ (see Ex. 6 in Exercises 1.4[A]). As k takes different values, the equation $3x - y = k$ represents different lines, but they are all parallel to one another; conversely, any line parallel to $3x - y + 7 = 0$ has an equation in the form $3x - y = k$. The equation $3x - y = k$ is therefore said to represent a family of parallel lines. What must be the value of k for the member of the family that goes through

(**a**) (2, 1)? (**b**) (0, 0)? (**c**) (2, 8)? (**d**) (−2, 1)?

3. Using the technique suggested by Ex. 2, write an equation for the line

(**a**) through (3, 7), parallel to $x + 2y - 4 = 0$.

Solution: An equation of the family of lines parallel to the given line is $x + 2y = k$. If a member of the family passes through (3, 7), we must have $3 + 2(7) = k$, or $k = 17$. The required equation is therefore $x + 2y - 17 = 0$.

(**b**) through (−1, 5), parallel to $2x + 3y - 11 = 0$.

(**c**) through (0, 0), parallel to $172x - 313y + 4478 = 0$.

(**d**) through (4, −3), parallel to $3x + 4y - 13 = 0$.

(**e**) through (x_1, y_1), parallel to $ax + by + c = 0$.

4. For any value of k, the line $x + 3y = k$ is perpendicular to the line $3x - y + 7 = 0$ (see Ex. 4 in Exercises 1.4[B]). What must be the value of k if $x + 3y = k$ goes through

(**a**) (3, −2)? (**b**) (1, −4)? (**c**) (0, 0)? (**d**) (−6, 2)?

5. Using the technique suggested by Ex. 4, write an equation of the line

(**a**) through $(2, 1)$, perpendicular to $2x + 3y - 5 = 0$.

(**b**) through $(-1, 6)$, perpendicular to $x + y + 4 = 0$.

(**c**) through $(0, 0)$, perpendicular to $43x - 192y + 471 = 0$.

6. (**a**) Write an equation of the line through $P(2, 6)$ perpendicular to the line L: $3x - 4y - 7 = 0$.

(**b**) Find the point Q in which this perpendicular meets L.

(**c**) Find PQ, the distance between P and L.

7. Using the technique suggested by Ex. 6, find the distance between

(**a**) $(3, 2)$ and $x - 2y + 2 = 0$.

(**b**) $(0, 0)$ and $2x + 5y - 1 = 0$.

(**c**) $(0, 0)$ and $ax + by + c = 0$.

8. For any value of m, the point-slope equation $y - 2 = m(x + 3)$ represents a line through $(-3, 2)$.

(**a**) What must be the value of m, and the resulting equation of the line, if the line also passes through (**i**) $(-2, 3)$? (**ii**) $(1, 6)$? (**iii**) $(0, 2)$? (**iv**) $(3, 5)$?

(**b**) The given equation does not represent the entire family of lines through $(-3, 2)$ because there is one such line whose equation cannot be written in the given form. What is the equation of that line?

9. Consider the two linear equations

$$L_1\colon\quad 2x + y - 3 = 0$$

and

$$L_2\colon\quad x + 5y + 4 = 0.$$

From them we can form a third equation

$$L_3\colon\quad (2x + y - 3) + k(x + 5y + 4) = 0,$$

or

$$(k + 2)x + (5k + 1)y + (4k - 3) = 0.$$

Note that (i) L_3 is the equation of a straight line (why?), and (ii) since any number-pair satisfying L_1 and L_2 must also satisfy L_3 (why?), the graph of L_3 passes through the intersection of L_1 and L_2. Hence the equation L_3 represents (almost; see part (b)) the family of lines through the intersection of L_1 and L_2.

(**a**) By determining the value of k and substituting that value in L_3, find the equation of that member of the family which (**i**) passes through $(0, -1)$; (**ii**) passes through $(0, 0)$; (**iii**) has slope 2; (**iv**) is parallel to the y-axis.

(**b**) What line through the intersection of L_1 and L_2 cannot have its equation written in the form L_3?

10. Using the technique suggested by Ex. 9, write an equation of the family of lines through the intersection of $x + y - 1 = 0$ and $2x + 3y - 1 = 0$, and determine from this the equation of that member of the family which (**a**) goes through $(-2, 1)$; (**b**) has slope 1; (**c**) is perpendicular to $x + 2y + 7 = 0$.

Exercises 1.4 [C]

1. Prove that the medians of any triangle are concurrent. (You are free, without loss of generality, to choose the axes in any position you wish. It is usually convenient to have as many zero coordinates as possible. Try placing the axes as suggested in Fig. 16.)

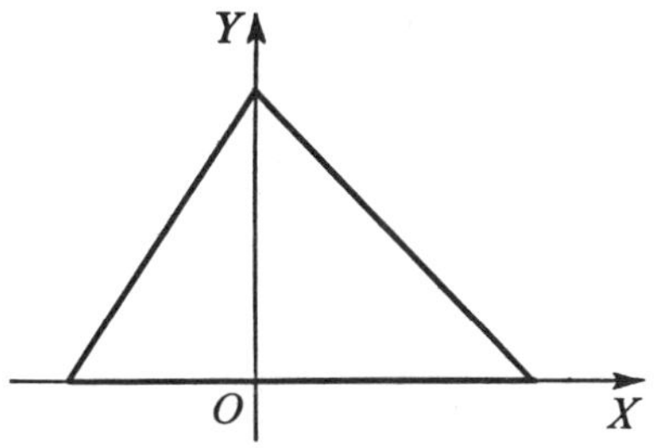

Fig. 16

2. Prove that the altitudes of any triangle are concurrent.

3. Prove that the perpendicular bisectors of the sides of any triangle are concurrent.

4. Prove that the points of concurrence found in the three preceding problems (the centroid, orthocenter, and circumcenter) are collinear.

5. Given the point $P(x_1, y_1)$ and the line L: $ax + by + c = 0$:

(**a**) Write an equation of the line M which passes through P and is perpendicular to L.

(**b**) Show that M meets L in the point

$$Q\left(\frac{b^2x_1 - aby_1 - ac}{a^2 + b^2}, \frac{a^2y_1 - abx_1 - bc}{a^2 + b^2}\right).$$

(**c**) Show that PQ, the perpendicular distance between P and L, is given by

$$\mathbf{PQ} = \frac{|\mathbf{a}x_1 + \mathbf{b}y_1 + \mathbf{c}|}{\sqrt{\mathbf{a}^2 + \mathbf{b}^2}}. \qquad (5)$$

6. An alternative derivation of (5) proceeds thus:

(a) If PS is perpendicular to the x-axis and crosses L at R, show that

$$PR = \left| \frac{ax_1 + by_1 + c}{b} \right|.$$

(b) If α is the inclination of L, show that

$$PQ = |PR \cos \alpha|.$$

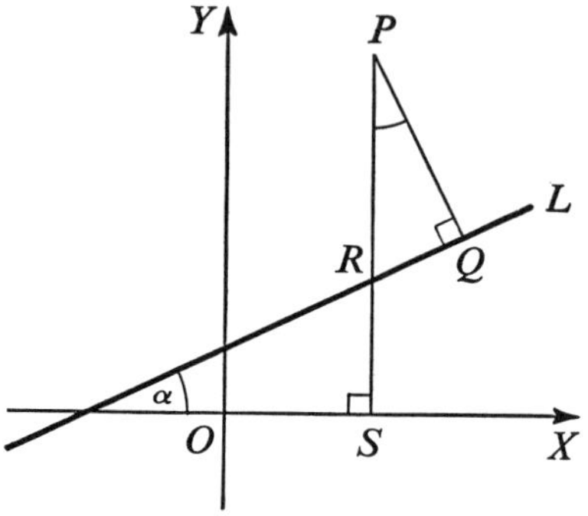

Fig. 17

(c) Show that $|\cos \alpha| = \left| \frac{b}{\sqrt{a^2 + b^2}} \right|.$

(d) Combine the results of (a), (b), and (c) to obtain (5).

(e) Show that this result is valid when α is right or obtuse, and for any relative positions of P and L.

7. Use (5) to obtain the distance between the given point and the given line:

(a) $(5, -2)$; $4x - 3y + 4 = 0$ (c) $(0, 0)$; $ax + by + c = 0$

(b) $(1, 4)$; $x + 3y + 7 = 0$ (d) $(-3, 1)$; $2x + y + 5 = 0$

8. Find the distance between the parallel lines:

(a) $2x + 3y - 7 = 0$, $2x + 3y + 3 = 0$

(b) $x - 5y + 1 = 0$, $2x - 10y - 3 = 0$

(c) $ax + by + c_1 = 0$, $ax + by + c_2 = 0$

9. (a) Show that if $|a| = |b|$, then either $a + b = 0$ or $a - b = 0$.

(b) Show that if $|x - 2| = |y + 3|$, then either $x + y + 1 = 0$ or $x - y - 5 = 0$.

(c) Write two linear equations that are equivalent to

$$|x + 2y - 3| = |3x - 5y|.$$

(**d**) Write two linear equations that are equivalent to

$$\frac{|x + y - 5|}{\sqrt{2}} = \frac{|2x - y + 3|}{3}.$$

10. (**a**) Use (5) to express in the form of an equation the fact that the distance between (X, Y) and $3x + 4y - 7 = 0$ is equal to the distance between (X, Y) and $5x - 12y + 4 = 0$, and then show that this equation is equivalent to the two equations $14X + 112Y - 111 = 0$ and $64X - 8Y - 71 = 0$.

(**b**) What is the locus of points equidistant from two intersecting lines?

(**c**) What are the equations of the bisectors of the angles between $3x + 4y - 7 = 0$ and $5x - 12y + 4 = 0$?

(**d**) Discuss the use of the capital letters (X, Y) in part (a).

11. Write equations of the bisectors of the angles between

(**a**) $x + y - 3 = 0$ and $x - y + 1 = 0$.

(**b**) $2x - y - 4 = 0$ and $2x + y + 7 = 0$.

(**c**) $x \cos \alpha + y \sin \alpha + p_1 = 0$ and $x \cos \beta + y \sin \beta + p_2 = 0$.

12. Find the area of the triangle with vertices

(**a**) $(1, 2), (2, 1), (-1, -1)$.

(**b**) $(1, 3), (3, 4), (0, 2)$.

(**c**) $(0, -7), (-4, 0), (6, 0)$.

(**d**) $(-3, -2), (3, 1), (5, 2)$.

(**e**) $(0, 0), (x_1, y_1), (x_2, y_2)$.

13. (**a**) Find the cosine of the angle subtended at the origin by the line segment which joins (x_1, y_1) and (x_2, y_2).

(**b**) Show that $(x_1x_2 + y_1y_2)^2 \leq (x_1^2 + y_1^2)(x_2^2 + y_2^2)$.

(**c**) Show that equality holds in (b) if $x_2 = kx_1$, $y_2 = ky_1$.

(**d**) Show that, if equality holds in (b), and if $x_1y_1 \neq 0$, then $x_2/x_1 = y_2/y_1 = k$.

(**e**) Interpret (c) and (d) geometrically.

Chapter Review

1. Find the points which are 10 units from both $(8, -4)$ and $(10, -2)$.

2. (**a**) Sketch the graph of $x + y + 2 = 0$.

(**b**) Indicate on your sketch the set of points for which $x + y + 2 > 0$ (this is the *graph* of $x + y + 2 > 0$).

3. Find the area of the triangle with sides $x - 2 = 0$, $x - y = 0$, and $3x - y - 8 = 0$.

4. (a) Express as an interval, and sketch, $\{y: |y - 2| \leq 3\}$. Is this interval open, closed, or neither?

   (b) Express as an interval $\{z: z^2 - 2z < 0\}$. Of what number is this a circular neighborhood?

5. Show that the triangle with vertices $(1, 4)$, $(-3, -2)$, and $(-5, 0)$ is isosceles, and find its angles.

6. Solve: $\dfrac{1}{x - 1} > 3$.

7. Write equations of the altitude and the median on the side AC of the triangle with vertices $A(-3, 2)$, $B(3, 3)$, and $C(5, -4)$.

8. Prove analytically: If the diagonals of a quadrilateral bisect each other, then the quadrilateral is a parallelogram.

2

Functions and Derivatives

2.1 Relations and functions

We were concerned, in § 1.4, with certain sets of points, the graphs of linear equations. Each point in a two-dimensional coordinate system is located by an ordered pair of numbers, its coordinates. If we shift our attention from the set of points to the set of coordinate pairs, we are led to consider a set whose elements are ordered pairs of numbers.

▶ **Definition.** A relation is a set of ordered pairs.

The sets $A = \{(2, 1), (2, 2), (1, 3)\}$ and $B = \{(2, 1), (3, 1), (-1, 2)\}$, for example, are relations. Most of the relations with which we shall be concerned contain infinitely many elements and must therefore be described by rule or formula. Using the set-builder notation, we have, for example, the relation $C = \{(x, y): x + 2y + 3 = 0\}$; in more familiar language, this is simply the set of coordinates of the points of the line whose equation is $x + 2y + 3 = 0$. Two other examples of relations described in this notation are $D = \{(x, y): x + 2y + 3 > 0\}$ and $E = \{(x, y): x^2 + y^2 = 4\}$.

This technical use of the word "relation" arises, as is often the case in mathematics, from a nontechnical use of the same word. If x is the brother of y, for example, we then say that they are *related* in a certain way. To determine whether two people are related in this way, we ordinarily find out whether the first of them is male and they have the same parents, but we could get the same information if we had a list of all pairs (x, y) such that x is the brother of y. Such a list would define a set $R = \{(x, y): x$ is the brother of $y\}$, and it is this set that we call the *relation* "is the brother of."

It is an easy step from graphs to relations, and the return step is just as easy:

▶ **Definition.** The graph of a relation is the set of points whose coordinates belong to the relation.

Note that if the relation is described by an equation, as are C and E, then this definition coincides with the definition given in § 1.4 for the graph of an equation. The present definition is, however, more general, since it also covers the graphs of relations like A and B, which are described by lists, and D, which is described by an inequality.

Each element of a relation is an ordered pair, but an ordered pair is itself made up of two elements, called the first coordinate and the second coordinate of the pair (although, as noted in § 1.3, the names *abscissa* and *ordinate* are traditional when we think of the ordered pair as a point in a Cartesian coordinate system). In the pair (s, t), for example, the first coordinate is s and the second coordinate is t. The relation A, above, contains two pairs, $(2, 1)$ and $(2, 2)$, which have the same first coordinate but different second coordinates, and the relation D has infinitely many pairs in which a given first coordinate is paired with different second coordinates: $(1, -1)$, $(1, \pi)$, $(1, 17)$, and $(1, 63)$ are a few instances. Similarly, E contains many instances in which a given first coordinate is paired with two second coordinates, such as $(1, \sqrt{3})$ and $(1, -\sqrt{3})$, $(0, 2)$ and $(0, -2)$, and so on. In relations B and C, however, no two pairs have the same first coordinate. This distinction is an important one, and a relation which resembles B and C in this respect is called a function.

Definition. A function is a relation in which no two distinct pairs have the same first coordinate.

More explicitly, if f is a function, and if $(x, y) \in f$ and $(x, z) \in f$, then $y = z$.

The distinction between functions and other sorts of relations has a specific graphical interpretation. If two (or more) number-pairs have the same first coordinate, then the corresponding points have the same abscissa and therefore lie on the same vertical line. It follows that a relation is a function if and only if no vertical line meets its graph in two or more points. The student should have this criterion in mind when examining the following figures. The diagrams labeled B and C in Fig. 1, and those labeled A, D, and E in Fig. 2, are graphs of the corresponding relations discussed above.

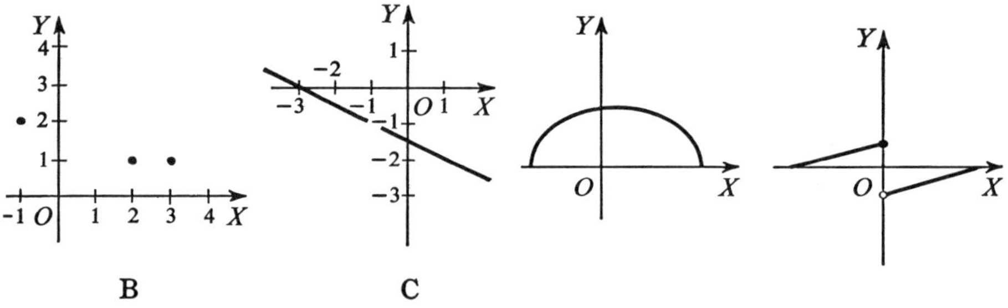

Fig. 1. Some relations that are functions

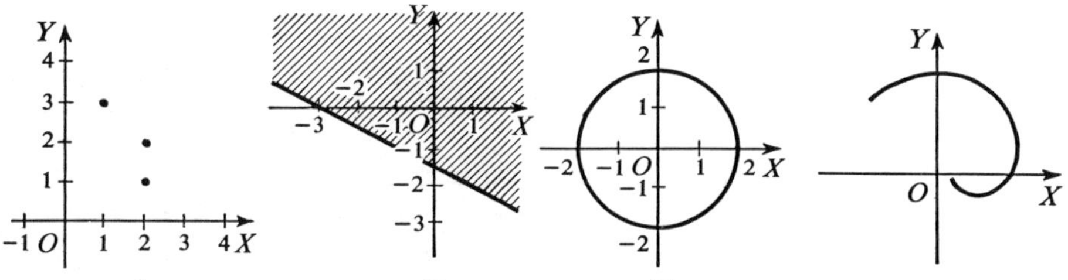

Fig. 2. Some relations that are not functions

In working with relations, it is often necessary to discuss the set of all first coordinates or the set of all second coordinates of the ordered pairs which belong to a particular relation. To facilitate this, we have the following definition:

Definition. The domain of a relation is the set of all first coordinates, and the range of a relation is the set of all second coordinates, of the ordered pairs which belong to the relation.

Hence the domain of A is $\{2, 1\}$ and the range is $\{1, 2, 3\}$; the domain of B is $\{-1, 2, 3\}$ and the range is $\{1, 2\}$. When a relation is described by a rule, the relation is not completely specified unless the domain is stated. An example is E: clearly, the relation $E_1 = \{(x, y) : x^2 + y^2 = 4, -2 \leq x \leq 0\}$ does not consist of the same pairs as $E_2 = \{(x, y) : x^2 + y^2 = 4, 1 \leq x \leq \sqrt{3}\}$, and therefore $E_1 \neq E_2$. It is, however, conventional and convenient to assume that, if a relation is described by a rule, then the domain of the relation is the most inclusive set of real numbers for which the rule makes sense, it being understood that the range, also, consists solely of real numbers. We therefore assume, in the absence of any explicit restriction, that the domain of E is the interval $[-2, 2]$, and that the range is the same interval. Similarly, given the function $F = \{(x, y) : y = x^2\}$, the same convention tells us that the domain of F is $(-\infty, \infty)$, the set of all real numbers, and that the range is therefore $[0, \infty)$.

It is important to note that when we describe sets by formula we can use as variables any symbols which have not already been given a particular meaning. Thus, the function F is a set of ordered pairs of real numbers; it has no x's or y's in it, and we could just as well describe it as $\{(s, t) : t = s^2\}$, or as $\{(p, q) : q = p^2\}$, or even as $\{(y, x) : x = y^2\}$, although the last choice runs counter to a habit which leads us to associate the letter x with elements of the domain, with the first coordinate, and the letter y with elements of the range, with the second coordinate. Note, however, that the relation

$\bar{F} = \{(x, y): x = y^2\}$ (read this "F-bar") is *not* the same as F and, indeed, is not even a function.

No matter what letters we use in describing a relation, it is conventional in graphing to let points of the domain correspond to points of the horizontal axis, and points of the range to points of the vertical axis.

If f is any function, then, for any element x of its domain, the symbol $f(x)$ (read "f at x" or "f of x") denotes the corresponding element of its range, the second coordinate that is paired with the first coordinate x. The number $f(x)$ is called the value of the function f at x. For the function $C = \{(x, y): x + 2y + 3 = 0\}$ we have $C(2) = -\frac{5}{2}$, $C(-5) = 1$, and, in general, $C(x) = -(x + 3)/2$ for each number x in the domain of C; similarly, for the function $F = \{(x, y): y = x^2\}$ we have $F(3) = 9$, $F(-8) = 64$, and, in general, $F(x) = x^2$ for each number x in the domain of F. Now, if we know (i) the domain of a function, and (ii) what element to pair with each element of the domain, we then have complete knowledge of the function. We may therefore describe F by saying "F is the function which has the value $F(x) = x^2$ at each real number x," and C by saying "C is the function which has the value $C(x) = -(x + 3)/2$ at each real number x." Such descriptions are common in mathematics and are often extremely convenient; they are frequently shortened by omitting an explicit description of the domain (in accordance with the convention described above), and even a name for the function, so that we find, for example, "the function $F(x) = x^2$," or "the function x^2," or, with an obvious implication of Cartesian coordinates, "the function $y = x^2$." Any and all of these may be used, as convenient, in what follows. The student should, however, bear in mind that $f(x)$ is not a *function*, but an element of the range of the function f, and therefore usually a *number;* the function is f, and $f(x)$ is its value at x. When the symbol x appears in this context, it is sometimes called the argument of the function f. Similarly, if we have occasion to write $f(2a)$, then $2a$ is the argument.

For obvious reasons (after all, f is for "function"), the letters f and F, and their Greek equivalents ϕ and Φ (small and capital phi), are often used as names of functions. More names than these are needed, however, and many letters, of many alphabets, are used as names of functions.

If we have an equation in two variables, such as x and y, it often happens that there exists a function f that takes the value $y = f(x)$ at the number x. In such a case, the pairs (x, y) which belong to f all satisfy the given equation, and we then say that the equation defines the function. If the equation has the form $y = f(x)$, if, that is, y stands alone on one side of the equation and the other side involves no variable other than x, we can then substitute values of x and compute the corresponding values of y by evaluation of this formula; in this case, we say that the equation defines the function explicitly. If the equation has some other form, it is still possible to determine whether a particular pair (x, y) satisfies the equation and therefore belongs to the function; in such a case, we say that the equation defines the function implicitly. In either case, it is customary to say that the equation *defines y as a function of x*. The equation $x + 2y + 3 = 0$, for example, defines y implicitly as a function of x; the function in question is precisely the function C defined above. Since the set $\overline{C} = \{(y, x): x + 2y + 3 = 0\}$ is also a function, it is also true, in this case, that the same equation defines x implicitly as a function of y.

In many cases, an equation defines a relation which is not a function but which can be broken up into functions. Thus, the equation $x^2 + y^2 = 4$ defines the relation E considered above, and this relation can be expressed as the union of two functions:

$$E = \{(x, y): y = \sqrt{4 - x^2}\} \cup \{(x, y): y = -\sqrt{4 - x^2}\}.$$

Relations as mappings. Each ordered pair of a relation associates an element of its domain with an element of its range, much as a map associates points of a geographical region with points on a piece of paper. It is sometimes helpful to exploit this analogy by thinking of a relation as mapping its domain onto its range. Thus, for example, the relation A mentioned at the beginning of this section maps 2 onto 1, 2 onto 2, and 1 onto 3; the points 1, 2, and 3 of the range are sometimes called the images under the mapping A of the points 2, 2, and 1 of the domain. Similarly, the function B maps 2 onto 1, 3 onto 1, and -1 onto 2. We can represent each of these mappings pictorially by putting dots on number lines to represent the domain and the range and drawing arrows to represent the mapping, as in Fig. 3. Note the important distinction, in the diagram, between A, which is not a function, and B, which is: in A, two arrows *start from* a single point; in B, this does not happen.

In general, we may regard any relation as a mapping from a set P (the domain) to a set Q (which includes but, conventionally, may exceed the

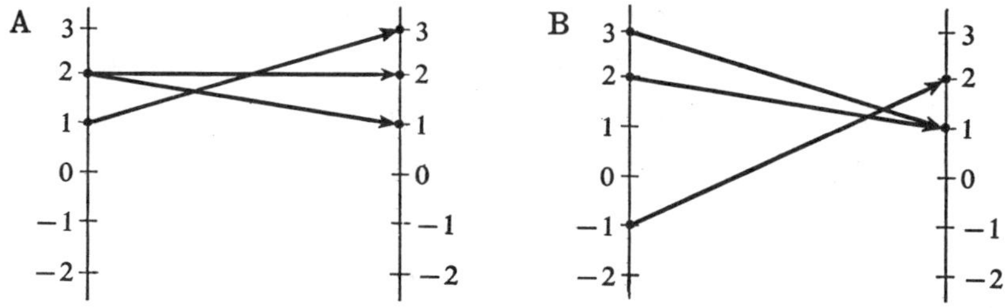

Fig. 3

range), and we may represent this pictorially by some such diagram as Fig. 4. (Note that this mapping is a function.) The objects in the sets P

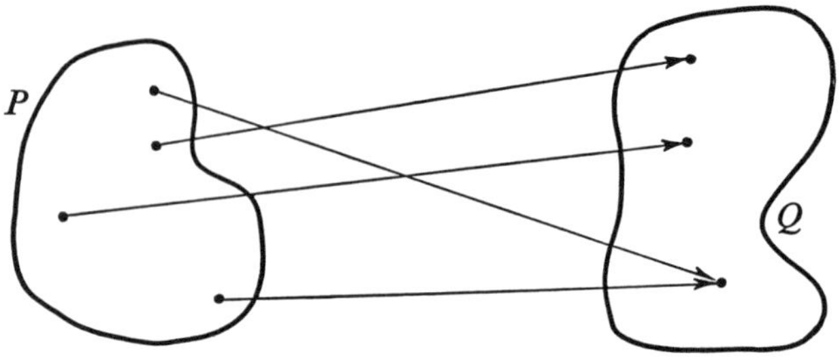

Fig. 4

and Q can be of any kind whatever, but we shall concern ourselves almost exclusively with cases in which they are real numbers.

We have defined functions in terms of sets of ordered pairs. It is, however, entirely possible to use this idea of mapping as the starting point and define functions in terms of mapping. The two definitions are logically equivalent, and both points of view are useful in calculus.

Exercises 2.1 [A]

1. For each of the following relations, (i) sketch a graph, (ii) determine the domain and the range, and (iii) state whether the relation is a function:
   (a) $\{(7, 3), (3, 7), (7, 14)\}$
   (b) $\{(x, y): y = x\}$
   (c) $\{(x, y): y > x\}$
   (d) $\{(x, y): y = 1\}$
   (e) $\{(x, y): x = 1\}$
   (f) $\{(s, t): t = 2s\}$

2. Determine the domain and the range of each of the following functions:
   (a) $f = \{(x, y): y = |x|\}$
   (b) g, if $g(t) = \sqrt{t - 1}$
   (c) $h(x) = \sqrt{4 - x^2}$
   (d) $F = \{(x, y): y = \log x\}$
   (e) $\left\{(x, f(x)): f(x) = \dfrac{x^2 - 4}{x - 2}\right\}$

3. Use the set-builder notation to describe the function f if
   (a) $f(x) = 1/x$. (b) $f(u) = u^2 - 5$. (c) $f(\theta) = \cos\theta$.

4. The domain of a function f is $\{1, 2, 3, 4, 5\}$. Describe the function by a rule, or formula, in the form "$f(x) = \cdots$," if
   (a) $f = \{(1, 2), (2, 4), (3, 6), (4, 8), (5, 10)\}$.
   (b) $f = \{(1, 0), (2, 1), (3, 2), (4, 3), (5, 4)\}$.
   (c) $f = \{(1, 1), (2, 4), (3, 9), (4, 16), (5, 25)\}$.

5. If $f = \{(x, y)\colon 2x + y - 1 = 0\}$, find $f(-2)$, $f(3)$, $f\left(\frac{1}{1+\sqrt{3}}\right)$, and $f(1 - a)$.

6. If $g = \{(u, v)\colon v = u(u - 2)\}$, find $\frac{1}{g(x)} - g\left(\frac{1}{x}\right)$.

7. If $f = \{(x, y)\colon 2x - y + 1 = 0\}$ and $g = \{(x, y)\colon y = x^2\}$, find expressions for $f(g(t))$ and $g(f(t))$, and solve the equation $f(g(t)) = g(f(t))$.

8. State precisely why the function $f = \left\{(x, y)\colon y = \frac{x^2 - 3x + 2}{x - 1}\right\}$ is not the same as the function $g = \{(x, y)\colon y = x - 2\}$.

9. If $f(x) = \begin{cases} 2x + 1 \text{ when } x \le 0 \\ x^2 \text{ when } x > 0, \end{cases}$ find $f(-5), f(0), f(1)$, and $f(3)$, and state the domain and the range of f.

Exercises 2.1 [B]

1. For each of the following relations, (i) sketch a graph, (ii) determine the domain and the range, and (iii) state whether the relation is a function:
   (a) $\{(-6, 2), (4, 4), (\pi, 4)\}$
   (b) $\{(x, y)\colon 2x + 3y - 6 = 0\}$
   (c) $\{(x, y)\colon 2x + 3y - 6 \le 0\}$
   (d) $\{(u, v)\colon u = v^2\}$
   (e) $\{(x, f(x))\colon f(x) = x^2 + 1\}$
   (f) $\{(t, v)\colon v = 32t - 8 \text{ and } t > 0\}$

2. Determine the domain and the range of each of the following functions:
   (a) $s(x) = |x|/x$
   (b) $S(x) = \sin x$
   (c) f, if $f(x) = \sqrt{1 - x^2} \cdot (1/x)$
   (d) $T(x) = \tan x$
   (e) $\phi = \left\{(x, y)\colon y = \frac{x^2 + 2x - 8}{x - 2}\right\}$

3. Use the set-builder notation to describe the function f if
   (a) $f(x) = (2x - 5)/3$. (b) $f(z) = z^3 - 5z$. (c) $f(r) = \log r$.

4. The domain of a function f is $\{2, 3, 5, 7\}$. Describe the function by a rule, or formula, in the form "$f(x) = \cdots$," if
   (a) $f = \{(2, -1), (7, -6), (3, -2), (5, -4)\}$.
   (b) $f = \{(2, \frac{1}{3}), (3, \frac{1}{4}), (5, \frac{1}{6}), (7, \frac{1}{8})\}$.
   (c) $f = \{(2, 3), (3, 5), (5, 9), (7, 13)\}$.

5. If $f(x) = x^2 + 2x - 1$, find $f(\sqrt{3} - 1)$ and $f(z + 1)$, and list the elements of $\{t\colon f(t) = 0\}$.

6. If $F = \left\{(x, y): y = x + \frac{1}{x}\right\}$, find $F(\sqrt{2} + 1) + F(\sqrt{2} - 1)$.

7. If $f(x) = \dfrac{ax + b}{cx - a}$ and $a^2 + bc \neq 0$, find and simplify an expression for $f(f(x))$.

8. State the distinction between the functions f and g if $f(h) = \dfrac{(h + 2)^2 - 4}{h}$ and $g(h) = h + 4$.

2.2 Circular functions: definitions and fundamental properties

At this point we consider in detail a family of functions which are of major importance in calculus: the *circular*, or *trigonometric, functions*. Since it is assumed that the student has already met these functions (indeed, they were used in Chapter 1), the following development, although self-contained, will proceed at a fairly rapid pace, and a number of important details will be left for the exercises.

We need to use a Cartesian coordinate plane, but because it is desirable to save the letters X and Y for another use, we shall label the axes U and V. In this plane, we consider the circle with center at the origin and unit radius, the circle $u^2 + v^2 = 1$. We wish to establish a scale on this circle. To this

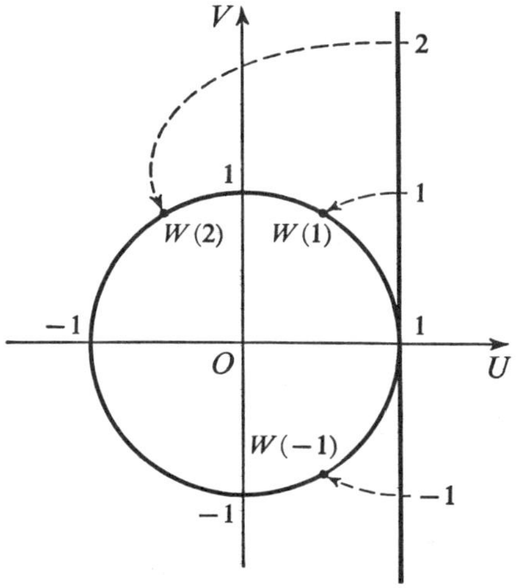

Fig. 5

end, we take the scale on either coordinate axis and reproduce it on the line $u = 1$, with the origin of this scale at the point $(1, 0)$. We then wrap this number line around our circle, in both directions, the negative part of the number line going around clockwise, the positive part counterclockwise, somewhat as an endless fishing line might be wound, both ways, around a reel. It is essential in this process that the fishing line not be stretched; the original distances between points on the number line must be reproduced as the same distances between points along the arc of the circle. In this wrapping process, each point x of the number line is carried onto a unique point P of the circle, and we thus see that the set $W = \{(x, P)\}$ is a function, called a wrapping function, whose domain is the set of all real numbers and whose range is the set $\{(u, v): u^2 + v^2 = 1\}$ of points on the unit circle.

Since the circumference of a unit circle is 2π, the wrapping function W maps points of the number line that are 2π units apart onto the same point of the unit circle. Thus, for example, the image under W of the number 0 is the point $(1, 0)$, that is, $W(0) = (1, 0)$. When the number line is wrapped around the circle, the number 2π also falls on $(1, 0)$, and hence $W(2\pi) = (1, 0)$. Similarly, $W(4\pi) = (1, 0)$, $W(-2\pi) = (1, 0)$, and generally, for each integer n, $W(2n\pi) = (1, 0)$. A similar property holds at each point of the unit circle: $W(\pi/2) = W(5\pi/2) = W(9\pi/2) = \cdots = (0, 1)$; $W(\pi) = W(3\pi) = W(5\pi) = \cdots = (-1, 0)$; and in general, for each real number x and for each integer n, $W(x + 2n\pi) = W(x)$. A function with this property is called *periodic*.

▶ **Definition.** A function f is periodic with period $k \neq 0$ if, whenever x is in the domain of f, then also $x + k$ is in the domain of f, and $f(x + k) = f(x)$.

The smallest positive k satisfying the conditions of this definition is the fundamental period of f. Hence W is periodic with fundamental period 2π.

Each element of the range of W is a point, $W(x) = (u, v)$. We now turn our attention from W itself to two functions whose values at x are the coordinates u and v of the point $W(x)$. These functions are the cosine and sine, abbreviated cos and sin.

▶ **Definition.** If $W(x) = (u, v)$, then $\cos x = u$ and $\sin x = v$.

It follows immediately that the domain of cos, and that of sin as well, is the same as the domain of W: the set of all real numbers. To find the range of cos, we note that $u^2 + v^2 = 1$, that is,

▶
$$\cos^2 x + \sin^2 x = 1 \tag{1}$$

("$\cos^2 x$" means "$(\cos x)^2$"). Hence $|u| \leq 1$, that is, $|\cos x| \leq 1$. Since, moreover, every $u \in [-1, 1]$ is the abscissa of at least one point on the unit circle, it follows that the range of cos is precisely the interval $[-1, 1]$. Similar considerations show that the range of sin is also $[-1, 1]$.

Certain further properties of cos and sin are of interest. Because W is periodic with period 2π, it follows that cos and sin are also periodic with period 2π; it happens that 2π is also the *fundamental* period of these two functions, but this result is not automatic, and the student will be asked in the exercises to supply proofs. If $W(x) = (u, v)$, then $W(-x) = (u, -v)$, as Fig. 6 suggests; a proof, using congruent triangles, is easy but rather long, since several different cases must be considered. It then follows from the definitions of cos and sin that

$$\begin{aligned} \cos(-x) &= \cos x, \\ \sin(-x) &= -\sin x. \end{aligned} \tag{2}$$

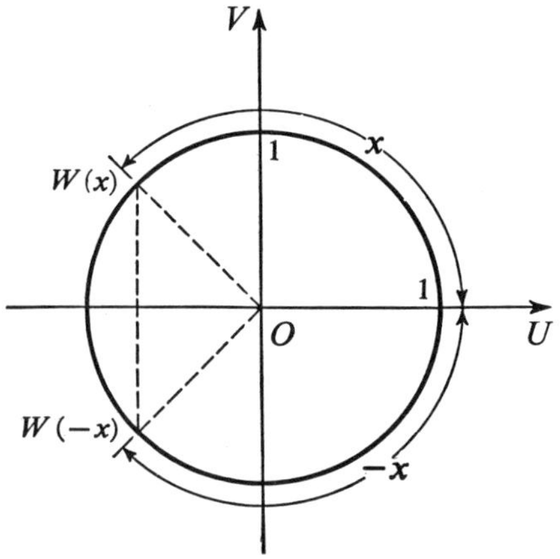

Fig. 6

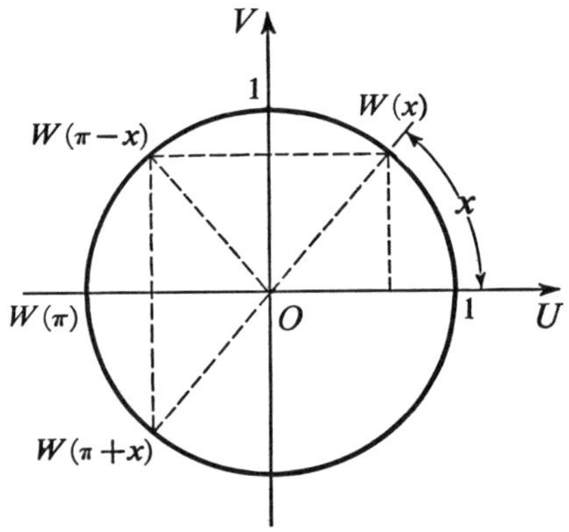

Fig. 7

Similarly, if $W(x) = (u, v)$, then $W(\pi - x) = (-u, v)$ and $W(\pi + x) = (-u, -v)$, as indicated in Fig. 7; hence

$$\begin{aligned} \cos(\pi - x) &= -\cos x, \\ \sin(\pi - x) &= \sin x, \\ \cos(\pi + x) &= -\cos x, \\ \sin(\pi + x) &= -\sin x. \end{aligned} \tag{3}$$

Four other circular functions will be useful. The functions tan (tangent) and sec (secant) are defined for all x such that $\cos x \neq 0$ by

$$\tan x = \frac{\sin x}{\cos x} \quad \text{and} \quad \sec x = \frac{1}{\cos x},$$

and the functions cot (cotangent) and csc (cosecant) are defined for all x such that $\sin x \neq 0$ by

$$\cot x = \frac{\cos x}{\sin x} \quad \text{and} \quad \csc x = \frac{1}{\sin x}.$$

It is left to the student in the exercises to determine the fundamental periods of tan, cot, sec, and csc.

Example 1. Simplify $\sin x + \cos x \cot x$.

Solution: $\sin x + \cos x \cot x = \sin x + \cos x \dfrac{\cos x}{\sin x}$ (by definition of cot)

$$= \frac{\sin^2 x + \cos^2 x}{\sin x}$$

$$= \frac{1}{\sin x} \quad \text{(by (1))}$$

$$= \csc x \quad \text{(by definition of csc).}$$

Example 2. Find all $x \in [0, 2\pi)$ such that $3 \cos x = 2 \sin^2 x$.

Solution: The problem will be simplified if we can obtain an equivalent equation involving only one of the circular functions. Equation (1) enables us to accomplish this; with its help, we can always express $\sin^2 x$ in terms of $\cos^2 x$, or vice versa. Hence

$$\begin{aligned} 3 \cos x &= 2 \sin^2 x \\ &= 2(1 - \cos^2 x) \quad \text{(by (1))} \\ &= 2 - 2 \cos^2 x; \\ 2 \cos^2 x + 3 \cos x - 2 &= 0, \\ (2 \cos x - 1)(\cos x + 2) &= 0, \\ \cos x &= \tfrac{1}{2}, -2. \end{aligned}$$

If $W(x) = (u, v)$, then $\cos x = u$ by definition. To find the values of x that correspond to $\cos x = \frac{1}{2}$, we note that the line $u = \frac{1}{2}$ meets the circle $u^2 + v^2 = 1$ at the points $(\frac{1}{2}, \frac{1}{2}\sqrt{3})$ and $(\frac{1}{2}, -\frac{1}{2}\sqrt{3})$; by elementary geometry, these points are seen to be $W(\pi/3)$ and $W(-\pi/3) = W(5\pi/3)$. There is no real number x such that $\cos x = -2$ because, as noted above, the range of cos is $[-1, 1]$. Hence the only solutions in the specified interval are $x = \pi/3, 5\pi/3$.

Example 3. Show that $\dfrac{1 + \cos x}{\sin x} = \dfrac{\sin x}{1 - \cos x}$ at all real numbers x at which the equation has meaning.

Solution: We note that if we clear the given equation of fractions, we get $1 - \cos^2 x = \sin^2 x$, which is an alternative form of (1) and therefore true. There is a temptation to conclude from this that the given equation is true, but what the argument in fact establishes is the converse: if the given equation is true, then so is (1). Since we already know (1) is true, we have accomplished nothing. What we must do is change the *form* of one or both sides of the original equation, without changing the *value* of either, until we have identical forms on both sides.

Using the fraction on the left, we may assume $\sin x \neq 0$, since otherwise the given equation is meaningless; this also implies $\cos^2 x \neq 1$. Hence,

$$\begin{aligned}\frac{1+\cos x}{\sin x} &= \frac{1+\cos x}{\sin x}\cdot\frac{\sin x}{\sin x}\\ &= \frac{\sin x(1+\cos x)}{\sin^2 x}\\ &= \frac{\sin x(1+\cos x)}{1-\cos^2 x} \qquad \text{(by (1))}\\ &= \frac{\sin x(1+\cos x)}{(1+\cos x)(1-\cos x)}\\ &= \frac{\sin x}{1-\cos x}.\end{aligned}$$

An equation such as the one given in Example 3, which is true whenever it is meaningful, is called an identity, and the process just described and illustrated is called proving the identity.

Miscellaneous Exercises 2.2 (1)

Fundamental Properties

1. Make a sketch of the circle $u^2 + v^2 = 1$ and use elementary geometry to complete a table like the one started below. Extend your table to include the following values of x: $\pi/3$, $\pi/2$, $2\pi/3$, $3\pi/4$, $5\pi/6$, π, $7\pi/6$, $5\pi/4$, $4\pi/3$, $3\pi/2$, $5\pi/3$, $7\pi/4$, $11\pi/6$, 2π. Keep your table for use in the problems that follow.

x	$W(x)$	sin x	cos x	tan x	cot x	sec x	csc x
0	(1, 0)	0	1	0	—	1	—
$\pi/6$	$(\frac{1}{2}\sqrt{3}, \frac{1}{2})$	$\frac{1}{2}$					
$\pi/4$	$(\frac{1}{2}\sqrt{2}, \frac{1}{2}\sqrt{2})$						

2. Using the table that you made in Ex. 1, sketch a graph of each of the six circular functions over the interval $[0, 2\pi]$.

3. Determine the domain and the range of **(a)** tan; **(b)** cot; **(c)** sec; **(d)** csc.

4. For the case $x \in \left(0, \frac{\pi}{4}\right)$, show that if $W(x) = (u, v)$ then $W\left(\frac{\pi}{2} - x\right) = (v, u)$ and $W\left(\frac{\pi}{2} + x\right) = (-v, u)$. (These results are actually valid

for all x.) Conclude that $\cos\left(\frac{\pi}{2} - x\right) = \sin x = -\cos\left(\frac{\pi}{2} + x\right)$, and $\sin\left(\frac{\pi}{2} - x\right) = \cos x = \sin\left(\frac{\pi}{2} + x\right)$.

5. A function f is periodic with period $\frac{2}{3}$. If $f(1) = 2$ and $f(2\frac{1}{6}) = -5$, find $f(1\frac{1}{2})$.

6. A square in the uv-plane has vertices at (0, 0), (1, 0), (1, 1), and (0, 1). A number line is formed using the line $u = 1$ and the same scale as on the coordinate axes, with the origin of the number line at the point $(u, v) = (1, 0)$. A wrapping function f then maps this scale onto the square, as suggested in Fig. 8. **(a)** Show that $f(1) = (1, 1)$, $f(2) = (0, 1)$, and $f(3) = f(-1) = (0, 0)$. **(b)** Find $f(5\frac{1}{2})$ and $f(\pi)$. **(c)** Show that f is periodic and determine the fundamental period. **(d)** Find $f(1{,}000{,}000)$.

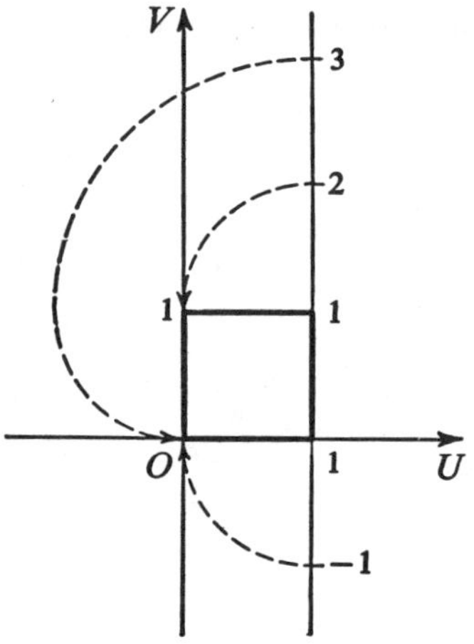

Fig. 8

7. The number line of Ex. 6 is mapped onto an equilateral triangle having vertices at (0, 0), (1, 0), and $(\frac{1}{2}, \frac{1}{2}\sqrt{3})$, by a wrapping function g. Show that g is periodic, determine the fundamental period, and find $g(2)$, $g(4\frac{1}{2})$, $g(5)$, and $g(686{,}147)$.

8. Prove that

▶

$$\sec^2 x = \tan^2 x + 1, \tag{4}$$

$$\csc^2 x = \cot^2 x + 1, \tag{5}$$

$$\tan x \cot x = 1 \tag{6}$$

at all points in the intersection of the domains of the functions involved.

9. Show that the fundamental period of cos is 2π. (*Hint:* If k is any period of cos, then $\cos(x + k) = \cos x$ for each real number x and in particular for $x = 0$. Find $\cos k$ and conclude that $k = 2n\pi$.)

10. Show that the fundamental period of sin is 2π.

11. Show that the fundamental period of tan and of cot is π.

12. Show that the fundamental period of sec and of csc is 2π.

13. A function f is periodic with period 2, $f(x) = x$ for all $x \in (-1, 1)$, and $f(1) = 0$. Sketch the graph of f.

14. Show that there is no function g with period 2 such that $g(x) = x$ for all $x \in [-1, 1]$.

15. Determine the fundamental period of f if (**a**) $f(x) = \sin 2x$; (**b**) $f(x) = \tan 3x$; (**c**) $f(x) = \sin kx$, $k > 0$.

Miscellaneous Exercises 2.2 (2)

Simplifications

Simplify each of the following expressions. The definitions of the circular functions, and equations (1), (4), and (5), will be especially useful.

1. $\cos x \tan x$
2. $\sin x \sec x \cot x$
3. $\cos x \tan x \csc x$
4. $\cos x + \sin x \tan x$
5. $\sin^3 x + \cos^2 x \sin x$
6. $\sin x + \dfrac{\cos^2 x}{\sin x}$
7. $\dfrac{1}{\cos^2 x} - \dfrac{1}{\cot^2 x}$
8. $\dfrac{\cos x}{1 - \sin x} \cdot \dfrac{\cos x}{1 + \sin x}$
9. $\sec x \csc x - \tan x$
10. $\csc x - \cot x \cos x$
11. $\sin x \tan x + \cos x$
12. $\dfrac{\sin x + \cos x \cot x}{\cot x}$
13. $\dfrac{(1 + \cos x)^2 + \sin^2 x}{1 + \cos x}$
14. $(\csc x + \cot x)(1 - \cos x)$
15. $\cos x(\tan x + \cos x \csc x)$
16. $1 + (\csc x - \cot x)(\csc x + \cot x)$
17. $\tan x(1 - \cot^2 x) + \cot x(1 - \tan^2 x)$
18. $(\sec x - \tan x)(1 + \csc x)$
19. $(\sec x + \tan x)(1 - \sin x)$
20. $\dfrac{\sin x \cot x + \cos x}{2 \cot x}$
21. $\dfrac{\cos x}{1 - \sin x} + \dfrac{1 + \sin x}{\cos x}$
22. $\dfrac{\sin x}{1 + \cos x} + \dfrac{1 + \cos x}{\sin x}$
23. $(1 + \tan x)\left(\dfrac{\sin x}{\sin x + \cos x}\right)$
24. $\tan x(\tan x + \cot x)$
25. $\dfrac{\cos x}{1 + \sin x} + \dfrac{1 + \sin x}{\cos x}$

26. $\cos^4 x + 2\sin^2 x \cos^2 x + \sin^4 x$

27. $(\sin x + \cos x)^2 + (\sin x - \cos x)^2$

28. $(\cos^2 x - \sin^2 x)^2 + (2 \sin x \cos x)^2$

29. $\dfrac{1}{1 + \tan^2 x} + \dfrac{1}{1 + \cot^2 x}$

30. $\dfrac{\cot^2 x}{1 + \cot^2 x}$

31. $(1 + \tan x + \sec x)(1 + \tan x - \sec x)$

32. $\dfrac{(\tan x - \sin x)(1 + \cos x)}{\sin^3 x}$

33. $\left(\dfrac{2 - \sec^2 x}{\sec^2 x}\right)\left(\dfrac{\cos x + \sin x}{\cos x - \sin x}\right) - \dfrac{2 \tan x}{1 + \tan^2 x}$

34. $3(\cos^4 x + \sin^4 x) - 2(\cos^6 x + \sin^6 x)$

35. $\dfrac{\sin(\pi - x)\sin(\pi + x) - \cos(\pi - x)\cos(\pi + x)}{\tan(\pi - x)\tan(\pi + x) - 1}$

Miscellaneous Exercises 2.2 (3)

Equations and Identities

1. With the aid of your table from Ex. 1, page 44, find all $x \in [0, 2\pi)$ such that

   (a) $\sin x = \cos x$. (*Hint:* Divide by $\cos x$. Why is this possible?)

   (b) $\sec^2 x = 2$.

   (c) $\sin^2 x = \sin x$.

   (d) $\sin^2 x = 1 + \cos x$.

   (e) $\sin^2 x - 3 \sin x + 2 = 0$.

   (f) $\cos x = 2\cos^2 x$.

   (g) $\cos x = 2 \sec x$.

   (h) $4 \sin x - 3 \csc x = 0$.

   (i) $\tan x = \cot x$.

2. Find the value at x of all six circular functions if

   (a) $\sin x = \frac{3}{5}$ and $x \in (0, \pi/2)$.

   (b) $\tan x = -2$ and $x \in (\pi/2, \pi)$.

   (c) $\cos x = \frac{5}{13}$ and $x \in (3\pi/2, 2\pi)$.

   (d) $\sec x = 1$.

   (e) $\sin x = -\frac{1}{2}$ and $\tan x > 0$.

   (f) $\tan x = \frac{7}{24}$ and $\cos x < 0$.

   (g) $\sin x = \dfrac{m^2 - n^2}{m^2 + n^2}$, $m > n > 0$, and $\cos x < 0$.

Show that statements 3–18 are true at each point x (and each point y) at which they have a meaning.

3. $\sin x + \cos x \cot x = \cot x \sec x$

4. $2\cos^2 x - 1 = 1 - 2\sin^2 x$

5. $\tan^2 x - \sin^2 x = \tan^2 x \sin^2 x$

6. $\dfrac{\cos^2 x}{1 + \sin x} = 1 - \sin x$

7. $\sec x + \csc x = \dfrac{1 + \tan x}{\sin x}$

8. $\dfrac{1 + \sin x}{\cos x} = \dfrac{\cos x}{1 - \sin x}$

9. $\dfrac{1}{\sec x + \tan x} = \sec x - \tan x$

10. $\sec^2 x + \csc^2 x = \sec^2 x \csc^2 x$

11. $\dfrac{\tan^2 x}{1 + \tan^2 x} = \sin^2 x$

12. $\dfrac{1}{\csc x - \cot x} = \csc x + \cot x$

13. $\dfrac{\cot x - \cos x}{\cot x \cos x} = \dfrac{\cot x \cos x}{\cot x + \cos x}$

14. $\dfrac{\tan x - \sin x}{\sin^3 x} = \dfrac{\sec x}{1 + \cos x}$

15. $(\sin x + \sec x)^2 + (\cos x + \csc x)^2 = (1 + \sec x \csc x)^2$

16. $\dfrac{\tan x + \tan y}{\cot x + \cot y} = \tan x \tan y$

17. $\cos^2 x \cos^2 y - \sin^2 x \sin^2 y = \cos^2 x - \sin^2 y$

18. $\sin^2 x \cos^2 y - \cos^2 x \sin^2 y = \sin^2 x - \sin^2 y$

19. Find all $x, y \in [0, 2\pi)$ such that

$$\frac{\sin x}{\sin y} = \sqrt{2} \quad \text{and} \quad \frac{\tan x}{\tan y} = \sqrt{3}.$$

20. Determine all $x \in [0, 2\pi]$ such that

(a) $\sin x > \cos x$. (b) $\sin^2 x > \cos^2 x$.

(*Hint:* A sketch graph should help.)

2.3 Circular functions: sum formulas

It is not generally true that $f(x + y) = f(x) + f(y)$; indeed, only a very special type of function has this property (it is an interesting problem, not excessively difficult, to find some functions that do have this property). It is generally of interest, however, to discover, for any function f, a formula that expresses $f(x + y)$ in terms of functions of x and y alone, and we shall here develop such formulas for some of the circular functions. It is convenient to start with a formula for $\cos(x - y)$.

We assume from plane geometry that in a given circle equal arcs have equal chords. In the elementary treatment, it is often implied, if not explicitly stated, that each of the arcs involved is less than a semicircle. We shall consider here that the assumption is valid for all arcs. It follows, referring to our unit circle $u^2 + v^2 = 1$ and our wrapping function W, that

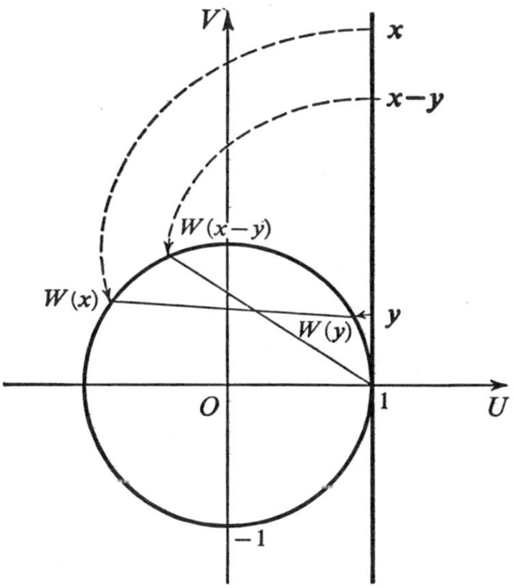

Fig. 9

the distance from $W(x) = (\cos x, \sin x)$ to $W(y) = (\cos y, \sin y)$ is the same as the distance from $W(x - y) = (\cos[x - y], \sin[x - y])$ to $W(0) = (\cos 0, \sin 0) = (1, 0)$. If we let d stand for each of these distances and use equation (1) of § 1.3, we have, using the coordinates of $W(x)$ and $W(y)$,

$$\begin{aligned} d^2 &= (\cos x - \cos y)^2 + (\sin x - \sin y)^2 \\ &= \cos^2 x - 2\cos x \cos y + \cos^2 y + \sin^2 x - 2\sin x \sin y + \sin^2 y \\ &= 2 - 2\cos x \cos y - 2\sin x \sin y, \end{aligned}$$

where we have twice made use of the relationship $\cos^2 x + \sin^2 x = 1$, which is equation (1) of § 2.2. Using the coordinates of $W(x - y)$ and $W(0)$, we have

$$\begin{aligned} d^2 &= [\cos (x - y) - 1]^2 + [\sin (x - y) - 0]^2 \\ &= \cos^2 (x - y) - 2 \cos (x - y) + 1 + \sin^2 (x - y) \\ &= 2 - 2 \cos (x - y). \end{aligned}$$

Equating these two expressions for d^2 and simplifying, we have

► $$\cos (x - y) = \cos x \cos y + \sin x \sin y. \tag{1}$$

We can now manipulate this result in such a way as to produce several other important formulas. First, replacing y by $-y$ in (1), we have

$$\cos [x - (-y)] = \cos x \cos (-y) + \sin x \sin (-y),$$

or, since $\cos (-y) = \cos y$ and $\sin (-y) = -\sin y$, by (2) of § 2.2,

► $$\cos (x + y) = \cos x \cos y - \sin x \sin y. \tag{2}$$

Putting $x = \pi/2$ in (1) and using $\cos \pi/2 = 0$, $\sin \pi/2 = 1$, we find

$$\cos \left(\frac{\pi}{2} - y\right) = \sin y; \tag{3}$$

and replacing y by $\frac{\pi}{2} - y$ in (3), we have

$$\sin \left(\frac{\pi}{2} - y\right) = \cos y. \tag{4}$$

Replacing y by $x + y$ in (3), we have

$$\sin (x + y) = \cos \left[\frac{\pi}{2} - (x + y)\right] = \cos \left[\left(\frac{\pi}{2} - x\right) - y\right],$$

or, by (1),

$$\sin (x + y) = \cos \left(\frac{\pi}{2} - x\right) \cos y + \sin \left(\frac{\pi}{2} - x\right) \sin y.$$

Using (3) and (4), we have

► $$\sin (x + y) = \sin x \cos y + \cos x \sin y. \tag{5}$$

It is left to the student as an exercise to show that

► $$\sin (x - y) = \sin x \cos y - \cos x \sin y. \tag{6}$$

Miscellaneous Exercises 2.3 (1)

Derivations and Evaluations

1. Use equations (5) of this section and (2) of § 2.2 to establish (6).

2. Show that (**a**) $\sin\left(\frac{\pi}{2} + x\right) = \cos x.$

(**b**) $\cos\left(\frac{\pi}{2} + x\right) = -\sin x.$

(**c**) $\sin(\pi - x) = \sin x.$

(**d**) $\cos(\pi - x) = -\cos x.$

3. Find (**a**) $\sin \pi/12$; (**b**) $\cos \pi/12$. (*Hint:* $\pi/12 = \pi/4 - \pi/6$; use (6) and (1), in conjunction with the values of sin and cos at $\pi/4$ and $\pi/6$ obtained in Ex. 1, p. 44.)

4. Find (**a**) $\sin 5\pi/12$. (*Hint:* $5\pi/12 = \pi/3 + \pi/12 = \pi/2 - \pi/12$.)

(**b**) $\cos 5\pi/12$.

(**c**) $\sin 7\pi/12$.

(**d**) $\cos 7\pi/12$.

(**e**) $\sin 11\pi/12$.

(**f**) $\cos 11\pi/12$.

(**g**) $\tan 5\pi/12$

5. If $\sin x = \frac{3}{5}$, $x \in (0, \pi/2)$, and $\cos y = \frac{12}{13}$, $y \in (3\pi/2, 2\pi)$, find (**a**) $\cos x$, $\sin y$; (**b**) $\sin(x + y)$; (**c**) $\cos(x + y)$.

6. If $\cos x = -\frac{7}{25}$, $x \in (\pi/2, \pi)$, and $\cos y = \frac{4}{5}$, $y \in (3\pi/2, 2\pi)$, find (**a**) $\sin(x - y)$; (**b**) $\cos(x - y)$.

7. (**a**) Show that $\cos x \cos(y - x) - \sin x \sin(y - x) = \cos y$.

(**b**) Simplify $\sin x \cos(x - y) - \cos x \sin(x - y)$.

(**c**) Simplify $\sin(x + y)\cos(x - y) + \cos(x + y)\sin(x - y)$.

8. Simplify: (**a**) $\sin(x + y) + \sin(x - y)$

(**b**) $\cos(x - y) - \cos(x + y)$

9. Prove:

(**a**) $\sin(x + y)\sin(x - y) = \sin^2 x - \sin^2 y = \cos^2 y - \cos^2 x$

(**b**) $\cos(x + y)\cos(x - y) = \cos^2 x - \sin^2 y = \cos^2 y - \sin^2 x$

10. Prove: (a) $\dfrac{\sin(x+y)}{\cos x \cos y} = \tan x + \tan y$

(b) $\dfrac{\sin(\pi/4 + x)}{\cos(\pi/4 + x)} = \dfrac{1 + \tan x}{1 - \tan x}$

(c) $\dfrac{\sin(x+y)}{\sin(x-y)} = \dfrac{\tan x + \tan y}{\tan x - \tan y}$

(d) $\dfrac{\cos(x+y)}{\cos(x-y)} = \dfrac{1 - \tan x \tan y}{1 + \tan x \tan y}$

11. Use the definition $\tan(x+y) = \dfrac{\sin(x+y)}{\cos(x+y)}$ to show that

▶ $$\tan(x+y) = \frac{\tan x + \tan y}{1 - \tan x \tan y} \quad (\cos x \cos y \neq 0). \tag{7}$$

12. Show that

▶ $$\tan(x-y) = \frac{\tan x - \tan y}{1 + \tan x \tan y} \quad (\cos x \cos y \neq 0). \tag{8}$$

13. Find (a) $\tan 5\pi/12$; (b) $\tan 7\pi/12$; (c) $\tan 11\pi/12$.

14. Use (7) to show that π is a period of tan.

15. If $\tan x = 2$ and $\tan y = 3$, find all $z \in [0, 2\pi)$ such that $\tan z = \tan(x+y)$.

16. If $\sin x = -\frac{4}{5}$, $x \in (\pi, 3\pi/2)$, and $\cos y = \frac{7}{25}$, $y \in (0, \pi/2)$, find $\tan(x-y)$.

17. Show that $\dfrac{\tan(x + \pi/4) - \tan x}{1 + \tan(x + \pi/4)\tan x} = 1.$

18. (a) Show that $\tan\left(\dfrac{\pi}{4} - x\right) = \dfrac{1 - \tan x}{1 + \tan x}$.

(b) Set $x = \pi/8$ in (a) and solve the resulting equation for $\tan \pi/8$.

(c) Find $\tan 3\pi/8$.

19. Show that

▶ $$\sin 2x = 2 \sin x \cos x. \tag{9}$$

$$\cos 2x = \cos^2 x - \sin^2 x. \tag{10a}$$

$$\tan 2x = \frac{2 \tan x}{1 - \tan^2 x}. \tag{11}$$

(*Hint:* Set x for y in (5), (2), (7), respectively.)

20. Using (10a), show that

$$\cos 2x = 2\cos^2 x - 1; \tag{10b}$$

$$\cos 2x = 1 - 2\sin^2 x. \tag{10c}$$

21. If $\sin x = \frac{3}{5}$ and $x \in (0, \pi/2)$, find
(a) $\cos x$. (b) $\sin 2x$. (c) $\cos 2x$. (d) $\sin 4x$.

22. (a) If $\tan y = \frac{1}{5}$, find $\tan 2y$ and $\tan 4y$.
(b) If also $y \in (0, \pi/2)$, conclude that y is very nearly equal to $\pi/16$.

23. Simplify and then evaluate:

(a) $2 \sin \dfrac{\pi}{12} \cos \dfrac{\pi}{12}$ (d) $\dfrac{2 \tan \pi/8}{1 - \tan^2 \pi/8}$

(b) $\sin \dfrac{\pi}{8} \cos \dfrac{\pi}{8}$ (e) $2 \cos^2 \dfrac{\pi}{12} - 1$

(c) $\cos^2 \dfrac{\pi}{12} - \sin^2 \dfrac{\pi}{12}$ (f) $1 - 2\sin^2 \dfrac{7\pi}{8}$

24. Simplify:

(a) $2 \sin \dfrac{x}{2} \cos \dfrac{x}{2}$ (c) $\dfrac{2 \tan z/6}{1 - \tan^2 z/6}$ (e) $2\cos^2 \dfrac{y}{2} - 1$

(b) $\sin 17x \cos 17x$ (d) $\cos^2 5t - \sin^2 5t$ (f) $1 - 2\sin^2 3x$

25. Use (10b) and (10c) to show that

$$\left|\sin \frac{x}{2}\right| = \sqrt{\frac{1 - \cos x}{2}}. \tag{12}$$

$$\left|\cos \frac{x}{2}\right| = \sqrt{\frac{1 + \cos x}{2}}. \tag{13}$$

26. (a) If $\cos x = \frac{7}{25}$ and $x \in (0, \pi/2)$, find $\sin x/2$ and $\cos x/2$.
(b) If $\cos x = \frac{7}{25}$ and $x \in (3\pi/2, 2\pi)$, find $\sin x/2$ and $\cos x/2$.

27. If $\cos y = -\frac{1}{3}$ and $y \in (\pi, 3\pi/2)$, find $\sin y/2$ and $\cos y/2$.

28. Find (a) $\sin \pi/8$; (b) $\cos \pi/8$.

29. Show that

$$\tan \frac{x}{2} = \frac{1 - \cos x}{\sin x}; \tag{14a}$$

$$\tan \frac{x}{2} = \frac{\sin x}{1 + \cos x}. \tag{14b}$$

Miscellaneous Exercises 2.3 (2)

Equations and Identities

1. Find all $x \in [0, 2\pi)$ such that

(a) $\tan 2x = 1$.

(*Hint:* There are 4 solutions.)

(b) $\sin 3x = 1$.

(c) $2 \cos 2x = \sqrt{3}$.

(d) $\sin 2x = \sin x$.

(e) $\cos 2x = \cos x$.

(f) $\cos 4x = \sin 2x$.

(g) $\cos 4x = 1 - 3 \cos 2x$.

(h) $\sec 2x = 2 \cos 2x$.

(i) $\tan 2x + \tan x = 0$.

(j) $\sin\left(\frac{\pi}{4} - x\right) \cos\left(\frac{\pi}{4} - x\right) = \frac{1}{4}$.

Prove the following identities:

2. $\left(\sin \frac{x}{2} + \cos \frac{x}{2}\right)^2 = 1 + \sin x$

3. $2 \cos^2\left(\frac{\pi}{4} - \frac{x}{2}\right) = 1 + \sin x$

4. $\sin\left(\frac{\pi}{4} - z\right) \cos\left(\frac{\pi}{4} - z\right) = \frac{1}{2} \cos 2z$

5. $\sin y \tan y + \cos 2y \sec y = \cos y$

6. $\dfrac{2 \tan x}{1 + \tan^2 x} = \sin 2x$

7. $\dfrac{1 - \tan^2 y}{1 + \tan^2 y} = \cos 2y$

8. $\dfrac{\sin x}{1 - \cos x} = \cot \dfrac{x}{2}$

9. $\dfrac{1 + \cos 2x}{\sin 2x} = \cot x$

10. $\dfrac{\sin 3x}{\sin x} - \dfrac{\cos 3x}{\cos x} = 2$

11. $\dfrac{\tan 2x \tan x}{\tan 2x - \tan x} = \sin 2x$

12. $\tan^2\left(\dfrac{\pi}{4} - \dfrac{x}{2}\right) = \dfrac{1 - \sin x}{1 + \sin x}$

13. $\sin 4x = 4 \sin x \cos x \cos 2x$

14. $\dfrac{1 - \cos 4x}{1 + \cos 2x} = 4 \sin^2 x$

15. $1 + \tan x \tan \dfrac{x}{2} = \sec x$

16. $\sec^2 x(1 + \sec 2x) = 2 \sec 2x$

17. $\dfrac{\cos 2x}{1 + \sin 2x} = \tan\left(\dfrac{\pi}{4} - x\right)$

18. $\tan\left(\dfrac{\pi}{4} - x\right) + \tan\left(\dfrac{\pi}{4} + x\right) = 2 \sec 2x$

19. $2 \sin (x + y) \sin (x - y) = \cos 2y - \cos 2x$

20. $\csc x - \cot x = \tan \dfrac{x}{2}$

21. $\dfrac{\tan 3y + \tan y}{\tan 3y - \tan y} = 2 \cos 2y$

Miscellaneous Exercises 2.3 (3)

Sum and Product Formulas

1. Use equations (5) and (6) to show that

$$\sin (a + b) + \sin (a - b) = 2 \sin a \cos b.$$

Then let $x = a + b$, $y = a - b$, solve for a and b in terms of x and y, and show that this result is equivalent to

► $$\sin x + \sin y = 2 \sin \frac{x + y}{2} \cos \frac{x - y}{2}. \tag{15}$$

2. Show that

► $$\sin x - \sin y = 2 \cos \frac{x + y}{2} \sin \frac{x - y}{2}. \tag{16}$$

3. Show that

► $$\cos x + \cos y = 2 \cos \frac{x + y}{2} \cos \frac{x - y}{2}. \tag{17}$$

4. Show that

► $$\cos x - \cos y = -2 \sin \frac{x + y}{2} \sin \frac{x - y}{2}. \tag{18}$$

5. Use (15) to simplify $\sin\left(\dfrac{\pi}{4} + x\right) + \sin\left(\dfrac{\pi}{4} - x\right)$.

6. Show that $\sin 3x + \sin 5x = 4 \cos x \cos 2x \sin 2x$.

7. Simplify $\dfrac{\cos 3x - \cos x}{\sin 3x - \sin x}$, using (16) and (18).

8. Simplify $\dfrac{\sin 2x + \sin 2y}{\cos 2x - \cos 2y}$.

9. Show that $\dfrac{\sin 5x - \sin x}{2 \cos 3x - \cos 5x - \cos x} = \cot x$.

10. Show that $\dfrac{\cos x - \cos (x + y)}{\sin (x + y) + \sin x} = \tan \frac{1}{2}y$.

11. Prove: $\cos x + \cos\left(x + \dfrac{2\pi}{3}\right) + \cos\left(x + \dfrac{4\pi}{3}\right) = 0$.

12. Prove: $\sin x + \sin\left(x + \dfrac{2\pi}{3}\right) + \sin\left(x + \dfrac{4\pi}{3}\right) = 0$.

13. Find the maximum value of $\cos\left(x + \dfrac{\pi}{3}\right) - \cos\left(x + \dfrac{2\pi}{3}\right)$.

14. Find all $x \in [0, 2\pi)$ such that

(a) $\sin x + \sin 3x = \sin 2x$. (*Hint:* Use (15).)

(b) $\cos x + \cos 2x + \cos 3x = 0$.

(c) $\sin 5x + \sin x = \sin 3x$.

15. Use (5) and (6) to show that

$$\sin x \cos y = \tfrac{1}{2}[\sin (x + y) + \sin (x - y)]. \tag{19}$$

16. Use (1) and (2) to show that

$$\cos x \cos y = \tfrac{1}{2}[\cos (x + y) + \cos (x - y)]; \tag{20}$$

$$\sin x \sin y = -\tfrac{1}{2}[\cos (x + y) - \cos (x - y)]. \tag{21}$$

17. Use (19) to show that $\sin \dfrac{5\pi}{12} \cos \dfrac{\pi}{12} = \frac{1}{2} + \frac{1}{4}\sqrt{3}$.

18. Show that (a) $\cos \dfrac{\pi}{12} \cos \dfrac{5\pi}{12} = \frac{1}{4}$.

(b) $\sin \dfrac{\pi}{12} \sin \dfrac{7\pi}{12} = \frac{1}{4}$.

(c) $\sin \dfrac{\pi}{12} \cos \dfrac{5\pi}{12} = \frac{1}{2} - \frac{1}{4}\sqrt{3}$.

19. Prove: $\sin\left(\frac{\pi}{4} - x\right) \sin\left(\frac{\pi}{4} + x\right) = \frac{1}{2} \cos 2x.$

20. Prove: $\sin x \sin\left(\frac{\pi}{3} - x\right) \sin\left(\frac{\pi}{3} + x\right) = \frac{1}{4} \sin 3x.$

21. Prove: $\cos x \cos\left(\frac{\pi}{3} - x\right) \cos\left(\frac{\pi}{3} + x\right) = \frac{1}{4} \cos 3x.$

22. Prove: $\dfrac{\sin 6x \cos 3x - \sin 8x \cos x}{\sin 3x \sin 4x - \cos 2x \cos x} = \tan 2x.$

23. Find all $x \in [0, 2\pi)$ such that

(a) $2 \cos x \cos\left(\frac{\pi}{3} - x\right) = 1.$ (b) $\sin\left(\frac{\pi}{3} - x\right) \cos\left(\frac{\pi}{3} + x\right) = \frac{1}{4}\sqrt{3}.$

2.4 Circular functions: angles and measurement

We turn now to the relationship between circular functions and angles, and to the use of these functions in the measurement of triangles. Strictly speaking, we must define a whole new set of functions; the functions we have been investigating have sets of real numbers as their domains, and we now require functions that have sets of angles as their domains. The relationship between the two sets of functions is so close, however, that we can often ignore the distinction (indeed, we shall even use the same names for functions in both sets), provided we make the appropriate choice of our unit of measure for angles.

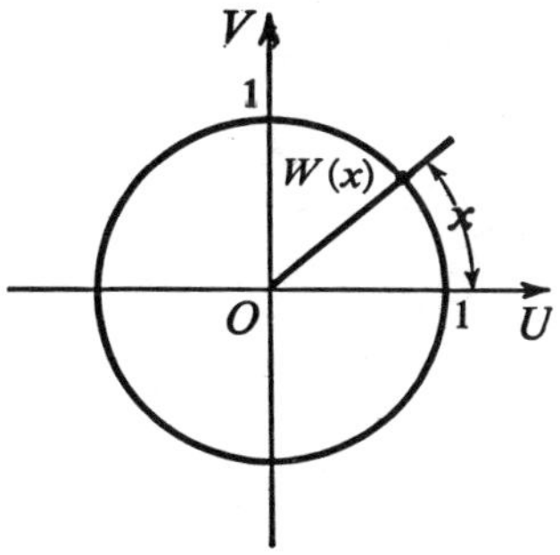

Fig. 10

To define these new functions, we consider an angle in the uv-plane, with its vertex at the origin and one side extending in the positive direction along the u-axis. The other side of the angle will then meet the unit circle at a point $W(x)$, where x is the measure of the arc intercepted by the sides of the angle. It is convenient to use this same number x as a measure of the angle; indeed, in calculus, any other measure would be wildly impractical. With this choice, the unit of measure is called a radian (abbreviated "rad."). Thus, in a circle of unit radius, a central angle of x rad. intercepts an arc x units long. The measure of an angle in radians is also called the circular measure of the angle. To obtain the relationship between circular and degree measure, we note that $W(\pi/2) = (0, 1)$ corresponds to a right angle, and therefore $90° = \pi/2$ rad., $1° = \pi/180$ rad., or approximately 0.01745 rad., and 1 rad. $= (180/\pi)°$, or approximately $57°18'$.

The definitions of circular (or trigonometric) functions of angles now follow easily. If $\angle A$ has circular measure x rad., we define $\sin \angle A$ by the equation $\sin \angle A = \sin x$, where $\sin x$ is the now-familiar sine function evaluated at the real number x; similarly, $\cos \angle A = \cos x$, $\tan \angle A = \tan x$ (provided x is in the domain of tan; if not, $\tan \angle A$ is not defined), and so on. The job is not quite done yet, however; we must also define a set of functions whose arguments are measured in degrees. To do this, we simply convert from degrees to radians: thus, for example, $\sin y° = \sin x$, where $x = \pi y/180$, and so on. For example, $\cos 60° = \cos \pi/3 = \frac{1}{2}$.

Tables and calculators. Exercise 1 on page 44 asks for a brief table of values of the circular functions, values that can be determined from the definitions of those functions with the help of a little geometry. We have now derived formulas that could be used to refine that table, inserting more values between those already found, but expressions thus found are clumsy. For example,

$$\sin \frac{\pi}{60} = \frac{1}{16}[(\sqrt{6} + \sqrt{2})(\sqrt{5} - 1) - 2(\sqrt{3} - 1)\sqrt{5 + \sqrt{5}}].$$

This exact expression has some theoretical interest, but in most applications it is less useful than the approximation

$$\sin \frac{\pi}{60} = 0.0523.$$

Such approximations are readily available from calculators or tables. Calculus provides the mathematical basis for the programs by which calculators produce these approximations, as well as efficient techniques for computing tables.

If, for example, you want to use a calculator to find an approximation to sin 2, you use the fact that the same number is the sine of an angle of *2 radians*. Thus, put the calculator in radian mode, press "2" and "sin", and get $\sin 2 = 0.9093$, correct to four significant digits. To use a table of values of the circular functions for the same purpose is a little harder. First, many such tables give angles measured only in degrees; in such a case you must convert radian measure to degree measure before you can use the table. Second, most such tables cover only the interval $[0, \pi/2]$,

The design of modern highways depends on the mathematical analysis of traffic flow. One problem is the orderly mixing of traffic from different sources. The length of time that it takes a driver to join the main stream is expressed as a function of the distribution of gaps between cars and of the anticipated performance of the individual car; this, in turn, determines the length of waiting lines on approach ramps. The engineer's goal is to maximize the safe carrying capacity of the road, a function of many variables.

that is, degree measure from 0° to 90°. Table 2, at the back of the book, covers the interval [0, 1.60]. This, however, is enough; as the student should show, using the formulas numbered (2) and (3) in § 2.2 and the periodicity property, if f is any one of the circular functions, if k is any integer, and if x is any point in the domain of f, then

$$|f(k\pi \pm x)| = |f(x)|. \tag{1}$$

Since the distance on the number line between any number x and the nearest integral multiple of π is not greater than $\pi/2$, if we determine this distance and use it to enter a table of values of f, equation (1) assures us that the number we find will be $|f(x)|$. Thus, to find $|\cos 17|$, we note that $17 - 5\pi = 1.2920$, approximately, and therefore $|\cos 17| = \cos 1.2920 = 0.2752$; similarly, $|\sin(-100)| = \sin(32\pi - 100) = \sin 0.5310 = 0.5064$. We still have to determine the sign of $f(x)$: this is best done by determining into what quadrant the wrapping function W maps the point x and then using the definition of f. Thus, since $5\pi < 17 < 5.5\pi$, it follows that $W(17)$ is in Quadrant III, where the cosine function takes only negative values, and therefore $\cos 17 = -0.2752$; and since $-32\pi < -100 < -31.5\pi$, it follows that $W(-100)$ is in Quadrant I, where all the circular functions take only positive values, and therefore $\sin(-100) = 0.5064$. The following partial summary is convenient:

If $W(x)$ is in Quadrant		I	II	III	IV
then	$\sin x$ is	+	+	−	−
	$\cos x$ is	+	−	−	+
and	$\tan x$ is	+	−	+	−

Miscellaneous Exercises 2.4 (1)

Evaluations

1. Convert the following circular measures to degrees: $\pi/5$, $5\pi/12$, $7\pi/6$, $10\pi/3$, $-\pi/4$, 20π, 2, -5.

2. Convert to circular measure: 50°, 150°, −300°, 450°, 27°10′.

3. Prove: If k is an odd integer, then (a) $|\sin(k\pi/2 \pm x)| = |\cos x|$; (b) $|\cos(k\pi/2 \pm x)| = |\sin x|$; (c) $|\tan(k\pi/2 \pm x)| = |\cot x|$.

4. Find a rational approximation to the distance between the point x and the nearest integral multiple of π for the following values of x:

(a) 1	(c) 3	(e) 20	(g) 16
(b) −5	(d) 7	(f) −9	(h) −12.4

5. Use tables to obtain rational approximations:

(a) $\sin 16\pi/3$ (f) $\sin 2$ (k) $\sec 12$

(b) $\tan -23\pi/5$ (g) $\tan 5$ (l) $\tan(-20)$

(c) $\cos 43\pi/6$ (h) $\sec(-3)$ (m) $\cos 3.6$

(d) $\sec 21\pi/4$ (i) $\csc 9$ (n) $\sin(-11.1)$

(e) $\cot 49\pi$ (j) $\tan(-10)$ (o) $\cos\sqrt{85}$

6. Find rational approximations for all $x \in [0, 2\pi)$ such that

(a) $\sin x = 0.6$ (f) $\csc x = 3$

(b) $\tan x = -2$ (g) $\tan x = -0.75$

(c) $\cos x = \frac{1}{3}$ (h) $\sin x = -0.2756$

(d) $\sec x = 1.7$ (i) $\csc x = 2.0211$

(e) $\cos x = -0.4$ (j) $\cos x = -0.91$

7. If $f(x) = x^2 + 2x \sin x$, find rational approximations to $f(-1)$ and $f(2)$.

8. Find rational approximations, correct to the nearest hundredth, to all $x \in [0, 2\pi)$ such that

(a) $2 \cos x = \sin x$. (h) $\sin^2 x - \sin x = \cos^2 x$.

(b) $\sin^2 x = 3 \sin x \cos x$. (i) $2(\tan x + \cot x) = 5$.

(c) $\sin x + 3 \cos x = 0$. (j) $\sec^2 x + 1 = 3 \tan x$.

(d) $\cot x - 2 \tan x = 0$. (k) $4 \cos^2 x + 3 \cos x = 0$.

(e) $5 \cos^2 x = 12 \sin x \cos x$. (l) $\sin^2 x + 2 \sin x - 1 = 0$.

(f) $6 \sin^2 x - 7 \sin x + 2 = 0$. (m) $\sec^2 x + 2 \tan x = 3$.

(g) $6 \sin^2 x + 5 \cos x - 7 = 0$. (n) $\sin x + 1 = 2 \cos x$.

9. Find rational approximations, correct to the nearest hundredth, to all $x \in [0, 2\pi)$ such that

(a) $\sin 2x = \sin x$. (h) $\tan 2x = \tan x$.

(b) $\cos 2x = \cos x$. (i) $\tan 2x - 2 \cot 2x = 1$.

(c) $\sin 3x = \cos 3x$. (j) $\sin x + \sin 3x = \sin 2x$.

(d) $\cos 2x = \sin x$. (*Hint:* Use equation (15) of § 2.3.)

(e) $\cos 2x = 1 - \sin x$. (k) $\cos x + \cos 3x = \cos 2x$.

(f) $\cos^2 2x = \sin^2 2x - \sin 2x$. (l) $\cos 2x - \cos x = \sin 2x - \sin x$.

(g) $3 \cos 2x + \cos x = 2$. (m) $\sin x + \sin 3x + \sin 5x = 0$.

10. Assume that, on a given circle, arc lengths are proportional to the central angles they subtend, and show that if on a circle of radius r an arc of length s subtends a central angle of circular measure θ, then

$$s = r\theta. \tag{2}$$

(*Hint:* What arc length s corresponds to $\theta = 2\pi$?)

11. The radius of a circle is 5 in. (**a**) What central angle is subtended by an 11-inch arc? (**b**) What length of arc does a central angle of 3 rad. intercept?

12. The radius of a flywheel is 18 in., and the wheel turns at 600 r.p.m. (revolutions per minute). Find the velocity, in feet per minute, of a point on the rim of the wheel.

13. Assume that, in a given circle, the areas of sectors are proportional to the central angles of the sectors. Prove: If in a circle of radius r a sector of angle θ radians has area K, then

$$K = \tfrac{1}{2}r^2\theta. \tag{3}$$

(*Hint:* See Ex. 10, above.)

14. Prove: A circular sector of radius r and arc length s has area $\frac{1}{2}rs$.

15. In a circle of radius 6 cm., find the area of a sector which has (**a**) a central angle of 2.5 rad.; (**b**) an arc 10 cm. long.

Miscellaneous Exercises 2.4 (2)

Right and Isosceles Triangles

1. A right triangle ABC is placed in the uv-plane with vertex A at the center of a unit circle, as indicated in Fig. 11.

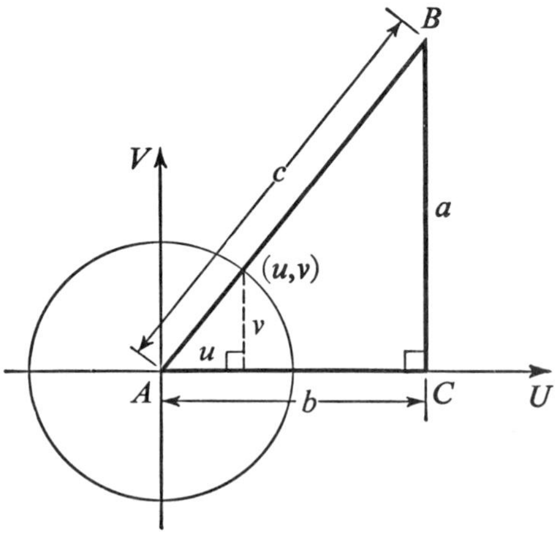

Fig. 11

Use similar triangles to show that

$$\sin A = \frac{a}{c} = \frac{\text{Leg opposite } \angle A}{\text{Hypotenuse}}$$

$$\cos A = \frac{b}{c} = \frac{\text{Leg adjacent to } \angle A}{\text{Hypotenuse}}$$

$$\tan A = \frac{a}{b} = \frac{\text{Leg opposite } \angle A}{\text{Leg adjacent to } \angle A}$$

$$\cot A = \frac{b}{a}$$

$$\sec A = \frac{c}{b}$$

$$\csc A = \frac{c}{a}.$$

(Here, and subsequently, we use the abbreviation "sin A" for "sin $\angle BAC$.")

2. Use the relations obtained in Ex. 1 to show that if $\triangle PQR$ has a right angle at R, then $p = r \sin P = q \tan P$, and obtain two similar expressions for q in terms of p or r and a circular function of Q.

3. Prove: If X and Y are the acute angles of a right triangle, then

   (a) $\sin X = \cos Y$. (b) $\tan X = \cot Y$. (c) $\sec X = \csc Y$.

4. Use tables to compute the missing dimensions of the given triangles. Find angles to the nearest *degree*, sides to two significant figures.

   (a) $\triangle ABC$: $a = 3.2$ cm., $b = 4.8$ cm., $C = 90°$

   (b) $\triangle XYZ$: $x = 0.13$ in., $y = 0.65$ in., $Z = 90°$

   (c) $\triangle PQR$: $P = 90°$, $Q = 43°$, $q = 40$ yd.

   (d) $\triangle ABC$: $A = 90°$, $B = 57°$, $a = 56$ ft.

   (e) $\triangle PQR$: $p = 5$ m., $q = 6$ m., $R = 90°$

   (f) $\triangle ABC$: $A = 90°$, $a = 16$ in., $b = 15$ in.

5. How high is a flagpole that casts a 30-foot shadow when the elevation of the sun is 53°?

6. A guy wire reaches to the ground from a point 20 ft. up a pole and makes an angle of 28° with the pole. How long is the wire? How far from the foot of the pole does it reach the ground?

7. The sides of a triangle are 4 in., 6 in., 6 in. Find the angles.

8. (a) Write a formula for the perimeter of a regular polygon of n sides if it is (i) inscribed in a circle of radius R; (ii) circumscribed about the same circle.

(b) Evaluate both formulas if $n = 60$, $R = \frac{1}{2}$.

9. If h is the perpendicular distance from A to BC in $\triangle ABC$, show that

$$h = \frac{a}{\cot B + \cot C}.$$

Consider (i) $0° < C < 90°$; (ii) $C = 90°$; (iii) $90° < C < 180°$.

10. Two observers are 1000 ft. apart, on opposite sides of a tower. One of them finds that the angle of elevation of the top of the tower is 43°, and the other finds that it is 52°. How high is the tower? (*Hint:* Use Ex. 9.)

11. A woman in a boat finds that the angle of elevation of the top of a lighthouse is 30° at one point and 40° at a point 100 yd. nearer the foot of the lighthouse. Use Ex. 9 to find the height of the lighthouse.

12. Areas. If h is the length of the altitude on side AC of $\triangle ABC$, then the area of the triangle is $K = \frac{1}{2}bh$. Show that, in either of the figures below, $h = a \sin C$ ($= a \sin \angle BCA$), and conclude that

$$\blacktriangleright \qquad K = \tfrac{1}{2}ab \sin C. \tag{4a}$$

Show also that

$$\blacktriangleright \qquad K = \tfrac{1}{2}bc \sin A = \tfrac{1}{2}ca \sin B. \tag{4b}$$

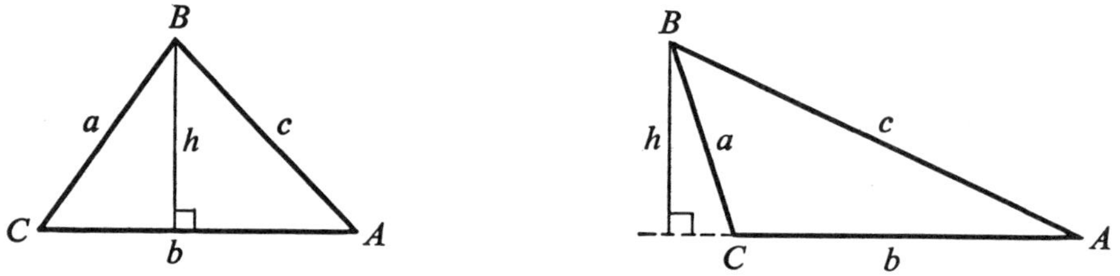

Fig. 12

13. Find the area of $\triangle ABC$, correct to two significant figures, if

(a) $a = 12$ ft., $b = 16$ ft., $C = 30°$.

(b) $a = 23$ in., $c = 42$ in., $B = 57°$.

(c) $b = 0.42$ m., $c = 1.3$ m., $A = 11°$.

(d) $a = 12$ in., $b = 5$ in., $C = 142°$.

14. Find the area of a parallelogram with sides of 6 in. and 7 in. and a 50° angle.

15. Find the area of a parallelogram if its diagonals are 8 in. and 10 in. long and meet at an angle of 46°.

16. Each side of a rhombus is 16 in. long, and one of its angles contains 37°. Find the lengths of the diagonals and the area.

17. Find the area of a regular polygon of 10 sides inscribed in a circle of radius 8 in.

18. (a) Write a formula for the area of a regular polygon of n sides, if it is (i) inscribed in a circle of radius R; (ii) circumscribed about the same circle.

(b) If $n = 90$, $R = 1$ in., (i) evaluate both formulas; (ii) compute the area of the circle.

19. A regular pentagon has a diagonal d in. long.

(a) Write a formula for the area in terms of d.

(b) Evaluate, if $d = 10$ in.

20. A circle of radius a is externally tangent to a circle of radius b $(a > b)$, and common external tangents to these circles meet at an angle θ. Show that $\sin\theta = \dfrac{4(a-b)\sqrt{ab}}{(a+b)^2}$.

Miscellaneous Exercises 2.4 (3)

Oblique Triangles

1. Law of sines. Use (4a) and (4b) to show that, in any $\triangle ABC$,

$$\frac{a}{\sin A} = \frac{b}{\sin B} = \frac{c}{\sin C}. \tag{5}$$

2. Use (5) to find the missing parts of $\triangle ABC$ if

(a) $a = 10$, $b = 12$, $B = 25°$.

(b) $A = 70°$, $B = 63°$, $a = 3.5$.

(c) $A = 112°$, $a = 17$, $b = 8$.

(d) $A = 53°$, $a = 0.26$, $c = 0.19$.

(e) $A = 143°$, $B = 11°$, $c = 2.4$.

3. In $\triangle PQR$, $P = 27°$, $p = 4$ in., $q = 7$ in. Show that there are two possible values of Q, and find the two corresponding values for r.

4. Do Ex. 3 if $P = 52°$, $p = 3.6$, $q = 4.0$.

5. The area of $\triangle XYZ$ is 48 sq. in., and $x = 16$ in., $y = 12$ in. Find Z.

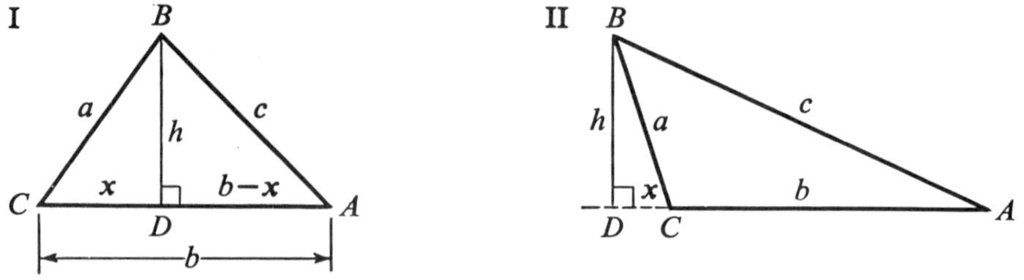

Fig. 13

6. Law of cosines.

(a) Use the Pythagorean theorem in $\triangle BCD$ and in $\triangle BAD$ to get two equations in a, b, c, h, x. Eliminate h between these equations and show that in Fig. 13, I, $c^2 = a^2 + b^2 - 2bx$, and in Fig. 13, II, $c^2 = a^2 + b^2 + 2bx$.

(b) Show that in Fig. 13, I, $x = a \cos C$ ($= a \cos \angle BCA$), and in Fig. 13, II, $x = -a \cos C$ ($= -a \cos \angle BCA$). (Because x denotes the distance between C and D, it is a non-negative number.)

(c) Conclude that, in any $\triangle ABC$,

$$c^2 = a^2 + b^2 - 2ab \cos C, \quad \text{and also}$$
$$a^2 = b^2 + c^2 - 2bc \cos A, \tag{6}$$
$$b^2 = c^2 + a^2 - 2ca \cos B.$$

7. Use (6) to find the third side of $\triangle ABC$, correct to the nearest tenth, if
(a) $b = 3$ in., $c = 4$ in., $A = 60°$. (c) $a = 6$ cm., $b = 10$ cm., $C = 43°$.
(b) $b = 3$ in., $c = 4$ in., $A = 120°$. (d) $c = 7$ in., $a = 4$ in., $B = 162°$.

8. Use (6) to find all three angles of a triangle whose sides are 4 in., 5 in., and 6 in. long.

9. Find the largest angle of a triangle with sides of 3 in., 4 in., and 6 in.

10. The sides of a triangle are 35 in., 43 in., and 48 in. Show that one of its angles is exactly 60°.

11. In $\triangle XYZ$, $y = 5$ cm., $z = 2$ cm., and $\cos X = \frac{13}{20}$. Use (6) to find the exact values of $\cos Y$ and $\cos Z$.

12. The sides of a parallelogram are 3 in. and 7 in., and one diagonal is 6 in. Use (6) to find the other diagonal, without the aid of tables.

13. Show that the sum of the squares of the diagonals of a parallelogram is equal to the sum of the squares of the sides.

14. If m is the length of the median from vertex A of $\triangle ABC$, show that $m^2 = \frac{1}{2}(b^2 + c^2 - \frac{1}{2}a^2)$.

15. Prove: In any triangle ABC, **(a)** $\sin C = \sin(A + B)$, **(b)** $\cos C = -\cos(A + B)$, **(c)** $\tan C = -\tan(A + B)$.

16. Prove: In any triangle ABC,

(a) $(b - c)\sin A + (c - a)\sin B + (a - b)\sin C = 0$.

(b) $2bc(1 + \cos A) = (b + c + a)(b + c - a)$.

(c) $b^2 + c^2 - a^2 = 4K \cot A$ (K is the area of $\triangle ABC$).

17. Prove: In any triangle ABC,

$$\tan A + \tan B + \tan C = \tan A \tan B \tan C.$$

2.5 Composite and inverse functions

Functions can be combined in various ways to yield new functions; there is, in fact, a whole algebra of functions. The *sum* of functions f and g, for example, is defined in a rather obvious way to be the function $f + g$ which has the value

$$(f + g)(x) = f(x) + g(x)$$

at each number x in the intersection of the domains of f and g. There are analogous definitions, which the student should have no trouble in supplying, for the difference $f - g$, the product $f \cdot g$, and the quotient f/g of f and g.

Most of the algebra of functions so closely parallels the familiar algebra of real numbers as to require no detailed investigation, but this algebra also includes an operation which is distinctive, which we shall need in some of what follows, and which we must therefore examine with some care.

▶ **Definition.** If f and g are functions, then the composite of f and g (in that order) is the function fg which has the value

$$(fg)(x) = f(g(x))$$

at each x in the domain of g such that $g(x)$ is in the domain of f.

(Note that we write the composite fg without a dot, the product $f \cdot g$ with a dot; to avoid confusing the two, the composite is sometimes written with an open dot, $f \circ g$.) If, for example,

$$f(x) = \sqrt{x} \quad \text{and} \quad g(x) = x - 10,$$

then

$$(fg)(x) = f(g(x)) = f(x - 10) = \sqrt{x - 10},$$

and the domain of fg is the interval $[10, \infty)$. We also have

$$(gf)(x) = g(f(x)) = g(\sqrt{x}) = \sqrt{x} - 10,$$

and the domain of gf is $[0, \infty)$. As this example suggests, we generally find that $fg \neq gf$, although, as we shall immediately see, there are exceptions.

A case of particular interest is that in which functions f and g have the properties

$$(fg)(x) = x \quad \text{and} \quad (gf)(x) = x.$$

In this case, we say that g is an *inverse* of f (or, equally, that f is an inverse of g). It turns out, as the student will be asked to prove in Ex. 8, page 75, that if a function has an inverse, then that inverse is unique. This permits us to speak of *the* inverse of a function f and to adopt for it the distinctive notation f^{-1}, a notation already familiar in such forms as $\tan^{-1}$ and $\sin^{-1}$; note that the -1 which appears in this notation is *not* an exponent, and f^{-1} should be read "f-inverse." We require that the domain of f^{-1} be the range of f; it follows that the range of f^{-1} is the domain of f. Summarizing, we make the following definition:

▶ **Definition.** Given a function f with domain D and range R, if there exists a function g with domain R and range D such that $(gf)(x) = x$ for each $x \in D$ and $(fg)(y) = y$ for each $y \in R$, then g is the inverse of f, and we write $g = f^{-1}$.

In the language of mappings, if f takes any element x of its domain into the element y of its range, then f^{-1} takes y right back into x. In the pictorial representation of § 2.1, this means that the arrows which represent a mapping f can be changed into the arrows which represent f^{-1} by simply reversing their direction.

If, for example, $f(x) = 2x + 3$ and $g(x) = \dfrac{x-3}{2}$,

then
$$(fg)(x) = f(g(x)) = f\left(\frac{x-3}{2}\right) = 2\left(\frac{x-3}{2}\right) + 3 = x,$$

and
$$(gf)(x) = g(f(x)) = g(2x + 3) = \frac{(2x + 3) - 3}{2} = x$$

at each real number x, and hence $g = f^{-1}$ (and also $f = g^{-1}$).

Each element of a function is an ordered pair, containing a first coordinate and a second coordinate. If a function pairs the first coordinate x with the second coordinate y, its inverse must reverse this order and pair y with x; for each pair (x, y) that belongs to a function, there is a pair (y, x) that belongs to its inverse. Thus, if

$$F = \{(1, 5), (2, -7), (3, 3)\},$$

then
$$F^{-1} = \{(5, 1), (-7, 2), (3, 3)\};$$

and if
$$f = \{(x, y): y = 3x + 4\},$$

then
$$f^{-1} = \{(y, x): y = 3x + 4\}.$$

This last set-builder, however, uses the letters x and y in an order that is directly contrary to a familiar pattern and is therefore somewhat objectionable. Hence, instead of interchanging x and y in the ordered pair (x, y), we can achieve the same result by interchanging them in the equation $y = 3x + 4$ that describes f, thus obtaining an equivalent description of f^{-1} in a more accustomed form:

$$f^{-1} = \{(x, y) : x = 3y + 4\}.$$

Finally, we can solve this last equation for y in terms of x, if that seems desirable, to obtain a third description of f^{-1}:

$$f^{-1} = \left\{(x, y) : y = \frac{x - 4}{3}\right\}.$$

You, the student, should make sure at this point that you understand all the notation in the preceding argument. You should satisfy yourself that, in each of the three forms, the function defined as f^{-1} is the same function and is indeed the inverse of f. You may find it helpful to verify this in a few specific numerical instances. For example, the function f contains the ordered pair $(2, 10)$, that is, $f(2) = 10$; hence the function f^{-1} must contain $(10, 2)$, that is, we must have $f^{-1}(10) = 2$. You should verify that this is the case, no matter which way we write f^{-1}. Other examples of the same nature can easily be devised.

We can infer from the preceding result a technique for finding the inverses of many functions. If the function is described by a rule in the form $y = f(x)$, we can commonly find a rule in the same general form for the inverse function (if it exists) by (i) interchanging x and y, thus getting $x = f(y)$, and then (ii) solving this equation for y in terms of x to obtain $y = f^{-1}(x)$. Thus, if

$$f = \{(x, y) : y = x^3\},$$

then

$$f^{-1} = \{(x, y) : x = y^3\} = \{(x, y) : y = \sqrt[3]{x}\}.$$

What conclusion can you draw about f and its inverse if, in step (ii) above, you find that more than one value of y corresponds to some value of x?

Since a function is changed into its inverse by interchanging coordinates in the ordered pairs of the function, it follows that the graph of a function can be changed into the graph of its inverse by merely changing labels on the coordinate axes. This, however, runs counter to the convention that the domain is associated with points of the horizontal axis and the range with points of the vertical axis; hence it is better to leave the axes alone and make an appropriate change in the graph. It turns out, as will be shown in § 8.1, that the appropriate change is a reflection of the graph in the line $y = x$. This transformation is illustrated in Fig. 14 at the top of page 70.

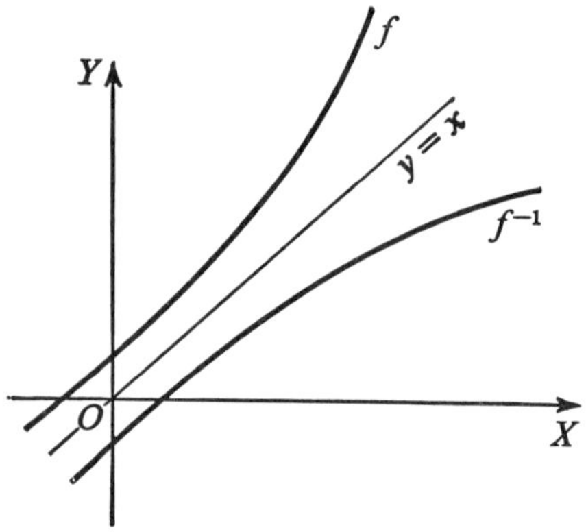

Fig. 14

We can take any function and reverse the order of the coordinates in each pair of the function. The result will certainly be a relation, but it will not necessarily be a function. If, for example, we execute this maneuver on the function $G = \{(1, 2), (-1, 2), (4, 17)\}$, we obtain the relation $\overline{G} = \{(2, 1), (2, -1), (17, 4)\}$, but this is not a function because it contains two different pairs that have the same first element. Hence the function G has no *inverse*, since an inverse must, by definition, be a function. The relation $\overline{G}$ is called the *converse* of G.

We have seen that a relation is a function if and only if no vertical line meets its graph in more than one point. Since an interchange of coordinates (equivalent to a reflection in the line $y = x$) changes vertical lines into horizontal lines, it follows that a function has an inverse if and only if no *horizontal* line meets its graph in more than one point. In the language of ordered pairs, this means that a function has an inverse if and only if no two distinct pairs which belong to the function have the same *second* coordinate. In the language of mapping, a function has an inverse if and only if, in the pictorial representation of § 2.1, no two of the arrows which represent the mapping *go to* the same point.

An example of a function which has no inverse is the familiar square function, $g = \{(x, y) : y = x^2\}$. If we try to find an inverse for g by the procedure outlined above, we interchange x and y, getting the equation $x = y^2$, and then solve for y. But for each $x > 0$ there are two values of y which satisfy this equation, and the set of all solutions (x, y) is therefore not a function. Hence g has a converse, as does every relation, but no inverse, and a glance at the graph of g will show that there are many horizontal lines which meet it in two points. We can, however, represent g as the union of two sets, each of which is a function and each of which has an inverse. In fact, this can be done in many ways, but perhaps the simplest

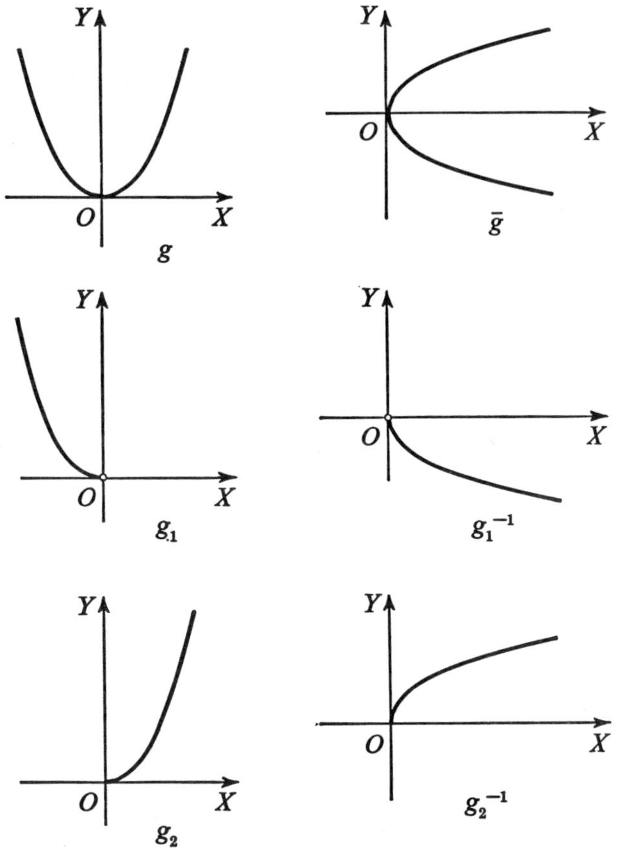

Fig. 15

and most natural way is to break the domain of g, the interval $(-\infty, \infty)$, into the two intervals $(-\infty, 0)$ and $[0, \infty)$, and then consider the two functions which are defined by the same rule as g but have these two intervals as their respective domains; each of these functions is called a *restriction* of g to the appropriate interval. We thus obtain the functions

$$g_1 = \{(x, y): y = x^2, x < 0\} \quad \text{and} \quad g_2 = \{(x, y): y = x^2, x \geq 0\},$$

which have the inverses

$$g_1^{-1} = \{(x, y): y = -\sqrt{x}\} \quad \text{and} \quad g_2^{-1} = \{(x, y): y = \sqrt{x}\}.$$

Furthermore, $g = g_1 \cup g_2$. Most of the functions that we shall meet can similarly be expressed as unions of functions that have inverses.

The circular functions involve especially important applications of this principle. They are periodic, as we pointed out on page 41, and every periodic function takes each of its values infinitely many times (if k is a period of f, then $f(x) = f(x + k) = f(x + 2k) = \cdots$). To be specific, $\sin x = \frac{1}{2}$ is satisfied by $x = (\pi/6) + 2n\pi$ or $x = (5\pi/6) + 2n\pi$ for any integer n. Any horizontal line that meets the graph of sin meets it infi-

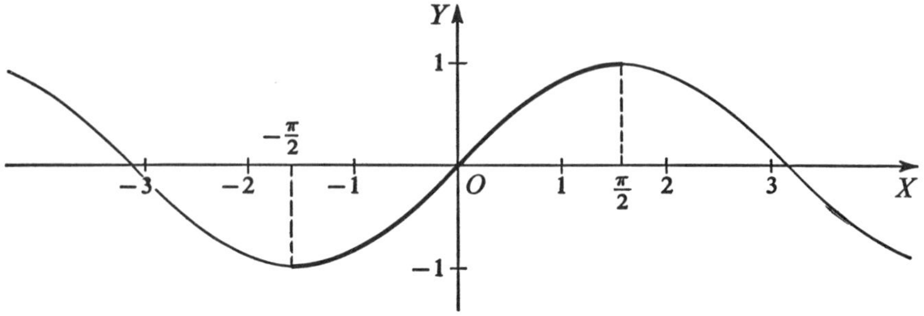

Fig. 16. Graph of sin

nitely often, and the function has no inverse. If, however, we restrict sin to the domain $[-\pi/2, \pi/2]$, the function thus restricted takes exactly once each value of the unrestricted sine and has an inverse, denoted "$\sin^{-1}$":

$$\blacktriangleright \quad \sin^{-1} = \{(x, y)\colon \quad x = \sin y \quad \text{and} \quad -\pi/2 \le y \le \pi/2\}.$$

Hence the domain of $\sin^{-1}$ is $[-1, 1]$, and the range is $[-\pi/2, \pi/2]$.

It must be emphasized that the function sin with which the student is familiar *has no inverse;* the symbol "$\sin^{-1}$" denotes only the inverse of a particular restriction of sin. Hence $\sin(\sin^{-1} x) = x$ for any x in the domain

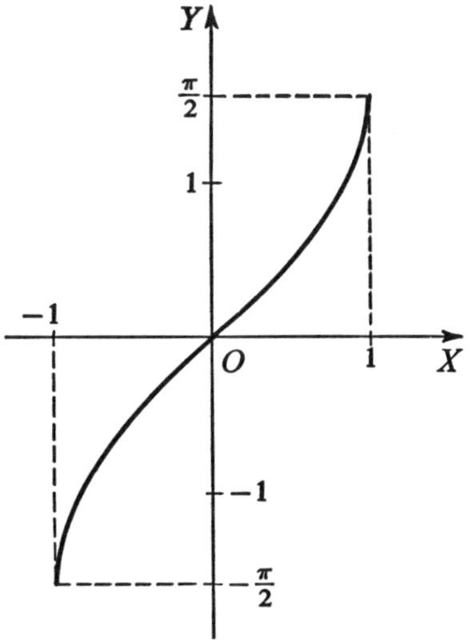

Fig. 17. Graph of $\sin^{-1}$

of $\sin^{-1}$, but $\sin^{-1}(\sin y) = y$ if and only if y is in the range of $\sin^{-1}$: $\sin^{-1}(\sin \pi/3) = \sin^{-1}(\sqrt{3}/2) = \pi/3$, but $\sin^{-1}(\sin 5\pi/3) = \sin^{-1}(-\sqrt{3}/2) = -\pi/3 \neq 5\pi/3$.

In restricting the sine function so as to obtain a function that has an inverse, the choice of restriction is to a considerable extent arbitrary. There are many intervals or unions of intervals, other than our choice of $[-\pi/2, \pi/2]$, on which sin takes each value only once and therefore has an inverse: $[\pi/2, 3\pi/2]$, for example, or $[-\pi/2, 0] \cup [\pi/2, \pi)$. The choice we made, which is standard, has a certain common-sense appeal, but, much more important, this choice gives the greatest possible simplicity to the use of $\sin^{-1}$ in calculus. One convenient feature of this choice can be demonstrated immediately: If we know that $y = \sin x$, then equation (1) of § 2.2 tells us that $\cos^2 x = 1 - y^2$ and hence either $\cos x = \sqrt{1 - y^2}$ or $\cos x = -\sqrt{1 - y^2}$; the ambiguity is unavoidable. If, however, we know that $x = \sin^{-1} y$, then we also know that $x \in [-\pi/2, \pi/2]$, which implies that $\cos x \geq 0$, and therefore $\cos x = \sqrt{1 - y^2}$, without ambiguity.

The inversion of cos and tan runs into the same difficulties as did the inversion of sin: each of these functions is periodic, hence each takes every value in its range infinitely often, and each must be restricted in order to obtain a function that has an inverse. The standard restriction of cos for this purpose is $[0, \pi]$, and that of tan is $(-\pi/2, \pi/2)$. Hence we make the following definitions:

$$\cos^{-1} = \{(x, y)\colon\ x = \cos y \ \text{ and } \ 0 \leq y \leq \pi\}.$$

$$\tan^{-1} = \{(x, y)\colon\ x = \tan y \ \text{ and } \ -\pi/2 < y < \pi/2\}.$$

We shall not need inverses of the other circular functions.

Exercises 2.5 [A]

1. Find formulas for the functions fg and gf, and state the domain and range of each, if
   (a) $f(x) = 2x + 3$; $g(x) = x - 5$.
   (b) $f(x) = ax$; $g(x) = bx$.
   (c) $f(x) = \log x$; $g(x) = x + 2$.
   (d) $f(x) = \sin x$; $g(x) = 2x$.
   (e) $f(x) = \sqrt{9 - x^2}$; $g(x) = 3x$.
2. Find the inverse f^{-1} of the function f if
   (a) $f(x) = x + 2$.
   (b) $f(x) = 2x - 7$.
   (c) $f(x) = |x|$, $x \geq 0$.
   (d) $f(x) = |x|$, $x < 0$.
   (e) $f = \{(1, 3), (2, 4), (3, 5)\}$.
   (f) $f(x) = (ax + b)/(cx - a)$.

3. Find functions f and g such that
   (a) $(fg)(x) = \sqrt{x-1}$.
   (b) $(fg)(x) = \sin(2x+3)$.
   (c) $(fg)(x) = 1/(x^2+3)$.
   (d) $(fg)(x) = |5x-7|$.
   (e) $(fg)(x) = x$.
   (f) $(fg)(x) = 2 \log x$.

4. (a) Sketch a graph of cos over the interval $[-2\pi, 2\pi]$.
   (b) Sketch a complete graph of $\cos^{-1}$.
   (c) Show that $\sin(\cos^{-1} x) \geq 0$, and conclude that $\sin(\cos^{-1} x) = \sqrt{1-x^2}$.

5. Evaluate:
   (a) $\sin^{-1}(\sin \pi)$
   (b) $\cos^{-1}(\cos \pi)$
   (c) $\tan^{-1}(\tan \pi)$
   (d) $\sin(\cos^{-1} \frac{1}{3})$
   (e) $\sin(2 \cos^{-1} \frac{1}{3})$
   (f) $\cos(2 \cos^{-1} \frac{1}{3})$
   (g) $\sin(\sin^{-1} \frac{1}{3} + \sin^{-1} \frac{2}{3})$
   (h) $\sin(\sin^{-1} \frac{1}{2} + \cos^{-1} \frac{1}{2})$

6. Prove: $\sin^{-1} x + \cos^{-1} x = \frac{\pi}{2}$. (*Hint:* Consider $\sin(\sin^{-1} x + \cos^{-1} x)$.)

Exercises 2.5 [B]

1. Find formulas for the functions fg and gf, and state the domain and range of each, if
   (a) $f(x) = 6x - 2$; $g(x) = \frac{1}{2}x$.
   (b) $f(x) = x^2$; $g(x) = x - 1$.
   (c) $f(x) = 3 \cos x$; $g(x) = 3x$.
   (d) $f(x) = \sqrt{25 - x^2}$; $g(x) = x^2$.
   (e) $f(x) = x^2$; $g(x) = \sqrt{-x}$.
   (f) $f(x) = -x^2$; $g(x) = \log x$.

2. Find the inverse f^{-1} of the function f if
   (a) $f(x) = 3x - 5$.
   (b) $f(x) = \log x$.
   (c) $f(x) = x^2 - 4x$, $x \geq 2$.
   (d) $f(x) = x^2 - 4x$, $x < 2$.
   (e) $f = \{(2, 7), (3, 3), (1, -5)\}$.
   (f) $f = \{(x, y): x + y + 2 = 0\}$.

3. Find functions f, g, h such that
   (a) $(fgh)(x) = \log \sin 5x$.
   (b) $(fgh)(x) = \sqrt{x^2+1}$.
   (c) $(fgh)(x) = 1/(2 + \sin x)$.
   (d) $(fgh)(x) = \tan^2 5x$.

4. (a) Sketch a graph of tan over the interval $[-2\pi, 2\pi]$.
   (b) Sketch a graph of $\tan^{-1}$.
   (c) Show that $\sec(\tan^{-1} x) = \sqrt{x^2+1}$.

5. Evaluate:
   (a) $\sin^{-1}(\sin(-\pi/2))$
   (b) $\cos^{-1}(\cos(-\pi/2))$
   (c) $\tan^{-1}(\tan(-\pi/2))$
   (d) $\tan(\sin^{-1}\frac{4}{5})$
   (e) $\tan(2\tan^{-1}\frac{1}{2})$
   (f) $\tan(2\sin^{-1}\frac{4}{5})$
   (g) $\tan(\tan^{-1}\frac{1}{2} - \tan^{-1}\frac{1}{3})$
   (h) $\tan(4\tan^{-1}\frac{1}{5})$
   (i) $\sin(2\sin^{-1}\frac{1}{3} + \cos^{-1}\frac{1}{3})$
   (j) $\cos(\tan^{-1}(-2) + \sin^{-1}\frac{1}{4})$
6. Show that if $x > 0$, then $\tan^{-1} x + \tan^{-1}(1/x) = \pi/2$, and find the value of $\tan^{-1} x + \tan^{-1}(1/x)$ when $x < 0$.

Miscellaneous Exercises 2.5

1. If $f(x) = k/x$, for what k does $f = f^{-1}$?
2. A function is *linear* if its graph is a straight line with slope $\neq 0$. Show that every linear function has an inverse which is also a linear function.
3. Prove: If f and g are linear functions, then (a) fg and gf are also linear functions; (b) $fg = gf$ if the graphs of f and g meet on the line $x - y = 0$.
4. If f, g, h are linear functions, show that $f(gh) = (fg)h$. (A similar relationship is true for any functions, linear or not. Can you prove it? This result makes it possible to refer, without ambiguity, to the function fgh.)
5. Prove: If a function f has an inverse f^{-1}, then f^{-1} has an inverse $(f^{-1})^{-1}$, and $(f^{-1})^{-1} = f$.
6. (a) If $f = \{(2, 7), (3, -4), (6, 2)\}$, list the elements of ff^{-1} and of $f^{-1}f$, and compare.
   (b) Repeat with $f = \{(1, 2), (2, 7), (7, 1)\}$.
   (c) State a necessary condition on the domain and the range of f in order that $ff^{-1} = f^{-1}f$.
7. Prove: If f and g have inverses f^{-1} and g^{-1}, then
   (a) fg has an inverse $(fg)^{-1}$, and $(fg)^{-1} = g^{-1}f^{-1}$;
   (b) $(fg)^{-1} \neq f^{-1}g^{-1}$.
   (*Hint:* For (a) use the definition of inverse function and the result of Ex. 4, above. For (b), find a counterexample.)
8. Using the following steps, prove that if a function has an inverse, then that inverse is unique. Let g be an inverse of f, and let h be an inverse of f.
   (a) What is the domain of g? of h? Conclude that g and h have the same domain.
   (b) If y is any element of the common domain of g and h, then there must be some x in the domain of f such that $y = f(x)$. Why? What is $g(y)$? $h(y)$? Conclude that $g(y) = h(y)$ and hence that $g = h$.

2.6 The slope of a curve

Slope, as a measure of steepness, seems intuitively applicable to a curve. It makes sense to speak of the steepness of a curve, and it seems reasonable to try to find a number, the slope, which will measure that steepness. Thus, in Fig. 18, a part of the graph of $y = x^2$, it seems that the slope

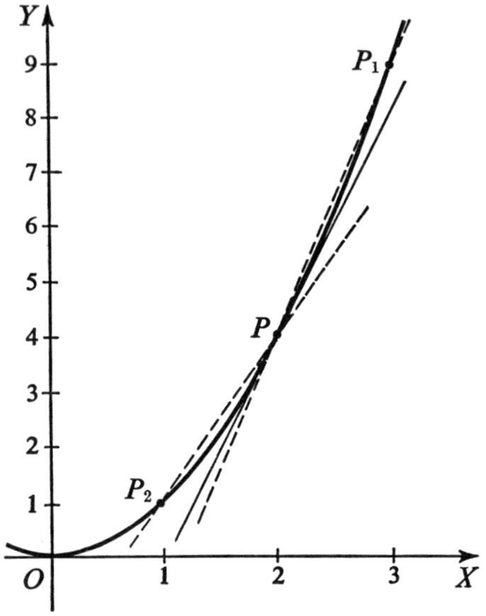

Fig. 18

at the origin should be 0, and that, as we trace the graph to the right, the slope should steadily increase, since we can see that the graph gets steeper and steeper. We shall try to find a number to measure that steepness at some specific point, say $P(2, 4)$. In order to do this, we shall find it desirable to introduce a totally new concept, the idea of a *limit*. This concept is the principal distinguishing feature of calculus.

We start by defining the average slope of an arc to be the slope of the chord of the arc, the straight line that connects the ends of the arc. Thus, if we take P as one end of an arc and take the other end at $P_1(3, 9)$, we have (see § 1.3)

$$\Delta x = 3 - 2 = 1, \quad \Delta y = 9 - 4 = 5,$$

and the average slope of the arc PP_1 is then

$$\frac{\Delta y}{\Delta x} = 5.$$

If the curve gets steeper as we go from P to P_1, as seems intuitively obvious, then the slope at P—which we have not yet defined—must be less than the average slope of PP_1, that is, less than 5. If, on the other hand, we consider an arc to the left of P, say from P to $P_2(1, 1)$, we find

$$\Delta x = 1 - 2 = -1, \quad \Delta y = 1 - 4 = -3,$$

and the average slope is

$$\frac{\Delta y}{\Delta x} = 3;$$

the figure suggests that the slope at P should be greater than this, since the curve at P is apparently steeper than the line PP_2. Hence we conclude that any definition we adopt for the slope of a curve should be one which in this particular instance will yield a number between 3 and 5.

From 3 to 5 is a long way; let us see if we can reduce this spread by taking points closer to P, by using shorter arcs. If, for example, we cut the values of Δx in half, we have, on the right of P,

$$x = 2.5, \quad y = (2.5)^2 = 6.25, \quad \Delta y = 2.25, \quad \text{and} \quad \frac{\Delta y}{\Delta x} = \frac{2.25}{0.5} = 4.5,$$

and on the left,

$$x = 1.5, \quad y = (1.5)^2 = 2.25, \quad \Delta y = -1.75, \quad \text{and} \quad \frac{\Delta y}{\Delta x} = \frac{-1.75}{-0.5} = 3.5.$$

Our intuitive notion of the slope of the curve at P now suggests that it must be a number between 3.5 and 4.5. If we try still smaller $|\Delta x|$, we get the results summarized in the following table:

x	y	Δx	Δy	$\frac{\Delta y}{\Delta x}$
3	9	1	5	5
2.5	6.25	0.5	2.25	4.5
2.1	4.41	0.1	0.41	4.1
2.01	4.0401	0.01	0.0401	4.01
...	...	...	...	...
1.99	3.9601	−0.01	−0.0399	3.99
1.9	3.61	−0.1	−0.39	3.9
1.5	2.25	−0.5	−1.75	3.5
1	1	−1	−3	3

As we take values of Δx closer and closer to 0, we get values of $\Delta y/\Delta x$ closer and closer to 4, and it looks as though $\Delta y/\Delta x$ could be made arbitrarily close to 4 just by making $|\Delta x|$ small enough. To see that this is indeed so, we consider the general case. If the point $Q(2 + \Delta x, 4 + \Delta y)$ is on the curve, its coordinates must satisfy the equation of the curve; hence

$$4 + \Delta y = (2 + \Delta x)^2 = 4 + 4\,\Delta x + (\Delta x)^2,$$
$$\Delta y = 4\,\Delta x + (\Delta x)^2,$$

and the average slope of the arc PQ is

$$\frac{\Delta y}{\Delta x} = \frac{4\,\Delta x + (\Delta x)^2}{\Delta x} = 4 + \Delta x,$$

a quantity which differs from 4 by an amount Δx and which can therefore be made arbitrarily close to 4 by taking Δx close enough to 0. This situation is described, in the language of calculus, by saying that the limit of $\Delta y/\Delta x$, as Δx approaches 0, is 4, or, in symbols,

$$\lim_{\Delta x \to 0} \frac{\Delta y}{\Delta x} = 4.$$

We now *define* the *slope* of this curve at this point P to be this limit, 4, and we also *define* the *tangent* to the curve at P to be the line through P with this slope.

At this point, an obvious question is: Since $\Delta y/\Delta x = 4 + \Delta x$, why not set $\Delta x = 0$ in the first place, and thus avoid all this fuss about limits? The answer should be equally obvious: Since division by 0 is impossible, $\Delta y/\Delta x$ does not exist when $\Delta x = 0$. The conclusion reached was based on reducing the fraction

$$\frac{4\,\Delta x + (\Delta x)^2}{\Delta x} \quad \text{to the form} \quad 4 + \Delta x,$$

and this reduction, in which numerator and denominator are divided by the common factor Δx, is valid only when $\Delta x \neq 0$. If we set $\Delta x = 0$, we get only the meaningless mark $\Delta y/\Delta x = 0/0$. We cannot set $\Delta x = 0$, nor is there any value of Δx for which $\Delta y/\Delta x = 4$. It was never expected that one of the average slopes which were computed would itself be the slope of the curve at P; rather, it was hoped that these values would point the way toward that slope. The successive reductions in $|\Delta x|$ led, as the table indicates, to the conclusion that the slope at P must lie between 3 and 5, between 3.5 and 4.5, between 3.9 and 4.1, and so on; the general case showed that it was possible to go on narrowing these intervals, in this way, as far as we liked. The only number contained in *all* such intervals is 4. The mathematical definition of limit, which will be stated and examined in detail in Chapter 5, does not here require that there be a value of Δx for which $\Delta y/\Delta x = 4$, and it specifically avoids the case $\Delta x = 0$, as it must, since division by 0 is impossible. To show that the limit of $\Delta y/\Delta x$ is 4, we are required merely to show that $\Delta y/\Delta x$ can be made arbitrarily close to 4, can be made to differ from 4 by less than any positive number that may be suggested, no matter how small it may be. Thus, if we are challenged to make $\Delta y/\Delta x$ within, say, 1/1,000,000 of 4, we can do this by making $|\Delta x| = 1/1{,}000{,}001$; if we are challenged with 1/1,000,000,000, we can take $|\Delta x| = 1/1{,}000{,}000{,}001$. We can meet any challenge of this sort; no matter how small a number is proposed, we can find values of $|\Delta x|$ which will make $\Delta y/\Delta x$ differ from 4 by less than this amount. This is the implication of the words "arbitrarily close to 4."

Let us find, by the same process, the slope of the curve $y = x^3 - 9x$ at $P(1, -8)$. Rather than go through all the numerical detail of the preceding example, however, we shall start by finding a general expression for the average slope, $\Delta y/\Delta x$, in terms of Δx. If the point $Q(1 + \Delta x, -8 + \Delta y)$ is also on the curve, then its coordinates must satisfy the equation of the curve:

$$-8 + \Delta y = (1 + \Delta x)^3 - 9(1 + \Delta x)$$
$$= -8 - 6\,\Delta x + 3(\Delta x)^2 + (\Delta x)^3;$$

hence
$$\Delta y = -6\,\Delta x + 3(\Delta x)^2 + (\Delta x)^3,$$

and the average slope of the arc PQ is then

$$\frac{\Delta y}{\Delta x} = \frac{-6\,\Delta x + 3(\Delta x)^2 + (\Delta x)^3}{\Delta x} = -6 + 3\,\Delta x + (\Delta x)^2$$

so long as $\Delta x \neq 0$. We can use this to find the numerical value of $\Delta y/\Delta x$ for any Δx (except, as always, 0):

If $\Delta x =$	0.1,	then	$\Delta y/\Delta x =$	-5.69
	0.01			-5.9699
	0.001			-5.996999
	-0.001			$-6.002999.$

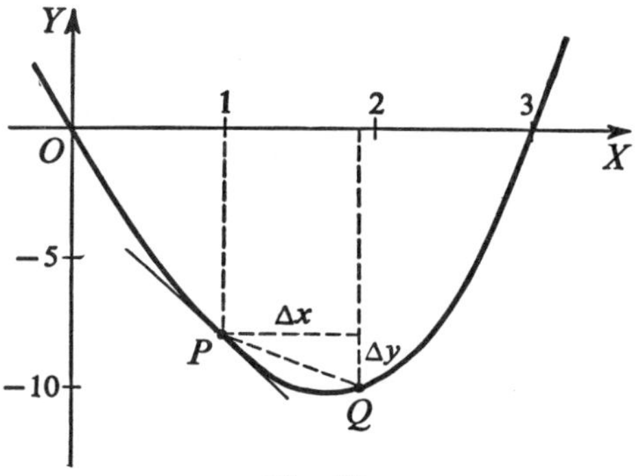

Fig. 19

Evidently, $\Delta y/\Delta x$ can be made arbitrarily close to -6 by making Δx close enough to 0, and we therefore define the slope of the curve at P to be

$$m = \lim_{\Delta x \to 0} \frac{\Delta y}{\Delta x} = -6.$$

Let us now return to our earlier example, the curve $y = x^2$, but to a more general question: What is the slope of this curve at an arbitrary point $P(x, y)$? The same procedure provides an answer. We take a second point, $Q(x + \Delta x, y + \Delta y)$ on the same curve, and find the average slope of the arc PQ. Since Q lies on the curve,

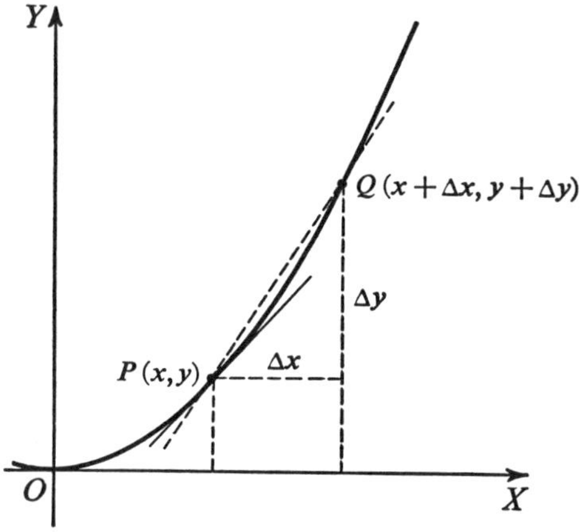

Fig. 20

$$\begin{aligned} y + \Delta y &= (x + \Delta x)^2 \\ &= x^2 + 2x \cdot \Delta x + (\Delta x)^2; \end{aligned}$$

since P lies on the curve,

$$y = x^2.$$

Hence

$$\Delta y = 2x \cdot \Delta x + (\Delta x)^2,$$

and the average slope of PQ is

$$\frac{\Delta y}{\Delta x} = \frac{2x \cdot \Delta x + (\Delta x)^2}{\Delta x} = 2x + \Delta x$$

if $\Delta x \neq 0$. Since this expression can evidently be made arbitrarily close to $2x$ by making $|\Delta x|$ small enough, the slope at P is

$$m = \lim_{\Delta x \to 0} \frac{\Delta y}{\Delta x} = 2x.$$

In other words, the slope of $y = x^2$ at any point is twice the abscissa of that point: at $(0, 0)$, the slope is $m = 2(0) = 0$; at $(1, 1)$, $m = 2(1) = 2$; at $(-3, 9)$, $m = 2(-3) = -6$, and so on.

Consider, finally, the general problem, the problem of finding the slope of the graph of an arbitrary function f at an arbitrary point $P(x, y)$ of that

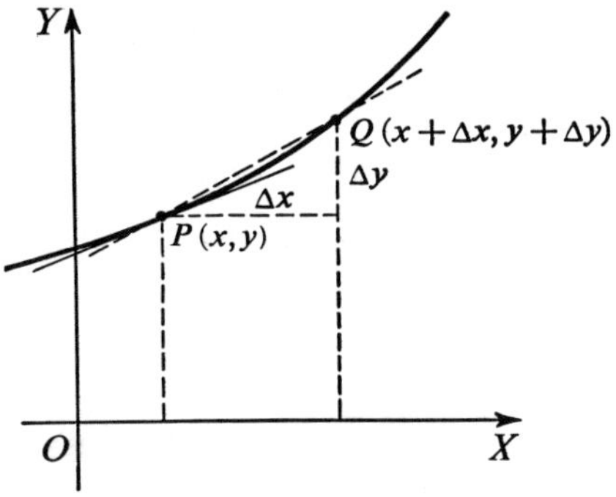

Fig. 21

graph. The procedure is exactly the same. We take a second point $Q(x + \Delta x, y + \Delta y)$ of the graph, find the average slope of the arc PQ, and define the slope at P to be the limit of this average slope as Δx approaches 0. Since Q is on the curve,

$$y + \Delta y = f(x + \Delta x);$$

since P is on the curve,

$$y = f(x).$$

Hence $$\Delta y = f(x + \Delta x) - f(x),$$

the average slope of PQ is

$$\frac{\Delta y}{\Delta x} = \frac{f(x + \Delta x) - f(x)}{\Delta x},$$

and we have the following definition:

▶ **Definition.** The <u>slope</u> of the graph of the function f at the point $(x, f(x))$ is

$$m = \lim_{\Delta x \to 0} \frac{f(x + \Delta x) - f(x)}{\Delta x}.$$

This definition does not, of course, give any indication of how to evaluate this limit in a particular instance; we shall study this problem at length in what follows. There are cases in which the limit, for one reason or another, simply does not exist; in these cases, we say the graph has no slope at that point.

To complete the work of this section, we have one more definition:

Definition. The line <u>tangent</u> to the graph of a function f at the point $(x, f(x))$ is the line through that point having the slope m defined above.

Exercises 2.6 [A]

1. The point $P(1, 2)$ is on the graph of $y = x^2 - 5x + 6$. If the point $Q(1 + \Delta x, 2 + \Delta y)$ is on the same curve:
   (a) Compute Δy and $\Delta y/\Delta x$, the average slope of PQ, for $\Delta x = \pm 1$, ± 0.1, ± 0.01.
   (b) Estimate $\lim_{\Delta x \to 0} \dfrac{\Delta y}{\Delta x}$.
   (c) Find an expression for Δy in terms of Δx, use it to compute $\Delta y/\Delta x$ in terms of Δx, and verify your answer to (b).
   (d) Sketch the curve and the tangent line at P.
2. Using the points $P(2, 0)$ and $Q(2 + \Delta x, 0 + \Delta y)$ on the graph of $y = x^2 - 4$, carry out the instructions for Ex. 1.
3. Proceed as in Ex. 1 with the points $P(2, -4)$ and $Q(2 + \Delta x, -4 + \Delta y)$ on $y = x^2 - 4x$.
4. Proceed as in Ex. 1 with $P(0, 0)$ and $Q(\Delta x, \Delta y)$ on $y = x^3 - 4x$.
5. Proceed as in Ex. 1 with $P(1, 1)$ and $Q(1 + \Delta x, 1 + \Delta y)$ on $y = \sqrt{x}$.

2.7 A rate problem

The formula $D = RT$, Distance = Rate × Time, is a common one in elementary algebra. It can be used when the rate, or velocity, is constant, and it can be used, even when the rate varies, to find an *average* velocity, as one sometimes does in dividing the total distance traveled by the total time elapsed, at the end of an automobile trip. Because most of the motion problems found in nature involve changing velocities, the formula can ordinarily be used only in this average sense, yet, even if the velocity changes, it seems reasonable that there must be some relationship among distance, rate, and time. In many important cases, calculus can profitably be used to explore and describe this relationship.

The distance s feet that a body falls in t seconds, for example, is given under certain circumstances by the formula $s = 16t^2$. As we shall shortly see, this implies a constantly changing velocity. What is that velocity at a particular time? What is it, for example, at the end of exactly 2 seconds? The question has no immediately accessible answer, for we do not even have a definition of velocity at a specific instant of time. The average velocity over the first 2 seconds is, of course, easily found: the elapsed time is 2 seconds, the total distance traveled is, by formula, $16(2)^2 = 64$ ft., and the average velocity is then $64/2 = 32$ feet per second (abbreviated ft./sec. or f.p.s.). But since the body starts from rest, that is, with zero velocity, it is evident that the velocity at the end of 2 seconds must be greater than the average velocity over those 2 seconds, greater, that is, than 32 ft./sec. It may be illuminating to find the average velocity in the one-second interval on either side of the instant $t = 2$. When $t = 1$, $s = 16$; when $t = 2$, $s = 64$. If we represent the change in time from $t = 2$ by $\Delta t = t - 2$ and the change in position by $\Delta s = s - 64$, we then have, for this interval, $\Delta t = 1 - 2 = -1$ and $\Delta s = 16 - 64 = -48$, and the average velocity is then $\Delta s/\Delta t = (-48)/(-1) = 48$. When $t = 3$, $s = 144$; the change in s in the 3d second of fall is $\Delta s = 144 - 64 = 80$, the change in t is $\Delta t = 3 - 2 = 1$, and therefore $\Delta s/\Delta t = 80$. In the second next before $t = 2$, the average velocity is 48 ft./sec.; in the second next after $t = 2$, the average velocity is 80 ft./sec.; intuitively, it seems that the average velocity at the instant $t = 2$ should lie between those two values. "Somewhere between 48 and 80" is not a very precise estimate; if we try to improve this estimate by decreasing the size of $|\Delta t|$, we get the following results:

t	s	Δt	Δs	$\frac{\Delta s}{\Delta t}$
1	16	-1	-48	48
1.9	57.76	-0.1	-6.24	62.4
1.99	63.3616	-0.01	-0.6384	63.84
...	...	...	...	...
2.01	64.6416	0.01	0.6416	64.16
2.1	70.56	0.1	6.56	65.6
3	144	1	80	80

As we take smaller and smaller values of $|\Delta t|$, we get average velocities closer and closer to 64, and it seems as though we could make the average velocity arbitrarily close to 64 by taking $|\Delta t|$ close enough to 0. To see that this is indeed so, we work the problem in terms of a general Δt: If the body has fallen $(64 + \Delta s)$ feet at the instant $(2 + \Delta t)$ seconds, we must have, by formula,

$$64 + \Delta s = 16(2 + \Delta t)^2 = 64 + 64\,\Delta t + 16(\Delta t)^2,$$

$$\Delta s = 64\,\Delta t + 16(\Delta t)^2,$$

and hence

$$\frac{\Delta s}{\Delta t} = \frac{64\,\Delta t + 16(\Delta t)^2}{\Delta t} = 64 + 16\,\Delta t \quad \text{if} \quad \Delta t \neq 0.$$

Evidently, this quantity can be made arbitrarily close to 64 by making $|\Delta t|$ small enough, and we are therefore led to *define* the velocity at $t = 2$ to be

$$v = \lim_{\Delta t \to 0} \frac{\Delta s}{\Delta t} = 64.$$

The same procedure can be used to find the velocity at any instant: If the body falls s feet in t seconds, $(s + \Delta s)$ feet in $(t + \Delta t)$ seconds, then we must have

$$s + \Delta s = 16(t + \Delta t)^2 = 16t^2 + 32t \cdot \Delta t + 16(\Delta t)^2.$$

Since $$s = 16t^2,$$

we have $$\Delta s = 32t \cdot \Delta t + 16(\Delta t)^2,$$

and therefore $$\frac{\Delta s}{\Delta t} = \frac{32t \cdot \Delta t + 16(\Delta t)^2}{\Delta t} = 32t + 16\,\Delta t \quad \text{if} \quad \Delta t \neq 0.$$

The term $16\,\Delta t$ can evidently be made close to 0 by making $|\Delta t|$ small enough, and the limit of the average velocity, as Δt approaches 0, is then seen to be $32t$. We define the limit to be the velocity v at time t:

$$v = \lim_{\Delta t \to 0} \frac{\Delta s}{\Delta t} = 32t.$$

Thus, when $t = 5$, the velocity is $32(5) = 160$ ft./sec.; when $t = 10$, the velocity is $32(10) = 320$ ft./sec., and so on.

We generalize these results in the following definition:

Definition. If the distance s which a body travels in time t is given by the value at t of a function f, $s = f(t)$, then the <u>velocity</u> of the body at time t is

$$v = \lim_{\Delta t \to 0} \frac{\Delta s}{\Delta t} = \lim_{\Delta t \to 0} \frac{f(t + \Delta t) - f(t)}{\Delta t}.$$

Exercises 2.7 [A]

1. If a stone is thrown straight up with a velocity of 96 ft./sec., its height t seconds later will be $h = 96t - 16t^2$ (feet). Using the method of the text, find the velocity at the end of 2 seconds, 4 seconds, and t seconds. For what value of t is the velocity 0? How high will the stone go?

2. A stone is thrown down, with a velocity of 24 ft./sec., from the top of a 160-foot cliff. Its height above the foot of the cliff t seconds later will be $h = 160 - 24t - 16t^2$ (feet).

   (a) How long does it take the stone to get to the bottom of the cliff?

   (b) With what velocity does it hit the ground?

3. If the distance an object travels in t minutes is s yards, where $s = 9t - t^3$, $0 \leq t \leq 3$, find its velocity when $t = 2$.

2.8 The derivative

In § 2.6, we defined the slope of the graph of the function f at the point $(x, f(x))$ to be

$$m = \lim_{\Delta x \to 0} \frac{f(x + \Delta x) - f(x)}{\Delta x}.$$

If f is a function such that the distance that a body has traveled up to time t is $f(t)$, we defined, in § 2.7, the velocity at t as

$$v = \lim_{\Delta t \to 0} \frac{f(t + \Delta t) - f(t)}{\Delta t}.$$

These two different problems have solutions in identical form: the only difference between the limit that defines the slope and the limit that defines the velocity is the letter used, x or t, a trivial distinction in this context. Limits in this particular form are of central importance in calculus, and we are therefore led to make the following definition:

▶ **Definition.** If f is a function, then the function f′ whose value at x is

$$f'(x) = \lim_{\Delta x \to 0} \frac{f(x + \Delta x) - f(x)}{\Delta x}$$

is called the derivative of f.

The term "derivative" is also often used to describe the limit $f'(x)$; the context will generally make it clear whether the function f' or its value $f'(x)$ is meant. If $y = f(x)$, then $f'(x)$ is sometimes called *the rate at which* y *changes with* x.

The student should note that the domain of f' is no greater than the domain of f, because the limit $f'(x)$ is defined only at values of x where $f(x)$ is also defined. As we shall see in Chapter 5, the limit $f'(x)$ may fail to exist even at some points where $f(x)$ is defined, and there may therefore be points of the domain of f which are not in the domain of f'.

The process of finding a derivative is called differentiation, and the branch of calculus that deals with derivatives and their applications is called the differential calculus. If a function f has a derivative at a point c, then f is said to be differentiable at c. If f is differentiable at all points of an interval, it is said to be differentiable on the interval.

Another symbol sometimes used for the derivative f' of a function f is Df, and if $y = f(x)$ is the value of f at x, then there are many symbols in use for the corresponding value $f'(x)$ of the derivative. Among them are

$$\frac{dy}{dx}, \quad y', \quad \frac{df(x)}{dx}, \quad D_x y, \quad D_x f(x), \quad \dot{y}.$$

We shall have some occasion to use the first three of these, but $f'(x)$ will suffice for a while.

The technical details of finding a derivative are conveniently summarized in the following four-step procedure:

1. Compute $f(x)$ and $f(x + \Delta x)$.
2. Subtract, obtaining the difference $f(x + \Delta x) - f(x)$.
3. Divide this difference by Δx.
4. Find the limit of this quotient as $\Delta x \to 0$.

The first three steps are mechanical. The fourth step can demand considerable skill and ingenuity, but the exercises at the end of this section will be confined to a few simple types.

Example 1. If $f(x) = x^2 - 3x + 5$, find $f'(x)$.

Solution: Following the steps listed above, we obtain

1.
$$\begin{aligned} f(x + \Delta x) &= (x + \Delta x)^2 - 3(x + \Delta x) + 5 \\ &= x^2 + 2x \cdot \Delta x + (\Delta x)^2 - 3x - 3\,\Delta x + 5 \\ f(x) &= x^2 \qquad\qquad\qquad\qquad\quad - 3x \qquad\quad + 5 \end{aligned}$$

2.
$$f(x + \Delta x) - f(x) = 2x \cdot \Delta x + (\Delta x)^2 - 3\,\Delta x$$

3.
$$\frac{f(x + \Delta x) - f(x)}{\Delta x} = \frac{2x \cdot \Delta x - 3\,\Delta x + (\Delta x)^2}{\Delta x}$$

$$= 2x - 3 + \Delta x \quad \text{if} \quad \Delta x \neq 0.$$

4. The limit of this last quantity, as $\Delta x \to 0$, is $2x - 3$, and therefore

$$f'(x) = 2x - 3.$$

Some further discussion of the last step is desirable. Consider the two functions defined, for each real number x, by

$$p = \{(\Delta x, z) \colon z = 2x - 3 + \Delta x \quad \text{and} \quad \Delta x \neq 0\}$$

and $q = \{(\Delta x, z) \colon z = 2x - 3 + \Delta x \quad \text{and} \quad \Delta x \text{ is any real number}\}$.

These functions are not the same, because the ordered pair $(0, 2x - 3)$ belongs to q but not to p. The last step in the preceding example is to find $\lim_{\Delta x \to 0} p(\Delta x)$. Now for all $\Delta x \neq 0$, it is true that $p(\Delta x) = q(\Delta x)$, and it follows that $\lim_{\Delta x \to 0} p(\Delta x) = \lim_{\Delta x \to 0} q(\Delta x)$, since both limits are defined in terms of values of p and q near but not at $\Delta x = 0$. To determine $\lim_{\Delta x \to 0} q(\Delta x)$ is easy; we need only compute $q(0) = 2x - 3$, which is the required limit not only of $q(\Delta x)$ but also of $p(\Delta x)$. The principle involved here is of general applicability, and we may therefore paraphrase as follows the fourth step in the instructions for finding a derivative: "Reduce the fraction found in Step 3 to a form which will have meaning if 0 is substituted for Δx, and make this substitution; the resulting number is $f'(x)$."

Example 2. If $f(x) = \sqrt{x + 2}$, find $f'(x)$ and compute $f'(7)$ and $f'(14)$.

Solution: Following the steps listed above, we obtain:

1.
$$f(x + \Delta x) = \sqrt{x + \Delta x + 2}$$
$$f(x) = \sqrt{x + 2}$$

2.
$$f(x + \Delta x) - f(x) = \sqrt{x + \Delta x + 2} - \sqrt{x + 2}$$

3.
$$\frac{f(x + \Delta x) - f(x)}{\Delta x} = \frac{\sqrt{x + \Delta x + 2} - \sqrt{x + 2}}{\Delta x}.$$

4. To carry out this step, we use a device which has proved useful in problems which involve radicals in this way; we rationalize the numerator of the fraction found in Step 3.

$$\frac{\sqrt{x + \Delta x + 2} - \sqrt{x + 2}}{\Delta x} \cdot \frac{\sqrt{x + \Delta x + 2} + \sqrt{x + 2}}{\sqrt{x + \Delta x + 2} + \sqrt{x + 2}}$$
$$= \frac{(x + \Delta x + 2) - (x + 2)}{\Delta x(\sqrt{x + \Delta x + 2} + \sqrt{x + 2})}$$
$$= \frac{1}{\sqrt{x + \Delta x + 2} + \sqrt{x + 2}},$$

and the limit of this, as $\Delta x \to 0$, is clearly

$$f'(x) = \frac{1}{2\sqrt{x + 2}}.$$

Hence $f'(7) = \frac{1}{6}$ and $f'(14) = \frac{1}{8}$.

Exercises 2.8 [A]

1. If $f(x) = 3x^2 - 7x$, find $f'(x)$ and hence evaluate $f'(2)$ and $f'(-3)$.
2. If $g = \{(x, y): y = x^3\}$, find $g'(x)$ and hence evaluate g' at -2, -1, 0, 1, and 2. What do these numbers tell you about the slope of the curve $y = g(x)$? Making use of what you know about the slope, sketch the graph of $y = g(x)$, $-2 \le x \le 2$.
3. If $F(x) = x^3 - 6x^2 + 5$, find $F'(x)$. For what values of x does $F'(x) = 0$? What implications does this have for the graph of F?
4. Find $f'(x)$ if: (a) $f(x) = x^4$; (b) $f(x) = x^5$.
5. If a stone is thrown into the air with a velocity of 48 ft./sec., its height t seconds later will be s feet, where $s = f(t) = 48t - 16t^2$. Find $f'(t)$ and hence find the velocity of the stone when $t = 1$ and $t = 2$.
6. If $\phi(x) = 1/(x - 1)$, find $\phi'(x)$ and thus evaluate $\phi'(0)$ and $\phi'(2)$.
7. Find $f'(x)$ if

   (a) $f(x) = \dfrac{x}{x + a}$.   (c) $f(x) = \dfrac{x - 1}{x + 1}$.

   (b) $f(x) = \dfrac{1}{x^2 - 4}$.   (d) $f(x) = \dfrac{x^2 - 1}{x^2 + 1}$.

8. If $F(x) = \sqrt{x}$, find $F'(x)$ (see Example 2). State the domains of F and F'. Find $F'(1)$, $F'(\frac{1}{4})$, $F'(\frac{1}{16})$, and $F'(0.01)$. What is the inclination of the line tangent at the origin to the graph of F?
9. If $g(x) = \sqrt{25 - x^2}$, find $g'(x)$ and thus evaluate $g'(3)$. Verify by elementary methods that the line through (3, 4) with slope $g'(3)$ is tangent to the semicircle $y = \sqrt{25 - x^2}$.

2.9 Rules for differentiation

It is awkward to have to use the process of § 2.8, which defines the derivative, each time we want to find a derivative; we can save ourselves quite a lot of work by establishing a few general rules, or formulas, for differentiation. Four of these follow. There are many more, as we shall see, but these four are of particular importance, and they will suffice for the work of the next few chapters.

If a function f takes the same value c at each real number x, then f is an example of a constant function. Since the graph of such a function is a straight line parallel to the x-axis, its slope is zero, and the following result is therefore entirely to be expected:

Rule 1. If $f(x) = c$ for all x in some interval, then $f'(x) = 0$ on that interval.

To derive Rule 1, we carry through the four steps listed in § 2.8:

1. $$f(x + \Delta x) = c$$

$$f(x) = c$$

2. $$f(x + \Delta x) - f(x) = 0$$

3. $$\frac{f(x + \Delta x) - f(x)}{\Delta x} = 0, \quad \Delta x \neq 0.$$

4. $$f'(x) = \lim_{\Delta x \to 0} \frac{f(x + \Delta x) - f(x)}{\Delta x}$$

$$= \lim_{\Delta x \to 0} 0 = 0.$$

Thus, for example, if $f(x) = 5$ at each real number x, then $f'(x) = 0$ at each real number x.

We have seen that

$$\begin{aligned} &\text{if } f(x) = x^2, \quad \text{then} \quad f'(x) = 2x \quad (\S\ 2.6),\\ &\text{if } f(x) = x^4, \quad \text{then} \quad f'(x) = 4x^3 \quad (\text{Ex. 4, page 88}),\\ \text{and } &\text{if } f(x) = x^5, \quad \text{then} \quad f'(x) = 5x^4 \quad (\text{Ex. 4, page 88}). \end{aligned}$$

These results suggest the following generalization:

Rule 2. If $f(x) = x^n$, then $f'(x) = nx^{n-1}$.

This rule is valid for any real exponent n. We shall, however, prove it here only for n a positive integer. In this case, expanding $f(x + \Delta x)$ by the binomial theorem and following the usual steps for finding the derivative, often called the "delta process," we have

1. $$\begin{aligned} f(x+\Delta x) &= (x+\Delta x)^n \\ &= x^n + nx^{n-1}\Delta x + \frac{n(n-1)}{2} x^{n-2}(\Delta x)^2 + \cdots + (\Delta x)^n \end{aligned}$$

$$f(x) = x^n$$

2. $$f(x+\Delta x) - f(x) = nx^{n-1}\Delta x + \frac{n(n-1)}{2} x^{n-2}(\Delta x)^2 + \cdots + (\Delta x)^n$$

3. $$\begin{aligned} \frac{f(x+\Delta x) - f(x)}{\Delta x} &= nx^{n-1} + \frac{n(n-1)}{2} x^{n-2}(\Delta x) + \cdots + (\Delta x)^{n-1} \\ &= nx^{n-1} + \Delta x\left[\frac{n(n-1)}{2} x^{n-2} + \cdots + (\Delta x)^{n-2}\right]. \end{aligned}$$

No matter what the value of the long expression in brackets on the right-hand side of this last equation, the term in which it appears can be made arbitrarily small by making $|\Delta x|$ small enough; the limit of the right-hand side must therefore be just the first term:

$$4. \quad f'(x) = \lim_{\Delta x \to 0} \frac{f(x + \Delta x) - f(x)}{\Delta x}$$

$$= \lim_{\Delta x \to 0} \left[nx^{n-1} + \Delta x \left(\frac{n(n-1)}{2} x^{n-2} + \cdots + (\Delta x)^{n-2} \right) \right]$$

$$= nx^{n-1}.$$

Thus, for example, the derivative of x^{17} is $17x^{16}$ (that is, if $f(x) = x^{17}$, then $f'(x) = 17x^{16}$); the derivative of x^{100} is $100x^{99}$; and, since we shall assume the validity of Rule 2 even when n is not a positive integer, the derivative of $x^{-7/5}$ is $-\frac{7}{5}x^{-12/5}$.

We now know how to differentiate all powers of a variable, but what do we do about $2x^3$ or $5x^{-2}$, terms in which we have a numerical coefficient other than 1? The following rule provides an answer:

▶ **Rule 3.** **If f and g are functions such that, for some constant c, $f(x) = cg(x)$, and if the derivative g' exists, then f' also exists, and $f'(x) = cg'(x)$.**

The proof of Rule 3 uses the familiar "delta process":

$$1. \quad f(x + \Delta x) = cg(x + \Delta x)$$

$$f(x) = cg(x)$$

$$2. \quad f(x + \Delta x) - f(x) = cg(x + \Delta x) - cg(x)$$

$$= c[g(x + \Delta x) - g(x)]$$

$$3. \quad \frac{f(x + \Delta x) - f(x)}{\Delta x} = c\,\frac{g(x + \Delta x) - g(x)}{\Delta x}.$$

$$4. \quad f'(x) = \lim_{\Delta x \to 0} \frac{f(x + \Delta x) - f(x)}{\Delta x}$$

$$= \lim_{\Delta x \to 0} c\,\frac{g(x + \Delta x) - g(x)}{\Delta x}$$

$$= c \lim_{\Delta x \to 0} \frac{g(x + \Delta x) - g(x)}{\Delta x}.$$

The last step, in which c is "factored out" of the limit process, is rather less simple than it seems; it will be justified in Chapter 5. The last limit in Step 4 is $g'(x)$ by definition, and Rule 3 is therefore established.

Rules 2 and 3 together now enable us to differentiate terms like cx^n directly and easily, without having to go back to the "delta process." Thus, if $f(x) = 2x^3$, we can take $g(x) = x^3$, whence $f(x) = 2g(x)$ and

$$\begin{aligned} f'(x) &= 2g'(x) \quad \text{(by Rule 3)} \\ &= 2\cdot 3x^2 \quad \text{(by Rule 2)} \\ &= 6x^2. \end{aligned}$$

Similarly, if $f(x) = 5x^{-2}$, we take $g(x) = x^{-2}$ and find $f'(x) = 5g'(x) = 5(-2x^{-3}) = -10x^{-3}$.

In summary, to differentiate a term like cx^n, we (i) find the new coefficient by multiplying the original coefficient by the original exponent, and (ii) find the new exponent by subtracting 1 from the original exponent.

Finally, we need a rule that will enable us to deal with expressions that contain two or more terms:

▶ **Rule 4.** If f, f_1, and f_2 are functions such that

$$f(x) = f_1(x) + f_2(x),$$

and if the derivatives f_1' and f_2' exist, then the derivative f' also exists, and

$$f'(x) = f_1'(x) + f_2'(x).$$

To prove Rule 4, we follow the usual four steps:

1. $$f(x + \Delta x) = f_1(x + \Delta x) + f_2(x + \Delta x)$$

2. $$\begin{aligned} f(x) &= f_1(x) + f_2(x) \\ \hline f(x + \Delta x) - f(x) &= f_1(x + \Delta x) - f_1(x) + f_2(x + \Delta x) - f_2(x) \end{aligned}$$

3. $$\frac{f(x + \Delta x) - f(x)}{\Delta x} = \frac{f_1(x + \Delta x) - f_1(x)}{\Delta x} + \frac{f_2(x + \Delta x) - f_2(x)}{\Delta x}.$$

4. $$\begin{aligned} \lim_{\Delta x \to 0} \frac{f(x + \Delta x) - f(x)}{\Delta x} &= \lim_{\Delta x \to 0} \left[\frac{f_1(x + \Delta x) - f_1(x)}{\Delta x} + \frac{f_2(x + \Delta x) - f_2(x)}{\Delta x}\right] \\ &= \lim_{\Delta x \to 0} \frac{f_1(x + \Delta x) - f_1(x)}{\Delta x} + \lim_{\Delta x \to 0} \frac{f_2(x + \Delta x) - f_2(x)}{\Delta x}. \end{aligned}$$

The last step, in which the limit of the sum of two terms is expressed as the sum of their limits, is the crucial one in the argument; it will be justified in Chapter 5. The three limits in this last expression are, by definition, $f'(x)$, $f_1'(x)$, and $f_2'(x)$, and the desired result has therefore been established.

Rule 4 can be extended, by repeated applications, to a function which is the sum of any number of functions. We therefore have formulas which permit us to write, at sight, the derivative of the sum of any number of terms of the form cx^n.

Example 1. Find $f'(x)$ if $f(x) = 3x^5 - x^{1/3}$.

Solution: By Rule 4, $f'(x)$ is the sum of the derivatives of $3x^5$ and $-x^{1/3}$. By Rules 2 and 3, the first of these derivatives is

$$3(5x^4) = 15x^4,$$

and by Rule 2 the second is $$-\tfrac{1}{3}x^{-2/3}.$$

Therefore $$f'(x) = 15x^4 - \tfrac{1}{3}x^{-2/3}.$$

Example 2. Differentiate $F(x) = \sqrt{x}\,(2x - 3)^3$.

Solution: We first express $F(x)$ as a sum of terms like cx^n by performing the indicated multiplications:

$$F(x) = 8x^{7/2} - 36x^{5/2} + 54x^{3/2} - 27x^{1/2},$$

and therefore $$F'(x) = 28x^{5/2} - 90x^{3/2} + 81x^{1/2} - \tfrac{27}{2}x^{-1/2}.$$

Example 3. Differentiate $g(t) = \sqrt{3t}$.

Solution: Since $$g(t) = \sqrt{3}\,(t^{1/2}),$$

we have $$g'(t) = \sqrt{3}(\tfrac{1}{2}t^{-1/2}) = \frac{1}{2}\sqrt{\frac{3}{t}}.$$

Exercises 2.9 [A]

1. Differentiate:

(a) $x^3 - 5$

(b) $3x^4 - 6x + 3$

(c) $7z^6 - 3z^{-2}$

(d) $3x^9 - 2x^4 + \dfrac{1}{5x}$

(e) $\frac{2}{7}x^{14} - x^3 + 0.23$

(f) $\frac{2}{3}x^{2/3} - 7x^{-7}$

(g) $2\sqrt{y} - 4\sqrt[3]{y}$

(h) $(2x)^{-3} - \dfrac{8}{x} + 5$

(i) $2x^{-3} - \sqrt[3]{x^4}$

(j) $\sqrt{5t} - \dfrac{5}{\sqrt{5t}}$

2. Write the first three terms in the expansion of

(a) $(x + y)^4$; (b) $(x + \Delta x)^7$; (c) $\left(t^2 - \dfrac{2}{t}\right)^5$; (d) $(1 + 1)^{10}$.

3. Multiply or divide as necessary, and then differentiate:

(a) $\dfrac{x^2 - 3}{x}$ (c) $\dfrac{x^4 - 5x^3 + 3x^2 - 7}{3\sqrt{x}}$

(b) $(2y - 1)^3$ (d) $\sqrt{2z}\left(z^2 - \dfrac{3}{5z}\right)$

4. If $f(x) = 3x^2 + 2x - 5$, find $f'(x)$ and use it to determine the slope of $y = f(x)$ at the places where $x = 3, -2$, and $-\frac{1}{3}$.

5. If a ball is thrown up from the top of a 10-meter building with a velocity of 24.5 m./sec., its height above the ground t seconds later will be s meters, where $s = f(t) = 10 + 24.5t - 4.9t^2$. Find $f'(t)$ and use it to find the velocity of the ball when $t = 1$ and when $t = 2$. What significance has the sign of $f'(t)$?

6. Find functions f which have

(a) $f'(x) = 3x^2$. (b) $f'(x) = 5x^4$. (c) $f'(x) = x^3$.

7. (a) Show that the derivative of $(x + a)^3$ is $3(x + a)^2$.

(b) Show that the derivative of $(x + a)^n$ is $n(x + a)^{n-1}$ (n a positive integer).

Exercises 2.9 [B]

1. Differentiate:

(a) $3x^4 - 7x + 2$ (f) $\dfrac{t^2 + 2t - 5}{\sqrt[3]{t}}$

(b) $4t^2 - 5t^{-2} + 7$ (g) $(\sqrt{6x} - 3)^2$

(c) $(z - 2)^3$ (h) $\dfrac{1}{\sqrt{2y}}(7y - 1)$

(d) $3(\sqrt{2y} - 5)^2$ (i) $x(x + 3)(x - 5)$

(e) $(x^2 + x^{-2})^2$ (j) $\dfrac{2 + z^{-2/3}}{z^{-3/2}}$

2. Each of the following equations defines y as a function of x. Solve for y, thus obtaining the form $y = f(x)$, and then find $f'(x)$.

(a) $2x + 3y - 7 = 0$ (c) $x^2y + x - 1 = 0$

(b) $y\sqrt{x} - 4 = 0$ (d) $\sqrt{x} + \sqrt{y} = \sqrt{a}$

3. Find the slope of $y = x^3 - 6x^2 + 1$ at the points where $x = -4, -2, 0, 2, 4$.

4. Differentiate (a) $x^2 + 7x - 3$, (b) $x^2 + 7x + 5$, and (c) $x^2 + 7x - 40$, and interpret geometrically the relationship among these derivatives.

5. A particle P moves on a number line in such a way that its distance from the origin at time t is determined by a function f such that $f(t) = 3t^2 - t^3$, $0 \leq t \leq 3$. Find the velocity of P when $t = 1, 2, 3$. Where is P when $t = 3$?

6. Find functions f which have (a) $f'(x) = 8x^7$; (b) $f'(x) = -x^{-2}$.

7. (a) Show that the derivative of $(ax + b)^3$ is $3a(ax + b)^2$.
   (b) Show that the derivative of $(ax + b)^n$ is $na(ax + b)^{n-1}$ (n a positive integer).

Miscellaneous Exercises 2.9

1. Differentiate:

(a) $2x^8 - 3x^2 + 7x$	(f) $\sqrt{5x}(x^2 - 5)$
(b) $(2x^2 - 5)^2$	(g) $(2x^{-1} + 7)^2$
(c) $\dfrac{x^5 - 4x^2 + 13}{x^3}$	(h) $(\sqrt[3]{3x} - 5)^2$
(d) $\dfrac{x^3 + 7x}{\sqrt{x}}$	(i) $(x^{2/3} - x^{-2/3})^2$
(e) $(t^{1/3} - 2)^3$	(j) $(2x)^{1/2} + (2x)^{-1/2}$

2. Solve for y, thus obtaining the form $y = f(x)$, and then find $f'(x)$:
   (a) $x^2 - xy = 2$ (b) $x^{1/3} + y^{1/3} = a^{1/3}$ (c) $\sqrt{xy} + 2x - 3 = 0$

3. If $F(x) = \frac{1}{3}x^3 - 2x^2 + 5$, find $F'(x)$. For what values of x is $F'(x) = 0$? What does this tell you about the graph of F?

4. A particle P moves on a number line in such a way that its distance from the origin at time t is $t^2 - 6t + 8$. Find its position and velocity when $t = 1, 2, 3$. Explain the signs of the results.

5. (a) Find the slope of $y = 3x^2 - 7x$ at the points where $x = -1, 0, 1, 2$.
   (b) Sketch the graph of $y = 3x^2 - 7x$, $-1 \leq x \leq 2$.
   (c) Where is the slope zero?
   (d) What is the minimum value of $3x^2 - 7x$?

6. If $f(x) = 2x^2 + 5x - 12$, where is $f'(x) > 0$? < 0? Where is the vertex of the parabola $y = 2x^2 + 5x - 12$?

Chapter Review

1. (a) What is the range of $f = \{(x, |x - 3|): 0 \leq x \leq 4\}$?
   (b) Either describe f^{-1} or show that it does not exist.

2. If $f(x) = x^2$ and $g(x) = \sqrt{1 - x^2}$, determine the domains of f, g, fg, and gf.

3. Find all $x \in [0, 2\pi)$ such that $\sin 2x = \sin 3x$.

4. (a) Simplify: $\dfrac{\sin 4x}{1 + 2\cos 2x + \cos 4x}$.
   (b) Find a relationship between h and y such that
$$\frac{\sin(x + h) - \sin x}{h} = \cos\left(x + \frac{h}{2}\right)\frac{\sin y}{y}.$$

5. Evaluate (a) $\lim\limits_{h \to 1} \dfrac{h^3 - 1}{h - 1}$; (b) $\lim\limits_{h \to 2} \dfrac{h^3 - 2h^2 - h + 2}{h^2 - 4}$.

6. A particle moves along the x-axis in such a way that its abscissa at time t is $x = t^3 - 6t^2 + 20$. Find its velocity when $t = 0, 1, 2, 3$.

7. Use the definition of derivative to find the derivative, at $x = 2a$, of
   (a) $\dfrac{x + a}{x - a}$.
   (b) $\sqrt{x^2 - a^2}$.

8. Differentiate:
   (a) $\dfrac{x^3 - 3}{3x}$
   (b) $(2t)^{-1/2}$
   (c) $\left(\sqrt{u} - \dfrac{1}{u}\right)^2$
   (d) $z(z + 1)(z - 1)$
   (e) $(y^2 - 1)^3$
   (f) $\dfrac{\sqrt{2x} + \sqrt{3x}}{\sqrt{5x}}$

9. Find the points on the graph of $y = x^3 - 3x^2 - 6x + 2$ where the slope is 3.

3

Applications of the Derivative

3.1 Curve tracing

The usual procedure in graphing a function is to compute the coordinates of several points on the graph, plot the points, and draw a smooth curve through them. A knowledge of the derivative can be very helpful in this process, as the following examples will show, and may bring out features in the graph which would otherwise be overlooked.

Example 1. Graph $f = \{(x, y): y = x^3 - 3x^2 - 9x + 5\}, x \in [-3, 5]$.

Solution: The table shows the values of f at all integers in the prescribed interval. (Synthetic division and the remainder theorem are convenient tools for this computation.)

x:	-3	-2	-1	0	1	2	3	4	5
$f(x)$:	-22	3	10	5	-6	-17	-22	-15	10

Points with these coordinates are plotted in Fig. 1. (Note that the scales on the

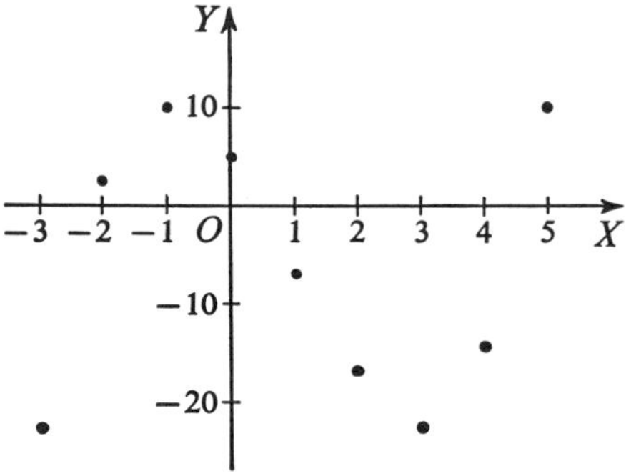

Fig. 1

two axes are not the same; this distortion has been introduced to make it easier to draw the graph.) It is now entirely possible to draw a smooth curve through these points, but a knowledge of the slope at each point not only will make the job easier but may make it a good deal more accurate. The derivative f' is defined by the formula

$$f'(x) = 3x^2 - 6x - 9,$$

and this formula gives the slope of our graph at each point in terms of its abscissa x. Expanding the table to include values of the slope, we have:

x:	-3	-2	-1	0	1	2	3	4	5
$f(x)$:	-22	3	10	5	-6	-17	-22	-15	10
$f'(x)$:	36	15	0	-9	-12	-9	0	15	36

If these values of the slope are now used to draw a short segment of the tangent line at each plotted point, these tangent segments outline the curve in sufficient detail to give a reasonably accurate impression of its general nature and make it easier to finish the job.

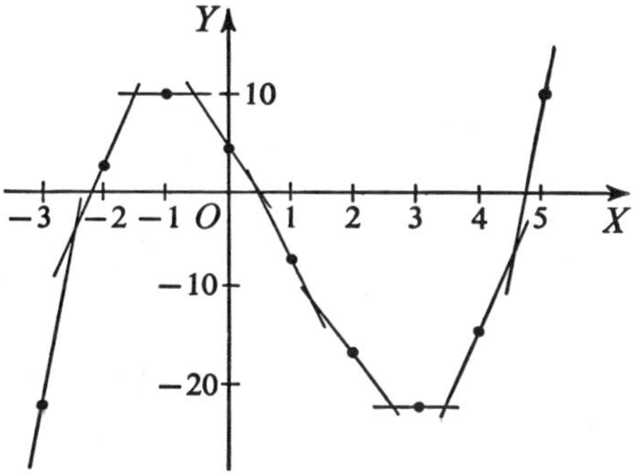

Fig. 2

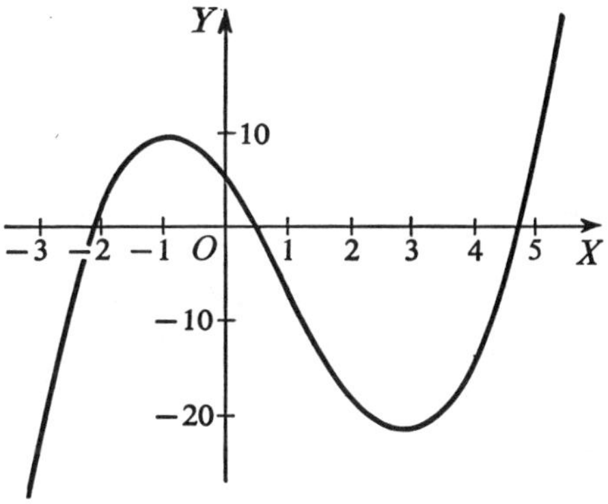

Fig. 3

Note the relationship between the slope and the appearance of the curve in the preceding example. When $x = -3$, the slope has the rather large value 36, and the curve climbs steeply as we trace it from left to right. (Actually, the climb is much steeper than the figure suggests; the curve has been flattened out by the use of different scales on the two axes.) As x increases, the curve climbs less and less steeply until, when $x = -1$, the tangent has turned parallel to the x-axis and the curve is horizontal. As x continues to increase, the curve turns to go down again, descending more and more steeply until $x = 1$, and then less steeply until, at $x = 3$, it again turns to go up, climbing ever more steeply as x increases. On an interval where an increase in x produces an increase in $f(x)$, in this case $(-\infty, -1]$

or $[3, \infty)$, the function f is said to be *increasing;* on an interval where an increase in x produces a decrease in $f(x)$, in this case $[-1, 3]$, the function f is said to be *decreasing.* More formally:

Definition. A function f is increasing on an interval I if, for any $x_1 \in I$ and $x_2 \in I$, $x_2 > x_1$ implies $f(x_2) > f(x_1)$.

A function f is decreasing on an interval I if, for any $x_1 \in I$ and $x_2 \in I$, $x_2 > x_1$ implies $f(x_2) < f(x_1)$.

The student should note that, in Example 1, if the derivative f' is defined and takes only positive values on an interval, then the function f is increasing on that interval, and if f' is defined and takes only negative values on an interval, then f is decreasing on that interval. This relationship between the behavior of the function and the sign of its derivative is quite general, as we shall prove in Chapter 5, and the sign of the derivative is therefore often a useful criterion in determining where a function increases and where it decreases. The criterion is unfortunately not perfect, since it does not tell us what the situation is at points where the derivative is zero or is not defined, but these cases will be exceptional in what follows, and they can generally be treated by examining the behavior of the function itself, or of its derivative, in a neighborhood of each such point.

In Example 1, the points of the graph where $x = -1$ and where $x = 3$ are of particular interest and importance. At each of them, $f'(x) = 0$. If we examine the behavior of f near the first of these, the point $(-1, 10)$, we find that f is increasing on the left of this point and decreasing on the right. It follows, as the graph illustrates, that this point is the highest in its immediate vicinity, or, in more precise language, that the domain of f includes a neighborhood of -1 on which the largest value that f takes is $f(-1) = 10$; the point $(-1, 10)$ is therefore called a *maximum point* of the graph, the function-value $f(-1) = 10$ is called a *maximum value* of f, and f is said to *have a maximum* at -1. Similarly, if we examine the behavior of f near $(3, -22)$, we find that f is decreasing on the left of this point, increasing on the right. It follows that the domain of f includes a neighborhood of 3 on which the smallest value of f is $f(3) = -22$; the point $(3, -22)$ is therefore called a *minimum point* of the graph, the function-value $f(3) = -22$ is called a *minimum value* of f, and f is said to *have a minimum* at 3. There are, of course, values of f greater than 10 and values less than -22; the terms "maximum" and "minimum" are used in a relative sense and refer only to the local behavior of the function (Example 3, on page 100, affords a striking example of this). We can summarize this discussion by the following definition:

Definition. A function f has a maximum (plural maxima) at a point a of its domain if there exists a neighborhood N of a such that $f(x) \leq f(a)$ at each point x which is in the intersection of N with the domain of f. In this case, $f(a)$ is a maximum value of f.

To define minimum (plural minima) and minimum value, we merely reverse the sign of inequality in the preceding definition.

Maxima and minima are known collectively as extreme values, or extrema, and points of the domain at which extrema are likely to occur are known as critical points. We shall postpone to § 7.3 a formal definition of "critical point," but we note here that the function just considered took its extreme values at the zeros of the derivative, and it follows that zeros of the derivative are critical points. There are critical points of other sorts, as we shall see in § 7.3, but until then we shall avoid them. The zeros of the derivative f' are known as stationary points of the domain of the function f, and the corresponding values of f are stationary values: if $f'(c) = 0$, then c is a stationary point of the domain of f, and $f(c)$ is a stationary value of f. The student should note that not all stationary values are extreme values; Example 4, on page 101, illustrates this.

Example 2. Graph $y = f(x) = x^4 + x^3 - 7x^2 - 8x - 8$, $x \in [-3, 3]$. Locate all stationary points, and determine the intervals on which f is increasing and those on which it is decreasing.

Solution: Noting that $f'(x) = 4x^3 + 3x^2 - 14x - 8$, we make the following table of values:

x:	−3	−2	−1	0	1	2	3
$f(x)$:	7	−12	−7	−8	−21	−28	13
$f'(x)$:	−47	0	5	−8	−15	8	85

The table shows that $f'(-2) = 0$, and -2 is therefore a stationary point. To locate the others, we complete the solution of the equation $f'(x) = 0$:

$$4x^3 + 3x^2 - 14x - 8 = 0,$$

$$(x + 2)(4x^2 - 5x - 4) = 0,$$

$$x = -2, \quad \frac{5 \pm \sqrt{89}}{8}, \quad \text{or} \quad -2, \quad -0.55, \quad 1.80$$

approximately. The stationary values of f are then $f(-2) = -12$, $f(-0.55) = -5.8$, and $f(1.80) = -28.8$, the last two being only approximate. The graph,

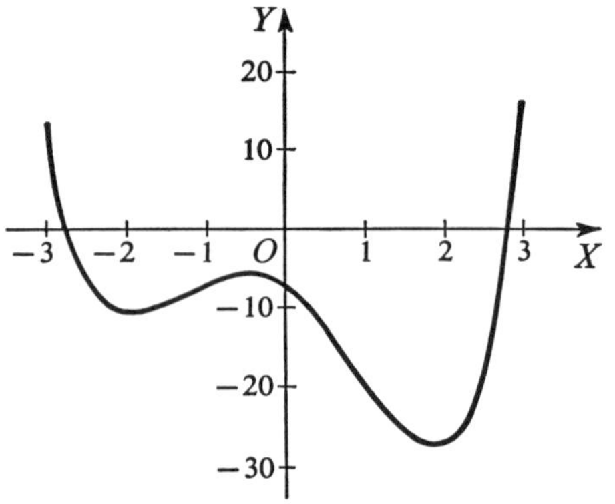

Fig. 4

drawn with the aid of this information, shows that f is increasing on $[-2, -0.55]$ and on $[1.80, \infty)$ and is decreasing on $(-\infty, -2]$ and on $[-0.55, 1.80]$.

Example 3. Graph $f(x) = x + x^{-1}$, $x \in [-4, 4]$. Find the stationary points, and determine the intervals on which f is increasing and those on which it is decreasing.

Solution: Since $f'(x) = 1 - x^{-2}$, we find

x:	-4	-3	-2	-1	0	1	2	3	4
$f(x)$:	$-4\frac{1}{4}$	$-3\frac{1}{3}$	$-2\frac{1}{2}$	-2	–	2	$2\frac{1}{2}$	$3\frac{1}{3}$	$4\frac{1}{4}$
$f'(x)$:	$\frac{15}{16}$	$\frac{8}{9}$	$\frac{3}{4}$	0	–	0	$\frac{3}{4}$	$\frac{8}{9}$	$\frac{15}{16}$

We note that $f(0)$ does not exist, but, aside from this, we know very little about the behavior of f on the interval $[-1, 1]$; we therefore expand the table and take a closer look:

x:	$-\frac{1}{2}$	$-\frac{1}{4}$	-0.1	0.1	$\frac{1}{4}$	$\frac{1}{2}$
$f(x)$:	$-2\frac{1}{2}$	$-4\frac{1}{4}$	-10.1	10.1	$4\frac{1}{4}$	$2\frac{1}{2}$
$f'(x)$:	-3	-15	-99	-99	-15	-3

We see that the curve gets closer and closer to the y-axis as x approaches 0, and gets closer and closer to the line $y = x$ (the dashed line in Fig. 5) as $|x|$ increases. (These lines are called asymptotes of the curve; we will look again at such lines in § 8.1.) The stationary points are -1, which yields $f(-1) = -2$, a maximum, and 1, which yields $f(1) = 2$, a minimum. Note that the value of f at the maximum is 4 units *less* than its value at the minimum. The function is increasing on $(-\infty, -1]$ and $[1, \infty)$, decreasing on $[-1, 0)$ and $(0, 1]$.

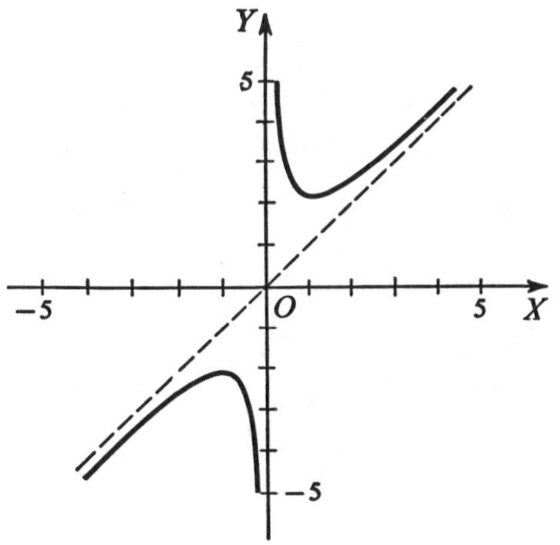

Fig. 5

Example 4. Determine the stationary values of f and the intervals on which it is increasing, if $f(x) = x^3 - 1$.

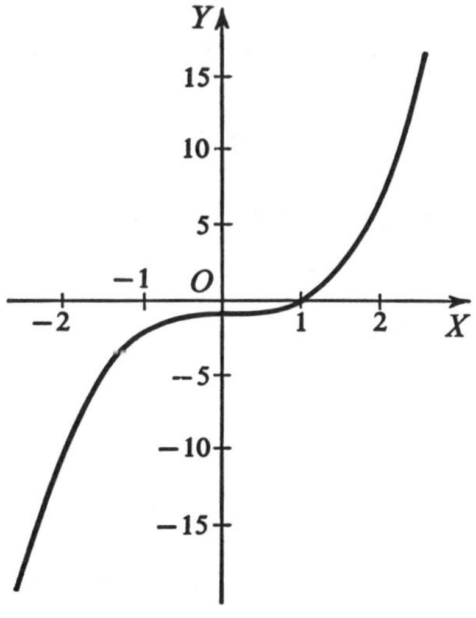

Fig. 6

Solution: Since $f'(x) = 3x^2$, the only stationary point is $x = 0$, but since $3x^2 > 0$ for all $x \neq 0$, the derivative f' indicates that f is increasing for all $x \neq 0$; hence the point $(0, -1)$ is neither a maximum nor a minimum. Since, moreover,

$$f(x_2) - f(x_1) = x_2^3 - x_1^3 > 0 \quad \text{if} \quad x_2 > x_1,$$

it follows that f is increasing throughout its domain (the set of all real numbers).

In many of the following exercises, small changes in the values assigned to the variable x produce relatively large changes in the value of the function. In such cases, the graph will be easier to draw if a given segment of the y-axis represents more units than an equal segment of the x-axis. The figures for the preceding examples may serve as indications of reasonable proportions to use.

Exercises 3.1 [A]

In Exs. 1–4, compute the values of f and of f' at each integer in the stated interval, and at such other points as may be needed. Locate all stationary points, graph, and state the intervals on which each function is increasing and those on which each is decreasing.

1. $f(x) = 2x^2 + 3x - 5,\ x \in [-4, 3]$
2. $f(x) = x^3 - 3x^2 + 2,\ x \in [-3, 4]$
3. $f(x) = x^4 - 8x^2 + 10,\ x \in [-3, 3]$
4. $f(x) = x^2 + 3x + \dfrac{1}{x},\ x \in [-5, 3]$
5. On one set of axes, draw the graphs of $y = x^2$, $y = 2x$, and $y = x^2 - 2x$, $x \in [-4, 4]$. Explain the relationship among the three graphs.

Exercises 3.1 [B]

Follow the directions for the first four exercises in Exercises 3.1 [A].

1. $f(x) = -x^2 + x - 1,\ x \in [-3, 3]$
2. $f(x) = 3x^3 + x + 1,\ x \in [-3, 3]$
3. $f(x) = 3x^4 - 4x^3 - 12x^2 + 5,\ x \in [-2, 3]$
4. $f(x) = x + \dfrac{10}{x^2},\ x \in [-4, 4]$
5. On one set of axes, draw the graphs of $y = x^2$, $y = \dfrac{1}{x}$, and $y = x^2 + \dfrac{1}{x}$, $x \in [-4, 4]$. Explain the relationship among the three graphs.

3.2 Extreme values; applications

The problem of finding the extreme values of a function is a common one in applications of mathematics (there are some examples below, others in § 7.3), and calculus is an almost indispensable tool in the solution of this problem. It is usually convenient to attack the problem in two stages: first, find the stationary points of the domain, and second, examine the behavior of the function on a neighborhood of each stationary point. For the moment, as noted in § 3.1, the first step boils down to finding the zeros of the derivative. When we have found a stationary point, there are several tests by which we can determine whether the corresponding function-value is a maximum, a minimum, or neither; three will be considered here, and a fourth in § 3.3.

1. By definition, $f(a)$ is a maximum value of f if $f(a) \geq f(x)$ for all x in some neighborhood of a. It is not often convenient to evaluate f at *all* points in a neighborhood of a, but a valid conclusion can be reached by comparing $f(a)$ with $f(x_1)$, where $x_1 < a$, and with $f(x_2)$, where $x_2 > a$, provided a is the only stationary point in the interval $[x_1, x_2]$ (see Fig. 7). The test for a minimum is similar.

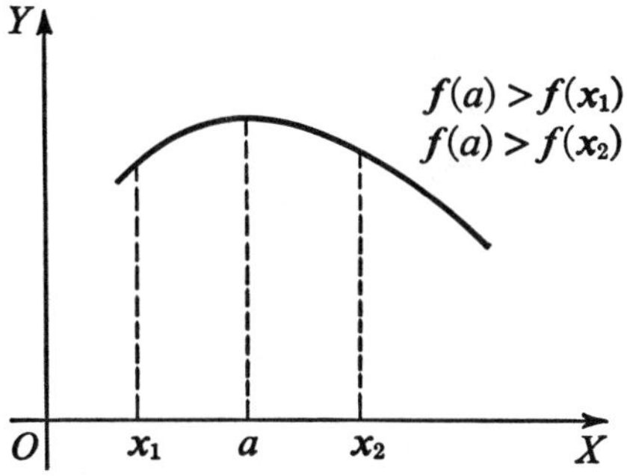

Fig. 7

2. If $f'(a) = 0$, then a is a stationary point for the function f. If there is a neighborhood of a on which $x < a$ implies $f'(x) > 0$ and $x > a$ implies $f'(x) < 0$, then f is an increasing function to the left of a, increases to a maximum $f(a)$ at a, and decreases to the right of a. This describes a graph which has, locally, an inverted "U" shape, with a maximum point $(a, f(a))$ (see Fig. 8). As with the first test, it is not often convenient to test *all*

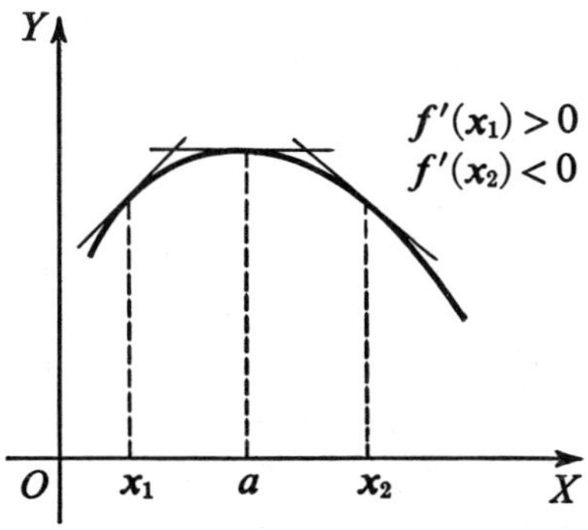

Fig. 8

values of the derivative in a neighborhood of a, but, as above, a single value on each side is enough to permit a valid conclusion if the values belong to an interval that contains no stationary point other than a. The modification of this argument for a minimum, a graph which has locally a right-side-up "U" shape, is left to the student.

3. In many cases, a consideration of the physical or geometrical problem involved or an acquaintance with the type of function that has been used to represent the problem will make it intuitively clear that one has a maximum or has a minimum. This proves, in practice, to be the sort of test most commonly used; it is best explained with reference to specific examples.

Example 1. Find the extreme values of f if $f(x) = x^3 - 4x^2 - 4x + 16$, and identify each as a maximum or as a minimum.

Solution: The value at x of the derivative f' is

$$f'(x) = 3x^2 - 8x - 4.$$

Since this must be zero at a stationary point, we have

$$3x^2 - 8x - 4 = 0,$$

or

$$x = \frac{8 \pm \sqrt{64 + 48}}{6} = -0.43 \text{ or } 3.10, \text{ approximately.}$$

These are the stationary points; we must now examine the behavior of f in a neighborhood of each. Any one of the above tests will give us the information we require; we shall, however, look at all three.

1. We find $f(-0.43) = 16.9$, approximately. To make the work as easy as possible, we evaluate f at the integers on each side of -0.43: $f(-1) = 15 < 16.9$, $f(0) = 16 < 16.9$. Since -0.43 is the only stationary point in the interval $[-1, 0]$, it follows that 16.9 is a maximum value of f. Similarly, $f(3.10) = -5.05$ is less than $f(3) = -5$ and $f(4) = 0$, 3.10 is the only stationary point in $[3, 4]$, and -5.05 is therefore a minimum value of f.

2. Since $f'(-1) = 7$ and $f'(0) = -4$, f is increasing to the left of -0.43 and decreasing to the right; $f(-0.43)$ is therefore a maximum. Similarly, $f'(3) = -1$, $f'(4) = 12$, f is therefore decreasing on the left of 3.10 and increasing on the right, and $f(3.10)$ is a minimum.

3. The student who has studied the forms of the graphs of polynomials will recall that the graph of a cubic with positive x^3 term, such as this, and with two distinct bends, or elbows, as indicated by the two distinct stationary points, comes with increasing x up from deep in the third quadrant to a maximum, goes around an S-curve to a minimum, and then goes on out into the far reaches of the first quadrant. The sign of x^3 is important in this analysis because, for large $|x|$, this term dominates the entire polynomial.

Figure 9 shows some of the quantities on which classification of the extreme values was based.

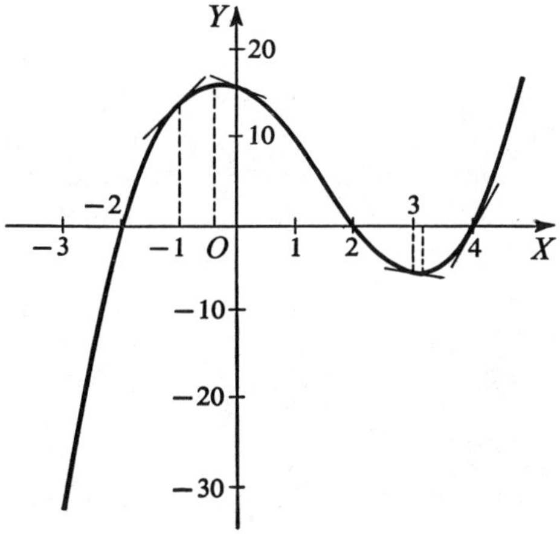

Fig. 9

Example 2. Equal squares are cut out of the corners of a 9-inch by 12-inch sheet of metal, and the sides of the remaining piece are then bent up to form an open-top box. How big should the squares be if the volume of the box is to be a maximum?

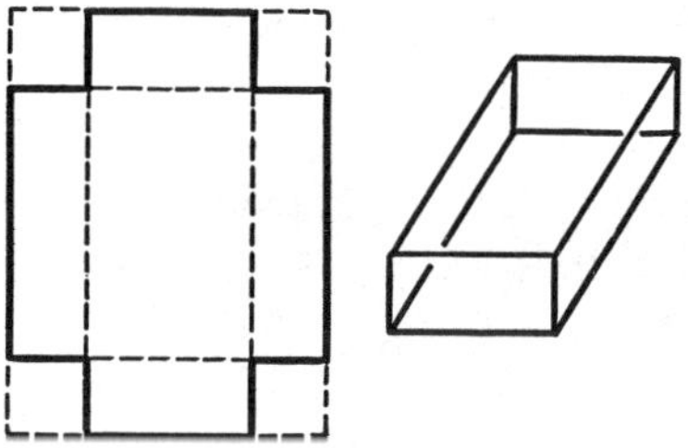

Fig. 10

Solution: If the squares are each x inches on a side, the number of cubic inches in the volume of the box will be

$$f(x) = x(9 - 2x)(12 - 2x) = 4x^3 - 42x^2 + 108x,$$

where the domain of the function f is $(0, 4.5)$ (why?). We find the stationary points of this domain as in Example 1:

$$f'(x) = 12x^2 - 84x + 108,$$

$$12x^2 - 84x + 108 = 0,$$

$$x = \frac{7 \pm \sqrt{49 - 36}}{2} = 1.70 \text{ or } 5.30, \text{ approximately.}$$

Since 5.30 is not in the domain of f, it must be discarded; this leaves 1.70 as the only stationary point. If $f(1.70) = 82^-$ is not the required maximum, the method has failed. A test by the third method is easy here: The volume of the box is zero for $x = 0$ and for $x = 4.5$, and positive for intermediate values of x, so the only stationary point in this interval must yield a maximum. The other two tests are also available, though rather laborious. The graph of f (Fig. 11) may help to make the argument clear.

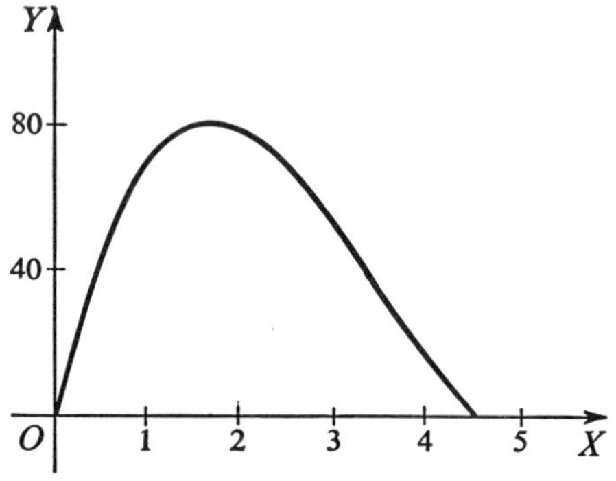

Fig. 11

Exercises 3.2 [A]

1. Find any extreme values of the functions defined by the following rules, and identify each as a (relative) maximum or minimum by comparing it with neighboring values of the function (the first test of the preceding text):

   (a) $f(x) = 5x^2 + 3x + 10$ (b) $f(x) = x^3 - 3x + 4$ (c) $g(t) = t + \frac{2}{t}$

2. Find any extreme values and classify each by observing the slope of the curve in the neighborhood of the extreme point (the second test of the preceding text):
   (a) $\phi(x) = x^4$
   (b) $F(x) = \frac{1}{3}x^3 + \frac{1}{2}x^2 - 2x + 6$
   (c) $f(t) = t^4 - 8t^2 + 5$

3. A one-foot square of metal is to be made into a box without a top by cutting equal squares out of the corners and turning up the edges. What dimensions will make the volume a maximum? What is the maximum volume?

4. Express 10 as the sum of two numbers whose product is a maximum.

5. Show that, among rectangles with a given perimeter P, the square has the greatest area.

6. The girth of a package is defined to be the shortest distance around the package. For a package to be accepted for shipment by a particular parcel delivery service, the sum of its length and girth must not exceed 72 in. If a package is a right square prism, express the volume in terms of a side of the square cross section and thus find the maximum volume which will meet this requirement. Consider two cases:

(**a**) The package is relatively long and narrow, so that the girth is the perimeter of the square cross section (a fishing rod).

(**b**) The package is relatively wide and flat, so that the girth is the perimeter of a rectangle which lies in a plane perpendicular to the square cross sections (a phonograph record).

Exercises 3.2 [B]

1. Find any extreme values and classify each by the first method of the text:

(**a**) $F(x) = -7 - x^2$

(**b**) $g(x) = x^3 + 2x^2 - 4x + 7$

(**c**) $f(x) = (x + 3)(x + 2)(x - 1)$ (approximate to the nearest tenth)

2. Find any extreme values and apply the second test of the text:

(**a**) $f(x) = x^3 + 5x^2 - 8x - 8$

(**b**) $F(x) = (x + 1)^2(x - 2)$

(**c**) $g(x) = x^3 \mid 3x^2 \mid 9x \quad 2$

3. Express 12 as the sum of two numbers such that the sum of their squares is a minimum.

4. Express 9 as the sum of two positive numbers such that the product of one by the square of the other is a maximum.

5. Show that, among rectangles with a given area A, the square has the least perimeter.

6. A strip of metal 20 in. wide is going to be bent to form an open gutter of rectangular cross section. What dimensions of the cross section will make the carrying capacity a maximum?

7. A farmer has 100 ft. of wire fencing with which to enclose a rectangular plot, using an existing stone wall as one side of the plot. What dimensions will maximize the area of the plot?

Miscellaneous Exercises 3.2

1. Find and classify all extreme values:

(a) $f(x) = x^3 + 3x^2 - 9x - 4$

(b) $F(x) = x^3 + 3x^2 + 9x + 4$

(c) $f(x) = x^{1/3} - 4x$

(d) $g(x) = x^4 - x^3 + x^2 - 3x + 1$

(e) $f(z) = z^2 + z^{-2}$

(f) $h(x) = 4x^3 - 19x^2 - 40x + 6$

(g) $r(x) = x^{1/2} + x^{-1/2}$

(h) $F(x) = \frac{1}{4}x^4 - \frac{1}{3}x^3 - x^2 + 3$

(i) $g(x) = x^4 + 2x^2 - 40x - 5$

(j) $f(x) = x^4 - 2x^3$

2. What number exceeds its square by the greatest amount?

3. Suppose a parcel delivery service requires that the sum of the length and the girth of a parcel not exceed 100 in. Work Ex. 6 of Exercises 3.2[A] under this condition.

4. A bead is being moved along a straight wire in such a way that t seconds after it starts its velocity in inches per second is $v = 24t^2 - t^3$ $(0 \leq t \leq 24)$.

(a) What is its greatest velocity?

(b) How long does it take to reach this velocity?

3.3 The second derivative

In § 3.1, we considered at some length the graph of $f = \{(x, y): y = x^3 - 3x^2 - 9x + 5\}$; that graph is reproduced here for further consideration, as the upper part of Fig. 12. Suppose the graph represents a section of highway, and you are driving over it in the direction of increasing x, from A toward E. As you go from A to B and on toward C, you are on the inside of the curve, the road is turning to your right; by the time you get to D, you are on the outside of the curve, turning to your left. Somewhere near C, the direction of turning must have changed. Precisely where did this change take place? To examine the same phenomenon from another point of view, observe that the curve is concave down, "sheds water," in the region from A to B to C, and is concave up, "holds water," in the region from C to D to E. Exactly where does the change in concavity occur?

These questions can be answered by an examination of the behavior of the derivative f' as the curve is traced from A to E. Since $f'(x) = 3x^2 - 6x - 9$, we find that, near A, f' has a relatively large positive value; for instance, $f'(-3) = 36$. If f' were to keep this value, the graph would be a straight line, the road would go straight ahead, but what actually happens, in this case, is that $f'(x)$ decreases as we go from A toward B, the graph of f

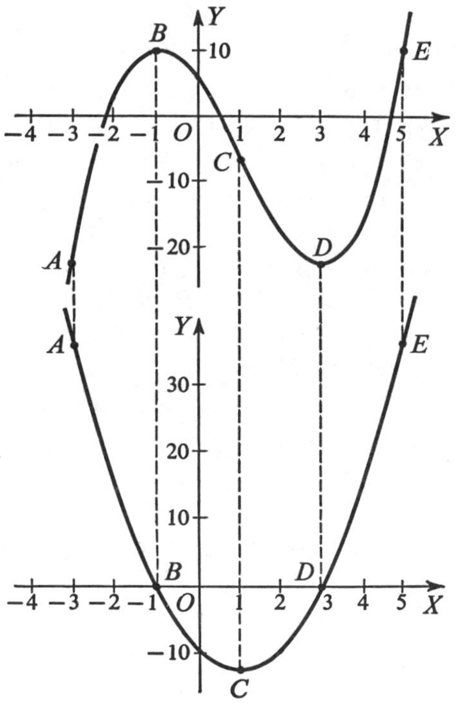

Fig. 12

rises less rapidly, the tangent line turns toward the horizontal. At B, the tangent line is horizontal, $f'(x) = 0$. As we continue to the right, the value of f' continues to decrease, the road continues to curve to the right. Somewhere near C, however, the derivative f' reaches its minimum value and then starts to increase, and the road starts to curve the other way, toward the left. By D, the value of f' has increased to zero, and it continues to increase as we go to E and beyond. All this is summarized by the graph of $y = f'(x)$ (lower part of Fig. 12), on which the points marked A, B, C, D, and E correspond to the points marked with the same letters on the graph of f.

Now, the point where the graph of f changes its direction of curvature is the point where the value of f' stops decreasing and starts increasing, that is, the point where f' takes its minimum value. But the problem of locating the minimum value of a function is a familiar one: we must find the stationary points of the domain of f'. The derivative of f' is $(f')'$, usually written f'', and we have $f''(x) = 6x - 6 = 6(x - 1)$, and $f''(1) = 0$. The graph of f therefore changes its direction of curvature at the point where $x = 1$, that is, the point $(1, -6)$.

The derivative of the derivative of a function, encountered here, is called the second derivative; if the function is f, then we shall denote the second derivative f'', read "f second" or "f double-prime." It follows from the definition of derivative that the second derivative is a function. In general, the second derivative will itself have a derivative, called the third derivative of the original function, and this process of finding successive derivatives is one which, in many cases, can go on forever.

As the preceding analysis indicates, negative values of the second derivative correspond to regions in which the slope is decreasing and the graph is therefore concave down, and positive values of the second derivative correspond to regions in which the slope is increasing and the graph is concave up. A point where the direction of curvature changes is called a point of inflection, or inflection point, of the curve and is evidently a point at which the sign of the second derivative changes. Such a change usually occurs at a zero of the second derivative (exceptions will be considered in § 7.3). To find inflection points, therefore, it suffices for the time being to examine the zeros of f'' and determine at which of them the value of f'' changes sign.

The second derivative can also be used, in many cases, to determine whether a stationary point corresponds to a maximum or to a minimum. When the second derivative is easily determined, this is usually the easiest test. Since a stationary point at which the graph of a function is concave down is a maximum, and a stationary point at which the graph is concave up is a minimum, this test can be stated as follows:

▶ **If $f'(a) = 0$ and $f''(a) < 0$, then f has a maximum at a. If $f'(a) = 0$ and $f''(a) > 0$, then f has a minimum at a.**

Note that the test fails if $f''(a) = 0$. This situation is considered in Examples 2 and 3, on pages 111 and 112.

Example 1. Find the maximum, minimum, and inflection points of the graph of

$$y = f(x) = 3x^4 + 8x^3 - 6x^2 - 24x + 10.$$

Solution: The extreme values will occur at zeros of f':

$$f'(x) = 12x^3 + 24x^2 - 12x - 24 = 12(x + 2)(x + 1)(x - 1).$$

Then $$f'(x) = 0 \quad \text{if} \quad x = -2, -1, \text{ or } 1.$$

If the corresponding values of f are maxima or minima, this will be shown by the sign of $f''(x)$:

$$f''(x) = 36x^2 + 48x - 12 = 12(3x^2 + 4x - 1).$$

Hence, when				
	$x =$	-2	-1	1
	$f(x) =$	18	23	-9
	$f'(x) =$	0	0	0
	$f''(x) =$	36	-24	72
and f therefore has a		minimum	maximum	minimum.

The inflection points will be at zeros of f'':

$$f''(x) = 12(3x^2 + 4x - 1) = 0,$$

$$x = \frac{-4 \pm \sqrt{16 + 12}}{6} = -1.55 \text{ or } 0.22, \text{ approximately.}$$

It is left to the student to show that f'' does indeed change sign at these points. The corresponding values of f are approximately $f(-1.55) = 20.4$ and $f(0.22) = 4.5$; because the slope at an inflection point is an aid to graphing, we also compute $f'(-1.55) = 7.6$ and $f'(0.22) = -25.3$. Figure 13 illustrates these findings.

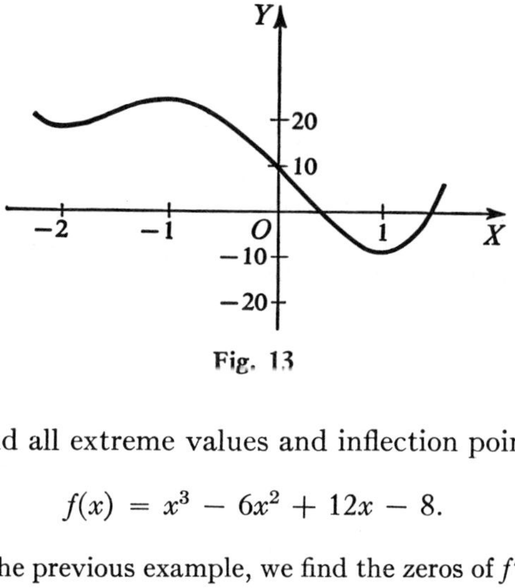

Fig. 13

Example 2. Find all extreme values and inflection points:

$$f(x) = x^3 - 6x^2 + 12x - 8.$$

Solution: As in the previous example, we find the zeros of f':

$$f'(x) = 3x^2 - 12x + 12 = 3(x - 2)^2;$$

$$f'(x) = 0 \quad \text{if} \quad x = 2.$$

Testing this result, we find $f''(x) = 6x - 12$ and hence $f''(2) = 0$. The second-derivative test for an extreme value therefore fails, and the result moreover suggests that $x = 2$ is the abscissa of a point of inflection. Since the linear function f'' is negative for all $x < 2$, positive for $x > 2$, there is, indeed, a change in the sign of $f''(x)$ at $x = 2$, and this therefore yields an inflection point of the graph of

f. The fact that $f'(x) = 3(x - 2)^2$ is positive for all $x \neq 2$ also shows that 2 cannot give an extreme value of f. Figure 14 illustrates the situation.

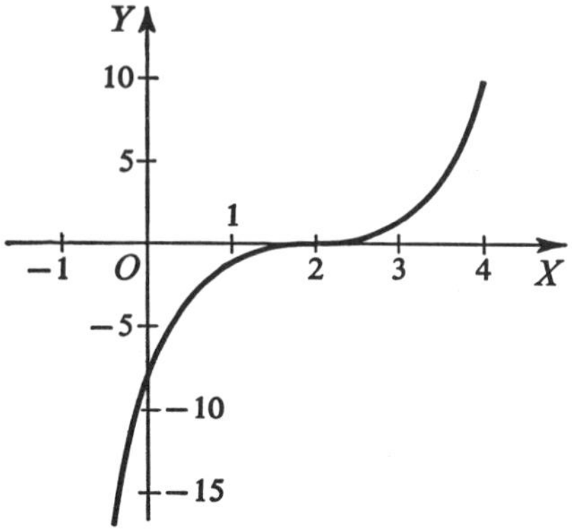

Fig. 14

Example 3. Find all extreme values and inflection points:

$$f(x) = x^4 - 5.$$

Solution: Proceeding as before, we find $f'(x) = 4x^3$, $f''(x) = 12x^2$. Hence 0, the only zero of f', is also a zero of f'', and the second-derivative test again fails. But $f''(x) \geq 0$ for all x, and therefore does not change sign; hence there is no inflection point. The fact that $f'(x) < 0$ for $x < 0$ and $f'(x) > 0$ for $x > 0$ shows that $f(0) = -5$ is a minimum, as does the fact that $f(x) > -5$ for all $x \neq 0$. The graph is shown in Fig. 15.

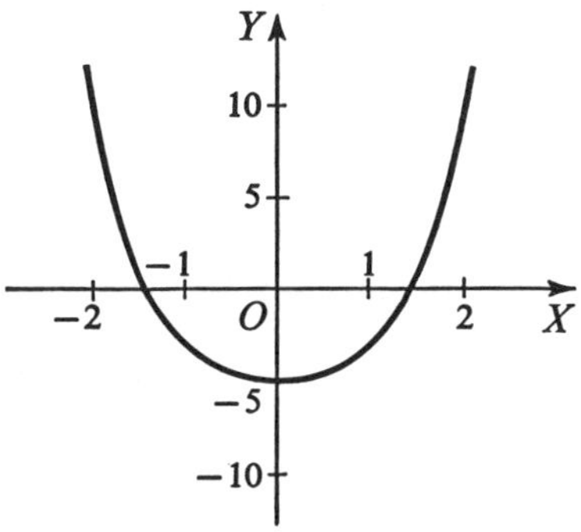

Fig. 15

Example 4. Find the coefficients a, b, c, and d if the graph of $f(x) = ax^3 + bx^2 + cx + d$ has a maximum at $(-2, 8)$ and an inflection point at $(1, 4)$.

Solution: Since the curve must go through the two given points, we know that $f(-2) = 8$ and $f(1) = 4$, that is

$$-8a + 4b - 2c + d = 8 \tag{1}$$

and

$$a + b + c + d = 4. \tag{2}$$

In addition, $f'(-2) = 0$. Since

$$f'(x) = 3ax^2 + 2bx + c,$$

this gives

$$12a - 4b + c = 0; \tag{3}$$

and since $f''(1) = 0$, we find $f''(x) = 6ax + 2b$,

and

$$6a + 2b = 0. \tag{4}$$

Equations (1)–(4) in the unknown coefficients can be solved by standard methods, giving

$$a = \tfrac{2}{27}, \quad b = -\tfrac{2}{9}, \quad c = -\tfrac{16}{9}, \quad d = \tfrac{160}{27},$$

and therefore

$$f(x) = \tfrac{2}{27}x^3 - \tfrac{2}{9}x^2 - \tfrac{16}{9}x + \tfrac{160}{27}, \quad \text{or} \quad \tfrac{2}{27}(x^3 - 3x^2 - 24x + 80).$$

Exercises 3.3 [A]

1. Find the extreme values of the following and use the second-derivative test to determine whether each is a maximum or is a minimum.

(a) $3x^2 - 7x + 13$

(b) $-6 + 3x - 2x^2$

(c) $x^2 - 5x - 3$

(d) $x^2 - 6x + 9$

2. Find the maximum, minimum, and inflection points of the following. Use the second-derivative test to identify each extreme value as a maximum or as a minimum.

(a) $y = 2x^3 - 3x^2 - 12x + 5$ (b) $y = \frac{1}{3}x^4 - 6x^2 + 2$ (c) $y = \dfrac{1}{x} - x$

3. On one set of axes, plot the maximum, minimum, and inflection points of the following curves, draw the tangents at the points of inflection, and sketch the curves.

(a) $y = x^3 - 3x^2 - 9x + 7$

(b) $y = x^3 - 3x^2 - 3x + 1$

(c) $y = x^3 - 3x^2 + 3x - 5$

(d) $y = x^3 - 3x^2 + 9x - 11$

4. The graph of a cubic function has a maximum at $(-2, 95)$ and an inflection point at $(1, -13)$. Find the function.

5. The graph of a quadratic function goes through $(-1, 3)$ and is tangent at $(2, 9)$ to the line $5x - y - 1 = 0$. Find the function.

6. Show that (a) no quadratic function has an inflection point on its graph; (b) every cubic function has an inflection point.

Exercises 3.3 [B]

1. Find the maximum, minimum, and inflection points of the following. Use the second-derivative test to identify each extreme value as a maximum or as a minimum.

   (a) $y = x^3 - 3x^2 - 9x - 4$

   (b) $y = x^4 + 4x^3 - 2x^2 - 12x + 7$

   (c) $y = \frac{1}{5}x^5 - \frac{4}{3}x^3 + 3$

   (d) $y = \dfrac{2}{x^2} + \dfrac{1}{x}$

2. On one set of axes, plot the maximum and minimum points of the following and sketch each curve:

   (a) $y = x^4 - 8x^2 + 5$

   (b) $y = x^4 - \frac{4}{3}x^3 - 4x^2 + 5$

   (c) $y = x^4 - \frac{8}{3}x^3 + 5$

3. The extreme values of a cubic function f are $f(-2) = 6$ and $f(0) = 12$. Find the function.

4. The graph of a cubic function has a point of inflection at the origin and a slope of 5 at that point. If the graph also passes through (2, 6), find the function.

5. (a) What condition must be satisfied by the coefficients of $y = ax^4 + bx^3 + cx^2 + dx + e$ if its graph has an inflection point?

   (b) Show that the graph of every polynomial function of odd degree (degree greater than 1) has at least one inflection point. (*Hint:* What is the degree of the equation which must be solved to determine the abscissa of any inflection point? What do you know about the nature of the roots of such an equation?)

3.4 Tangent and normal

An equation in point-slope form (§ 1.4, equation (3)) of the line which passes through $P(x_1, y_1)$ and has slope m is

$$y - y_1 = m(x - x_1). \tag{1}$$

If P lies on the graph of a function f, we must then have $y_1 = f(x_1)$; if, in addition, the line is tangent to this graph, then $m = f'(x_1)$. In these circumstances, (1) becomes

$$y - f(x_1) = f'(x_1)(x - x_1), \tag{2}$$

which is therefore an equation of the line tangent to the graph of f at the point P. If, therefore, P is known and f can be differentiated, the problem of writing an equation of the tangent line offers no great difficulty. If, on the other hand, P is not given, the problem may be somewhat more challenging (see Example 2, below).

Another line of some importance in geometry is the *normal.*

▶ **Definition.** If T is the line tangent to a graph at P, then the normal to the graph at P is the line through P perpendicular to T.

The condition for perpendicular lines (§ 1.3) gives us the slope of the normal: if the point P is (x_1, y_1), then the slope of the tangent at P is $f'(x_1)$ and the slope of the normal at P is therefore $-1/f'(x_1)$. Hence an equation in point-slope form of the line normal at P to the graph of f is

▶
$$y - f(x_1) = -\frac{1}{f'(x_1)}(x - x_1). \tag{3}$$

Example 1. Find equations of the tangent and the normal to the graph of $y = x^3 - 5x - 3$ at the point where $x = -1$.

Solution: If $\quad f(x) = x^3 - 5x - 3, \quad$ then $\quad f(-1) = 1.$

Also, $\quad f'(x) = 3x^2 - 5, \quad$ and therefore $\quad f'(-1) = -2.$

Hence, (2) gives $\quad y - 1 = -2(x + 1), \quad$ or $\quad 2x + y + 1 = 0,$

as the required equation of the tangent, and (3) gives

$$y - 1 = \tfrac{1}{2}(x + 1), \quad \text{or} \quad x - 2y + 3 = 0,$$

for the normal.

Example 2. Find an equation of the line through $(\frac{3}{2}, 1)$ tangent to the graph of $f = \{(x, y) : y = x^2 + 2x + 2\}$.

Solution: This problem is very different from the preceding one because the given point, $(\frac{3}{2}, 1)$, is not on the given curve; we do not know the point of tangency, and hence we do not know what number x_1 to substitute in (2). Let us, however, assume that a point of tangency exists (if this is not so, the fact will appear as we try to work the problem), and call it, say, (r, s). Since this point lies on the graph of f, we must have

$$s = f(r) = r^2 + 2r + 2,$$

and, in terms of the yet-unknown r, the slope of the required tangent is

$$f'(r) = 2r + 2.$$

Hence, in these terms, (2) becomes

$$y - (r^2 + 2r + 2) = (2r + 2)(x - r). \tag{4}$$

This is, in a sense, the required equation, but the result is not satisfactory because it depends on the unknown abscissa r. What we now need is a further condition on the tangent line which will permit us to find the value of r. We have such a condition at hand: the line must pass through $(\frac{3}{2}, 1)$. Substituting these coordinates in (4), we find

$$1 - (r^2 + 2r + 2) = (2r + 2)(\tfrac{3}{2} - r),$$

or

$$r^2 - 3r - 4 = 0.$$

Hence

$$r = 4, \quad -1$$

and

$$s = 26, \quad 1.$$

The point of tangency (r, s) is therefore either $(4, 26)$ or $(-1, 1)$, and the required tangent is either

$$10x - y - 14 = 0$$

or

$$y - 1 = 0.$$

These lines are shown in Fig. 16.

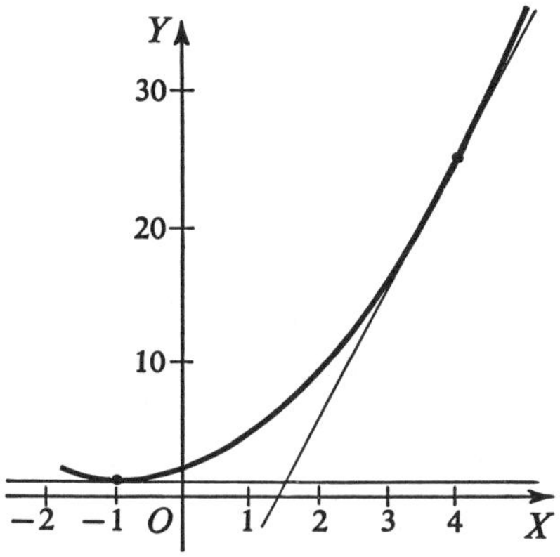

Fig. 16

3.5 Angle between two curves

Before we can talk constructively about an angle between two curved lines, we must specify what we mean. We therefore make a definition:

▶ **Definition.** An angle between two curves at a common point P is an angle between the lines tangent to each at P.

Since there are two supplementary angles between the tangent lines, there are also two supplementary angles between two intersecting curves. If it is necessary to distinguish between them, this distinction can generally be made (as in § 1.3) on the basis of a sketch.

Finding an angle between two graphs then breaks down into three steps:

1. Find the point or points of intersection.
2. Find the slope of each curve at each intersection.
3. Find the angle. Ordinarily, this involves the use of § 1.3, equation (5), and an answer in the form "$\tan^{-1} n$" is usually satisfactory.

If the angle between two graphs is either zero or a right angle, the situation involves certain special terminology.

▶ **Definition.** Two curves are said to be tangent if they are tangent to the same line at the same point.

▶ **Definition.** Two curves are said to be orthogonal if they meet at right angles.

Example 1. Find the angles of intersection of

$$y = f_1(x) = 4 - x^2 \quad \text{and} \quad y = f_2(x) = x^2 - 4.$$

The curves are shown in Fig. 17. See the following page for a solution.

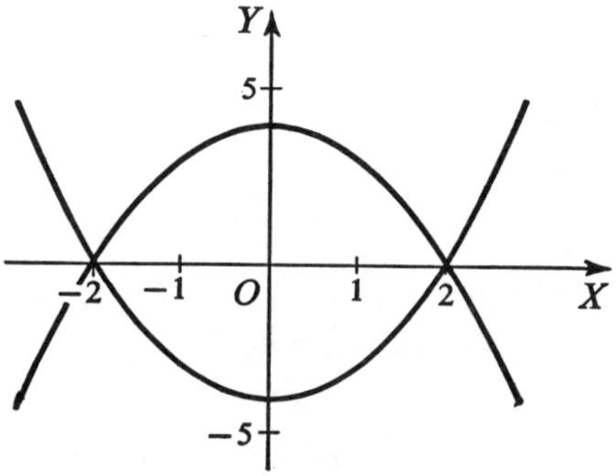

Fig. 17

Solution: Since $y = 4 - x^2$ and also $y = x^2 - 4$, we have

$$4 - x^2 = x^2 - 4,$$
$$2x^2 = 8,$$
$$x^2 = 4,$$
$$x = 2, \quad -2,$$
$$y = 0, \quad 0.$$

Hence the intersections are at $P_1(2, 0)$ and at $P_2(-2, 0)$. The values of the derivatives are given by the formulas

$$f_1'(x) = -2x, \qquad f_2'(x) = 2x,$$

so the slopes at P_1 are

$$f_1'(2) = -4, \qquad f_2'(2) = 4,$$

and the angles between the graphs at P_1 are

$$\tan^{-1} \frac{-4 - 4}{1 + (-4)(4)} = \tan^{-1} \frac{8}{15}$$

and its supplement. The same result is obtained at P_2. Note that the use of different scales on the two axes distorts these angles in Fig. 17.

Example 2. Show that every member of the family of hyperbolas $xy = k^2$ is orthogonal to the line $x - y = 0$.

Solution: Eliminating y, we have $x^2 = k^2$, $x = k, -k$. The intersections are therefore at (k, k) and $(-k, -k)$. Each hyperbola is the graph of a function f defined by

$$y = f(x) = k^2x^{-1};$$

hence $\quad f'(x) = -k^2x^{-2} \quad$ and $\quad f'(k) = f'(-k) = -1;$

that is, the slope of each hyperbola of the family is -1 at the points where it intersects the line. But the slope of the line is $+1$, and the two are therefore orthogonal, by § 1.3.

Exercises 3.5 [A]

1. Find equations of the tangent and the normal to the given curve at the point with given abscissa:
   (a) $y = 3x^2 - 7x + 2$, $x = 1$
   (b) $y = x^3 + 2x^2 + 5x + 2$, $x = -1$

2. Find equations of the tangents from $(0, -4)$ to $y = x^2 + 4x$, and sketch.
3. Find an angle, at each intersection, between $y = x^2 - 2x + 1$ and $y = 7 + 2x - x^2$.
4. If the normal at $P(2, -2)$ to the parabola $y = x^2 - x - 4$ meets the parabola again at Q, find the length PQ.

5. (a) Show that an equation of the line tangent to $y = ax^2$ at $P(x_1, y_1)$ is $y + y_1 = 2ax_1x$. (*Hint:* Remember, $y_1 = ax_1^2$.)
   (b) If this tangent meets the y-axis at Q, show that the x-axis bisects PQ.

Exercises 3.5 [B]

1. Find equations of the tangent and the normal to the given curve at the point with given abscissa:
   (a) $y = x^2 - 2x + 5$, $x = 1$ (b) $y = x - \dfrac{2}{x}$, $x = -2$
2. Write equations of the tangents from (1, 1) to $y = x^2 + 4$, and sketch.
3. Find an angle, at each intersection, between $y = x^3 - 9x$ and $y = x^2 - 9$.
4. Show that $x^2 = 8y$ and $x^2 + 4y = 12$ are orthogonal.
5. (a) Show that an equation of the line tangent to $xy = k^2$ at (x_1, y_1) is $y_1x + x_1y = 2k^2$. (*Hint:* $x_1y_1 = k^2$.)
   (b) Show that the triangle which this tangent line makes with the coordinate axes has the area $2k^2$, independent of the point of tangency.

Miscellaneous Exercises 3.5

1. Find equations of the tangent and the normal to the given curve at the point with given abscissa:
   (a) $y = x^3 - 3x^2 - 9x + 8$, $x = 3$
   (b) $y = x^{1/2} + x^{-1/2}$, $x = 4$
2. Find equations of the tangents to the given curve from the given point, and sketch:
   (a) $y = x^2 + 3x + 1$, (2, 2) (b) $xy = 9$, (1, 5)
3. Find the coordinates of the points of contact of the tangents from the given point to the given curve:
   (a) $(-1, 5)$, $y = x^3 - 4x^2 + x + 2$
   (b) $(2, -5)$, $y = x^3 - 6x^2 + 4x + 1$
4. Find an angle, at each intersection, between
   (a) $y = x^2$ and $x = y^2$.
   (b) $xy = 2k^2$ and $x^2 = 4ky$.
5. A line is tangent at (1, 1) to $y = x^3$. At what point does this line again meet the curve?

6. The normal to $y = ax^2$ at $P(x_1, y_1)$ meets the y-axis at N. Show that the length of the projection of PN onto the y-axis is independent of the choice of P.

7. The equation $y = \dfrac{x^2}{4k} - k$ represents a family of parabolas, a different parabola for each $k \neq 0$.
   (a) Show that, for fixed $x = x_1$ and $y = y_1$, there are two values of k, and that there are, therefore, two members of the family through each point (x_1, y_1) $(x_1 \neq 0)$.
   (b) Show that the two parabolas through (x_1, y_1) are orthogonal.

Exercises 3.5 [C]

1. Show that the sum of the x-intercept and the y-intercept of the line tangent at (x_1, y_1) to $\sqrt{x} + \sqrt{y} = \sqrt{a}$ is a constant, independent of the particular choice of (x_1, y_1).
2. The points $P(x_1, y_1)$ and $Q(x_2, y_2)$ are on the parabola $y = ax^2 + bx + c$, and the line tangent to the parabola at (x_3, y_3) is parallel to the chord PQ. Show that $x_3 = (x_1 + x_2)/2$.
3. If x_1 and x_2 have the same sign, show that the lines tangent at (x_1, y_1) and (x_2, y_2) to $y = x^2$ meet at $\left(\dfrac{x_1 + x_2}{2}, \sqrt{y_1 y_2}\right)$. At what point do the tangents meet if x_1 and x_2 have opposite signs?
4. P, Q, and R are three points on $y = x^2$ such that their abscissas are in geometric progression. Show that the tangents at P and R meet at a point which has the same ordinate as Q.
5. Show analytically that there are two, one, or no tangents from (h, k) to the parabola $y = ax^2 + bx + c$ $(a > 0)$ according as $ah^2 + bh + c$ is greater than, equal to, or less than k. Interpret geometrically.

3.6 Extreme values: further applications

The extreme-value problems already considered (§ 3.2) were such that each could readily be stated mathematically in terms of a single variable. There are, however, many such problems which cannot easily be formulated in mathematical terms without using two or more independent variables. The methods already developed can be applied to many such problems, but, if it is necessary to use more than one independent variable, then we shall need to find conditions on these variables which will permit eliminating all but one of them at a convenient stage of the work. If there is not enough information available to make this elimination possible, then there is not, in general, a complete specific solution to the problem. The technique is most conveniently exhibited in examples.

Example 1. The sum of the height and the radius of the base of a right circular cone is 9 in. Find the maximum volume of the cone.

Solution: If the height is h inches and the radius is r inches, then the volume in cubic inches is given by

$$V = \tfrac{1}{3}\pi r^2 h.$$

Here V depends on two variables, but we also know that

$$h + r = 9,$$

and we can therefore eliminate one of them:

$$h = 9 - r,$$

and therefore $$V = f(r) = \tfrac{1}{3}\pi r^2(9 - r) = \tfrac{1}{3}\pi(9r^2 - r^3);$$

hence $$f'(r) = \tfrac{1}{3}\pi(18r - 3r^2) = \pi(6r - r^2).$$

For a stationary value, we must have $f'(r) = 0$:

$$\pi(6r - r^2) = 0,$$

$$r = 0 \quad \text{or} \quad 6.$$

Clearly $r = 0$ does not give a maximum; if $r = 6$, then $V = f(6) = 36\pi$ (cu. in.). Since $f(0) = f(9) = 0$, and since $f(r) > 0$ for $0 < r < 9$, it is clear that the only stationary value in this interval must give a maximum; alternatively, $f''(r) = \pi(6 - 2r)$, and $f''(6) = -6\pi < 0$ tells us that we have the required maximum.

Example 2. Find the dimensions of the largest cylinder that can be cut from a solid sphere of radius a.

Solution: The volume of the cylinder,

$$V = \pi r^2 h,$$

depends on two variables, r and h; we must find a condition on them which will permit the elimination of one. As can be seen in the figure, an axial section, the

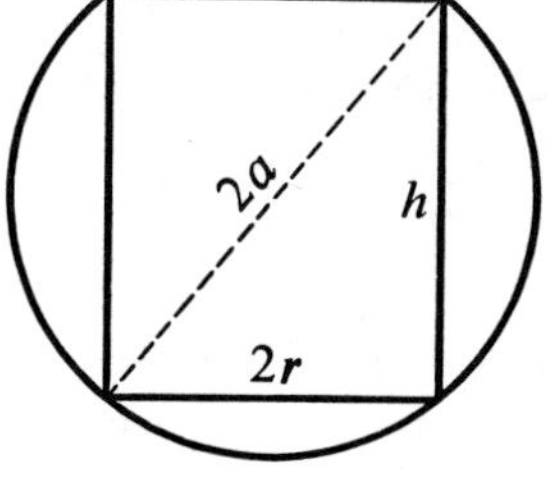

Fig. 18

theorem of Pythagoras can be applied to give such a condition, for

$$(2r)^2 + h^2 = (2a)^2.$$

Hence
$$r^2 = a^2 - \tfrac{1}{4}h^2,$$

and
$$V = f(h) = \pi(a^2 - \tfrac{1}{4}h^2)h$$
$$= \pi a^2 h - \tfrac{1}{4}\pi h^3.$$

Then
$$f'(h) = \pi a^2 - \tfrac{3}{4}\pi h^2,$$

and, if $f'(h) = 0$,
$$h = \frac{2a\sqrt{3}}{3}.$$

The corresponding value of r is
$$r = \frac{a\sqrt{6}}{3}.$$

That $f\left(\frac{2a\sqrt{3}}{3}\right)$ is a maximum is readily verified. The second derivative, $f''(h) = -\frac{3}{2}\pi h$, is negative for all positive h, and this establishes the desired fact. Also, an informal argument can be based on a consideration of what happens to V in the extreme cases $h = 0$ and $h = 2a$.

It is also possible to eliminate h from the expression for V. Since

$$h^2 = 4a^2 - 4r^2,$$
$$h = 2\sqrt{a^2 - r^2},$$

and
$$V = g(r) = 2\pi r^2\sqrt{a^2 - r^2}.$$

This does express V in terms of but a single variable, but the expression $g(r)$ cannot be differentiated by the methods we have developed so far. If this situation arises in some of the exercises that follow, as it may, it will then be necessary to back up and try eliminating some other variable.

Example 3. A cylindrical can must be designed to hold a given volume. What proportions will use the least amount of metal?

Solution: If the radius of the can is r and the height is h, then the total area S, the quantity to be minimized, is given by

$$S = 2\pi r^2 + 2\pi rh. \tag{1}$$

Here S depends on two variables, r and h; there must be some condition on r and h which will permit one of them to be eliminated. This condition is implied in the statement that the can must hold a given volume: the volume

$$V = \pi r^2 h \tag{2}$$

has a constant, fixed value V_1. Hence

$$h = \frac{V_1}{\pi r^2},$$

and, substituting this in (1), we find

$$S = f(r) = 2\pi r^2 + \frac{2V_1}{r},$$

an expression in which the only independent variable is r. Then

$$f'(r) = 4\pi r - \frac{2V_1}{r^2},$$

and if

$$f'(r) = 0,$$

a necessary condition for a stationary value, we find

$$r = \sqrt[3]{V_1/2\pi},$$

the only real zero of f'. The corresponding value of h is

$$h = 2\sqrt[3]{V_1/2\pi}.$$

Hence the stationary value of S is attained when $h = 2r$, a result which is conveniently expressed by saying that the can is just as high as it is wide. That $f(\sqrt[3]{V_1/2\pi})$ is actually a *minimum* value of S is readily determined. We may observe that S can be made arbitrarily large by taking r either very large or very near 0, and the only stationary value of f on the interval $(0, \infty)$ must therefore be a minimum. Alternatively, the second derivative,

$$f''(r) = 4\pi + \frac{4V_1}{r^3},$$

is positive for all positive r, and the stationary value is therefore a minimum by the second-derivative test.

The same problem can also be approached by eliminating r, instead of h, from (1). When $V = V_1$, (2) gives

$$r = \sqrt{V_1/\pi h}.$$

Substituting this expression in (1) and simplifying, we find

$$S = g(h) = \frac{2V_1}{h} + 2\sqrt{V_1 \pi h}.$$

Hence

$$g'(h) = -\frac{2V_1}{h^2} + \sqrt{V_1\pi/h},$$

and, as before, the only real zero of g' is

$$h = 2\sqrt[3]{V_1/2\pi}.$$

Exercises 3.6 [A]

1. A rectangular plot is to be enclosed by a fence and then divided into two plots by another fence joining the midpoints of two opposite sides. If the total fenced area is to be 24 sq. yd., what dimensions will minimize the length of fence used?
2. A strip of metal 24 in. wide is to be bent into an open gutter with a cross section in the shape of an isosceles trapezoid. If the sides of the trapezoid make an angle of 120° with the base, what dimensions will maximize the carrying capacity of the gutter?
3. A rectangular sheet of cardboard, 2 ft. by 4 ft., is to be cut as indicated in Fig. 19 and folded along the dashed lines to form a rectangular solid. What will be the dimensions of the solid if its volume is a maximum?

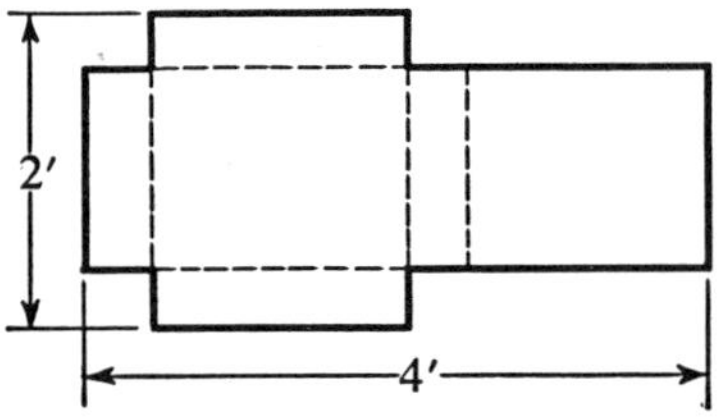

Fig. 19

4. An open tank with a square base and vertical sides is to have a capacity of 32 cu. ft. Find the dimensions that will minimize the cost of lining the tank with copper.
5. The lower side of a rectangle is on the x-axis, and the two upper vertices are on the curve $y = 4 - x^2$. What is the maximum area of the rectangle?
6. Find the maximum volume of a cone inscribed in a sphere of radius a in. (*Hint:* Let x be the distance from the center of the sphere to the base of the cone, as indicated in Fig. 20, and express all other dimensions in terms of a and x.)

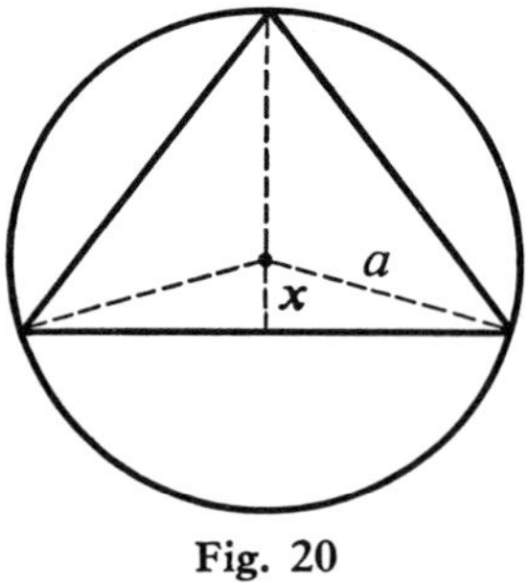

Fig. 20

Exercises 3.6 [B]

1. A roll of wire 50 ft. long is used to fence a rectangular plot of ground and to divide it into two plots by a fence joining the midpoints of two opposite sides. What dimensions maximize the total area enclosed?
2. Find the dimensions of the largest rectangular box with a square base and an open top that can be made with 675 sq. in. of material.
3. Find the coordinates of a point on the x-axis if the sum of the squares of its distances from (1, 2) and (4, 3) is a minimum.
4. A rectangle lies in the first quadrant, with one vertex at the origin, two sides along the coordinate axes, and the fourth vertex on the line $x + 2y - 6 = 0$. Find the maximum area of the rectangle.
5. A printed page must contain 20 sq. in. of printed matter, with a 2-inch margin at either side and at the top and a 3-inch margin at the bottom. What dimensions must the page have to minimize the amount of paper used? (*Hint:* Let the printed matter be x in. wide, and express all other dimensions in terms of x.)
6. A box is to be 3 times as long as it is wide and is to hold a fixed volume. If the material for the sides and bottom costs 20 cents a square foot and the top costs 10 cents a square foot, what proportions are the most economical?

Miscellaneous Exercises 3.6

1. The sides of a triangle are 10 in., 10 in., and 12 in. A rectangle is inscribed in the triangle, with one side in the 12-inch side of the triangle and the other two vertices on the 10-inch sides. Find the maximum area of the rectangle.
2. The figure of Ex. 1 is rotated about the altitude on the 12-inch side of the triangle, thus generating a cylinder inscribed in a cone. Find the maximum volume of the cylinder.
3. An isosceles triangle has base B in. and altitude H in. A second isosceles triangle, inverted, is inscribed in the first, with its apex at the center of the base of the first. Find the maximum area of the inscribed triangle.
4. The figure of Ex. 3 is rotated about its axis of symmetry, thus generating one right circular cone inscribed in another. Find the maximum volume of the inscribed cone.
5. A right circular cone has altitude of 12 in. and diameter of 12 in. Find the dimensions of the inscribed right circular cylinder of greatest volume. (The axis of the cylinder lies on the axis of the cone.)

6. Find the dimensions of the largest rectangle that can be inscribed in a triangle with sides 3 in., 4 in., and 5 in., (**a**) if one side of the rectangle lies in the 3-inch side and one in the 4-inch side of the triangle; (**b**) if one side of the rectangle lies in the 5-inch side of the triangle. Compare the areas of the two rectangles.

7. A rectangle lies in the first quadrant, with one vertex at the origin, two sides on the coordinate axes, and the fourth vertex on the first-quadrant arc of $y = -x^3 + 2x^2 - x + 3$. Find its maximum area.

8. A rectangular bin with square base and a cover is to be built so as to contain 1000 cu. ft. If the cost per square foot is $27\frac{1}{2}$ cents for the bottom, 40 cents for the top, and 10 cents for the sides, what are the dimensions for minimum cost?

9. Find the point on the line $x - y = 0$ such that the sum of the squares of its distances from (1, 0), (2, 0), and (0, 3) is a minimum.

10. The hypotenuse of a right triangle is 10 in., but the lengths of the two legs are variable. The triangle is rotated about its hypotenuse, thus generating a top-shaped solid consisting of two cones with a common base and with altitudes whose sum is 10 in. Find the maximum volume of this solid.

11. A right circular cone is inscribed in a sphere whose radius is 6 in. If the cone is h in. high, express the radius of the base in terms of h and find the value of h which makes the volume of the cone a maximum.

12. A manufacturer finds that his profit on the sale of n articles per week is $(50n - 0.3n^2 - 200)$ dollars.

    (**a**) What weekly production will maximize his total profit?

    (**b**) What weekly production will maximize his profit per item?

13. If 30 people hire a bus for an outing, they will be charged 10 dollars each, but this fare will be cut by 25 cents apiece for all passengers for each passenger over the 30; thus, 40 passengers would pay \$7.50 each. What number of passengers will produce the greatest income for the bus company?

14. A ship cruising at v knots uses $(0.3 + 0.001v^3)$ tons of fuel per hour. Find in terms of v the amount of fuel used on a 1000-mile trip, and thus determine the most economical speed.

15. The fuel cost per hour for running a ship is proportional to the cube of the speed, and there are additional fixed costs of \$864 per hour at any speed. If the fuel cost at 8 knots is \$128 per hour, find the most economical speed for a 500-mile trip.

Exercises 3.6 [C]

1. The amount of metal to be used in the construction of a cylindrical can is fixed. What proportions will make the volume of the can a maximum? Compare with Example 3, pages 122–123.
2. If, in Example 3, the top and bottom are each cut from a square and the waste is a total loss, what then will be the most economical proportions?
3. If an open-top cylindrical can is designed to hold a fixed volume, what proportions will use the least metal?
4. A window is in the form of a rectangle surmounted by a semicircle. What proportions will give the minimum perimeter for a given area?
5. The slant height of a cone is b in. What is the maximum volume of the cone?
6. Find the equation of that tangent to the first-quadrant arc of $y = 10 - x^2$ which forms with the axes a triangle of minimum area.
7. The strength of a rectangular beam is proportional to its width and the square of its depth. Find the shape of the strongest beam that can be cut from a cylindrical log.

Chapter Review

1. Locate all maximum, minimum, and inflection points on the graph of f, and determine the intervals on which f is increasing, if
   (a) $f(x) = x^3 + 3x^2 - 24x + 10$. (c) $f(x) = x^3 - 6x^2 + 6x + 2$.
   (b) $f(x) = x^3 + 3x^2 + 3x - 7$. (d) $f(x) = x^4 + 4x^3 - 3$.
2. Find a, b, and c if the graph of $y = ax^2 + bx + c$ passes through $(3, 3)$ and is tangent at $(1, 2)$ to the graph of $3x + 2y - 7 = 0$.
3. The curve $y = -x^2 + 6x - 16$ has two tangents which pass through the origin. Find their equations.
4. A square 4 in. on a side is inscribed in an isosceles triangle. Find the least possible area of the triangle.
5. (a) Find, in terms of m, an angle between $y = mx$ and $xy = 1$.
   (b) For what values of m do these curves meet at 45°?
6. Show that $y = ax^3 + bx^2 + cx + d$ has a maximum and a minimum if and only if $b^2 - 3ac > 0$.
7. If the tangent at $P(x_1, y_1)$ to $xy = 1$ meets the x-axis at A and the y-axis at B, show that $AP = PB$.

During the space-shuttle launch, computers perform approximately six integrations per second on an acceleration function. The results are used to keep the shuttle continually on course with the correct position and velocity relative to an earth-fixed coordinate system. Some time after liftoff, the shuttle switches to a star-oriented coordinate system. This helps to establish and keep the correct orbit until the shuttle and crew are ready to return home.

4

Integration

4.1 The definite integral

Areas of polygons can be computed by the methods of elementary geometry, but the problem of finding areas of regions with curved boundaries (even circles) properly belongs to calculus. Consider, then, the area A of the region enclosed by the continuous curve $y = f(x)$, the lines $x = a$ and $x = b$, and the x-axis. We assume for the moment that f is non-negative and non-decreasing on $[a, b]$. We divide $[a, b]$ into n subintervals of length $\Delta x = (b - a)/n$, and label the points of division $x_1, x_2, \ldots, x_{n-1}$; for symmetry of notation we also set $x_0 = a$, $x_n = b$. Fig. 1 shows the case $n = 5$. (The use here of equal subintervals is neither necessary nor always desirable, but

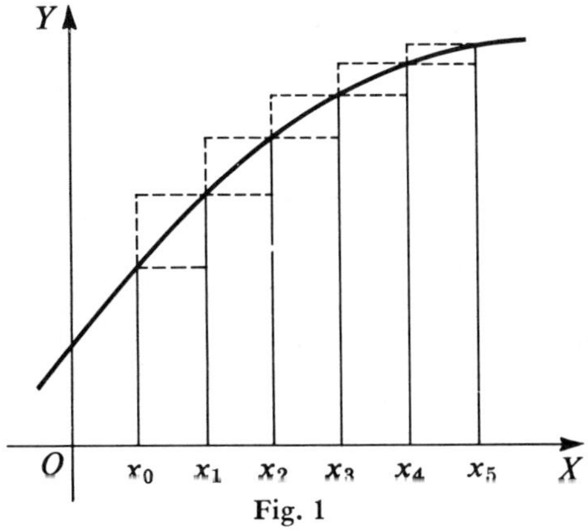

Fig. 1

it simplifies the argument.) Our assumptions about f then imply

$$0 \leq f(x_0) \leq f(x_1) \leq \cdots \leq f(x_n).$$

We next construct two rectangles on each of these subintervals: a lower rectangle, whose height is the least ordinate on its subinterval, and an upper rectangle, whose height is the greatest. The region we are considering then includes the union of all the lower rectangles and is included in the union of the upper rectangles. If ΔA_1 is the area under the curve between the lines $x = x_0$ and $x = x_1$, it then lies between the areas of the corresponding rectangles. In this case the height of the lower rectangle is $f(x_0)$, its width is Δx, and its area is therefore $f(x_0)\Delta x$. Similarly, the area of the corresponding upper rectangle is $f(x_1)\Delta x$. Hence

$$f(x_0)\Delta x \leq \Delta A_1 \leq f(x_1)\Delta x.$$

Similarly, if ΔA_2 is the corresponding area between $x = x_1$ and $x = x_2$, and so on, we have

$$f(x_1)\Delta x \le \Delta A_2 \le f(x_2)\Delta x,$$
$$\vdots$$
$$f(x_{n-1})\Delta x \le \Delta A_n \le f(x_n)\Delta x.$$

If we add these three columns of inequalities, the sum of the middle column is

$$A = \Delta A_1 + \Delta A_2 + \cdots + \Delta A_n.$$

The sum of the left-hand column, which we shall call the *lower sum*, is

$$S_L = f(x_0)\Delta x + f(x_1)\Delta x + \cdots + f(x_{n-1})\Delta x,$$

and the sum of the right-hand column, the *upper sum*, is

$$S_U = f(x_1)\Delta x + f(x_2)\Delta x + \cdots + f(x_n)\Delta x.$$

By the rules for addition of inequalities, we then have

$$S_L \le A \le S_U. \tag{1}$$

For obvious reasons, each of S_L and S_U is called a *rectangular approximation* to A. If we use S_L as an approximation, the error incurred is $A - S_L$. If we use S_U, the error is $S_U - A$. It follows from (1) that the sum of the errors is greater than either error and is equal to the difference between S_U and S_L, or

$$(S_U - A) + (A - S_L) = S_U - S_L.$$

But in the difference $S_U - S_L$, the terms $f(x_1)\Delta x$, $f(x_2)\Delta x$, . . ., $f(x_{n-1})\Delta x$ all vanish, leaving only

$$\begin{aligned} S_U - S_L &= f(x_n)\Delta x - f(x_0)\Delta x \\ &= [f(b) - f(a)]\Delta x. \end{aligned}$$

Since the difference $f(b) - f(a)$ is constant for a given a and b, $S_U - S_L$ can be made arbitrarily small by making Δx small enough. It can be shown that if f is continuous on $[a, b]$ then both S_U and S_L have limits, and it then follows from the preceding argument that these limits are equal. Hence

$$A = \lim_{\Delta x \to 0} S_L = \lim_{\Delta x \to 0} S_U.$$

The sums S_L and S_U were formed by choosing, respectively, the least and greatest values of f in each subinterval. (Such values always exist if f is continuous on $[a, b]$.) If we choose *any* element x_i^* in the ith subinterval, for $1 \leq i \leq n$, and use $f(x_i^*)$ as the height of a rectangle on that subinterval, we get a sum

$$S = f(x_1^*)\Delta x + f(x_2^*)\Delta x + \cdots + f(x_n^*)\Delta x$$

such that

$$S_L \leq S \leq S_U.$$

This follows directly from a term-by-term comparison of the three sums. Hence S has the same limit as do S_L and S_U. This common limit is called a *definite integral*. It is designated

$$\int_a^b f(x)dx.$$

(Read this "the definite integral of $f(x)dx$ from a to b.")

▶ **Definition.** Let f be a function continuous on $[a, b]$. Divide $[a, b]$ into subintervals of length $\Delta x = (b - a)/n$ by the points $x_0 = a, x_1, x_2, \ldots, x_n = b$. For each integer i, $1 \leq i \leq n$, let x_i^* be any point in $[x_{i-1}, x_i]$, and let

$$S = f(x_1^*)\Delta x + f(x_2^*)\Delta x + \cdots + f(x_n^*)\Delta x. \tag{2}$$

Then the definite integral of f from a to b is the number

$$\lim_{\Delta x \to 0} S = \int_a^b f(x)\, dx. \tag{3}$$

The function f, or the expression $f(x)$, is the integrand, and the numbers a and b are the limits of integration.

The symbol "$\int$" is called an *integral sign*; it is a stylized "S," reminding us that it represents the limit of a sum. The "dx," which seems here to serve no useful purpose, exists partly as a historical accident but principally, as we shall see in Chapter 7, because it can be very useful indeed. When the limit (3) exists the function f is said to be *integrable* on $[a, b]$, and the preceding argument shows that a function which is continuous on a *closed bounded* interval is integrable on that interval.

The study and application of integrals is a branch of calculus called *integral calculus*, as distinguished from differential calculus. While we began this section with a discussion of area, our definition of the definite integral makes no mention of area, nor does it require that the function f be either non-negative or non-decreasing. If $f(x) \geq 0$ for all $x \in [a, b]$, then $\int_a^b f(x)dx$ gives the area of the region shown in Figure 1, page 129, but the integral is an important mathematical object in its own right. One application of definite integrals is the computation of areas, but we shall see that there are many others.

In the preceding discussion we have required $a < b$. It is convenient to free ourselves of this restriction. In defining the sum S (Equation (2)) we have taken $\Delta x = (b - a)/n$. The effect of interchanging a and b is therefore to change the sign of each term in S and hence the sign of its limit. We therefore define

$$\int_b^a f(x)dx = -\int_a^b f(x)dx. \tag{4}$$

This in turn suggests (to see why, set $a = b$ in (4)) the further definition

$$\int_a^a f(x)dx = 0. \tag{5}$$

We also note that if $a < c < b$ then the sum S can be rearranged to yield two sums, one over the interval $[a, c]$ and the other over $[c, b]$, from which

$$\int_a^c f(x)dx + \int_c^b f(x)dx = \int_a^b f(x)dx. \tag{6}$$

Furthermore, it follows from (4) and (5) that (6) holds even if c is not between a and b; a demonstration of this is left as an exercise.

Two final facts about the definite integral will be useful in applications. If f and g are both integrable on $[a, b]$, then so are the functions $f + g$ and cf, where c is any constant, and

$$\int_a^b [f(x) + g(x)]dx = \int_a^b f(x)dx + \int_a^b g(x)dx, \tag{7}$$

$$\int_a^b cf(x)dx = c\int_a^b f(x)dx. \tag{8}$$

Miscellaneous Exercises 4.1

1. **(a)** Use a division of $[1, 2]$ into 10 equal subintervals to obtain rectangular approximations S_L and S_U to $\int_1^2 2x\,dx$.

(b) Draw a graph of $y = 2x$, and use elementary geometry to compute the area of the region bounded by the graph, the lines $x = 1$, $x = 2$, and the x-axis. Compare this area with the average of S_L and S_U.

2. Use the table of square roots, p. 436, and a division of $[1, 4]$ into 6 equal subintervals, to obtain rectangular approximations S_L and S_U to $\int_1^4 \sqrt{x}\,dx$. Compare the average of these approximations to the correct value, which is $4\frac{2}{3}$.

3. Use the table of reciprocals, p. 436, to compute a sum S approximating $\int_1^3 \frac{1}{x}\,dx$. Use a division of $[1, 3]$ into 10 equal subintervals, and take x_i^* at the middle of the ith subinterval, $1 \leq i \leq 10$. (The correct value of this integral is 1.099, to four significant figures.)

4. Given that Equation (6),

$$\int_a^c f(x)dx + \int_c^b f(x)dx = \int_a^b f(x)dx,$$

is true when $a < c < b$, use Equation (4),

$$\int_b^a f(x)dx = -\int_a^b f(x)dx,$$

and Equation (5),

$$\int_a^a f(x)dx = 0,$$

to show that (6) is also true when **(a)** $c < a$; **(b)** $c = a$; **(c)** $c = b$; **(d)** $c > b$.

5. Use the definition of the definite integral to prove that if f and g are both integrable on $[a, b]$, then

$$\int_a^b [f(x) + g(x)]dx = \int_a^b f(x)dx + \int_a^b g(x)dx \tag{7}$$

and

$$\int_a^b cf(x)dx = c\int_a^b f(x)dx \tag{8}$$

for any constant c.

6. Given that $\int_0^2 x^2\,dx = \frac{8}{3}$ and that $\int_0^2 x^3\,dx = 4$, evaluate

(a) $\int_0^2 3x^2\,dx$ **(b)** $\int_0^2 (x^2 + x^3)dx$ **(c)** $\int_0^2 (2x^2 - 5x^3)dx$.

7. Given that $\int_0^1 x^2\,dx = \frac{1}{3}$ and that $\int_0^2 x^2\,dx = \frac{8}{3}$, evaluate $\int_1^2 x^2\,dx$.

8. (For students with access to a computer.) Computations like those in Exercises 1–3 are most easily done by computer. Use a computer to obtain sums S_L and S_U approximating each of the following integrals. In each case, use enough subintervals so that $S_U - S_L < 0.001$.

(a) $\int_1^2 \frac{1}{x}\,dx$ **(b)** $\int_2^4 \frac{1}{x}\,dx$ **(c)** $\int_4^8 \frac{1}{x}\,dx$

(d) $\int_0^{\pi/2} \sin x\,dx$ **(e)** $\int_0^2 \sqrt{1 + x^4}\,dx$.

4.2 The fundamental theorems; indefinite integrals

It is surely one of the more incredible facts in mathematics that two such dissimilar objects as the derivative and the integral should be as closely related as the two sides of a coin. We shall explore that relationship in this section. Observe first that the definite integral, $\int_a^b f(x)dx$, is a name for a number, a numeral. This number is completely determined by the integrand f and the limits of integration, a and b. In particular, the "x" in this notation is a dummy variable, like the "x" in the set-builder $\{x\colon\ 0 \leq x \leq 1\}$, and can be replaced by any other letter not already in use. Thus,

$$\int_1^2 x^2\,dx = \int_1^2 y^2\,dy = \int_1^2 u^2\,du; \tag{1}$$

all three integrals represent the same number.

On the other hand, given an integrand f, the value of the integral does depend on the limits of integration. We have no reason to believe that $\int_1^3 x^2\,dx$ or $\int_0^2 x^2\,dx$ has the same value as the integrals in (1), and indeed neither one does. Consider, then, what happens if we fix the function f and the lower limit of integration a, and let the upper limit be a variable, say t. We then have a mathematical object that we designate

$$\int_a^t f(x)dx.$$

A specific instance is $\int_1^t x^2\,dx$. Because the square function is integrable on every interval, it follows that this integral represents a real number for every real t, and therefore that the equation

$$F(t) = \int_1^t x^2\,dx$$

defines a function F. More generally, if a function f is continuous on $[a, b]$, and c is an arbitrary but fixed point of $[a, b]$, then a function G with domain $[a, b]$ is defined by the rule

$$G(t) = \int_c^t f(x)dx.$$

We now make the simplifying assumption that f is non-negative and increasing on $[a, b]$; the results obtained will in fact be independent of both restrictions. One consequence of this assumption is that we can interpret our definite integrals as areas. Now suppose that t and $t + \Delta t$ both belong to the interval $[a, b]$. Then

$$G(t+\Delta t) - G(t) = \int_c^{t+\Delta t} f(x)dx - \int_c^t f(x)dx \qquad \text{(definition of } G\text{)}$$

$$= \int_t^c f(x)dx + \int_c^{t+\Delta t} f(x)dx \qquad \text{(Eqn. (4), § 4.1)}$$

$$= \int_t^{t+\Delta t} f(x)dx. \qquad \text{(Eqn. (6), § 4.1)}$$

When $\Delta t > 0$, this integral is the area ΔA of the region shown in Fig. 2. This region contains a rectangle of height $f(t)$ and base Δt, and is contained

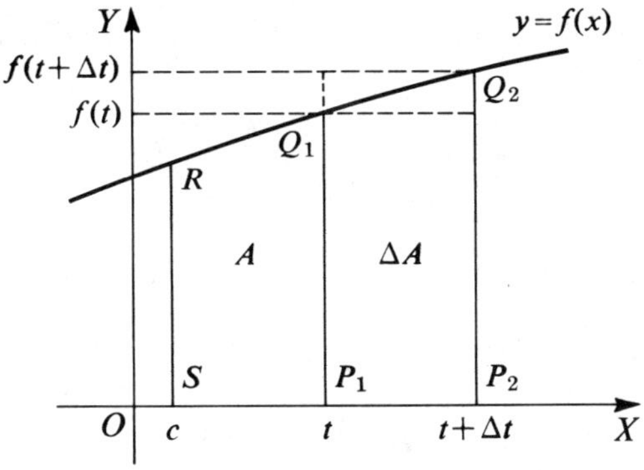

Fig. 2

in a rectangle of height $f(t+\Delta t)$ and base Δt. Hence

$$f(t)\Delta t \leq G(t+\Delta t) - G(t) \leq f(t+\Delta t)\Delta t,$$

or

$$f(t) \leq \frac{G(t+\Delta t) - G(t)}{\Delta t} \leq f(t+\Delta t). \tag{2}$$

If $\Delta t < 0$ we must reverse the inequalities in (2), and corresponding results can also be obtained if f is decreasing or negative. Consider now what happens in any of these cases as $\Delta t \to 0$. The expression on the left of (2), $f(t)$, does not change and therefore has $f(t)$ as its limit. Since the function f is continuous at t, that is, since there is no jump or gap at t in the graph of f, then the quantity on the right also has $f(t)$ as its limit. Hence the fraction in the middle, caught between two quantities that have a common limit, must have the same limit. But

$$\lim_{\Delta t \to 0} \frac{G(t+\Delta t) - G(t)}{\Delta t} = G'(t)$$

by definition. We therefore conclude that

$$G'(t) = f(t);$$

the derivative of the integral function G is the integrand f. This result is known as the First Fundamental Theorem of Calculus.

Theorem 4.2-1 (First Fundamental Theorem): Let f be continuous on $[a, b]$, and let c be any point of $[a, b]$. Then the function

$$G(t) = \int_c^t f(x)\, dx$$

has the derivative

$$G'(t) = f(t)$$

for each $t \in [a, b]$.

Now, given a function f, let F be any function such that $F' = f$. For instance, if $f(x) = x^2$, then it is easy to see that $F(x) = \frac{1}{3}x^3$ is suitable. Let $G(t) = \int_a^t f(x)\, dx$. Then $G' = f$ by the First Fundamental Theorem, and therefore $F' = G'$ by substitution. We shall show in Chapter 5 that if two functions have the same derivative then their difference is a constant. It follows that there exists a constant C such that

$$F(t) = G(t) + C.$$

But

$$G(t) = \int_a^t f(x) dx,$$

so

$$F(t) = \int_a^t f(x) dx + C.$$

When $t = a$ this yields

$$F(a) = \int_a^a f(x) dx + C = 0 + C = C,$$

and when $t = b$ we have

$$F(b) = \int_a^b f(x) dx + C$$

$$= \int_a^b f(x) dx + F(a),$$

or

$$\int_a^b f(x) dx = F(b) - F(a).$$

This result is the Second Fundamental Theorem of Calculus.

▶ **Theorem 4.2-2** (Second Fundamental Theorem): If F is any function such that $F' = f$ on $[a, b]$, then

$$\int_a^b f(x)dx = F(b) - F(a). \tag{3}$$

This theorem is the key to evaluating definite integrals. It tells us that to evaluate the integral in (3), we must

(1) find, if possible, a function F such that $F' = f$, and

(2) compute the difference $F(b) - F(a)$.

This difference is often denoted

$$F(x)\Big|_a^b, \quad \text{or} \quad [F(x)]_a^b,$$

and the function F is called an indefinite integral (or antiderivative, or primitive) of the function f.

Example 1. Evaluate $\int_1^2 x^2\,dx$.

Solution: We have noted that if $F(x) = \frac{1}{3}x^3$, then $F'(x) = x^2$. Hence

$$\int_1^2 x^2\,dx = \tfrac{1}{3}x^3\Big|_1^2 = \tfrac{8}{3} - \tfrac{1}{3} = \tfrac{7}{3}.$$

This is not the only possible choice of an indefinite integral F. If, for instance, $F(x) = \frac{1}{3}x^3 + 17$, then again $F'(x) = x^2$, and

$$\int_1^2 x^2\,dx = [\tfrac{1}{3}x^3 + 17]_1^2 = (\tfrac{8}{3} + 17) - (\tfrac{1}{3} + 17) = \tfrac{7}{3}$$

as before; the 17's vanish. More generally, we see that $F(x) = \frac{1}{3}x^3 + C$ is an indefinite integral of $f(x) = x^2$ for any value of the constant C.

Example 2. Evaluate $\int_{-1}^3 (x^2 - 4)dx$.

Solution: It is easy to check that $\frac{1}{3}x^3 - 4x$ is an indefinite integral for this integrand. Hence

$$\begin{aligned}\int_{-1}^3 (x^2-4)dx &= [\tfrac{1}{3}x^3 - 4x]_{-1}^3 \\ &= [\tfrac{1}{3}(3)^3 - 4(3)] - [\tfrac{1}{3}(-1)^3 - 4(-1)] \\ &= -3 - 11/3 = -20/3.\end{aligned}$$

As these examples suggest, the Second Fundamental Theorem makes the problem of evaluating definite integrals primarily a search for indefinite integrals. This process is commonly called *integration*. Given a function f, the standard notation for the family of indefinite integrals of f is

$$\int f(x)dx. \tag{4}$$

Thus Example 1 suggests that

$$\int x^2\,dx = \tfrac{1}{3}x^3 + C,$$

where C may be any constant whatever. Similarly, Example 2 yields

$$\int (x^2 - 4)dx = \tfrac{1}{3}x^3 - 4x + C.$$

Somewhat confusingly, (4) is commonly referred to as *the* indefinite integral of f, though strictly speaking it represents not one but the family of *all* indefinite integrals of f.

We note that by definition of the indefinite integral, the derivative of $\int f(x)dx$ is $f(x)$. The following basic properties then follow from the corresponding properties of derivatives:

▶ **Rule 1.** $\int [f(x) + g(x)]dx = \int f(x)dx + \int g(x)dx.$

▶ **Rule 2.** For any constant c, $\int cf(x)dx = c\int f(x)dx.$

▶ **Rule 3.** $\int x^n\,dx = \dfrac{1}{n+1}\,x^{n+1} + C, n \neq -1.$

Example 3. $\int \sqrt{x}\,dx = \int x^{1/2}\,dx = \dfrac{1}{3/2}\,x^{3/2} + C = \dfrac{2}{3}\,x^{3/2} + C.$

Example 4. $\int \dfrac{x^3 - 5}{x^2}\,dx = \int (x - 5x^{-2})dx = \dfrac{1}{2}\,x^2 + 5x^{-1} + C.$

Exercises 4.2 [A]

Integrate or evaluate, as indicated.

1. $\int (3x^2 - 5x + 1)dx$
2. $\int (2x - 5 + 3x^{-2})dx$
3. $\int \left(x^3 + 4 - \dfrac{5}{x^2}\right) dx$
4. $\int \left(x + \dfrac{1}{3x^3}\right) dx$

5. $\int \left(\sqrt{x} + \frac{1}{\sqrt{x}}\right) dx$

6. $\int \frac{s^2 + 2s - 1}{(\sqrt{s})^3}\, ds$

7. $\int (\sqrt[3]{5x^2} - 3)dx$

8. $\int \frac{(t + 2)(t - 5)}{t^4}\, dt$

9. $\int (ax^2 + bx + c)dx$

10. $\int_0^2 (x^2 - 2x)dx$

11. $\int_{-1}^2 (4 - y^2)dy$

12. $\int_0^4 (2x^{1/2} - \frac{1}{4}x^2)dx$

Exercises 4.2 [B]

Integrate or evaluate, as indicated.

1. $\int (x - a)^2\, dx$

2. $\int \sqrt{ax}\, dx$

3. $\int \frac{3x^2 - 5}{2x^2}\, dx$

4. $\int \frac{x^2 - 3x - 1}{\sqrt{x}}\, dx$

5. $\int \frac{y - 2}{y^{2/3}}\, dy$

6. $\int_0^3 (x^2 - 1)dx$

7. $\int_{-1}^3 x^4\, dx - \int_{-1}^3 (x^4 - 2x)dx$

8. $\int_0^2 \sqrt{x}\, dx + \int_2^4 \sqrt{x}\, dx$

9. $\int_0^h (h^2 - x^2)^2 dx$

10. $\int_0^{a\sqrt{2}} (x^2 - a^2)dx$

4.3 Areas of plane regions

We return now to the problem with which we started this chapter, the problem of computing areas of regions with curved boundaries. The argument used in § 4.1 led us to conclude that the area A of the region shown in Fig. 1, p. 129, is the limit of a certain sum (a rectangular approximation), and we represented this limit by a definite integral. Hence we conclude that the area is

$$A = \int_a^b f(x)dx; \quad f(x) \geq 0. \tag{1}$$

This conclusion is actually a definition. Elementary geometry fails to define an area for such a region as this, but an argument motivated by elementary geometry, based on approximating rectangles, suggests that this is a reasonable definition. Some of the exercises at the end of this section will show that (1) gives results consistent with elementary geometry in cases to which both apply.

Example 1. Find the area of the region bounded by $y = x^3 - 8x + 10$, the x-axis, and the lines $x = 1$, $x = 2$ (Fig. 3).

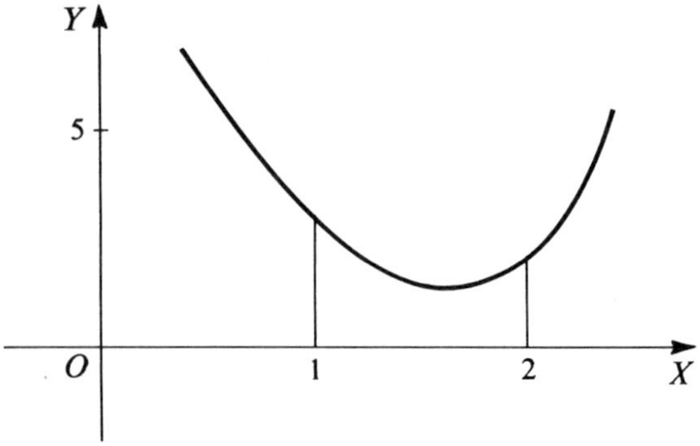

Fig. 3

Solution: Using (1), we have

$$\int_1^2 (x^3 - 8x + 10)dx = [\tfrac{1}{4}x^4 - 4x^2 + 10x]_1^2$$
$$= (4 - 16 + 20) - (\tfrac{1}{4} - 4 + 10)$$
$$= 1\tfrac{3}{4} \text{ (square units).}$$

The argument that led to (1) required that f be non-negative on $[a, b]$. If f is negative, a slight modification is required. In this case, as Fig. 4

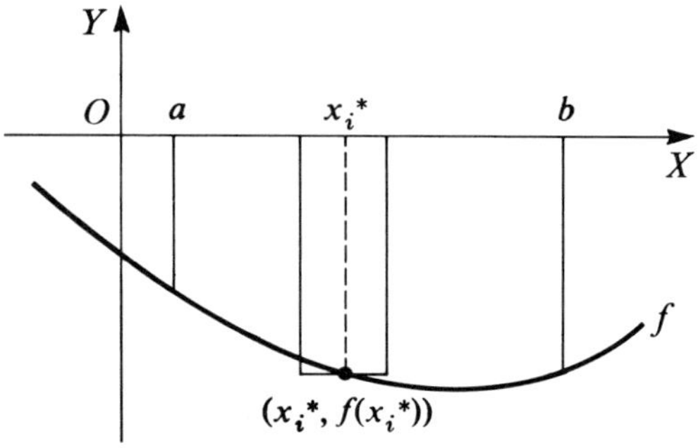

Fig. 4

indicates, the height of the ith approximating rectangle is $|f(x_i^*)| = -f(x_i^*)$, and its area is therefore $-f(x_i^*)\Delta x$. Hence a rectangular approximation to the required area is

$$S = -f(x_1^*)\Delta x - f(x_2^*)\Delta x - \cdots - f(x_n^*)\Delta x,$$

and
$$\lim_{\Delta x \to 0} S = \int_a^b -f(x)dx = -\int_a^b f(x)dx. \tag{2}$$

We define this to be the required area when f is negative.

Example 2. Find the area of the region enclosed by $y = x^2 - 5x - 6$ and the x-axis (Fig. 5).

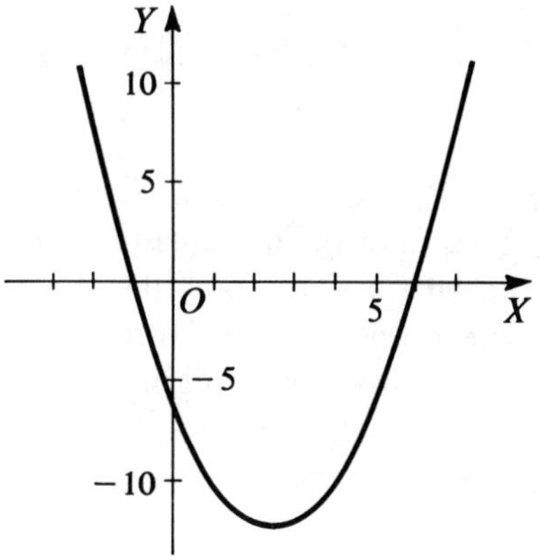

Fig. 5

Solution: If $x^2 - 5x - 6 = 0$, then $x = -1$ or 6. Hence the area is

$$-\int_{-1}^{6} (x^2 - 5x - 6)dx = -\left[\tfrac{1}{3}x^3 - \tfrac{5}{2}x^2 - 6x\right]_{-1}^{6}$$
$$= -[(72 - 90 - 36) - (-\tfrac{1}{3} - \tfrac{5}{2} + 6)]$$
$$= 57\tfrac{1}{6} \text{ (square units).}$$

A final complication arises when the integrand changes sign in the interval of integration. In this case it is necessary to divide the interval into subintervals on which it does not change sign, as in our next example.

Example 3. Find the area of the region enclosed by the x-axis and the loops of the curve $y = 4x - x^3$.

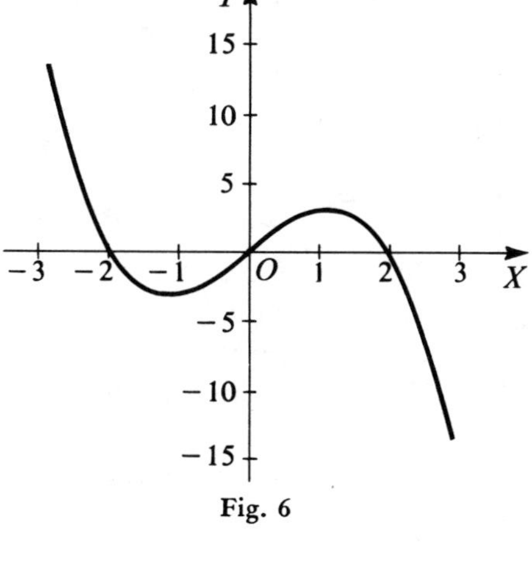

Fig. 6

Solution: The x-intercepts of the curve are -2, 0, and 2. As Fig. 6 shows, the region under consideration is actually two symmetric regions, one below the x-axis and one above. Hence the area is

$$-\int_{-2}^{0}(4x-x^3)dx+\int_{0}^{2}(4x-x^3)dx=-[2x^2-\tfrac{1}{4}x^4]_{-2}^{0}+[2x^2-\tfrac{1}{4}x^4]_{0}^{2}$$

$$=4+4=8 \text{ (square units)}.$$

A simpler way to carry out this computation would be to exploit the symmetry of the curve by finding the area of the first-quadrant region and then doubling.

A danger which is present in situations like this is that of not observing that the curve crosses the x-axis and therefore ignoring the fact in the integration. In this case such a procedure might give

$$\int_{-2}^{2}(4x-x^3)dx=[2x^2-\tfrac{1}{4}x^4]_{-2}^{2}=4-4=0$$

for the required area, a result that is not only wrong, even though the integration is correct, but manifest nonsense.

The preceding derivations are easily modified to give the area of a region enclosed between the graphs of two continuous functions, f and g, and the lines $x=a$ and $x=b$. In this case, if $f(x)\geq g(x)$ for $a\leq x\leq b$, we take approximating rectangles with their upper bases on the graph of f and their lower bases on the graph of g. The area of the ith rectangle is then

$$[f(x_i^*)-g(x_i^*)]\Delta x,$$

as shown in Fig. 7, and if we replace the function f by the function $f-g$

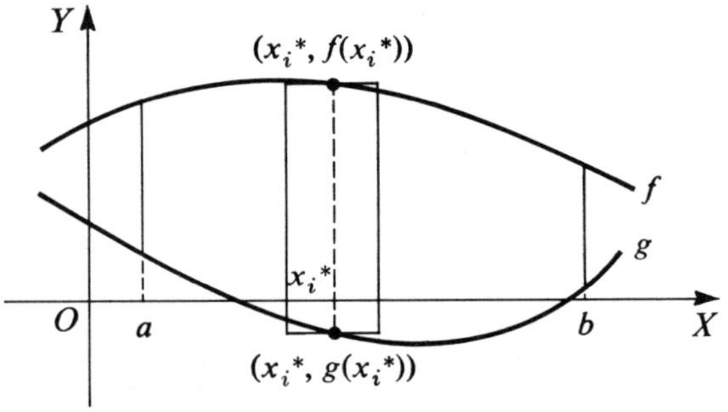

Fig. 7

in the argument that led to (1) we have, in this case,

$$A=\int_{a}^{b}[f(x)-g(x)]dx \tag{3}$$

as a definition of the required area. Note that this includes all the cases considered earlier. Thus, if the lower boundary of the region is the x-axis, we take $g(x) = 0$ to obtain (1), and if the upper boundary is the x-axis, we take $f(x) = 0$ to obtain a result equivalent to (2).

Example 4. Find the area of the region enclosed by $y = 11 - x^2$ and $y = x^2 - 4x + 5$.

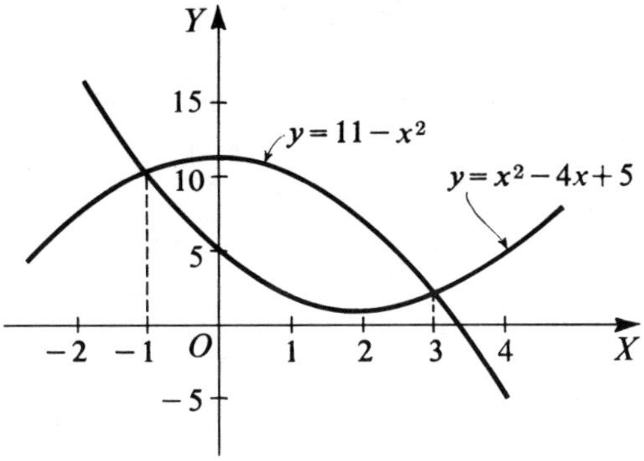

Fig. 8

Solution: The simultaneous solution of the two equations shows that the curves intersect at (−1, 10) and (3, 2), as shown in Fig. 8. Hence, by (3), the required area is

$$\int_{-1}^{3} [(11 - x^2) - (x^2 - 4x + 5)]dx = \int_{-1}^{3} (6 + 4x - 2x^2)dx$$
$$= \left[6x + 2x^2 - \tfrac{2}{3}x^3\right]_{-1}^{3}$$
$$= 18 - (-3\tfrac{1}{3}) = 21\tfrac{1}{3} \text{ (square units).}$$

Exercises 4.3 [A]

1. Sketch the region enclosed by the x-axis and the given curve and lines, and find its area:

   (a) $y = 3x^2 + 2, \quad x = 1, \quad x = 3$

   (b) $y = 1/x^3, \quad x = 1, \quad x = 2$

   (c) $y = x^2 - x - 2, \quad x = 0, \quad x = 4$

   (d) $y = x^2 - x - 2, \quad x = 2, \quad x = 4.$ Compare with part (c).

2. Sketch and find the area of the region enclosed by the given curve and the x-axis:

   (a) $y = 4 - x^2$ (b) $y = x^3 + 5x^2$ (c) $y = x^4 - 5x^2 + 4$

3. Find the area of the region enclosed by the parabola $y = x^2$ and the line $y = x$.
4. Find the area enclosed by $y = 8 + 2x - x^2$ and $x - y + 2 = 0$.
5. Use a definite integral to compute the area of the rectangle bounded by the coordinate axes and the lines $x = b$, $y = h$, where $b > 0$, $h > 0$. Compare with the value given by elementary geometry.

Exercises 4.3 [B]

1. Sketch the region enclosed by the x-axis and the given curve and lines, and find its area:

   (a) $y = 8 - x^2, \quad x = 0, \quad x = 2$

   (b) $y = x^2 - 5x - 6, \quad x = 0, \quad x = 4$

   (c) $y = 2\sqrt{x}, \quad x = 0, \ x = 9$

   (d) $x^2y - x^3 + 3 = 0, \quad x = 2, \quad x = 10.$

2. Sketch and find the area of the region enclosed by the given curve and the x-axis:

   (a) $y = x^2 - 4x - 5$ (b) $y = x^3 - 9x$ (c) $y = x^4 - 8x^2 + 16$

3. Find the area of the region enclosed by the parabola $y = x^2$ and the line $3x - y + 4 = 0$.
4. The equation $\sqrt{x} + \sqrt{y} = \sqrt{a}$ represents an arc of a parabola. Find the area of the region bounded by this arc and the coordinate axes.
5. Find by integration the area of a right triangle with legs a and b units long.

Miscellaneous Exercises 4.3

1. Find the area of the region enclosed by $x^2 = 4y$ and $y^2 = 4x$.
2. Find the area of the region enclosed by $y = x^3 - 4x + 4$ and $x + y - 6 = 0$.
3. Find the area of the region enclosed by $y^2 = 4px$ and $x = a$ $(p > 0, a > 0)$.
4. Find the area of the region between the parabola $y = (x - 2)^2$ and the chord which joins $(-1, 9)$ and $(3, 1)$.
5. Find the area of the region between the curve $y = x^2 + 3x + 3$ and the chord joining $(-2, 1)$ and $(1, 7)$.
6. Find the area of the region enclosed by $y = x^3 + 2$ and $y = x^2 + x + 1$.
7. (a) Find the area of the region enclosed by the x-axis, the curve $x^2y = 1$,

and the lines $x = 1$ and $x = k$, $k > 1$.

(b) Find the area of the region enclosed by the x-axis, the curve $y = x^{-1/2}$, and the lines $x = 1$ and $x = k$, $k > 1$.

(c) Discuss what happens to each of these areas as k takes larger and larger values.

8. Sketch an appropriate graph and use a geometrical argument to show that

(a) $\int_{-\pi}^{\pi} \sin x \, dx = 0$ (b) $\int_{-\pi/2}^{\pi/2} \cos x \, dx = 2\int_{0}^{\pi/2} \cos x \, dx.$

4.4 Elementary differential equations

Although, as we have pointed out, the derivative was introduced as a technique for finding the slope of a curve or the velocity of a falling body, we know that it has many applications in which the notions of slope and velocity do not explicitly appear. In this chapter the integral has been introduced as a device for computing areas, but we shall see (especially in Chapter 9) that it can be applied to a great variety of problems that have no obvious connection with areas. In this book we shall be principally concerned with applications of the derivative and the integral, but we should also recognize that these objects, considered as mathematical abstractions, have an existence and a meaning quite independent of any physical or geometrical interpretations.

Given a function f, its derivative f' and its integral F are also functions, where F is any function with the property $F' = f$. In most of the cases that we shall encounter here, a function f has a derivative f', which has a derivative f'', and so on; similarly, most of the functions that we shall meet here have integrals given by such formulas as $F(x) = \int f(x)dx$, and a function F thus determined has an integral given by $\int F(x)dx$, and so on; the chain can often be extended in both directions without end. The great importance of calculus arises from the fact that all these functions can be given a great variety of interpretations and can hence be used in the solution of a great variety of problems.

Many of the problems of physics can be stated in the following form (or some elaboration of it): Given a function f, find a function F such that

$$F'(x) = f(x). \tag{1}$$

An equation in this form, or any equation which involves derivatives, is called a differential equation, and a function F which satisfies (1) is called a solution of the differential equation. Such a solution is not completely determined by (1), for, as we have seen, if F_1 is any solution whatever of (1), then the function F given by

$$F(x) = F_1(x) + C \tag{2}$$

is also a solution, for any value of the constant C. We have in (2) the *general* solution, a family of solutions to (1). To pick out a particular member of the family (2), that is, a *particular* solution of (1), we must specify additional conditions which can be satisfied by only one value of C. Such conditions are commonly called *initial conditions.* Thus, the differential equation

$$F'(x) = 3x^2 + 1 \tag{3}$$

has a family of solutions

$$F(x) = \int (3x^2 + 1)dx$$

$$= x^3 + x + C.$$

Figure 9 shows a few members of this family. If to (3) we add the initial condition

$$F(2) = 5, \tag{4}$$

we find $$5 = 2^3 + 2 + C, \quad C = -5,$$

and the particular solution of (3) which also satisfies (4) is

$$F(x) = x^3 + x - 5.$$

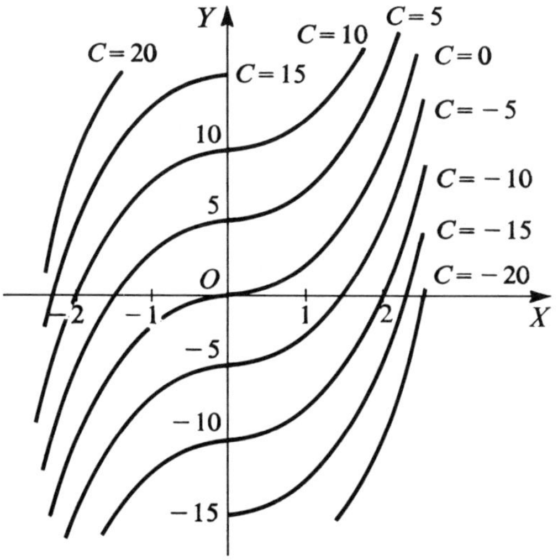

Fig. 9

Example 1. Find the general solution of the differential equation

$$f'(x) = \frac{2}{\sqrt{x}}$$

and the particular solution for which $f(1) = 0$.

Solution: The general solution is

$$\begin{aligned} f(x) &= 2\int x^{-1/2}\,dx \\ &= 4x^{1/2} + C. \end{aligned}$$

If $f(1) = 0,$

we find $0 = 4(1)^{1/2} + C$

and $C = -4.$

The required particular solution is therefore $f(x) = 4x^{1/2} - 4$.

Example 2. A curve has the slope $m = x^2 - 6x + 2$ at each point (x, y) through which it passes, and it also passes through $(2, -1)$. Find an equation of the curve.

Solution: If the equation of the curve is $y = F(x)$, then the first condition of the problem states, in effect, that

$$F'(x) = x^2 - 6x + 2,$$

and therefore $F(x) = \int (x^2 - 6x + 2)dx = \frac{1}{3}x^3 - 3x^2 + 2x + C$

for some value of the constant C. To find C, we use the condition that the curve must pass through $(2, -1)$, that is, that $F(2) = -1$:

$$F(2) = \tfrac{1}{3}(2)^3 - 3(2)^2 + 2(2) + C = -\tfrac{16}{3} + C = -1,$$

whence $C = \frac{13}{3}$.

Then $y = F(x) = \frac{1}{3}x^3 - 3x^2 + 2x + \frac{13}{3}$

is the equation of the curve.

Example 3. If $g'(t) = t^2 - 1$ and $g(3) = 0$, find $g(1)$.

Solution: We have $g(t) = \int (t^2 - 1)dt = \frac{1}{3}t^3 - t + C.$

Therefore $g(3) = 9 - 3 + C = 6 + C;$

but we were given $g(3) = 0,$

and therefore $C = -6$

and $g(t) = \frac{1}{3}t^3 - t - 6$

and $g(1) = -6\frac{2}{3}.$

Example 4. Find $f(t)$ if $f''(t) = 3t$, $f(0) = 8$, and $f(4) = 0$.

Solution: A differential equation which contains a second derivative, as does this one, is said to be of the *second order*. As we shall see, a solution to a second-order equation involves two constants of integration and may therefore be required, as in this problem, to meet two initial conditions.

Since $f''(t)$ is the derivative of $f'(t)$, we have

$$f'(t) = \int 3t\,dt = \tfrac{3}{2}t^2 + C_1,$$

and therefore
$$f(t) = \int (\tfrac{3}{2}t^2 + C_1)dt = \tfrac{1}{2}t^3 + C_1t + C_2.$$

We cannot combine C_1 and C_2, because one is a multiple of t and the other is not. The initial conditions require, however, that

$$8 = 0 + 0 + C_2$$

and
$$0 = 32 + 4C_1 + C_2.$$

Hence
$$C_1 = -10, \quad C_2 = 8,$$

and
$$f(t) = \tfrac{1}{2}t^3 - 10t + 8.$$

Exercises 4.4 [A]

1. Find the particular solutions of the following differential equations that correspond to the stated initial conditions:
   (a) $f'(x) = 2x + 3;\quad f(1) = -1$
   (b) $F'(y) = \sqrt{5y};\quad F(5) = \frac{25}{3}$
   (c) $g'(x) = 3x^2 - 4x + 7;\quad g(-1) = 13$
   (d) $f'(t) = (3t - 5)^2;\quad f(0) = 0$
   (e) $F'(x) = \dfrac{x^3 - 1}{3x^2};\quad F(2) = \frac{1}{12}$
2. The slope of a curve at the point (x, y) is given by $m = 2x^{-3}$, and the curve goes through $(2, 4)$. Find an equation of the curve.
3. Find an equation of a curve that goes through the origin if its slope is given, for any value of x, by the value of $3x^2 - 6x + 5$.
4. A family of curves is defined by the differential equation $f'(x) = x + 2$. Find an equation of the member of the family that is tangent to the x-axis.
5. The slope of a curve is given by $f'(x) = 3x^2 - 19$, and the curve has a y-intercept of 30. Where does the curve cross the x-axis?
6. Find $F(x)$ if $F''(x) = 12x$, $F(2) = 2$, and $F(-1) = 5$.

Exercises 4.4 [B]

1. Find the particular solutions of the following differential equations that correspond to the stated initial conditions:
   (a) $F'(x) = x^2 - 2x + 5; \quad F(3) = 1$
   (b) $h'(y) = (y - 1)^4; \quad h(0) = 13$
   (c) $f'(x) = \dfrac{5x + 2}{3x^{1/3}}; \quad f(-8) = 5$
2. If $\phi'(x) = x^2 - 3x + 5$ and $\phi(0) = 0$, find $\phi(2) - \phi(-1)$.
3. If $f'(x) = (\sqrt{a} - \sqrt{x})^2$ and $f(a) = 0$, find $f(0)$.
4. If $F'(x) = 3x^2 + x^{-2}$ and $F(-1) = 0$, for what values of x does $F(x) = 0$?
5. Find a function f that is positive only on the interval $(-1, 5)$ and that has a derivative $f'(x) = 4 - 2x$.
6. Describe geometrically the family of curves $y = f(x)$ for which $f''(x) = 0$.
7. If $f'(x) = 3x^2 - 2x - 4$ and $f(0) = -2$, find the zeros of f.
8. Find $g(x)$ if $g''(x) = x^{-1/2}$, $g(1) = -\frac{8}{3}$, and $g(9) = 40$.
9. Find the equation $y = f(x)$ of a curve if $f''(x) = 3x$ and the curve has slope 2 at the origin.

4.5 Volumes of solids of revolution

Consider a region in the xy-plane bounded by the continuous curve $y = f(x)$, the x-axis, and the lines $x = a$, $x = b$. If this region is rotated about the x-axis, a solid of revolution will be formed, and the problem of computing the volume V of this solid leads, like the area problem of § 4.1, to a definite integral. As in our analysis of the area problem, we divide the interval $[a, b]$ of the x-axis into n subintervals of length $\Delta x = (b - a)/n$, pick an arbitrary point in each subinterval, say x_i^* in the ith subinterval, and on this subinterval as base construct a rectangle of height $|y_i| = |f(x_i^*)|$. When this

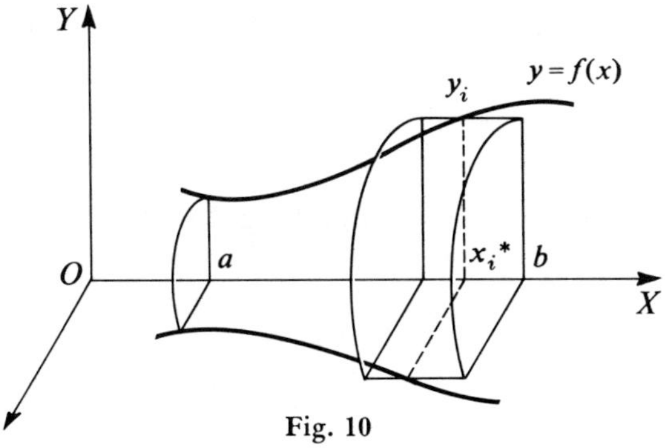

Fig. 10

rectangle is rotated about the x-axis it generates a cylinder of radius $|y_i|$ and height Δx, and the volume of this cylinder, given by the formula $V = \pi r^2 h$ (which we assume, as an axiom), is then $\pi y_i^2 \Delta x$. The sum of the volumes of all these cylinders is then a *cylindrical approximation*,

$$S = \pi y_1^2 \Delta x + \pi y_2^2 \Delta x + \cdots + \pi y_n^2 \Delta x,$$

to the volume V. It can be shown that if f is continuous on $[a, b]$, then the sum S has the same limit, as $\Delta x \to 0$, no matter how the points x_i^* are chosen. We define this limit to be the volume V. Since the sum S has precisely the same form as the sum discussed in § 4.1, with the function πf^2 replacing the function f, it follows that its limit is the definite integral

$$V = \pi \int_a^b y^2 \, dx.$$

Example: Find the volume of the solid generated by rotating about the x-axis the region bounded by $y = x^2$, the x-axis, and the lines $x = 1$, $x = 2$.

Solution: $V = \pi \int_1^2 y^2 \, dx = \pi \int_1^2 x^4 \, dx = \frac{1}{5} \pi x^5 \Big|_1^2 = \frac{32\pi}{5} - \frac{\pi}{5} = \frac{31\pi}{5}.$

Exercises 4.5 [A]

1. Find the volume of the solid formed by rotating about the x-axis the region bounded by the x-axis and
   (a) $x + 3y - 6 = 0, \quad x = 0, \quad x = 6$. Check by elementary geometry.
   (b) $y = x^2 + 2, \quad x = -1, \quad x = 1$.
   (c) $y = 1/x, \quad x = 1, \quad x = 5$.
2. Find the volume of the solid generated by rotating about the x-axis the region bounded by the x-axis and
   (a) $y = 9 - x^2$. (b) $y = x^3 - 4x^2$.
3. Find, by integration, the volume of the sphere generated by rotating about the x-axis the circle $x^2 + y^2 = 9$.
4. The region enclosed by $x^2 - y^2 = 4$ and $x = 5$ is rotated about the x-axis. Find the volume generated.
5. Find, by integration, the volume of a right circular cone of radius r and altitude h.
6. A cup-shaped solid is formed by rotating about the x-axis the region enclosed by the y-axis, the line $y = 3$, and the curve $y = \sqrt{x}$. Find the volume. (*Hint*: Find this as a difference of two volumes.)

7. Find the volume of the solid formed by rotating about the x-axis the region enclosed by $y = x^2 + 1$ and $y = 9 - x^2$.

8. Obtain, by integration, a formula for the volume of a conical frustum of height h and radii r_1 and r_2. Check by setting $r_2 = 0$ and comparing with Ex. 5, and by setting $r_1 = r_2$ and comparing with the formula for the volume of a cylinder.

Exercises 4.5 [B]

1. Find the volume of the solid formed by rotating about the x-axis the region bounded by the x-axis and
   (a) $y = \sqrt{x}, \quad x = 0, \quad x = 9.$
   (b) $y = x^2 - 5, \quad x = -2, \quad x = 1.$

2. Find the volume of the solid generated by rotating about the x-axis the region enclosed by the x-axis and $y = x^2 - 2x - 3$.

3. Find, by integration, the volume of a sphere of radius r.

4. Find the volume generated by rotating about the x-axis the ellipse $4x^2 + y^2 = 36$.

5. Find the volume generated by rotating about the x-axis the region bounded by $y^2 = 4x$ and the chord joining (0, 0) and (9, 6).

6. Find the volume generated by rotating about the x-axis the region bounded by $y^2 = a^2$ and $x^2 - y^2 = a^2$.

7. Find the volume of the solid generated by rotating about the x-axis the region bounded by $y = x^2 + 2x + 2$ and $y = 6 - x^2$.

8. That portion of the circle $x^2 + y^2 = r^2$ which is included between the lincs $x = a$ and $x - b$ $(-r \leq a \leq b \leq r)$ is rotated about the x-axis. Find the volume generated, and check by giving to a and b values which will make this volume equal to that of the whole sphere.

4.6 Accelerated motion

If the distance s which a body travels in time t is a function f of t, that is, if $s = f(t)$, then the average velocity over an interval of time Δt is (§ 2.7)

$$\frac{\Delta s}{\Delta t} = \frac{f(t + \Delta t) - f(t)}{\Delta t},$$

and the instantaneous velocity v at any instant t is defined (§ 2.7) to be the limit of this average velocity as Δt approaches zero:

$$v = \lim_{\Delta t \to 0} \frac{\Delta s}{\Delta t} = \lim_{\Delta t \to 0} \frac{f(t + \Delta t) - f(t)}{\Delta t}.$$

By the definition of derivative (§ 2.8), this limit is precisely the value at t of the derivative f'; hence $v = f'(t)$.

Motion with constant velocity rarely occurs. In most of the situations actually found in nature, velocity is continually changing, and it is therefore of some practical importance to examine the rate of change of the velocity, the acceleration. The average change in the velocity over any interval of time of length Δt, beginning at any time t in the domain of f' is

$$\frac{\Delta v}{\Delta t} = \frac{f'(t + \Delta t) - f'(t)}{\Delta t},$$

and it is therefore natural to define the acceleration a at time t to be the limit of this average as Δt approaches zero:

$$a = \lim_{\Delta t \to 0} \frac{\Delta v}{\Delta t} = \lim_{\Delta t \to 0} \frac{f'(t + \Delta t) - f'(t)}{\Delta t}.$$

But this is, by definition, the value at t of the derivative f'' of the velocity function f'; hence $a = f''(t)$, and the acceleration function is thus seen to be the second derivative of the distance function.

A velocity measures the distance traveled, or the change of position, in a unit time; the appropriate unit in which to measure a velocity is therefore distance per unit time, such as feet per second (abbreviated f.p.s. or ft./sec.) or miles per hour (m.p.h. or mi./hr.). Similarly, an acceleration measures the change of velocity in a unit time, and the appropriate unit is therefore velocity per unit time, such as feet per second *per second* (abbreviated ft./sec.^{2}).

If the distance function is known, it is then possible to obtain the velocity and acceleration functions by successive differentiations. Since, moreover, integration is, in a sense, the inverse of differentiation, it is also possible to reverse this process: if the acceleration function f'' is known, it may be possible to obtain the velocity function f' as a solution to the differential equation $a = f''(t)$ and then, in turn, the distance function f as a solution to the differential equation $v = f'(t)$. This implies the introduction of two constants of integration, and a problem which has a fully determined answer must, therefore, contain enough information to permit the evaluation of these constants.

Example 1. An object near the surface of the earth is attracted toward the earth by a force sufficient to give it a downward acceleration of approximately 32 ft./sec.^{2} A ball is thrown straight up, from the top of a 60-foot building, with a velocity of 20 ft./sec. With what velocity will the ball hit the ground?

Solution: Before the motion can be described in mathematical terms, it is necessary to choose a coordinate system. This choice is, to a considerable extent, arbitrary, but one reasonable choice is to take the origin at the surface of the earth and the positive direction upward. Then the ball will be thrown from a point at which $s = +60$, with a velocity $v = +20$, and throughout its flight it will be subject to an acceleration $a = -32$, *negative* because it is in the direction of *decreasing* s. We then have

$$a = f''(t) = -32; \tag{1}$$

hence
$$v = f'(t) = \int -32\, dt = -32t + C_1.$$

To evaluate the constant C_1, we note that the value of v at the beginning of the motion is known: the ball is thrown up with a velocity of 20 ft./sec. If we agree that this corresponds to $t = 0$, if, that is, we agree to start our clock at the instant the ball is released, we then have

$$20 = -32(0) + C_1,$$
$$C_1 = 20,$$

and therefore
$$v = -32t + 20. \tag{2}$$

If the initial velocity had any other value, say v_0, then the relation (2) would be

$$v = -32t + v_0,$$

and it is therefore common to use the symbol v_0, representing the velocity when $t = 0$, for the particular constant of integration C_1.

Now,
$$s = f(t) = \int (-32t + 20)dt = -16t^2 + 20t + C_2,$$

and since we know that $s = 60$ when $t = 0$, we have

$$60 = -16(0) + 20(0) + C_2$$

and
$$s = -16t^2 + 20t + 60. \tag{3}$$

The constant C_2 is the value of s when $t = 0$ and is therefore often designated s_0.

The motion of the ball is completely described by equations (1), (2), and (3), which give the acceleration, velocity, and position at any instant. To complete the solution of our problem, we note that the ball strikes the ground when $s = 0$, that is,

$$-16t^2 + 20t + 60 = 0.$$

Solving for t, we have $t = -1.41$ or 2.66, approximately.

The value $t = -1.41$ has no meaning in the physical problem and is therefore discarded. Setting $t = 2.66$ in (2), we have

$$v = f'(2.66) = -32(2.66) + 20 = -65.1, \text{ approximately,}$$

and we conclude that the ball strikes the ground with a (downward) velocity of about 65.1 ft./sec.

It is instructive to work the same problem in a different coordinate system. Let us take the origin at the top of the building and the positive direction downward. Then

$$a = f''(t) = +32, \tag{1a}$$

$$v = \int 32\,dt = 32t + v_0 = 32t - 20, \tag{2a}$$

and

$$s = \int (32t - 20)dt = 16t^2 - 20t + s_0 = 16t^2 - 20t. \tag{3a}$$

When the ball hits the ground, $s = 60$:

$$60 = 16t^2 - 20t,$$

$$t = -1.41 \text{ or } 2.66, \text{ approximately,}$$

and the required velocity is $f'(2.66) = +65.1$. The student should note and explain the differences between equations (1), (2), and (3), and their counterparts (1a), (2a), and (3a).

Example 2. A particle moves along the x-axis in such a way that its acceleration at time t is

$$a = 6t - 2 \quad (t \geq 0).$$

The particle is at the origin when $t = 1$ and again when $t = 3$. With what velocity does it pass through the origin?

Solution: If $x = f(t)$, we have

$$a = f''(t) = 6t - 2 \quad (t \geq 0),$$

and therefore

$$v = f'(t) = \int (6t - 2)dt = 3t^2 - 2t + v_0.$$

We have no information that will permit the direct evaluation of the initial velocity v_0, but we can find x in terms of v_0:

$$x = f(t) = \int (3t^2 - 2t + v_0)dt = t^3 - t^2 + v_0 t + x_0.$$

We know that $f(1) = f(3) = 0$; that is,

$$f(1) = 1 - 1 + v_0 + x_0 = 0,$$

and

$$f(3) = 27 - 9 + 3v_0 + x_0 = 0.$$

Solving these two equations for v_0 and x_0, we have

$$v_0 = -9, \quad x_0 = 9.$$

Then

$$x = f(t) = t^3 - t^2 - 9t + 9$$

gives the position of the particle at time t. Actually, all that is needed in this problem is the velocity,

$$v = f'(t) = 3t^2 - 2t - 9,$$

and we find $f'(1) = -8$ and $f'(3) = +12$ units per unit of time as the required velocities. The student should be sure he understands the meaning of the sign in each case.

Exercises 4.6[A]

NOTE: Near the surface of the earth, the acceleration due to gravity is approximately 32 ft./sec.^{2}.

1. A stone is dropped from the top of an 80-foot tower. **(a)** How long will it take the stone to reach the ground? **(b)** How fast will it be going when it hits?
2. If the stone of Ex. 1 is thrown down with an initial velocity of 40 ft./sec., with what velocity will it then hit the ground?
3. With what upward velocity must a girl throw a ball to get it up to a window 60 ft. above her head?
4. If a body moves on an incline which makes an angle θ with the horizontal, the gravitational acceleration is $32 \sin \theta$ ft./sec.^{2} A long ramp makes a 30° angle with the ground, and a ball is rolled up the ramp with a speed of 40 ft./sec. **(a)** How far up the slope does it go? **(b)** How high above its starting point does it go? **(c)** How long does it take the ball to go up and back?
5. A particle moves in a straight line with acceleration $a = 3t$ at time t. If it starts from a point P with velocity 8 when $t = 0$, how far from P is it when $t = 2$?
6. The velocity at time t of a particle moving in a line is $v = t^2 - 2$ (units per second). **(a)** Find the acceleration when $t = 6$. **(b)** How far does the particle go from $t = 3$ to $t = 4$?
7. A particle moving in a straight line starts from rest at the time $t = 0$ with an acceleration $a = 12 - 6t$. **(a)** Show that it is again at rest when $t = 4$. **(b)** Find how far from its starting point it then is.
8. A particle moving in a line will be s units from a fixed point P of the line at time t, where $s = 2t^2 - \sqrt{t} + 10$, $t > 0$. For what values of t will **(a)** the velocity, and **(b)** the acceleration, be zero?

Exercises 4.6[B]

1. A bullet fired vertically reaches a height of 3456 ft. in 3 sec. **(a)** How high will it go? **(b)** How long will it be in the air?

2. A helicopter is 200 ft. high and rising at a rate of 4 ft./sec. when a passenger drops a pen overboard. With what velocity will the pen hit the ground?

3. Coal is being discharged down a 40-foot chute which makes an angle of 60° with the horizontal. (See Prob. 4 of Exercises 4.6 [A].) How fast is the coal going at the bottom of the chute?

4. A particle moves on the x-axis in such a way that its abscissa at time t is $x = t^3 - 7t^2 + 10t$. Find when it goes through the origin and what its velocity is at that time.

5. A particle moves in a straight line with acceleration $a = 2t + 6$, $t \geq 0$. If it starts at $t = 0$ from a fixed point O with a velocity of -12 units/sec, show that it will return to O and find with what velocity it does return.

6. A particle moves on a straight line with acceleration $a = 8 - 6t$ at time t. If it is 3 units from a fixed point P of the line when $t = 1$ and 12 units from P (in the same direction) when $t = 4$, find its velocity when $t = 2$.

7. The acceleration of a particle is proportional to the square root of the time t, $t \geq 0$. If the particle starts from rest when $t = 0$ and is going 12 units/sec. at the end of 4 seconds, how far will it go in 9 seconds?

Chapter Review

1. Evaluate the following integrals:

(a) $\displaystyle\int_1^4 \frac{2x + 3}{\sqrt{x}}\, dx$ (b) $\displaystyle\int_{-3}^{2} (x^2 - 6)dx$ (c) $\displaystyle\int_a^0 (a^{3/2} - x^{3/2})^2\, dx$

2. Describe geometrically the family of curves $y = f(x)$ which have $f'(x) = 3$.

3. Find the area of the region enclosed by the x-axis and $y = h^2 - x^2$.

4. Find the particular solution $g(t)$ of the differential equation $g''(t) = 32$ for which $g(0) = g(3) = 0$.

5. The region cut out of the first quadrant by the line $x = k$ and the curve $y = x^2 - 4x + 9$ has an area of 18 square units. What is the value of k?

6. The region enclosed by $x^2 - 4y^2 = 16$ and $x = 6$ is rotated about the x-axis. Find the volume generated.

7. A stone is thrown straight up, with a velocity of 72 ft./sec., from the top of a 175-foot tower. With what velocity will it hit the ground?

8. The region above the line $y = 1$ and inside the ellipse $4x^2 + y^2 = 9$ is revolved about the x-axis. Find the volume of the solid generated.

5

Continuity and Limits

5.1 Continuity

Calculus is an enormous and diverse mathematical discipline, the creation of many people working over many centuries. It is not possible to name a particular mathematician at a specific time and say "Here is the beginning of calculus." A few names and two important times do stand out, however. The first time is approximately 250 B.C., and the first name is that of Archimedes, who lived from 287 to 212 B.C. in the Greek colony of Syracuse, in Sicily. One of Archimedes' many interests was the problem of computing the areas of plane regions with curved boundaries, such as circles and parabolic segments, and computing volumes of solids with curved surfaces, such as spheres, cylinders, and cones. In solving this problem he was led to create the branch of calculus now known as integral calculus, which we examined in Chapter 4.

The second time is almost 2,000 years later, and though many more names deserve attention, the two most outstanding are Isaac Newton (1642–1727), an Englishman, and Gottfried Wilhelm Leibniz (1646–1716), a German. These men were concerned with two problems: first, the problem of determining tangent lines to curves, and second, the problem of describing the motion of falling objects. As we saw in Chapter 2, these two apparently different problems have the same abstract formulation and therefore the same solution, the derivative. In obtaining this solution, Newton and Leibniz each created the second branch of calculus, differential calculus, and established the two fundamental theorems that relate it to the integral calculus of Archimedes.

There followed a period of tremendous productivity, as mathematicians and scientists rushed to exploit this new tool. They showed great originality in applying calculus to an enormous variety of problems, but much of their work was deplorably sloppy, by modern standards, because they paid scant attention to the conditions under which their results were valid. Indeed, it was over 100 years after the time of Newton and Leibniz before the root idea of calculus, the limit, received a clear and careful definition.

To some extent we have recapitulated this historical process in the preceding chapters. We have made extensive use of limits, and of the associated idea of continuity, without ever stating exactly what these concepts are. Instead, we have relied on your intuition to fill in some of the gaps and on your good will to tolerate the others. This somewhat untidy procedure made it possible to get quickly to applications of calculus and to see, with a minimum of formality, some of the things that can be done with this tool. The time has now come to replace our vague and intuitive statements with precise mathematical definitions and theorems. We shall begin with continuity.

Continuity is such a pervasive feature of familiar functions that to understand and appreciate it one must look at some functions that lack this property, some discontinuous functions.

Example 1. We have informally described a continuous function as one whose graph can be traced without lifting pencil from paper. The function f defined by the rule

$$f(x) = \begin{cases} 0 \text{ if } x < 0 \\ 1 \text{ if } x \geq 0 \end{cases}$$

is discontinuous since, as Fig. 1 shows, its graph cannot be traced across the

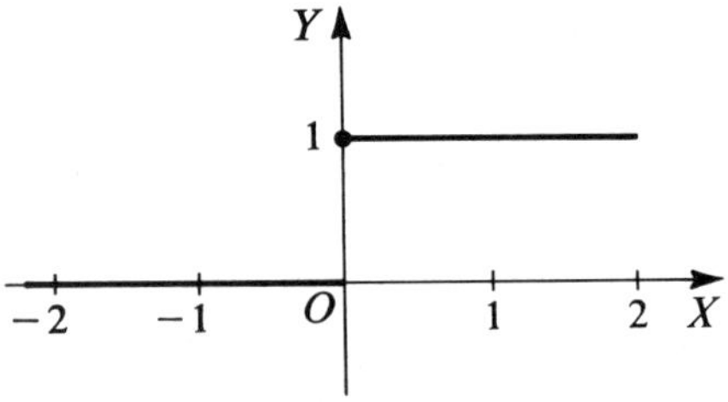

Fig. 1

point $x = 0$ without lifting pencil from paper.* Notice, however, that if we restrict the domain of f to either of the intervals $(-\infty, 0)$ or $[0, \infty)$, each of these restricted functions is continuous. The fact that f is discontinuous arises solely from its behavior at the single point $x = 0$ where its graph has a jump. This is our first important observation about continuity: it is strictly a local matter, a property of a function at a point of its domain by virtue of the behavior of the function in the immediate vicinity of that point.

Example 2. A discontinuous function with many applications in mathematics is the *greatest integer function*. For each real number x, its value is defined to be the greatest integer not exceeding x, and is denoted $[x]$. Thus $[2.7] = 2$, $[\pi] = 3$, $[4] = 4$, and $[-4.01] = -5$. As Fig. 2 shows, there are

* Functions like this are encountered in the study of electrical circuits; $f(x)$, for instance, could be the amount of current in a circuit that is energized at time $x = 0$, after which a current of 1 unit flows.

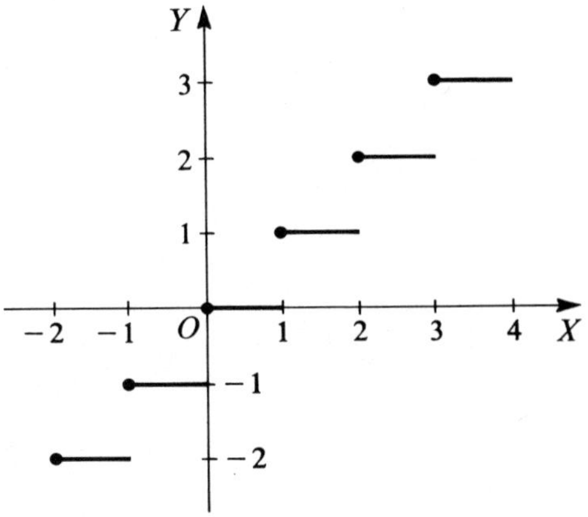

Fig. 2

one-unit jumps in the value of this function at each integer (the heavy dots indicate the values at the integers), and this function is therefore discontinuous at every integer. Thus we have a function with infinitely many points of discontinuity, but again the issue is a local one: $[x]$ is continuous on any interval that does not contain an integer, or even an interval such as $[1, 2)$ or $[2, 3)$ that has an integer as its left-hand endpoint.

Example 3. Let f be the function defined by

$$f(x) = \begin{cases} 3 - x & \text{if } x \neq 1 \\ 4 & \text{if } x = 1. \end{cases}$$

As shown in Figure 3, on page 160, the graph of f is a line with the point at $x = 1$ displaced, and the function is therefore discontinuous at $x = 1$. This time, however, the discontinuity is unlike our earlier examples; because it can be removed by redefining f at 1, thus:

$$f(x) = \begin{cases} 3 - x & \text{if } x \neq 1 \\ 2 & \text{if } x = 1. \end{cases}$$

It then follows that $f(x) = 3 - x$ for all x, and f is continuous (its graph is an unbroken line). Discontinuities of this type are called *removable discontinuities*. Despite their contrived appearance, they occur naturally and rather frequently in calculus.

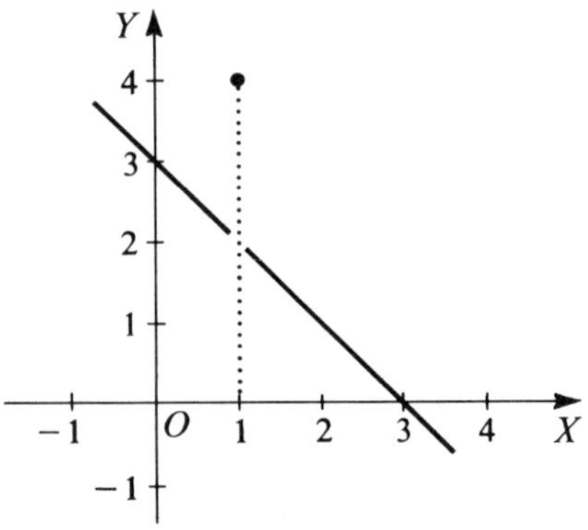

Fig. 3

In each of the preceding examples the discontinuities arise because the value of the function at a point of discontinuity is not close to the values of the function at *all* nearby points. This behavior is typical of a function at a point of discontinuity, and stands in sharp contrast to the behavior of a function at a point of continuity. For instance, in *every* neighborhood of 1 the greatest integer function assumes the values 0 *and* 1, and from this information alone we can infer that $[x]$ is discontinuous at 1. This rather obvious remark furnishes the key to a definition of continuity: a function is continuous at a point c of its domain if the values of f at all points near c are close to $f(c)$. If this condition is not met, the function is discontinuous at c.

We see that a definition of continuity must be worded to accommodate the following facts:

(1) Continuity is a local matter; a function can be continuous at some points, discontinuous at others. We must therefore define continuity *at a point*.

(2) A function f is continuous at a point c of its domain if $f(x)$ is near $f(c)$ when x is near c.

The second of these statements is almost the definition we require; the only trouble with it is that vague word "near." What remains, therefore, is to formulate some criterion of nearness. To do this, we turn to the idea of distance. The distance between $f(x)$ and $f(c)$ is by definition $|f(x) - f(c)|$. The distance between x and c is $|x - c|$. Hence $f(x)$ is "near" $f(c)$ if $|f(x) - f(c)|$ is small, and x is near c if $|x - c|$ is small. Admittedly, this substitutes one vague word for another, "small" for "near," but it is a forward step.

At this point we follow tradition by introducing as variables two Greek letters, δ (delta) and ϵ (epsilon). In view of the context, one often thinks of these variables as representing "small" numbers, but the only restriction explicitly placed on them is that, because they represent distances between distinct points, each of them must be positive. Our criterion of nearness is stated in terms of δ and ϵ:

Definition. A function f is continuous at a point c of its domain if, given any $\epsilon > 0$, there exists a $\delta > 0$ such that

$$|f(x) - f(c)| < \epsilon$$

whenever

$$|x - c| < \delta$$

and x is in the domain of f. A function that is not continuous at a point c of its domain is said to be discontinuous at c.

It is illuminating to think of this definition in terms of a contest. You claim that f is continuous at c. Your claim is challenged. The challenger produces a number, an ϵ. It may be any number: 17, π, 10^{-6}, just so long as it is positive. To support your claim, you must now match this ϵ with a δ such that $|x - c| < \delta$ implies $|f(x) - f(c)| < \epsilon$. You have no control over the choice of ϵ, but you may pick any positive δ that will do the job. Evidently, your choice of δ will generally depend on the function f, on the point c, and on the ϵ with which you are challenged. If you can demonstrate your ability to produce a δ to meet any ϵ challenge, then your claim of continuity at c is established.

In terms of the graph of f, our definition has the interpretation shown in Fig. 4. Given an ϵ, the values of f that satisfy the condition $|f(x) - f(c)| < \epsilon$ correspond to points of the graph that lie between the lines $y = f(c) - \epsilon$ and $y = f(c) + \epsilon$; these are the horizontal dashed lines in Fig. 4. The values of x that satisfy $|x - c| < \delta$ lie between the vertical lines $x = c - \delta$ and $x = c + \delta$. Continuity of f at c is established by finding a δ such that *all*

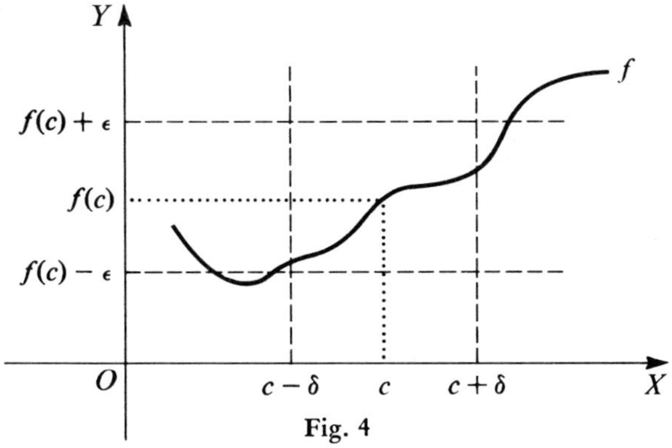

Fig. 4

points of the graph that lie between the vertical lines also lie between the horizontal lines.

Although continuity is a local property, many functions enjoy this property at each point of their domains. Indeed, this is true of most familiar functions. We thus have the following definitions:

Definition. A function is said to be continuous if it is continuous at every point of its domain; otherwise it is said to be discontinuous. A function that is continuous at every point of an interval is said to be continuous on the interval. A continuous function whose domain is the set of all real numbers is said to be everywhere continuous.

Miscellaneous Exercises 5.1

1. Sketch graphs and thus determine the points (if any) at which the following functions are discontinuous.

(a) $f(x) = \begin{cases} 1/x, & x \neq 0 \\ 0, & x = 0. \end{cases}$

(b) $f(x) = \begin{cases} 0, & x < 0 \\ 2x, & x \geq 0. \end{cases}$

(c) $f(x) = \begin{cases} -1, & x < 0 \\ 0, & x = 0 \\ 1, & x > 0. \end{cases}$

(d) $f(x) = \begin{cases} x + 1, & x < 0 \\ x - 1, & x \geq 0. \end{cases}$

(e) $f(x) = \begin{cases} 0, & x < \pi \\ \sin (x - \pi), & x \geq \pi. \end{cases}$

(f) $f(x) = 2^{[x]}$.

(g) $f(x) = \cos \pi[x]$.

(h) $f(x) = \sin \pi[x]$.

(i) $f(x) = |[x]|$.

(j) $f(x) = [x^2]$.

(k) $f(x) = [2x]$.

2. Where is f discontinuous if

(a) $f(x) = \begin{cases} \sin 1/x, & x \neq 0 \\ 0, & x = 0? \end{cases}$

(b) $f(x) = \begin{cases} x \sin 1/x, & x \neq 0 \\ 0, & x = 0? \end{cases}$

(c) $f(x) = \begin{cases} 0, & x \text{ rational} \\ 1, & x \text{ irrational?} \end{cases}$

(d) $f(x) = \begin{cases} 0, & x \text{ rational} \\ x, & x \text{ irrational?} \end{cases}$

3. For what x is $|[x] - 2| < 2$? <1? <0.1?

4. Show that $|x - a| < \epsilon$ if and only if $a - \epsilon < x < a + \epsilon$.

5. Show that if $|x - y| < \epsilon$ for every $\epsilon > 0$, then $x = y$.

5.2 Continuous functions

We propose now to support our earlier claim that most familiar functions are continuous. Because there are infinitely many of them, we cannot do this by analyzing them one at a time. Our strategy, therefore, will be to show that a few basic functions are continuous, and then to indicate how these functions can be put together to construct all familiar functions, in ways that preserve continuity.

▶ **Theorem 5.2-1.** The following functions are everywhere continuous:

(i) the constant functions, $f(x) = k$;
(ii) the identity function, $g(x) = x$;
(iii) the sine and cosine functions.

Proof: (i) This proof is left as an exercise. (ii) Recall the definition of continuity: g is continuous at c if, given any $\epsilon > 0$, we can find a $\delta > 0$ such that $|x - c| < \delta$ implies $|g(x) - g(c)| < \epsilon$. But for the identity function g we have

$$|g(x) - g(c)| = |x - c|.$$

Whatever ϵ may be, the definition will be satisfied if we take $\delta = \epsilon$.

(iii) The proof is essentially the same for both sine and cosine. We therefore consider only the sine, leaving the proof that the cosine is continuous to the exercises. We observe that

$$|\sin x - \sin c| = |2 \cos \tfrac{1}{2}(x + c) \sin \tfrac{1}{2}(x - c)| \tag{1}$$

by Equation (16) of § 2.3. But

$$|\cos \tfrac{1}{2}(x + c)| \leq 1$$

because the cosine takes only values between -1 and 1, and we shall show in §6.5 that

$$|\sin \tfrac{1}{2}(x - c)| \leq |\tfrac{1}{2}(x - c)|.$$

Substituting these values in (1), we have

$$|\sin x - \sin c| \leq |x - c|.$$

Once again, therefore, given any $\epsilon > 0$, we may take $\delta = \epsilon$, for in this case

$$|x - c| < \delta \quad \text{implies} \quad |\sin x - \sin c| < \epsilon,$$

as required.

Our next theorem states some ways in which continuous functions can be combined to make other continuous functions.

▶ **Theorem 5.2-2.** If f and g are continuous at c, then so are

(i) $af + bg$, for arbitrary constants a and b;
(ii) $f \cdot g$; and
(iii) f/g, provided $g(c) \neq 0$.

We omit a proof of this theorem. Observe, however, what it does for us. Every polynomial function can be built up from the identity function and constant functions by addition and multiplication of functions; it then follows from parts (i) and (ii) of the theorem that all polynomial functions are everywhere continuous. A *rational function* is the quotient of two polynomial functions; part (iii) then establishes the continuity of all rational functions. And since the sine and cosine are continuous, the continuity of the other circular functions follows from part (iii).

The operations of composition and inversion of functions also preserve continuity. The details are set forth in our next two theorems, which we also state without proof.

Theorem 5.2-3. (Chain Rule for Continuous Functions). If f is continuous at c and g is continuous at $f(c)$, then the composite function gf is continuous at c.

Theorem 5.2-4. If f has an inverse and is continuous at c, then f^{-1} is continuous at $f(c)$.

It follows from these theorems that functions given by such rules as

$$f(x) = 2x^3 + \sin\sqrt{x-1},$$
$$g(x) = \tan^{-1}\left(\frac{x}{1-x^2}\right),$$

and so on, are all continuous. Surely this list includes most familiar functions, as claimed.

Our informal discussion of continuity, in earlier chapters, said in effect that the graph of a function is an unbroken curve on any interval where the function is continuous. The following theorem, which is illustrated in Fig. 5, asserts that this is in fact the case.

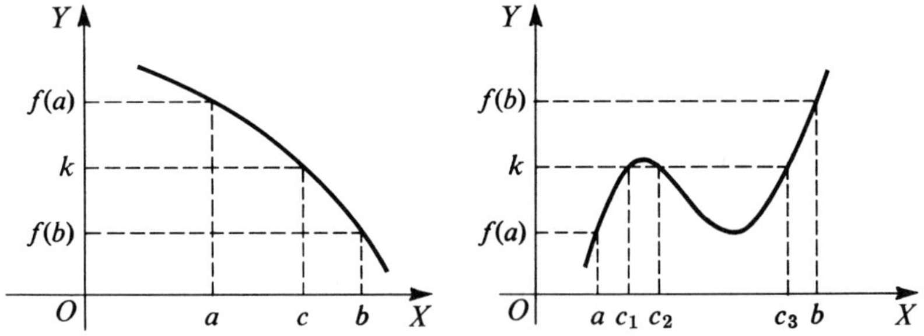

Fig. 5

Theorem 5.2-5. (Intermediate-value Theorem). Let f be continuous on an interval $[a, b]$, with $f(a) \neq f(b)$, and let k be any number between $f(a)$ and $f(b)$. Then there exists a number $c \in (a, b)$ such that $f(c) = k$.

As Fig. 5 shows, there may be more than one such c. A special case of the Intermediate-value Theorem is widely used in locating roots of equations: if f is continuous on $[a, b]$, and if $f(a)$ and $f(b)$ have opposite signs, then the equation $f(x) = 0$ has a root in (a, b).

The Intermediate-value Theorem is an example of an *existence* theorem: it asserts the existence, under specified conditions, of a point c with a certain property. The following theorem is also an existence theorem. It guarantees the existence, under certain circumstances, of solutions to extreme-value problems.

Theorem 5.2-6 (Extreme-value Theorem). If f is continuous on a *closed* interval I, then f has an absolute maximum and an absolute minimum on I.

Note carefully that the interval must be closed. For instance, $f(x) = 1/x$ is continuous on the half-open interval $(0, 1]$ but has no maximum there.

The proofs of Theorems 5.2-5 and 5.2-6 are beyond the scope of this book.

Miscellaneous Exercises 5.2

1. Prove part (i) of Theorem 5.2-1.
2. Prove that the cosine is everywhere continuous.
3. Show in detail how the function $F(x) = x^3 - 2x + 7$ can be built up from the functions $f(x) = 1$ and $g(x) = x$, using only the operations of Theorem 5.2-2.
4. Sketch a graph of a function, defined on $[0, 1]$, that has
   (a) a maximum but no minimum.
   (b) a maximum and a minimum.
   (c) neither a maximum nor a minimum.

   In which cases is there a *continuous* function satisfying the given conditions?
5. Note that $\tan \pi/4 = 1$ and $\tan 3\pi/4 = -1$. (a) Does the equation $\tan x = 0$ have a root in $[\pi/4, 3\pi/4]$? Explain, with reference to Theorem 5.2-5. (b) Does the tangent have a maximum or a minimum in $[\pi/4, 3\pi/4]$? Explain, with reference to Theorem 5.2-6.
6. Using only the definition of continuity (and not any of the theorems of this section), prove that $f(x) = 2x + 3$ is continuous at $x = 5$.

7. For what x is it true that $0 < |x|$? that $0 < |x - 2|$? that $0 < |x - c|$?

8. Let $f(x) = 2x + 3$ when $x \neq 1$, and let $f(1) = 0$.
   (a) Sketch a graph of f for $0 \leq x \leq 2$.
   (b) Draw on your graph the lines $y = 4$, $y = 6$.
   (c) For what values of x does the graph of f lie between the lines drawn in (b)?
   (d) Find a δ, if one exists, such that $|f(x) - 5| < 1$ whenever $0 < |x - 1| < \delta$.
   (e) Find a δ, if one exists, such that $|f(x) - 5| < 1$ whenever $|x - 1| < \delta$.

9. Let $f(x) = 5 - 3x$ when $x \neq 1$; $f(1)$ is not defined.
   (a) Sketch a graph of f for $0 \leq x \leq 2$.
   (b) For what x is it true that $|f(x) - 2| < 1$?
   (c) For what δ, if any, is it true that $|f(x) - 2| < \frac{1}{2}$ whenever $0 < |x - 1| < \delta$?
   (d) For what δ, if any, is it true that $|f(x) - 2| < \frac{1}{2}$ whenever $|x - 1| < \delta$.
   (e) What is the relationship between δ and ϵ if $|f(x) - 2| < \epsilon$ whenever $0 < |x - 1| < \delta$?

10. A function f is everywhere continuous. Define a function F by the rule

$$F(h) = \frac{f(3 + h) - f(3)}{h}.$$

For what h is F continuous?

11. (a) Prove that if $|f(x) - L| < \epsilon$ whenever $|x - c| < \delta$, then $|f(x) - L| < \epsilon$ whenever $0 < |x - c| < \delta$.
    (b) Show by an example that the converse of (a) is false. (*Hint:* Fig. 3 of § 5.1.)

5.3 Limits

Having considered continuity at some length, we now turn to the problem of defining precisely what we mean by a limit. Our first mention of a limit was in § 2.6, where we considered the slope

$$m = \lim_{\Delta x \to 0} \frac{4\Delta x + (\Delta x)^2}{\Delta x}.$$

We decided there that because $[4\Delta x + (\Delta x)^2]/\Delta x$ can be made arbitrarily close to 4 by taking Δx close enough to 0, provided $\Delta x \neq 0$, the value of this limit is 4. This is very nearly the same language we used in § 5.1 in connection with continuity. In fact we can give an informal description of the limit of a function f at a point c in almost exactly the same words we used for an informal description of continuity at c:

A function f has the limit L at a point c if, for all x *near to but different from* c, $f(x)$ is near L. Very little needs to be done to transform this into a formal rigorous definition. We must take care of two matters:

(1) There is obviously no point in talking about values of f at points near c if there aren't any. This would be the case if some neighborhood of c contained no points of the domain of f. *Therefore, in what follows we shall assume that the domain of f contains a punctured neighborhood of c.* Whether c itself belongs to the domain of f is immaterial; it's all right if it does, but in at least one important case, the derivative of a function f,

$$Q(\Delta x) = \frac{f(x + \Delta x) - f(x)}{\Delta x},$$

we cannot require that it does. That is, $\Delta x = 0$ is not in the domain of Q.

(2) We must again make specific the vague word "near." For this purpose we use exactly the same ϵ, δ language we used in connection with continuity: given any $\epsilon > 0$, we seek a $\delta > 0$ such that $|x - c| < \delta$ implies $|f(x) - L| < \epsilon$, where L represents the limit in question. To this, however, we must add the requirement $x \neq c$, or, equivalently, $0 < |x - c|$. When we combine this with the requirement $|x - c| < \delta$, we have $0 < |x - c| < \delta$. Thus we have the following definition.

Definition. The function f is said to have the limit L at c if, given any $\epsilon > 0$, there exists a $\delta > 0$ such that

$$|f(x) - L| < \epsilon \quad \text{whenever} \quad 0 < |x - c| < \delta.$$

In this case we write $\lim_{x \to c} f(x) = L$, and say that f(x) <u>converges</u> to L as x approaches c.

The notation "$\lim_{x \to c} f(x)$" is read "the limit of $f(x)$ as x approaches c," and

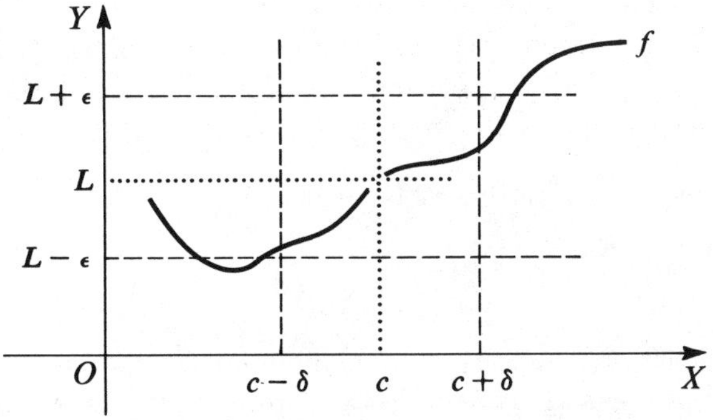

Fig. 6

"$x \to c$" alone is read "x approaches c."

Fig. 4 showed a graphical interpretation of the definition of continuity at a point c. A small modification of this figure, shown as Fig. 6, serves as an interpretation of the definition of the limit at c. The values of f that satisfy the condition $|f(x) - L| < \epsilon$ correspond to points of the graph that lie between the horizontal lines $y = L - \epsilon$ and $y = L + \epsilon$. The values of x that satisfy $0 < |x - c| < \delta$ correspond to points that lie between the lines $x = c - \delta$ and $x = c + \delta$ *but not on the line* $x = c$. The fact that L is the limit of f at c is established by finding a number δ such that all points of the graph between the vertical lines, except $(c,f(c))$ if it exists, also lie between the horizontal lines. Whether $f(c)$ exists, and if it does exist, whether $(c,f(c))$ lies between the horizontal lines, is immaterial. In fact, our definition of limit was designed to make this issue immaterial.

Consider, for example, the functions

$$f_1(x) = 3 - x,\ x \neq 1, \qquad f_2(x) = \begin{cases} 3 - x, & x \neq 1 \\ 4, & x = 1, \end{cases} \qquad f_3(x) = 3 - x,$$

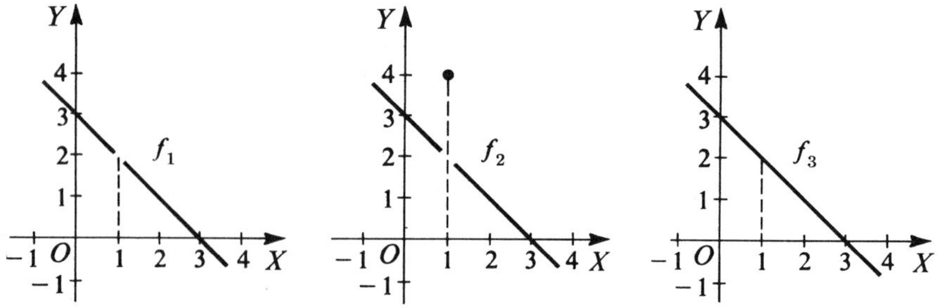

Fig. 7

graphed in Fig. 7. Since these functions are identical everywhere save at $x = 1$, they *must* have the same limit at 1. This limit is clearly 2, since values of x near 1 give functional values near 2 in each case. Note, however, that f_1 is not even defined at 1, and that $f_2(1)$, though defined, is different from the limit of $f_2(x)$ as x approaches 1. This illustrates the point made a moment ago: the value of $\lim_{x \to c} f(x)$, if it exists, is quite independent of the value, if any, of $f(c)$. Finally, in the case of f_3, we see that not only does $f_3(1)$ exist but

$$\lim_{x \to 1} f_3(x) = f_3(1).$$

This equality exists because f_3 is continuous at 1, and exemplifies the principal relationship between continuity and limits. We state this relationship in our next theorem.

Theorem 5.3-1. Let f be a function and c a number such that

(i) $f(c)$ exists, and
(ii) $\lim_{x \to c} f(x)$ exists.

Then f is continuous at c if and only if these two are equal:

$$\lim_{x \to c} f(x) = f(c). \tag{1}$$

Proof: If equation (1) is given, then by the definition of a limit, we know that for any $\epsilon > 0$ there exists a $\delta > 0$ such that $0 < |x - c| < \delta$ implies $|f(x) - f(c)| < \epsilon$. This falls short of establishing continuity at c because of the condition $0 < |x - c|$; it does not tell us what happens when $x = c$. When $x = c$, however, $|f(x) - f(c)| = |f(c) - f(c)| = 0$, which is certainly less than ϵ. Therefore f is continuous at c.

Conversely, suppose that f is continuous at c. It then follows from the definition of continuity that for any $\epsilon > 0$ there exists a $\delta > 0$ such that $|x - c| < \delta$ implies $|f(x) - f(c)| < \epsilon$. Now suppose that $0 < |x - c| < \delta$. Then $|x - c| < \delta$, and it follows that $|f(x) - f(c)| < \epsilon$. Thus (1) holds, and the proof is complete.

It follows from this theorem that the problem of finding the limit of a function at a point of continuity is trivial: just evaluate the function at that point. For instance,

$$\lim_{x \to 3} (x^2 - 2x + 1) = 3^2 - 2(3) + 1 = 4.$$

In view of the close connection between continuity and limits expressed in Theorem 5.3-1, it should come as no surprise that many theorems about limits closely resemble theorems about continuity. For example:

Theorem 5.3-2. If $\lim_{x \to c} f(x) = L$ and $\lim_{x \to c} g(x) = M$, then

(i) $\lim_{x \to c} [af(x) + bg(x)] = aL + bM$, for arbitrary constants a, b;
(ii) $\lim_{x \to c} f(x)g(x) = LM$; and
(iii) $\lim_{x \to c} f(x)/g(x) = L/M$, provided $M \neq 0$.

We omit the proof of this theorem.

When $a = b = 1$ in (i) we have $\lim_{x \to c} [f(x) + g(x)] = L + M$, an equation which asserts that *the limit of a sum is the sum of the limits.* Similarly, we see in part (ii) of the theorem that *the limit of a product is the product of the limits,* and in part (iii) that *the limit of a quotient is the quotient of the limits* (provided the latter quotient exists). This theorem plays an essential part in the proof of our next theorem, which links continuity and differentiability.

▶ **Theorem 5.3-3.** **If f is differentiable at c, that is, if $f'(c)$ exists, then f is also continuous at c.**

Proof: It suffices, by Theorem 5.3-1, to prove that

$$\lim_{x \to c} f(x) = f(c),$$

or, equivalently that

$$\lim_{\Delta x \to 0} [f(c + \Delta x) - f(c)] = 0$$

(see Exercise 1). But

$$\begin{aligned}\lim_{\Delta x \to 0} [f(c + \Delta x) - f(c)] &= \lim_{\Delta x \to 0} \Delta x \left(\frac{f(c + \Delta x) - f(c)}{\Delta x}\right) \\ &= (\lim_{\Delta x \to 0} \Delta x)\left(\lim_{\Delta x \to 0} \frac{f(c + \Delta x) - f(c)}{\Delta x}\right) \\ &= 0 \cdot f'(c) = 0,\end{aligned}$$

and the proof is complete.

Observe that this theorem gives us a new proof that all polynomials are continuous, and when we see how to differentiate rational and circular functions, it will provide an independent proof of their continuity as well. It also tells us that all functions supplied by the First Fundamental Theorem are continuous.

Differential calculus begins with the definition of the derivative:

$$f'(x) = \lim_{\Delta x \to 0} \frac{f(x + \Delta x) - f(x)}{\Delta x}.$$

The problem of how to evaluate this limit in specific instances is therefore of fundamental importance. In § 2.8 the following instructions for carrying out this evaluation were given: "Reduce the fraction found in Step 3 to a form which will have meaning if 0 is substituted for Δx, and make this substitution; the resulting number is $f'(x)$." The justification for this procedure is contained in our next theorem.

▶ **Theorem 5.3-4.** **If $f(x) = g(x)$ on some punctured neighborhood N of c, and if g is continuous at c, then $\lim_{x \to c} f(x)$ exists and is equal to $g(c)$:**

$$\lim_{x \to c} f(x) = g(c). \qquad (1)$$

Proof: Throughout the proof we consider only values of x in N. Let $\epsilon > 0$ be given. Then, since g is continuous at c, there exists a $\delta > 0$ such that

$$|g(x) - g(c)| < \epsilon \tag{2}$$

whenever

$$|x - c| < \delta. \tag{3}$$

Now suppose that $0 < |x - c| < \delta$. In this case not only does (3) hold, but $f(x) = g(x)$. Thus we can substitute $f(x)$ for $g(x)$ in (2) to obtain

$$|f(x) - g(c)| < \epsilon$$

whenever

$$0 < |x - c| < \delta,$$

and (1) follows by definition of limit.

Our next theorem is an extremely useful generalization of the preceding one. We state it without proof.

▶ **Theorem 5.3-5.** (Pinching Theorem). If f, g, h are functions such that

$$f(x) \leq g(x) \leq h(x)$$

on some punctured neighborhood of c, and if $\lim_{x \to c} f(x)$ and $\lim_{x \to c} h(x)$ exist and are equal,

$$\lim_{x \to c} f(x) = L = \lim_{x \to c} h(x),$$

then the limit of g at c exists and is equal to L:

$$\lim_{x \to c} g(x) = L.$$

Still another theorem about limits that closely resembles one about continuity is closely analogous to the Chain Rule for Continuous Functions. We state it without proof.

▶ **Theorem 5.3-6.** (Chain Rule for Limits). If $\lim_{x \to c} f(x) = L$, and if g is continuous at L, then

$$\lim_{x \to c} (gf)(x) = g(L) = g(\lim_{x \to c} f(x)).$$

We conclude this section by considering a different type of limit, the limit of $f(x)$ as $x \to \pm\infty$, that is, as x increases or decreases without bound. For example, consider $f(x) = 1/x$. What happens to the values of f as x increases? Since $f(100) = 0.01$, $f(1{,}000) = 0.001$, $f(10^6) = 10^{-6}$, it appears that as x increases $f(x)$ gets closer and closer to 0, and that we can make $f(x)$ arbitrarily close to 0 by making x big enough. Much of this description echoes the pattern of our earlier discussion of limit at a point; certainly, "$f(x)$ arbitrarily close to 0" suggests an inequality of the form $|f(x) - 0| < \epsilon$.

The new element here is the notion of making x "big enough." Where we have so far satisfied an ϵ-condition by making x close enough to some number c, we must now satisfy it by making x "close enough to infinity," that is, very large. To look again at $f(x) = 1/x$, we see that

$$|f(x) - 0| = |1/x - 0| = 1/x \quad \text{if} \quad x > 0,$$

and that, given any $\epsilon > 0$, $1/x < \epsilon$ is equivalent to $x > 1/\epsilon$. Hence

$$|f(x) - 0| < \epsilon \quad \text{whenever} \quad x > 1/\epsilon.$$

This discussion generalizes easily:

Definition. The function f is said to have the limit L at infinity if, given any $\epsilon > 0$, there exists an N such that $|f(x) - L| < \epsilon$ whenever $x > N$. In this case we write

$$\lim_{x\to\infty} f(x) = L, \quad \text{or} \quad \lim_{x\to+\infty} f(x) = L,$$

and say that $f(x)$ *converges* to L as x increases without bound.

Clearly, this definition is meaningful only if there are values of x greater than N in the domain of f; we therefore require that the domain have no upper bound.

It is left to the student as an exercise to define what is meant by

$$\lim_{x\to-\infty} f(x) = L.$$

The notations "$x \to \infty$," "$x \to +\infty$," and "$x \to -\infty$," by themselves, are usually read "x approaches infinity," "x approaches positive infinity," and "x approaches negative infinity," respectively.

The obvious modifications of Theorems 5.3-2, 5.3-5, and 5.3-6 also hold for limits at $\pm\infty$.

Example 1. Evaluate $\displaystyle\lim_{x\to\infty} \frac{2x - 17}{3x}$.

Solution: We observe that when x is very large the numerator of this fraction is dominated by the term $2x$; the -17 is negligible. Hence for large x the fraction acts like $2x/3x = \frac{2}{3}$, and we therefore guess that the limit is $\frac{2}{3}$. To support this conjecture with proof, we observe that

$$\left|\frac{2x - 17}{3x} - \frac{2}{3}\right| = \left|-\frac{17}{3x}\right| = \frac{17}{3x} \quad \text{for} \quad x > 0,$$

and that, given any $\epsilon > 0$,

$$\frac{17}{3x} < \epsilon \quad \text{if} \quad x > \frac{17}{3\epsilon}.$$

(Check this!) Hence our conjecture is proved by taking $N = 17/3\epsilon$ in the definition of limit at infinity. Note that this function has the same limit at negative infinity:

$$\lim_{x \to -\infty} \frac{2x - 17}{3x} = \frac{2}{3}.$$

Example 2. In any polynomial, the term of highest degree dominates when the absolute value of the variable is large. Thus, if

$$f(x) = \frac{x^3 - 53}{5x^2 + 11x - 2},$$

then when $|x|$ is large the numerator is dominated by the term x^3 and the denominator by the term $5x^2$. Hence, for large $|x|$, $f(x)$ acts like $x^3/5x^2 = x/5$. It follows that as x increases without bound, so does $f(x)$, and we conclude that f has no limit at infinity. We indicate the situation encountered here by writing

$$\lim_{x \to \infty} f(x) = \infty.$$

Similarly,

$$\lim_{x \to -\infty} f(x) = -\infty.$$

The student should have little difficulty defining a function f for which

$$\lim_{x \to \infty} f(x) = -\infty \quad \text{and} \quad \lim_{x \to -\infty} f(x) = \infty.$$

Miscellaneous Exercises 5.3

1. Show that $\lim_{\Delta x \to 0} [f(c + \Delta x) - f(c)] = 0$ is equivalent to $\lim_{x \to c} f(x) = f(c)$.

2. (a) Give a reasonable definition of $\lim_{x \to -\infty} f(x) = L$.

   (b) Use your definition to prove that $\lim_{x \to -\infty} 1/x = 0$.

3. (a) Suppose that (i) $\lim_{x \to c} f(x) = L$, and (ii) $f(x) > 0$ on a punctured neighborhood of c. Prove that $L \geq 0$.

   (b) Show by an example that the conclusion in (a) cannot be strengthened to $L > 0$.

   (c) What conclusion can be drawn if in (ii) we have $f(x) \geq 0$? $f(x) = 0$? $f(x) < 0$?

4. It is important to be aware of some of the things that Theorem 5.3-2 does *not* say. Give examples of functions f and g such that neither has a limit at 0 but

   (a) $\lim_{x \to 0} [f(x) + g(x)]$ exists. (b) $\lim_{x \to 0} f(x)g(x)$ exists.

5. Find, if it exists, the limit as $x \to 0$ of each of the following. (*Hint*: Rationalize numerators.)

(a) $\dfrac{1 - \sqrt{1 - x}}{x}$ (c) $\dfrac{1 - \sqrt{1 - x}}{x^2}$

(b) $\dfrac{\sqrt{1 + x} - \sqrt{1 - x}}{x}$ (d) $\dfrac{1 - \sqrt{1 - x^2}}{x^2}$

6. Evaluate: (a) $\lim_{x \to \infty} (\sqrt{x + 1} - \sqrt{x})$

(b) $\lim_{x \to \infty} (2\sqrt{x} - 3)(\sqrt{x + 1} - \sqrt{x})$

(c) $\lim_{x \to \infty} \sqrt{x + 2}(\sqrt{x + 1} - \sqrt{x})$

7. Prove: (a) $\lim_{x \to \infty} \dfrac{\sin x}{x} = 0$. (*Hint:* $|\sin x| \leq 1$.)

(b) $\lim_{x \to 0} x \sin 1/x = 0$.

8. Show by an example that the converse of Theorem 5.3-3 is false.

5.4 The mean-value theorem

A major reason for proof in mathematics is to confirm that something we believe is really true. In previous chapters we have made extensive use of derivatives to analyze and draw a variety of conclusions about functions, but all this has been done in an informal, intuitive manner, without benefit of proof. The principal tool for supporting our beliefs about derivatives is called the Mean-Value Theorem. This theorem, one of the most important in calculus, makes a very simple assertion: somewhere on an arc such as

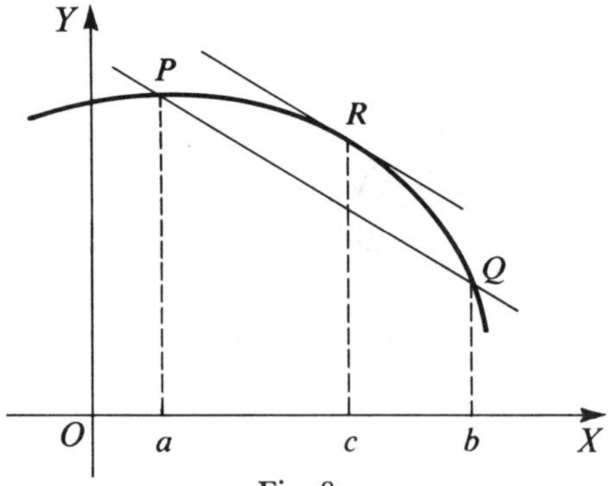

Fig. 8

PQ in Fig. 8, there is a point at which the tangent line is parallel to the secant line PQ. Since the slope of line PQ is

$$\frac{f(b) - f(a)}{b - a},$$

the theorem then asserts the existence of a point c between a and b such that

$$f'(c) = \frac{f(b) - f(a)}{b - a}.$$

Consider first an intuitive argument. Imagine moving the secant in Fig. 8 upward without changing its slope, so that it stays always parallel to its original position. The points P and Q will then move closer and closer together, and when they coincide the secant will have become the required tangent, parallel to the original secant PQ. This argument also suggests that the point of tangency R is the point where the vertical distance between the arc and the secant is a maximum. This conjecture is not only true but also useful in framing a proof of the theorem.

We start with a subsidiary theorem, a special case of the Mean-Value Theorem in which the secant PQ is the x-axis. Such a subsidiary theorem is often called a *lemma*. Since the slope of the x-axis is 0, the lemma requires us to prove the existence of a point on the arc where the slope is 0.

Lemma 5.4-1. (Rolle's Theorem). If a function f is continuous on the closed interval $[a, b]$ and differentiable on the open interval (a, b), and if $f(a) = f(b) = 0$, then there exists a point c, with $a < c < b$, such that $f'(c) = 0$.

There may of course be more than one possible choice of c, as Fig. 9 suggests; Rolle's Theorem merely asserts the existence of at least one.

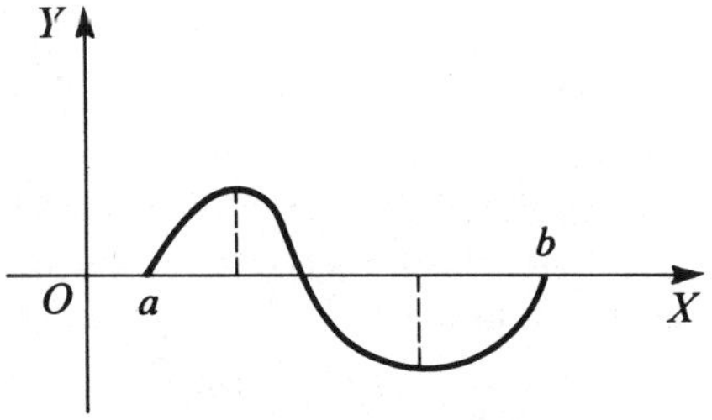

Fig. 9

Proof: If $f(x) = 0$ for all $x \in [a, b]$, then also $f'(x) = 0$ on $[a, b]$, and we can take any $c \in (a, b)$ to complete the proof. If $f(x) \neq 0$ for some $x \in [a, b]$, then for this x either $f(x) > 0$ or $f(x) < 0$; we consider first the case $f(x) > 0$. By the Extreme-Value Theorem (Theorem 5.2-6), f has a maximum value at some point $c \in [a, b]$, and since $f(a) = f(b) = 0$, it follows that $a \neq c \neq b$

and therefore $a < c < b$. We shall complete the proof of this case by showing that $f'(c) = 0$.

Since $f(c)$ is a maximum, we know that

$$f(c) \geq f(c + \Delta x),$$

or

$$f(c + \Delta x) - f(c) \leq 0, \tag{1}$$

for all sufficiently small $|\Delta x|$. We define a function Q by the rule

$$Q(\Delta x) = \frac{f(c + \Delta x) - f(c)}{\Delta x} \tag{2}$$

for $\Delta x \neq 0$. By definition,

$$f'(c) = \lim_{\Delta x \to 0} Q(\Delta x).$$

We know from (1) that the numerator of the fraction in (2) is non-positive. If $\Delta x > 0$, it follows that $Q(\Delta x) \leq 0$, and therefore that the limit of $Q(\Delta x)$, as Δx approaches 0 through positive values, is non-positive. If on the other hand $\Delta x < 0$, then $Q(\Delta x) \geq 0$, and the limit as Δx approaches 0 through negative values must therefore be non-negative. But the existence of $f'(c)$ requires that these two "one-sided" limits be the same. This forces

$$\lim_{\Delta x \to 0} Q(\Delta x) = 0,$$

or $f'(c) = 0$, as required. The proof of the case $f(x) < 0$ is left as an exercise.

Our proof of Rolle's Theorem establishes a corollary which we have already used many times in our work with extreme values:

▶ **Corollary 5.4-2.** If a function f is differentiable on an open interval I and has an extreme value at a point $c \in I$, then $f'(c) = 0$.

We turn now to the Mean-Value Theorem itself:

▶ **Theorem 5.4-3.** (Mean-Value Theorem). If a function f is continuous on the closed interval $[a, b]$ and differentiable on the open interval (a, b), then there exists a point c, with $a < c < b$, such that

$$f'(c) = \frac{f(b) - f(a)}{b - a}. \tag{3}$$

Proof: We return to our conjecture that the point c is the abscissa at which the vertical distance between the arc PQ and the secant PQ (Fig. 8) is a maximum. An equation of the line PQ is

$$y - f(a) = \frac{f(b) - f(a)}{b - a}(x - a),$$

or
$$y = f(a) + \frac{f(b) - f(a)}{b - a}(x - a).$$

An equation of the arc PQ is $y = f(x)$, and the vertical distance between the arc and the chord is then the difference of these two y's:

$$F(x) = f(x) - f(a) - \frac{f(b) - f(a)}{b - a}(x - a).$$

(The distance is the negative of this quantity if the arc lies below the secant.) The student should now check that, as the geometry of the problem should lead him to expect, $F(a) = F(b) = 0$. The continuity and differentiability of F follow from the corresponding properties of f, and F therefore satisfies the conditions of Rolle's Theorem. Hence there is a point $c \in (a, b)$ such that $F'(c) = 0$. But

$$F'(x) = f'(x) - \frac{f(b) - f(a)}{b - a}.$$

(Check this!) Thus

$$F'(c) = f'(c) - \frac{f(b) - f(a)}{b - a} = 0,$$

and
$$f'(c) = \frac{f(b) - f(a)}{b - a},$$

as required.

Example 1. If $f(x) = x^2$, then f is continuous and differentiable on any interval $[a, b], f'(x) = 2x$, and (3) therefore gives

$$2c = \frac{b^2 - a^2}{b \quad a} = b + a,$$

$$c = \tfrac{1}{2}(b + a).$$

Hence the line tangent to the parabola $y = x^2$ at the point on the curve where $x = \frac{1}{2}(a + b)$ is parallel to the secant joining the points (a, a^2) and (b, b^2).

Example 2. If you average 50 mi./hr. on an automobile trip, then the Mean-Value Theorem says that at some instant in the trip you must have been going exactly 50 mi./hr. To see this, suppose that the distance s traveled in time t is given by a function f: $s = f(t)$. Then the velocity at time t is $v = f'(t)$. Physical considerations suggest that f is continuous and differentiable. The average velocity from time a to time b is the distance traveled, $f(b) - f(a)$, divided by the elapsed time, $b - a$, and we are told that this average is 50 mi./hr.:

$$\frac{f(b) - f(a)}{b - a} = 50.$$

It follows from the Mean-Value Theorem that there is a time c between a and b at which the velocity function takes the value 50:

$$v = f'(c) = 50.$$

We turn now to the defense of the statement, made earlier in this section, that the Mean-Value Theorem is the instrument whereby we confirm most of our intuitive judgments about derivatives and their interpretation. We start with the question of increasing and decreasing functions.

▶ **Theorem 5.4-4.** A differentiable function f is increasing on an interval I if $f'(x) > 0$ for each $x \in I$.

Proof: We must prove that $f(x_2) > f(x_1)$ for every choice of x_1 and x_2 in I such that $x_2 > x_1$. Since f is differentiable on I, it is continuous (Why?) and differentiable on the subinterval $[x_1, x_2]$. Hence, by the Mean-Value Theorem,

$$f(x_2) - f(x_1) = f'(c)(x_2 - x_1) \tag{4}$$

for some $c \in (x_1, x_2)$. But $f'(c) > 0$ because $c \in I$; moreover, $x_2 > x_1$ implies $x_2 - x_1 > 0$. Hence

$$f'(c)(x_2 - x_1) > 0,$$

and it follows that

$$f(x_2) - f(x_1) > 0,$$

or

$$f(x_2) > f(x_1).$$

We can strengthen the preceding theorem slightly, as follows:

▶ **Corollary 5.4-5.** If f is continuous on $[a, b]$ and differentiable on (a, b), and if $f'(x) > 0$ for all $x \in (a, b)$, then f is increasing on $[a, b]$.

In other words, it doesn't matter what f' does at the endpoints, a and b. It can be zero there, or even fail to exist.

Proof: Since the Mean-Value Theorem applies to f on every subinterval of $[a, b]$, the argument used in the proof of Theorem 5.4-4 applies unchanged in this case also.

Example 3. Let $f(x) = 12x - x^3$. Then

$$f'(x) = 12 - 3x^2,$$

and $f'(x) = 0$ if and only if $x = \pm 2$. If $-2 < x < 2$, then $f'(x) > 0$. Since f is everywhere continuous, it follows from Corollary 5.4-5 that f is increasing not only on the open interval $(-2, 2)$ but also on the closed interval $[-2, 2]$. Since $f(-2) = -16$ and $f(2) = 16$, this means that the arc of the graph of f joining the points $(-2, -16)$ and $(2, 16)$ has the property that a line joining any two of its points (including either or both endpoints) has a positive slope. It can also be shown (see Exercise 8 in the next set) that f is decreasing on $(-\infty, -2]$ and $[2, \infty)$. The fact that 2 (and -2 as well) belongs both to an interval on which f is increasing and to an interval on which f is decreasing involves no contradiction: the words "increasing" and "decreasing" describe the behavior of functions on *intervals*, not at single points.

In the exercises that follow, the student is invited to state and prove the counterparts of the preceding theorem and corollary for decreasing functions, and also to prove our next two theorems.

▶ **Theorem 5.4-6.** If $f'(x) = 0$ for each x in an interval I, then f is constant on I.

▶ **Theorem 5.4-7.** If f_1 and f_2 are functions such that

$$f_1'(x) = f_2'(x)$$

for each x in some interval I, then there exists a constant C such that

$$f_1(x) = f_2(x) + C$$

for each $x \in I$.

Observe that Theorem 5.4-7 substantiates the assertion, made in Chapter 4, that the difference between any two indefinite integrals of the same function is a constant.

Miscellaneous Exercises 5.4

1. Determine all possible values of c in

$$f'(c) = \frac{f(b) - f(a)}{b - a}$$

and illustrate with a sketch graph, if

(a) $f(x) = 1/x$, $a = 1$, $b = 4$;
(b) $f(x) = x^3$, $a = -3$, $b = 3$;
(c) $f(x) = x^2 - 2x$, $a = -1$, $b = 5$.

2. Find c in terms of a and b, in

$$f'(c) = \frac{f(b) - f(a)}{b - a}, \quad \text{if}$$

(a) $f(x) = 1/x$ $(a > 0, b > 0)$;
(b) $f(x) = 1/x$ $(a < 0, b < 0)$;
(c) $f(x) = x^2 + px + q$.

3. Explain why the Mean-Value Theorem does not apply to
   (a) $x^{2/3}$ on $[-1, 8]$;
   (b) $|x|$ on $[-1, 1]$;
   (c) x^{-1} on $[-1, 1]$.
4. Complete the proof of Rolle's Theorem by treating the case $f(x) < 0$.
5. Use the Mean-Value Theorem to prove that
   (a) $\sqrt{26} - \sqrt{25} < 0.1$;
   (b) $\lim_{x \to \infty} (\sqrt{x+1} - \sqrt{x}) = 0$.
6. (a) Let f be continuous on $[a, b]$, and define a function F by the rule

$$F(t) = \int_a^t f(x)\, dx.$$

   Show that F satisfies the conditions of the Mean-Value Theorem on $[a, b]$.

   (b) Show that there exists a $c \in (a, b)$ such that

$$\int_a^b f(x)\, dx = f(c)(b - a).$$

7. State and prove the counterpart of Theorem 5.4-4 for decreasing functions.
8. State and prove the counterpart of Corollary 5.4-5 for decreasing functions.
9. Prove Theorem 5.4-6. (*Hint*: Use (4)).
10. Prove Theorem 5.4-7. (*Hint*: Define a function $F = f_1 - f_2$. What can you conclude about F'?)
11. Prove

▶ **Theorem 5.4–8.** (Cauchy Mean-Value Theorem.) If functions f and g are continuous on $[a, b]$ and differentiable on (a, b), then there exists a point c, with $a < c < b$, such that

$$f'(c)[g(b) - g(a)] = g'(c)[f(b) - f(a)].$$

(*Hint:* Look at $F(x) = f(x)[g(b) - g(a)] - g(x)[f(b) - f(a)]$.)

12. Prove that Theorem 5.4-3 is a special case of Theorem 5.4-8.

5.5 One-sided and infinite limits; L'Hôpital's Rule

The issues of continuity and limits become slightly more complicated at the endpoints of certain intervals than they are at interior points.

Consider, for example, the greatest-integer function, $f(x) = [x]$. (See Fig. 2, page 159, for a graph.) This function has no limit as $x \to 1$; it takes both the value 0 and the value 1 on every neighborhood of $x = 1$. But the restriction of f to the interval $[1,2)$ has the constant value 1 and therefore the limit 1 as we approach 1 from the right, which we write $x \to 1^+$. Thus

$$\lim_{x \to 1+} [x] = 1$$

Similarly, the restriction of f to $[0, 1)$ has the constant value 0, so we say

$$\lim_{x \to 1-} [x] = 0$$

where "$x \to 1^-$" is read "x approaches 1 from the left."

To formalize this discussion, we modify our previous definition of limit at a point (page 167) as follows. If $x > c$, then $|x - c| = x - c$, and the condition $0 < |x - c| < \delta$ becomes $0 < x - c < \delta$, while if $x < c$, then $|x - c| = c - x$, and $0 < |x - c| < \delta$ becomes $0 < c - x < \delta$. We therefore state our definitions of one-sided limits as follows.

► **Definition.** The right-hand limit of f at c is L, and we write

$$\lim_{x \to c+} f(x) = L,$$

if for any $\epsilon > 0$ there exists a $\delta > 0$ such that

$$|f(x) - L| < \epsilon$$

whenever

$$0 < x - c < \delta.$$

The left-hand limit of f at c is L, and we write

$$\lim_{x \to c-} f(x) = L,$$

if for any $\epsilon > 0$ there exists a $\delta > 0$ such that

$$|f(x) - L| < \epsilon$$

whenever

$$0 < c - x < \delta.$$

Evidently, if

$$\lim_{x \to c} f(x) = L,$$

then

$$\lim_{x \to c+} f(x) = \lim_{x \to c-} f(x) = L,$$

and conversely, but the existence of a right-hand and a left-hand limit at c does not guarantee a limit at c, as the case of $[x]$ illustrates.

By a similar argument, we arrive at the following definitions of one-sided continuity.

▶ **Definition.** A function f is continuous from the right at a point c of its domain if, for any $\epsilon > 0$, there exists a $\delta > 0$ such that

$$|f(x) - f(c)| < \epsilon$$

whenever

$$0 \leq x - c < \delta.$$

It is continuous from the left at c if, for any $\epsilon > 0$, there exists a $\delta > 0$ such that

$$|f(x) - f(c)| < \epsilon$$

whenever

$$0 \leq c - x < \delta.$$

Observe that $[x]$ is continuous from the right but not from the left at $x = 1$.

Infinite limits. Consider the behavior of $1/x$ near $x = 0$. As $x \to 0^+$, $1/x$ increases without bound. We describe this by writing

$$\lim_{x \to 0^+} 1/x = \infty$$

commonly read "$1/x$ goes to infinity as x goes to 0 from the right." The precise meaning of this statement can be expressed as follows.

▶ **Definition.** If for each number N there exists a number $\delta > 0$ such that

$$f(x) > N$$

whenever

$$0 < x - c < \delta,$$

then

$$\lim_{x \to c^+} f(x) = \infty.$$

It is left to the student to formulate similar definitions for

$$\lim_{x \to c^-} f(x) = \infty, \lim_{x \to c} f(x) = \infty, \lim_{x \to \infty} f(x) = \infty,$$

and other such statements in which $-\infty$ replaces ∞. Observe that

$$\lim_{x \to 0^-} 1/x = -\infty.$$

L'Hôpital's Rule. Recall part (iii) of Theorem 5.3-2: If $\lim_{x \to c} f(x) = L$ and $\lim_{x \to c} g(x) = M$, *and if* $M \neq 0$, then $\lim_{x \to c} f(x)/g(x) = L/M$. If $M = 0$ and $L \neq 0$, then $f(x)/g(x)$ has no limit at c. But if $L = M = 0$, a situation represented symbolically as the indeterminate form $0/0$, there may be a limit, and the following theorem can often be used to find it. A similar result holds for each of the cases $x \to c^+$, $x \to c^-$, $x \to \infty$, and $x \to -\infty$. The conclusion of the theorem also applies to the case $L = M = \infty$, in the sense of the preceding discussion of infinite limits; this yields the <u>indeterminate form</u> ∞/∞. We start with the form $0/0$, and the case $x \to c^+$.

Theorem 5.5-1. (L'Hôpital's Rule.*) Let f and g be functions continuous on the interval $[c, b)$ and such that

(i) $f(c) = g(c) = 0$,

(ii) f' and g' are defined on (c,b), and

(iii) $g'(x) \neq 0$ for all $x \in (c, b)$.

Then if $\lim_{x\to c+} f'(x)/g'(x)$ exists, it follows that $\lim_{x\to c+} f(x)/g(x)$ also exists, and

$$\lim_{x\to c+} \frac{f(x)}{g(x)} = \lim_{x\to c+} \frac{f'(x)}{g'(x)}. \tag{1}$$

Proof: By the Cauchy Mean-Value Theorem (Theorem 5.4-8), for any $x \in (c, b)$ there exists a $t \in (c, x)$, such that

$$g'(t)[f(x) - f(c)] = f'(t)[g(x) - g(c)],$$

or

$$g'(t)f(x) = f'(t)g(x) \tag{2}$$

since $f(c) = g(c) = 0$. Moreover, since $g(c) = 0$ we must have $g(x) \neq 0$, because otherwise Rolle's Theorem (Lemma 5.4-1) would require $g'(s) = 0$ for some $s \in (c, x)$, thus contradicting part (iii) of our hypothesis. Since $g'(t) \neq 0$, we may divide (2) by $g(x)g'(t)$ to obtain

$$\frac{f(x)}{g(x)} = \frac{f'(t)}{g'(t)}.$$

Now let $x \to c^+$; this forces $t \to c^+$, and the proof is complete.

L'Hôpital's Rule as stated and proved here can easily be modified for a limit from the left,

$$\lim_{x\to c-} \frac{f(x)}{g(x)} = \lim_{x\to c-} \frac{f'(x)}{g'(x)}, \tag{3}$$

and if the limits in Equations (1) and (3) are the same, we have a two-sided version,

$$\lim_{x\to c} \frac{f(x)}{g(x)} = \lim_{x\to c} \frac{f'(x)}{g'(x)}.$$

We omit proofs of the cases involving limits at infinity or infinite limits.

* This theorem was discovered in 1694 by the Swiss mathematician Jean Bernoulli, at a time when he was employed by a French nobleman, Marquis G. F. A. de L'Hôpital.

Example 1. Evaluate $\lim_{x\to 1} \dfrac{x^2 - 3x + 2}{x^2 - 1}$.

Solution: Observe that substituting $x = 1$ gives the indeterminate form $0/0$. Thus it is legitimate to try L'Hôpital's Rule:

$$\lim_{x\to 1} \frac{x^2 - 3x + 2}{x^2 - 1} = \lim_{x\to 1} \frac{2x - 3}{2x} = \frac{2 \cdot 1 - 3}{2 \cdot 1} = -\frac{1}{2}.$$

Example 2. Evaluate $\lim_{x\to \infty} \dfrac{5x - 1}{3x + 2}$.

Solution: Here we have the form ∞/∞, so L'Hôpital's Rule is applicable.

$$\lim_{x\to \infty} \frac{5x - 1}{3x + 2} = \lim_{x\to \infty} \frac{5}{3} = \frac{5}{3}.$$

In applying L'Hôpital's Rule, if a single differentiation of numerator and denominator again leads to one of the forms $0/0$ or ∞/∞, the process can be repeated, again and again if necessary. But you must be careful to stop this process when you get to an expression that does not yield $0/0$ or ∞/∞.

Example 3. Evaluate $\lim_{x\to 2} \dfrac{x^3 - 6x^2 + 12x - 8}{x^3 - 3x^2 + 4}$.

Solution: Substituting $x = 2$ gives $0/0$; use the rule:

$$\lim_{x\to 2} \frac{x^3 - 6x^2 + 12x - 8}{x^3 - 3x^2 + 4} = \lim_{x\to 2} \frac{3x^2 - 12x + 12}{3x^2 - 6x}.$$

Substituting $x = 2$ in the result again gives $0/0$, so use the rule again:

$$\lim_{x\to 2} \frac{3x^2 - 12x + 12}{3x^2 - 6x} = \lim_{x\to 2} \frac{6x - 12}{6x - 6} = \frac{6 \cdot 2 - 12}{6 \cdot 2 - 6} = \frac{0}{6} = 0.$$

Note that L'Hôpital's Rule does not apply to

$$\lim_{x\to 2} \frac{6x - 12}{6x - 6},$$

because this fraction does not give $0/0$ when $x = 2$.

Example 4.

$$\lim_{x\to 0+} \frac{x^2 - 3x}{x + 2\sqrt{x}} = \lim_{x\to 0+} \frac{2x - 3}{1 + 1/\sqrt{x}} \text{ (by L'Hôpital's Rule)}$$

$$= \lim_{x\to 0+} \frac{\sqrt{x}(2x - 3)}{\sqrt{x} + 1} \text{ (Simplifying)}$$

$$= \frac{0(2 \cdot 0 - 3)}{0 + 1}$$

$$= 0.$$

Warning! In L'Hôpital's Rule we obtain a new fraction $f'(x)/g'(x)$ by differentiating the numerator and denominator of the original fraction $f(x)/g(x)$. This has no connection with the operation of differentiating a quotient.

Miscellaneous Exercises 5.5

1. For each of the functions described in Exercise 1 of § 5.1, determine, if it exists, (i) $\lim_{x\to 0^-} f(x)$; (ii) $\lim_{x\to 0^+} f(x)$; (iii) $\lim_{x\to 0} f(x)$.
2. Prove that (a) $\lim_{x\to 0^-} |x|/x = -1$; (b) $\lim_{x\to 0^+} |x|/x = 1$.
3. Prove that if both limits exist, then
   (a) $\lim_{x\to\infty} f(x) = \lim_{t\to 0^+} f(1/t)$ (b) $\lim_{x\to -\infty} f(x) = \lim_{t\to 0^-} f(1/t)$.
4. (a) Prove that if $\lim_{x\to c^+} f(x) = L$ and $f(c)$ both exist, then f is continuous from the right at c if and only if $L = f(c)$.
   (b) State and prove a corresponding theorem for continuity from the left.
5. Comment on $\lim_{x\to\infty} x \sin x$. (Is it ∞? $-\infty$?)
6. Use L'Hôpital's rule to evaluate these limits.

   (a) $\lim_{x\to 2} \dfrac{x^2 - 4}{x^2 - 3x + 2}$ (f) $\lim_{x\to 0+} \dfrac{\sqrt{x+1} - 1}{\sqrt{x}}$

   (b) $\lim_{x\to 1} \dfrac{x^3 + 6x - 7}{x^3 + 5x^2 - 6}$ (g) $\lim_{x\to 0} \dfrac{\sqrt{x+1} - 1}{\sqrt{x+4} - 2}$

   (c) $\lim_{x\to 1} \dfrac{x^5 - 1}{x^8 - 1}$ (h) $\lim_{x\to 4} \dfrac{\sqrt[3]{x+4} - 2}{\sqrt{x} - 2}$

   (d) $\lim_{x\to\infty} \dfrac{x^3 + 2x}{5x^2 + 3}$ (i) $\lim_{x\to -\infty} \dfrac{\sqrt{1-x}}{x}$

   (e) $\lim_{x\to\infty} \dfrac{x^2 + 2x}{5x^2 + 3}$ (j) $\lim_{x\to 1-} \dfrac{|x - 1|}{x^2 - 1}$

7. (a) Give a sensible definition for right- and left-hand derivatives of a function f at a pont c of its domain.
   (b) Find, if they exist, the right- and left-hand derivatives of the following functions at 0: (i) $f(x) = |x|$; (ii) the function of Example 1, § 5.1; (iii) $f(x) = |x|/x$.
   (c) What conditions on the right- and left-hand derivatives of f at c guarantee the existence of the derivative $f'(c)$?

Chapter Review

1. Evaluate the following limits:

(a) $\lim_{x\to\infty} \dfrac{x^2 + 3x + 7}{4x^2 - 1}$

(b) $\lim_{x\to 0} \dfrac{3x^3}{2x^2}$

(c) $\lim_{x\to 0} \dfrac{3x^3 + 0.001}{2x^2}$

(d) $\lim_{x\to\infty} (\sqrt{x + 2} - \sqrt{x})$

(e) $\lim_{x\to 1} \dfrac{x^3 - x^2 + 2x - 2}{x - 1}$

(f) $\lim_{x\to 0} \dfrac{1 - \cos x}{\sin x}$

(g) $\lim_{x\to 1+} \dfrac{\sqrt{x} - 1}{x - 1}$

(h) $\lim_{x\to 2} \dfrac{x^3 - x - 6}{x^2 + x - 6}$

2. Use the definition of limit to prove that $\lim_{x\to 1} (2x + 3) = 5$.

3. Why is it true that if $\lim_{x\to\infty} f(x)$ exists, then $\lim_{x\to\infty} [f(x)]^3 = \left[\lim_{x\to\infty} f(x)\right]^3$?

4. Use the definition of continuity to prove that, if $f(x) = x^2$, then f is continuous at 3. (*Hint*: Since the issue here is the nature of f near 3, you can take $|x - 3| < 1$, and this implies that $|x + 3| < 7$.)

6

The Differential

6.1 The differential

A curve and its tangent line not only coincide at the point of tangency but also commonly lie close together for some distance on either side of that point. Over a short interval, therefore, a tangent to a curve can often be used as a fair approximation to the curve itself. If the curve is the graph of a function f, then an equation of the line tangent at (x_1, y_1) is (§ 3.4)

$$y - y_1 = f'(x_1)(x - x_1).$$

If x changes from x_1 to $x_2 = x_1 + \Delta x$, then the corresponding change in y along the curve is

$$\Delta y = y_2 - y_1 = f(x_1 + \Delta x) - f(x_1).$$

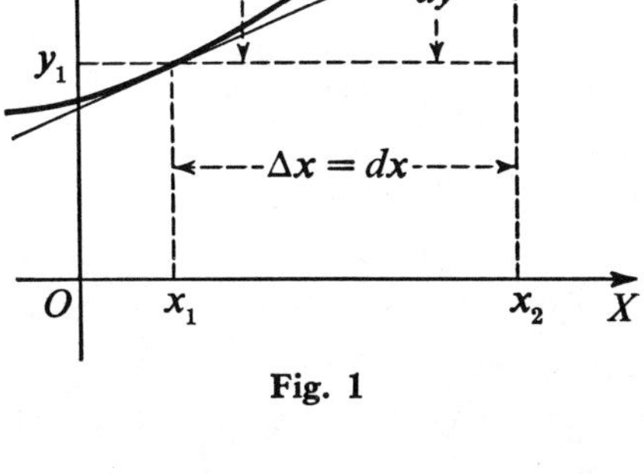

Fig. 1

If we use the tangent line as an approximation to the curve, then the change in y along this line will generally be different from Δy. From the equation of this line, we find

$$y_3 = y_1 + f'(x_1)(x_2 - x_1) = y_1 + f'(x_1)\Delta x,$$

and the change in y along the line is therefore

$$y_3 - y_1 = f'(x_1)\Delta x.$$

This quantity is important, and it is therefore desirable to have a symbol for it. We cannot use Δy, which describes the change along the curve, but because $y_3 - y_1$ is related to Δy we use a d, the Roman equivalent of the Greek Δ, and call the change along the tangent line dy. Also, for both historical and substantial practical reasons, the variable Δx is commonly denoted dx in this equation. Finally, since x_1 is any point in the domain of f', we drop the subscript. Thus we have

$$\mathbf{dy = f'(x)dx.} \tag{1}$$

Though this equation was reached by a geometrical argument, it has a meaning independent of any geometrical considerations: for a given differentiable function f and a given x in the domain of f', (1) describes a function $\{(dx, dy)\}$ whose domain is the set of all real numbers dx.* As Fig. 1 suggests, however, dy is generally a good approximation to Δy only when dx is small; we shall investigate this issue in § 7.2.

We have so far glossed over one difficulty. We have treated dx as an independent variable. But suppose x, in turn, is a function of a third variable, say, $x = g(t)$. We can then define the differential of x by the equation

$$dx = g'(t)dt$$

which corresponds to (1) in this case. Thus we seem to have a conflict of interpretation: is dx an independent variable, or is it a differential, and hence a dependent variable? We shall show in § 6.3 that there is, in fact, no conflict; the two interpretations are compatible.

For all $dx \neq 0$, (1) yields

$$\mathbf{f'(x) = \frac{dy}{dx},} \tag{2}$$

and the right member of this equation, called the differential quotient, may therefore be used as *an alternative notation for the derivative.* This notation has many convenient features and will therefore be used extensively in all that follows. Note, however, that strictly speaking the differential quotient dy/dx denotes not the derivative, a function, but the value at x of the deriva-

* Strictly speaking, the differential dy is a function of three variables: the function f, the variable x, and the variable dx. If we agree to talk about a particular function f, then, subject to this agreement, dy depends on, and is a function of, the two variables x and dx. If we further agree to a particular x in the domain of f, then dy is a function of dx alone. The study of functions of more than one variable is a major element in more advanced courses in calculus.

tive, that is, a number, just as y itself denotes not the function f but the value $f(x)$ which f takes at the number x.

Since $y = f(x)$, (1) may also be written in the form

$$d(f(x)) = f'(x)dx,$$

and (2) in the form $\dfrac{d(f(x))}{dx} = f'(x)$, or $\dfrac{d}{dx}f(x) = f'(x)$.

The symbols d and $\dfrac{d}{dx}$ are examples of operational symbols, or operators; they give instructions to perform, respectively, the operation "find the differential" and the operation "find the derivative." Because of the close connection between these two operations, implied by (1), we shall, in what follows, interpret "differentiate" to mean either.

The result of twice performing an operation is commonly indicated, by analogy with algebra, by an exponent. Thus, if $\dfrac{d}{dx}$ means "differentiate," then $\left(\dfrac{d}{dx}\right)^2$ means "differentiate twice"; this is usually written $\dfrac{d^2}{dx^2}$. If $y = f(x)$, we therefore write the second derivative $f''(x)$ as

$$\frac{d^2y}{dx^2}, \quad \text{or} \quad \frac{d^2}{dx^2}f(x),$$

the third derivative as

$$\frac{d^3y}{dx^3}, \quad \text{or} \quad \frac{d^3}{dx^3}f(x),$$

and so on.

One of the great advantages of these new symbols is that they permit a new economy of notation. Thus, for example, the statement

$$\text{If} \quad f(x) = 7x^5, \quad \text{then} \quad f'(x) = 35x^4$$

can be written $\quad \text{If} \quad y = 7x^5, \quad \text{then} \quad \dfrac{dy}{dx} = 35x^4,$

or, more concisely, $\quad \dfrac{d}{dx}(7x^5) = 35x^4.$

Expressing the same statement in the language of differentials, we have

$$\text{If} \quad y = 7x^5, \quad \text{then} \quad dy = 35x^4\,dx,$$

or, finally and most concisely,

$$d(7x^5) = 35x^4\,dx.$$

Note also that

$$\frac{d^2y}{dx^2} = \frac{d^2}{dx^2}(7x^5) = 140x^3,$$

$$\frac{d^3y}{dx^3} = \frac{d^3}{dx^3}(7x^5) = 420x^2,$$

and so on.

We listed, in § 2.9, four rules for differentiation. Each of these has its counterpart in differential notation. They are:

Rule 1a. $dc = 0.$

Rule 2a. $d(x^n) = nx^{n-1}\,dx.$

Rule 3a. $d[cg(x)] = cd[g(x)].$

Rule 4a. $d[f_1(x) + f_2(x)] = df_1(x) + df_2(x).$

Exercises 6.1 [A]

1. Find dy in terms of x and dx if
   (a) $y = x^2 - 5x + 6.$
   (b) $y = 3x^{-2} + \sqrt{7x}.$
   (c) $y = \dfrac{x^2 - 3}{\sqrt{x}}.$
   (d) $y = (x^2 - 2)^3.$
2. If $y = x^3 - 5x$, find the value of dy
   (a) when $x = 3$ and $dx = -10.$
   (b) when $x = 1$ and $dx = 0.01.$
3. If $s = at^2 + bt + c$, show that $t\dfrac{d^2s}{dt^2} - \dfrac{ds}{dt} + b = 0$ for all values of t.
4. If $u = \dfrac{1}{z}$, find $\dfrac{du}{dz}$, $\dfrac{d^2u}{dz^2}$, and $\dfrac{d^3u}{dz^3}$, and from these special cases guess a formula for $\dfrac{d^nu}{dz^n}$ for any positive integer n. Can you prove that your guess is correct?
5. If $y = x^2 - 4x$, for what values of x will a positive value of dy correspond to a negative value of dx?
6. If $y = f(x)$ and if $f'(x)$ is defined for all real numbers x,
   (a) what can you say about dy when $dx = 0$?
   (b) for what values of x, if any, does $dy = 0$ no matter what the value of dx?

7. Find all possible expressions for y if $dy = 8x^3\, dx$.

8. If $y = x^n$ for some positive integer n, what are $\dfrac{d^n y}{dx^n}$ and $\dfrac{d^{n+1}y}{dx^{n+1}}$?

Exercises 6.1 [B]

1. Find dy if
   (a) $y = x^3 + 3x^2 - 7$.
   (b) $y = \sqrt{2x} - \dfrac{2}{\sqrt{x}}$.
   (c) $y = (x^{1/2} + x^{-1/2})^2$.
   (d) $y = \dfrac{x^{2/3} - 3}{x^{-1/3}}$.

2. What is y if $dy = (3x^2 - 2)dx$?

3. If $y = 3x^2$, find dy, Δy, and $\dfrac{\Delta y}{dy}$ for $x = 1$ and $dx = 10$, 1, 0.1, and 0.01.

   What seems to be $\lim\limits_{\Delta x \to 0} \dfrac{\Delta y}{dy}$?

4. If $s = t - \dfrac{1}{t}$, find $\dfrac{d^2 s}{dt^2}\left(s^2 - \dfrac{ds}{dt}\right)$.

5. Find $\dfrac{d^2 y}{dx^2}$ if $\dfrac{d^3 y}{dx^3} = 2x$.

6. If k and n are positive integers, and $k < n$, find and simplify an expression for $\dfrac{d^k}{dx^k}\left(\dfrac{x^n}{n!}\right)$. Can you prove that this result is correct?

7. If $\dfrac{dy}{dx} = -y$, show that $\dfrac{d^2 y}{dx^2} = y$.

8. If $y = x^4 - 6x^2$, show that y, $\dfrac{dy}{dx}$, and $\dfrac{d^2 y}{dx^2}$ are all negative on the interval $(0, 1)$, but that $\dfrac{d^3 y}{dx^3}$ is positive on the same interval.

6.2 Derivatives of products and quotients

The rules for differentiation that we derived in Chapter 2 enable us to differentiate only functions described by sums of terms like cx^n. In the rest of this chapter we shall develop additional formulas and techniques that will permit us to differentiate many other types of functions and therefore to solve a greatly expanded variety of problems. In this section we shall see how to differentiate the product and the quotient of differentiable functions.

Products. Let g and h be functions differentiable at a point x, and let $f = g \cdot h$ be the product function described by the formula

$$f(x) = g(x) \cdot h(x). \tag{1}$$

We wish to determine how the derivative f' is related to the derivatives g' and h'.

The argument can be somewhat simplified if we introduce letters, called auxiliary variables, to stand for the function-values involved. Let

$$y = f(x), \quad u = g(x), \quad v = h(x); \tag{2}$$

then in terms of these auxiliary variables (1) becomes

$$y = uv. \tag{3}$$

Returning to the four steps for finding the derivative listed in § 2.8, we compute, using (1),

$$y + \Delta y = f(x + \Delta x) = g(x + \Delta x) \cdot h(x + \Delta x). \tag{4}$$

We have from (2)

$$u + \Delta u = g(x + \Delta x) \quad \text{and} \quad v + \Delta v = h(x + \Delta x); \tag{5}$$

hence (4) becomes

$$y + \Delta y = (u + \Delta u)(v + \Delta v) = uv + u\,\Delta v + v\,\Delta u + \Delta u\,\Delta v. \tag{6}$$

Subtracting (3) from (6), we have

$$\Delta y = u\,\Delta v + v\,\Delta u + \Delta u\,\Delta v,$$

and dividing by Δx, we have

$$\frac{\Delta y}{\Delta x} = u\frac{\Delta v}{\Delta x} + v\frac{\Delta u}{\Delta x} + \Delta u\frac{\Delta v}{\Delta x}. \tag{7}$$

By definition,
$$\frac{dy}{dx} = f'(x) = \lim_{\Delta x \to 0} \frac{\Delta y}{\Delta x}.$$

Taking limits on both sides of (7) and making generous use of Theorem 5.3-2, we find

$$\begin{aligned}\frac{dy}{dx} = \lim_{\Delta x \to 0}\frac{\Delta y}{\Delta x} &= \lim_{\Delta x \to 0}\left(u\frac{\Delta v}{\Delta x}\right) + \lim_{\Delta x \to 0}\left(v\frac{\Delta u}{\Delta x}\right) + \lim_{\Delta x \to 0}\left(\Delta u\frac{\Delta v}{\Delta x}\right)\\ &= u\lim_{\Delta x \to 0}\frac{\Delta v}{\Delta x} + v\lim_{\Delta x \to 0}\frac{\Delta u}{\Delta x} + \lim_{\Delta x \to 0}\Delta u\lim_{\Delta x \to 0}\frac{\Delta v}{\Delta x}\\ &= u\frac{dv}{dx} + v\frac{du}{dx} + 0 \cdot \frac{dv}{dx}\end{aligned}$$

since the existence of g' implies $\lim_{\Delta x \to 0} \Delta u = 0$ as was shown in the proof of Theorem 5.4-6. Hence we have

▶ **Rule 5.** $f'(x) = g(x)h'(x) + h(x)g'(x)$.

In differential-quotient notation, this becomes (using (2))

$$\frac{dy}{dx} = u\frac{dv}{dx} + v\frac{du}{dx}. \tag{8}$$

If we multiply (8) by the differential dx and use the definitions

$$dy = f'(x)dx = \frac{dy}{dx}\,dx,$$

$$du = g'(x)dx = \frac{du}{dx}\,dx,$$

$$dv = h'(x)dx = \frac{dv}{dx}\,dx,$$

we have

▶ **Rule 5a.** $dy = u\,dv + v\,du$, or $d(uv) = u\,dv + v\,du$.

There are similar expressions for the derivative of the product of three or more functions. If, for example,

$$f = g\cdot h\cdot i,$$

then
$$f' = g\cdot h\cdot i' + g\cdot h'\cdot i + g'\cdot h\cdot i,$$

or, setting $y = f(x),\quad u = g(x),\quad v = h(x),\quad w = i(x),$

$$dy = uv\,dw + uw\,dv + vw\,du.$$

If $y \neq 0$, we can put this last in a particularly neat form by dividing by $y = uvw$:

$$\frac{dy}{y} = \frac{du}{u} + \frac{dv}{v} + \frac{dw}{w}.$$

Example 1. Differentiate as a product: $y = (x^3 + 5)\left(1 + \frac{2}{x}\right)$.

Solution: If we set $u = x^3 + 5,\quad v = 1 + \frac{2}{x},$

we have $du = 3x^2\,dx,\quad dv = -\frac{2}{x^2}\,dx,$

and Rule 5a then gives $dy = (x^3 + 5)\left(-\frac{2}{x^2}\,dx\right) + \left(1 + \frac{2}{x}\right)(3x^2\,dx)$

$$= \left(3x^2 + 4x - \frac{10}{x^2}\right) dx.$$

It is, of course, possible first to carry out the indicated multiplication and then differentiate by the methods of Chapter 2, but the student will soon meet problems where this option does not exist.

Quotients. Let g and h be, as above, functions differentiable at a point x, and let us impose the additional condition $h(x) \neq 0$. We define a new function f, the quotient function $f = g/h$, described by the formula

$$f(x) = \frac{g(x)}{h(x)}. \tag{9}$$

In terms of the substitutions (2), (9) takes the form

$$y = \frac{u}{v}. \tag{10}$$

Returning once again to the method of § 2.8, we find

$$y + \Delta y = \frac{g(x + \Delta x)}{h(x + \Delta x)};$$

in terms of (5), this becomes

$$y + \Delta y = \frac{u + \Delta u}{v + \Delta v}. \tag{11}$$

Subtracting (10) from (11), we have

$$\Delta y = \frac{u + \Delta u}{v + \Delta v} - \frac{u}{v} = \frac{v\,\Delta u - u\,\Delta v}{v(v + \Delta v)},$$

and dividing by Δx, we get

$$\frac{\Delta y}{\Delta x} = \frac{v\,\dfrac{\Delta u}{\Delta x} - u\,\dfrac{\Delta v}{\Delta x}}{v(v + \Delta v)}. \tag{12}$$

If we now find $\lim_{\Delta x \to 0}$ of both sides of (12) we get, with considerable help from Theorems 5.3-2 and 5.3-3,

$$\frac{dy}{dx} = \frac{v\,\dfrac{du}{dx} - u\,\dfrac{dv}{dx}}{v^2},$$

or

▶ **Rule 6.** $f'(x) = \dfrac{h(x)g'(x) - g(x)h'(x)}{[h(x)]^2}$.

The details of this last step are left to the student as an exercise. Using the substitutions (5), and translating into differentials, we get

▶ **Rule 6a.** $dy = \dfrac{v\,du - u\,dv}{v^2}$, or $d\left(\dfrac{u}{v}\right) = \dfrac{v\,du - u\,dv}{v^2}$.

Again, the details are left as an exercise.

Example 2. Differentiate $y = \dfrac{x^2}{x + 1}$.

Solution: Setting $u = x^2$, $v = x + 1$, we have $du = 2x\,dx$, $dv = dx$, and therefore

$$dy = \frac{(x + 1)(2x\,dx) - (x^2)(dx)}{(x + 1)^2} = \frac{x^2 + 2x}{(x + 1)^2}\,dx.$$

A matter of technique. The rules of this section have been stated in terms of specific letters. The student should, however, learn to use these rules without relying on specific letters. To this end, Rules 5 and 6 should be learned and used in somewhat the following form:

Rule 5. The derivative of the product of two functions is the first function times the derivative of the second, plus the second times the derivative of the first.

Rule 6. The derivative of the quotient of two functions is the denominator times the derivative of the numerator, minus the numerator times the derivative of the denominator, all divided by the square of the denominator.

Example 3. In differentiating $\dfrac{3x^4}{x^2 - 3}$, the student should have in mind the pattern of Rule 6 and should fill that pattern in directly by writing

$$\frac{(x^2 - 3)(3 \cdot 4x^3) - 3x^4(2x)}{(x^2 - 3)^2}$$

without preliminary reference to any u or v or other auxiliary variable. This result must, of course, be simplified.

Exercises 6.2 [A]

1. Complete the derivation of Rule 6.
2. Show in detail how Rule 6a follows from Rule 6.
3. Differentiate by Rule 5 or 5a; do not perform the indicated multiplications:
   (a) $(x^2 - 2)(x + 3)$
   (b) $(x^2 - 3x)\left(x^2 - \dfrac{3}{x}\right)$
   (c) $(x^2 + x)(3x^5 - 7)$
   (d) $x(x - 1)(x + 3)$
4. Differentiate:
   (a) $\dfrac{t}{t + 1}$
   (b) $\dfrac{z^2 - z + 3}{2z - 1}$
   (c) $\dfrac{\sqrt{x}}{x^2 + x - 3}$
   (d) $\dfrac{x^2 + 2x - 3}{x^2 - x - 12}$
5. Find the first and second derivatives of $\dfrac{x - a}{x + a}$.
6. Show that $f = \left\{(x, y) : y = \dfrac{x}{x^2 - 4}\right\}$ is a decreasing function throughout its domain.
7. Write equations of the tangent and normal to $y = \dfrac{x}{x + 1}$ at $(1, \frac{1}{2})$.

Exercises 6.2 [B]

1. Show how the formula for differentiating x^n, for any integer n (Rule 2), can be derived *without* using the binomial theorem, as follows:
   (a) Use the procedure of § 2.8 to show that
   $$\frac{d}{dx}(x^1) = 1 \cdot x^0 = 1.$$
   (b) Set $u = x$, $v = x^k$, and show by Rule 5 that $\dfrac{d}{dx}(x^k) = k \cdot x^{k-1}$ implies $\dfrac{d}{dx}(x^{k+1}) = (k + 1)x^k$ for any positive integer k. Why does this establish the required result for any *positive* integer n?
   (c) Set $u = 1$, $v = x^n$, and show by the immediately preceding result and Rule 6 that the required formula is valid for any *negative* integer. (*Hint:* $x^{-n} = 1/x^n = u/v$.)
   (d) What integer n has been omitted? Prove the formula in this case.

2. Differentiate, using the product rule (5 or 5a):

(a) $(x^2 + 3x - 5)(x^3 - 17)$
(c) $(ax + b)(ax + b)(ax + b)$
(b) $(\sqrt{x} - 5)(x^2 + 3x^{-2})$
(d) $(ax^2 + bx + c)(ax^2 + bx + c)$

3. Differentiate:

(a) $\dfrac{2t + 3}{5t - 1}$ (b) $\dfrac{x^2 - 5}{x + 4}$ (c) $\dfrac{\sqrt{x} - 1}{\sqrt{x} + 1}$ (d) $\dfrac{1}{ax + b}$

4. Differentiate:

(a) $\dfrac{1}{(ax + b)^2}$ (b) $\dfrac{1}{ax^2 + bx + c}$

5. From Exercises 2(c), 3(d), and 4(a) can you guess a rule for differentiating $(ax + b)^n$?

6. Find the extreme values of $\dfrac{2x - 3}{x^2 + 2x + 7}$ and classify each as a maximum or as a minimum.

7. If $y = \dfrac{8}{x^2 + 4}$:

(a) Find all extreme values and inflection points.
(b) Determine all points at which $y < 0$, $y = 0$, $y > 0$.
(c) Find $\lim_{x \to \infty} y$ and $\lim_{x \to -\infty} y$.
(d) Sketch the graph of the equation.

6.3 Derivatives of composite functions

We have now applied the process of differentiation to most of the algebra of functions (§ 2.5). In § 2.9 we saw how to differentiate a multiple of a differentiable function and a sum of differentiable functions, and in § 6.2 we saw how to differentiate a product and a quotient of differentiable functions. In this section we shall derive a rule for differentiating a composite of differentiable functions. The result is of crucial importance in both differentiation and integration.

A typical problem is to find $f'(x)$ if $f(x) = \sqrt{3x^2 + 5}$. If we define functions g and h by the rules

$$g(x) = \sqrt{x} \quad \text{and} \quad h(x) = 3x^2 + 5, \tag{1}$$

then f is the composite of g and h: $f = gh$. Moreover, we know the derivatives g' and h':

$$g'(x) = \tfrac{1}{2}x^{-1/2} = 1/2\sqrt{x}, \quad h'(x) = 6x. \tag{2}$$

We shall see that this knowledge of g' and h' can be used, in a very simple way, to find f'.

It is convenient to introduce a new symbol, or auxiliary variable, to denote $h(x)$:

$$u = h(x) = 3x^2 + 5; \tag{3}$$

and also one for $f(x)$:

$$y = f(x) = g[h(x)] = g(u) = \sqrt{u}. \tag{4}$$

We then have

$$\begin{aligned} \Delta y &= f(x + \Delta x) - f(x) \\ &= g[h(x + \Delta x)] - g[h(x)] \\ &= g(u + \Delta u) - g(u) \\ &= \sqrt{u + \Delta u} - \sqrt{u}, \end{aligned}$$

where, following (3), we have set

$$\begin{aligned} \Delta u &= h(x + \Delta x) - h(x) \\ &= 3(x + \Delta x)^2 + 5 - (3x^2 + 5) \\ &= 6x \cdot \Delta x + 3(\Delta x)^2. \end{aligned} \tag{5}$$

We note for future reference that, for each real number x, (5) defines Δu as a function of Δx, and that this function $\{(\Delta x, \Delta u)\}$ is continuous at $\Delta x = 0$.

The derivative we are looking for is, by definition, $\lim_{\Delta x \to 0} \Delta y/\Delta x$. So long as $\Delta u \neq 0$—and this will commonly be the case, as it is in the particular example we are considering, if $\Delta x \neq 0$—we may write

$$\begin{aligned} \frac{\Delta y}{\Delta x} &= \frac{\Delta y}{\Delta u} \cdot \frac{\Delta u}{\Delta x} \\ &= \frac{g(u + \Delta u) - g(u)}{\Delta u} \cdot \frac{h(x + \Delta x) - h(x)}{\Delta x} \end{aligned}$$

from (4) and (3). Hence

$$\begin{aligned} f'(x) &= \lim_{\Delta x \to 0} \frac{\Delta y}{\Delta x} = \lim_{\Delta x \to 0} \left(\frac{\Delta y}{\Delta u} \cdot \frac{\Delta u}{\Delta x} \right) \\ &= \lim_{\Delta x \to 0} \frac{\Delta y}{\Delta u} \cdot \lim_{\Delta x \to 0} \frac{\Delta u}{\Delta x} \end{aligned} \tag{6}$$

by Theorem 5.3-2, if both the last limits exist. The second of these offers no difficulty: we assume that the derivative h' is defined at x, and therefore $\lim_{\Delta x \to 0} \frac{\Delta u}{\Delta x} = h'(x) = \frac{du}{dx}$ by definition. The first, $\lim_{\Delta x \to 0} \frac{\Delta y}{\Delta u}$, offers a little more difficulty; if only it were $\lim_{\Delta u \to 0} \frac{\Delta y}{\Delta u}$, things would be much more pleasant. But, in fact, the existence of the derivative h' assures us, thanks to Theorem 5.3-3, that Δu goes to zero, and we therefore have

$$\begin{aligned}\lim_{\Delta x \to 0} \frac{\Delta y}{\Delta u} &= \lim_{\Delta u \to 0} \frac{\Delta y}{\Delta u} \\ &= \frac{dy}{du} \\ &= g'(u) \\ &= g'[h(x)].\end{aligned}$$

Hence (6) becomes

$$\frac{dy}{dx} = \frac{dy}{du} \cdot \frac{du}{dx}, \tag{7}$$

or, eliminating the auxiliary variable u which has now served its purpose,

$$f'(x) = g'[h(x)] \cdot h'(x).$$

Thus, in the particular example with which we began this discussion,

$$\begin{aligned}f'(x) &= \tfrac{1}{2}u^{-1/2} \cdot 6x \\ &= \tfrac{1}{2}(3x^2 - 5)^{-1/2} \cdot 6x \\ &= 3x(3x^2 + 5)^{-1/2}.\end{aligned}$$

We thus have a new rule for differentiation:

▶ **Rule 7. (Chain Rule for Derivatives). If $f(x) = g[h(x)]$, then**

$$f'(x) = g'[h(x)] \cdot h'(x).$$

There are a few exceptional cases in which the preceding proof is not valid. The difficulty arises in connection with (6), which requires $\Delta u \neq 0$. If there is some punctured neighborhood of x on which $h(x + \Delta x) \neq h(x)$, and therefore $\Delta u \neq 0$, we can restrict our argument to that neighborhood, and all is well. But unfortunately it is possible for h and x to be such that *every* punctured neighborhood of x contains a point $x + \Delta x$ such that $h(x + \Delta x) = h(x)$. For a proof that also covers this exceptional case, see, for instance, p. 176 of *Calculus*, Vol. 1, Second Edition, by T. M. Apostol.

We saw in §6.1 that if $y = f(x)$, then the differential dy is defined by the equation

$$dy = f'(x)dx, \tag{8}$$

in which dx is an independent variable, with domain the set of all real numbers. If x in turn is a function of a new variable t, say $x = g(t)$, then a differential dx is defined by

$$dx = g'(t)dt. \tag{9}$$

We thus have the symbol dx with totally different meanings in (8) and (9), and a cloud of confusion seems to loom. Happily, the chain rule saves the day and reconciles these two uses of dx. For if $y = f(x)$ and $x = g(t)$, then we can form the composite function $F = fg$, and obtain

$$y = f[g(t)] = F(t).$$

In this case we have

$$dy = F'(t)dt. \tag{10}$$

But by the chain rule

$$F'(t) = f'[g(t)] \cdot g'(t). \tag{11}$$

Hence (10) becomes

$$\begin{aligned} dy &= f'[g(t)] \cdot g'(t)dt \\ &= f'(x)dx \end{aligned}$$

by virtue of (9). Hence (8) holds *regardless of the presence of the variable t.* We may therefore, without confusion, call dx the differential of x, and regard (8) as a statement about two differentials.

The chain rule thus reconciles the use of dx in (8) and (9), but the important fact, indeed the major reason for our concern with this special notation for differentials and derivatives, is that the favor is reciprocated: differential notation reduces the use of the chain rule to an exercise in elementary algebra. For if $dx \neq 0$ and $dt \neq 0$, we may write (8), (9), and (10) as

$$\frac{dy}{dx} = f'(x), \quad \frac{dx}{dt} = g'(t), \quad \text{and} \quad \frac{dy}{dt} = F'(t),$$

respectively. Since $f'[g(t)] = f'(x)$, the statement of the chain rule in (11) becomes

$$\frac{dy}{dt} = \frac{dy}{dx} \cdot \frac{dx}{dt}$$

in differential notation. Cancelling the dx's then reduces this to the identity

$$\frac{dy}{dt} = \frac{dy}{dt}.$$

We therefore restate the chain rule in terms of differentials. If y is a function of u, and if u, in turn, is a function of x, we then have:

▶ **Rule 7a.** $\dfrac{dy}{dx} = \dfrac{dy}{du} \cdot \dfrac{du}{dx}.$

Note again that if we interpret the quantities in Rule 7a as differentials, the du's cancel and the equation reduces to the identity $dy/dx = dy/dx$.

It is this consequence of the chain rule that makes the differential notation the great convenience that it is, and the student will therefore probably find Rule 7a easier to remember and use than Rule 7.

The chain rule is of value, evidently, only when it is possible to introduce the auxiliary variable u in such a way that the derivatives dy/du and du/dx are readily determined. In most cases, the choice of auxiliary variable is obvious. There may, however, be instances in which the two-stage procedure outlined here fails to break an expression down into units which we can differentiate; in such cases, the chain rule can be applied repeatedly, thus extending the chain to three or more "links." Thus, if $y = f(x)$, it may be convenient to write $y = g(v)$, $v = h(u)$, $u = k(x)$. In this case, the chain rule becomes

$$\frac{dy}{dx} = \frac{dy}{dv} \cdot \frac{dv}{du} \cdot \frac{du}{dx}.$$

Note that, in terms of differentials, this, too, reduces to an identity.

Example 1. Differentiate $y = (3x^2 - 1)^{100}$.

Solution: It would be possible to expand this expression, using the binomial theorem, and then differentiate the resulting 101 terms one at a time, but the method has little to recommend it. Instead, we set

$$u = 3x^2 - 1,$$

whence $$du = 6x\,dx, \quad y = u^{100},$$

and $$dy = 100u^{99}\,du = 100(3x^2 - 1)^{99}(6x\,dx) = 600x(3x^2 - 1)^{99}\,dx.$$

Example 2. Differentiate $y = (3x^2 - 5)\sqrt{2x + 3}$.

Solution: We regard this, in the first instance, as a product (Rule 5), and set

$$u = 3x^2 - 5, \quad v = \sqrt{2x + 3}.$$

The differential $du = 6x\,dx$ offers no difficulty, but to find dv we must turn to the chain rule:

$$w = 2x + 3, \quad v = w^{1/2},$$

whence $$dw = 2\,dx, \quad dv = \frac{1}{2}w^{-1/2}\,dw = \frac{dx}{\sqrt{2x + 3}},$$

and finally $$dy = u\,dv + v\,du = \frac{3x^2 - 5}{\sqrt{2x + 3}}dx + (6x\sqrt{2x + 3}\,)dx$$

$$= \frac{15x^2 + 18x - 5}{\sqrt{2x + 3}}\,dx.$$

Example 3. Differentiate $y = \dfrac{2x}{\sqrt{x-4}}$.

Solution: We start by treating this as a quotient (Rule 6a). Setting

$$u = 2x, \quad v = \sqrt{x-4},$$

we find $du = 2\,dx$ with ease, but for dv we must use the chain rule:

$$w = x - 4, \quad v = w^{1/2},$$

$$dw = dx, \quad dv = \frac{1}{2} w^{-1/2}\,dw = \frac{dx}{2\sqrt{x-4}}.$$

By Rule 6a, $$dy = \frac{\sqrt{x-4}\cdot 2\,dx - 2x\cdot\dfrac{dx}{2\sqrt{x-4}}}{x-4} = \frac{(x-8)dx}{(x-4)^{3/2}}.$$

The chain rule: a corollary. The only use that has been made of the chain rule in the preceding examples has been to differentiate $y = u^n$, where the derivative du/dx is known. We shall make other use of the chain rule, especially starting in § 6.5, but this particular case is of such common occurrence that it merits special emphasis, and we therefore state it as a special rule in the usual two forms:

▶ **Rule 8.** $\dfrac{d}{dx}(u^n) = nu^{n-1}\dfrac{du}{dx}.$

Rule 8a. $d(u^n) = nu^{n-1}\,du.$

The student would be well advised to learn this rule not in terms of particular letters, as u and x, but rather in some such form as:

Rule 8. The derivative of the nth power of a function is n, times the $(n-1)$th power of the function, times the derivative of the function.

The important thing to learn here is a pattern of procedure, not a rule. Because facility in this particular procedure is essential to what follows, an unusually large number of exercises is provided at the end of this section.

Example 4. In differentiating $\sqrt{4-7x} = (4-7x)^{1/2}$, the student should recognize that, by Rule 8a,

$$d(\quad)^{1/2} = \tfrac{1}{2}(\quad)^{-1/2}\,d(\quad)$$

is a valid statement if any differentiable expression whatever is used to fill all three parentheses. In this case,

$$d(4 - 7x)^{1/2} = \frac{1}{2}(4 - 7x)^{-1/2}\,d(4 - 7x)$$

$$= \frac{1}{2}(4 - 7x)^{-1/2}(-7\,dx) = \frac{-7\,dx}{2\sqrt{4 - 7x}}.$$

There may be instances in which y is given directly in terms of a variable u, and u, in turn, is given in terms of x. In such cases we can use (7) directly to find dy/dx, without troubling to eliminate u, as the following example shows.

Example 5. If $y = u^{2/3}$ and $u = x^2 + 2x - 5$, then

$$\frac{dy}{du} = \frac{2}{3}u^{-1/3}, \quad \frac{du}{dx} = 2x + 2,$$

and (7) then gives
$$\frac{dy}{dx} = \frac{2}{3}u^{-1/3}(2x + 2).$$

Exercises 6.3 [A]

1. Find dy/du and du/dx and use these to determine dy/dx if
   (a) $y = u^9, \quad u = x^2 - 7x + 3.$
   (b) $y = 2u^3 - u, \quad u = x^2 + x^{-2}.$

2. Differentiate: (a) $\sqrt{y + 2}$ (c) $(t^3 + 3t^2 - 6)^{-2/3}$
   (b) $(x^2 - 7x)^{10}$ (d) $\left(x - \frac{1}{x}\right)^5$

3. Differentiate: (a) $x\sqrt{2 + x}$ (c) $\left(\frac{2x - 1}{x - 1}\right)^2$
   (b) $\frac{x^2 + 2x}{\sqrt{x + 1}}$ (d) $\sqrt{\frac{x + 1}{x - 1}}$

4. Find the second derivative of (a) $(x^2 + 2x + 2)^5$; (b) $\frac{t}{\sqrt{t + 2}}$.

5. Differentiate $(x + \sqrt{a^2 - x^2}\,)^6$.

6. Find equations of the tangent and the normal to $y = \sqrt{x^2 + 9}$ at $(4, 5)$.

7. Show that the maximum area of a rectangle with diagonal d is $d^2/2$.

Exercises 6.3 [B]

1. Find dy/dx and dx/dt and use these to determine dy/dt if

 (a) $y = \frac{1}{x}, \quad x = 2t - \frac{2}{t}.$

 (b) $y = x^{10} + 13x^2 + x^{-3}, \quad x = t^3 + 4t.$

2. Differentiate: (a) $\sqrt{2 - 4x^3}$ (b) $(x^2 + 3x - 5)^8$ (c) $(ax^2 + bx + c)^n$ (d) $(x^{1/2} + x^{-1/2})^{1/2}$

3. Differentiate: (a) $(x + 3)\sqrt{x - 1}$ (b) $\left(\frac{1 - t^2}{1 + t^2}\right)^3$ (c) $\frac{x}{\sqrt{a^2 - x^2}}$ (d) $(\sqrt{x + 2} + x^2)^6$

4. Find an angle of intersection of $y = \sqrt{25 - x^2}$ and $y = \sqrt{x^2 - 7}$.

5. If f is a function such that $\frac{df(x)}{dx} = \frac{1}{x}$, show that $\frac{df(ax)}{dx} = \frac{1}{x}$ for any constant $a \neq 0$.

6. Show that the quotient rule (6 or 6a) can be derived from the product rule and the chain rule by differentiating $u \cdot v^{-1}$. (This result may help the student remember the order of the terms in the numerator of the quotient rule.)

7. Show that every normal to $y = \sqrt{a^2 - x^2}$ passes through the origin.

Miscellaneous Exercises 6.3

Differentiate:

1. $(ax + b)^n$
2. $(x^2 - 5)^{10}$
3. $(2x^3 - 7x + 3)^5$
4. $\sqrt{x^2 - a^2}$
5. $\sqrt{x^2 + a^2}$
6. $\sqrt{6x + x^2}$
7. $\sqrt{ax^2 + bx + c}$
8. $(x + 1)^3(5x - 3)$
9. $(7x^2 + 2x + 3)(x + 2)^2$
10. $(x + 1)^2(2x - 3)^3$
11. $(x + 3)\sqrt{2x - 1}$
12. $(x + 1)\sqrt{x + 2}$
13. $(x^2 + 5)(x - 3)^{2/3}$
14. $(2x + 1)^3(3x - 1)^5$
15. $(x + 3)^2(2x + 5)^3$
16. $(1 + 3x^2)^3(x - 1)^2$

17. $(4x - x^2)^{-3}$

18. $\dfrac{x + 3}{x - 5}$

19. $\dfrac{2x - 5}{3x + 4}$

20. $\dfrac{3 + t}{4 - 7t}$

21. $\dfrac{px + q}{rx + s}$

22. $\dfrac{x^2 + 2x + 1}{2x - 3}$

23. $\dfrac{x^2}{x - 1}$

24. $\dfrac{2y + 3}{y^2 + 4}$

25. $\dfrac{x^3 + 7x}{x^3 + 4}$

26. $\dfrac{x^5}{9 - x}$

27. $\dfrac{1}{\sqrt{a^2 - x^2}}$

28. $\dfrac{(x + 1)\sqrt{x}}{x - 3}$

29. $\dfrac{2x - 3}{(x + 2)\sqrt{x - 1}}$

30. $\sqrt{\dfrac{x + a}{x - a}}$

31. $\sqrt{\dfrac{2x + 3}{7x - 5}}$

32. $\sqrt{\dfrac{2x + 1}{x^2 - 1}}$

33. $\sqrt{1 + \sqrt{x}}$

34. $\sqrt{\dfrac{x^2 + a^2}{x^2 - a^2}}$

35. $\dfrac{x - 1}{\sqrt{x^2 - 1}}$

36. $\sqrt{x}(2x - 3)^4$

37. $\dfrac{\sqrt{x^2 + a^2}}{x}$

38. $\dfrac{x}{x + \sqrt{a^2 - x^2}}$

39. $\dfrac{x + \sqrt{a^2 - x^2}}{x}$

40. $\left(\sqrt{a^2 + x^2} + \sqrt{a^2 - x^2}\right)^3$

6.4 Implicit functions

An equation such as

$$2x + 3y - 6 = 0 \tag{1}$$

is said (see also § 2.1) to define y as a function of x because $f = \{(x, y): 2x + 3y - 6 = 0\}$ is a function. It is possible to solve (1) for y in terms of x,

$$y = 2 - \tfrac{2}{3}x. \tag{1a}$$

In the form (1a), in which y stands alone on one side, the equation is said to define y *explicitly* as a function of x, whereas in any other form, such as (1), the function is said to define y *implicitly* as a function of x. We know, of course, how to find the derivative dy/dx from (1a), but it is also possible, and often much more convenient, to find the same derivative directly from the form (1). We differentiate (1) term by term, with respect to x, obtaining

$$\frac{d}{dx}(2x) + \frac{d}{dx}(3y) - \frac{d}{dx}(6) = \frac{d}{dx}(0), \tag{2}$$

or

$$2 + 3\frac{dy}{dx} - 0 = 0, \tag{2a}$$

or

$$\frac{dy}{dx} = -\frac{2}{3}.$$

This procedure is known as differentiating *implicitly*.

The equation

$$x^2 + y^2 = a^2 \tag{3}$$

has elements in common with (1), though the relation $\{(x, y): x^2 + y^2 = a^2\}$ is not a function because there are two values of y corresponding to each $x \in (-a, a)$. We can, however, solve (3) for y, obtaining

$$y = \sqrt{a^2 - x^2} \tag{3a}$$

and

$$y = -\sqrt{a^2 - x^2}, \tag{3b}$$

and each of these equations does define y as a function of x. We may, however, find the derivative dy/dx for *each* of these functions directly from (3). As with (1), we differentiate implicitly:

$$\frac{d}{dx}(x^2) + \frac{d}{dx}(y^2) = \frac{d}{dx}(a^2). \tag{4}$$

The first and last terms of (4) are easily differentiated, but the second is a little more complicated, in that it requires use of the chain rule (in the special form, Rule 8). If y is defined implicitly as a function of x, then y^2 is, in turn, a (composite) function of x, and we therefore have

$$\frac{d}{dx}(y^2) = \frac{d}{dy}(y^2) \cdot \frac{dy}{dx} = 2y\frac{dy}{dx}.$$

Hence (4) becomes

$$2x + 2y\frac{dy}{dx} = 0, \tag{4a}$$

or
$$\frac{dy}{dx} = -\frac{x}{y}. \tag{5}$$

The same result can be obtained, with considerably more labor, from the explicit forms (3a) and (3b):

If
$$y = \sqrt{a^2 - x^2}, \tag{3a}$$

then
$$dy = \tfrac{1}{2}(a^2 - x^2)^{-1/2}\, d(a^2 - x^2) = \tfrac{1}{2}(a^2 - x^2)^{-1/2}(-2x\, dx)$$
$$= \frac{-x}{\sqrt{a^2 - x^2}}\, dx = -\frac{x}{y}\, dx,$$

and if
$$y = -\sqrt{a^2 - x^2}, \tag{3b}$$

then
$$dy = -\tfrac{1}{2}(a^2 - x^2)^{-1/2}\, d(a^2 - x^2) = -\tfrac{1}{2}(a^2 - x^2)^{-1/2}(-2x\, dx)$$
$$= \frac{x}{\sqrt{a^2 - x^2}}\, dx = -\frac{x}{y}\, dx.$$

Thus the single formula (5) gives, for the appropriate choice of y, the derivative of each of the functions described by (3a) and (3b). The implicit method clearly gives a much greater return for the effort invested.

The equation
$$x^5 + x^2y^3 + y^5 = 43, \tag{6}$$

like the two cases just considered, defines y implicitly as one or more functions of x, in that there is at least one function $\{(x, y)\}$ each of whose pairs satisfies (6). In this case, however, it is impossible to solve explicitly for y, and the method of implicit differentiation is not merely the most convenient method for finding dy/dx, it is the only method. Applying it, we have

$$\frac{d}{dx}(x^5) + \frac{d}{dx}(x^2y^3) + \frac{d}{dx}(y^5) = \frac{d}{dx}(43). \tag{7}$$

The first term, $d(x^5)/dx = 5x^4$, is elementary. The second term, $d(x^2y^3)/dx$, must be treated as a product, using Rule 5:

$$\frac{d}{dx}(x^2y^3) = x^2\frac{d}{dx}y^3 + y^3\frac{d}{dx}(x^2) = x^2\left(3y^2\frac{dy}{dx}\right) + y^3(2x)$$

since
$$\frac{d}{dx}(y^3) = \frac{d}{dy}(y^3)\cdot\frac{dy}{dx} = 3y^2\frac{dy}{dx}$$

by the chain rule. Also by the chain rule, the third term is

$$\frac{d}{dx}(y^5) = \frac{d}{dy}(y^5) \cdot \frac{dy}{dx} = 5y^4 \frac{dy}{dx}.$$

The last term, $d(43)/dx = 0$, offers no difficulty. Hence (7) becomes

$$5x^4 + 2xy^3 + 3x^2y^2 \frac{dy}{dx} + 5y^4 \frac{dy}{dx} = 0, \tag{7a}$$

or

$$\frac{dy}{dx} = -\frac{5x^4 + 2xy^3}{3x^2y^2 + 5y^4}.$$

It should, with practice, be possible to write the form (7a) directly from (6), without writing any intervening steps.

Equation (1) not only defines y as a function f of x, it also defines x as a function of y, and this function is, in fact, the inverse of f, $f^{-1} = \{(y, x): 2x + 3y - 6 = 0\}$ (§ 2.5). Just as $y = f(x)$, so $x = f^{-1}(y)$, and just as the derivative of f is given by the formula dy/dx, so the derivative of f^{-1} is given by a formula dx/dy. We can solve (1) for x in terms of y,

$$x = 3 - \tfrac{3}{2}y, \tag{1b}$$

and then find dx/dy from this explicit form, but it is also possible to differentiate (1) implicitly with respect to y, obtaining

$$2\frac{dx}{dy} + 3 - 0 = 0, \tag{2b}$$

or

$$\frac{dx}{dy} = -\frac{3}{2}.$$

In the same way, we may regard (3) as defining x as one or more functions of y. Differentiating implicitly with respect to y, we find

$$2x\frac{dx}{dy} + 2y = 0, \tag{4b}$$

or

$$\frac{dx}{dy} = -\frac{y}{x}.$$

Similarly, differentiating (6) with respect to y, we obtain

$$5x^4 \frac{dx}{dy} + 3x^2y^2 + 2xy^3 \frac{dx}{dy} + 5y^4 = 0, \tag{7b}$$

or

$$\frac{dx}{dy} = -\frac{3x^2y^2 + 5y^4}{5x^4 + 2xy^3}.$$

If equation (2a) is multiplied by the differential dx, we find

$$2\,dx + 3\frac{dy}{dx}\,dx = 0. \tag{2c}$$

But by the definition of § 6.1 we know that $\frac{dy}{dx}\,dx = dy$, and equation (2c) is therefore equivalent to

$$2\,dx + 3\,dy = 0. \tag{2d}$$

Similarly, multiplying (2b) by dy gives the same result, since the roles of x and y are interchanged in (2a) and (2b). In many cases, (2d) is the most convenient form in which to have our information about the derivatives of the variables x and y (strictly, the derivatives of the functions f and f^{-1}) in (1). Like (1), (2d) can be interpreted to mean either that x is a function of y or that y is a function of x, and the corresponding derivatives can be found from (2d) by division: by dx, to find dy/dx, or by dy, to find dx/dy. Similar forms can be obtained in the other two cases we have considered here. If (4a) is multiplied by dx, or (4b) by dy, we get

$$2x\,dx + 2y\,dy = 0,$$

or

$$x\,dx + y\,dy = 0, \tag{4c}$$

and (7a) or (7b) will similarly yield

$$5x^4\,dx + 3x^2y^2\,dy + 2xy^3\,dx + 5y^4\,dy = 0. \tag{7c}$$

The forms (2d), (4c), and (7c), involving differentials instead of derivatives, can be obtained directly from the original equations, and they have much in the way of convenience to recommend them.

Second derivatives. If y is an implicitly defined function of x, then the derivative dy/dx will usually contain both x and y. In finding the second derivative d^2y/dx^2, and derivatives of higher order, it is then important to remember that they are obtained by differentiating with respect to x, and that wherever y is encountered in this differentiation there will result a term containing the first derivative dy/dx. The value of dy/dx already obtained must then be introduced, as the examples will show.

It is also important to bear in mind that the variables x and y which are used must satisfy the original equation. This fact can often be used to effect simplifications, as the examples will also show. Note also that a value of x and a value of y must both be given before we can evaluate a derivative of any order.

Example 1. If $x^2 + y^2 = a^2$, find $\dfrac{d^2y}{dx^2}$ and $\dfrac{d^3y}{dx^3}$.

Solution: Returning to equation (5), we have $\dfrac{dy}{dx} = -\dfrac{x}{y}$. Using the quotient rule, we find

$$\frac{d^2y}{dx^2} = \frac{d}{dx}\left(-\frac{x}{y}\right) = -\frac{y - x\dfrac{dy}{dx}}{y^2} = -\frac{y - x\left(-\dfrac{x}{y}\right)}{y^2}$$

$$= -\frac{y^2 + x^2}{y^3} = -\frac{a^2}{y^3}$$

since $x^2 + y^2 = a^2$. Similarly,

$$\frac{d^3y}{dx^3} = \frac{d}{dx}\left(-\frac{a^2}{y^3}\right) = \frac{d}{dx}(-a^2y^{-3}) = 3a^2y^{-4}\frac{dy}{dx} = \frac{3a^2}{y^4}\left(-\frac{x}{y}\right) = -\frac{3a^2x}{y^5}.$$

Example 2. If $x^2 + xy + y^2 = 5$, find $\dfrac{d^2y}{dx^2}$.

Solution: Differentiating implicitly, we find $2x\,dx + x\,dy + y\,dx + 2y\,dy = 0$, and therefore $\dfrac{dy}{dx} = -\dfrac{2x + y}{x + 2y}$.

Hence

$$\frac{d^2y}{dx^2} = -\frac{(x + 2y)\left(2 + \dfrac{dy}{dx}\right) - (2x + y)\left(1 + 2\dfrac{dy}{dx}\right)}{(x + 2y)^2}$$

$$= -\frac{(x + 2y)\left(2 - \dfrac{2x + y}{x + 2y}\right) - (2x + y)\left(1 - 2\dfrac{2x + y}{x + 2y}\right)}{(x + 2y)^2}$$

$$= -\frac{(x + 2y)(3y) - (2x + y)(-3x)}{(x + 2y)^3} = -\frac{6(x^2 + xy + y^2)}{(x + 2y)^3}$$

$$= -\frac{30}{(x + 2y)^3}.$$

Example 3. We have proved that $d(x^n)/dx = nx^{n-1}$ for any integer n. (See Exercises 6.2 [B], Ex. 1.) The methods of the present section can be used to extend this result to the case of any rational exponent n. We let $y = x^{p/q}$, where p and q are integers, $q \neq 0$. Then $y^q = x^p$,

and
$$\frac{d}{dx}(y^q) = \frac{d}{dx}(x^p),$$

or
$$qy^{q-1}\frac{dy}{dx} = px^{p-1}.$$

Multiplying by y gives
$$qy^q\frac{dy}{dx} = pyx^{p-1};$$

substituting x^p for y^q,
$$qx^p\frac{dy}{dx} = pyx^{p-1};$$

dividing by qx^p,
$$\frac{dy}{dx} = \frac{p}{q}\cdot\frac{y}{x} = \frac{p}{q}x^{(p/q)-1}.$$

Exercises 6.4 [A]

1. Find $\dfrac{dy}{dx}$ if $y^2 - 8y + x^2 = 5$.
2. Find $\dfrac{dy}{dx}$ if $x^2 + 2y^2 = 7$.
3. Find $\dfrac{dy}{dx}$ if $x^3 + y^3 = a^3$.
4. Find $\dfrac{dy}{dx}$ if $x^2y^3 = 6$.
5. Find $\dfrac{dy}{dx}$ if $y\sqrt{x} + x\sqrt{y} = a\sqrt{a}$.
6. Find $\dfrac{d^2y}{dx^2}$ if $y^2 = 4x$.
7. Find $\dfrac{d^2y}{dx^2}$ if $3x^2 - 5y^2 = 7$.
8. Find an equation of the line tangent at $(-1, 2)$ to $x^2 + 3xy + 2y^2 - 2x + y - 7 = 0$.
9. (a) The derivative dy/dx is the trigonometric tangent of the inclination of the tangent line. Give a similar geometrical interpretation of the derivative dx/dy.

   (b) What is the inclination of the line tangent to a curve at a point where $dy/dx = 0$? where $dx/dy = 0$?

   (c) At what angles does $x^4 + y^4 = 1$ cross the coordinate axes?
10. Show that the hyperbolas $xy = \sqrt{2}$ and $x^2 - y^2 = 1$ are orthogonal.

Exercises 6.4 [B]

1. Find $\frac{dy}{dx}$ if $x^2y^2 = k^4$.

2. Find $\frac{dy}{dx}$ if $x^{1/2} + y^{1/2} = a^{1/2}$.

3. Find $\frac{dy}{dx}$ if $x^3 + y^3 - 3axy = 0$.

4. Find $\frac{dy}{dx}$ if $x^3 + x^2y + xy^2 + y^3 = 10$.

5. Find $\frac{dy}{dx}$ if $y^2 = \frac{x^2}{x - 4}$.

6. Find $\frac{d^2y}{dx^2}$ if $ay^2 = x^3$.

7. Find $\frac{d^2y}{dx^2}$ if $x^2 + 3y^2 - 2x = 7$.

8. Find the slope of $x^2 - 2xy - 3y^2 = 5$ at the points where $x = 2$.

9. Find the points on the graph of $x^2 + 4xy + 13y^2 = 9$ at which the tangent is horizontal, and determine the nature of each.

10. Show that the ellipse $x^2 + 10y^2 = 10$ and the hyperbola $x^2 - 8y^2 = 8$ are orthogonal.

6.5 Circular functions

To differentiate the function $f = \{(x, y): y = \sin x\}$, we return to first principles and compute

$$\frac{\Delta y}{\Delta x} = \frac{f(x + \Delta x) - f(x)}{\Delta x} = \frac{\sin (x + \Delta x) - \sin x}{\Delta x}.$$

If we use equation (16) of § 2.3, this becomes

$$\frac{\Delta y}{\Delta x} = \frac{2 \cos \left(x + \frac{\Delta x}{2}\right) \sin \frac{\Delta x}{2}}{\Delta x} = \cos \left(x + \frac{\Delta x}{2}\right) \frac{\sin \frac{\Delta x}{2}}{\frac{\Delta x}{2}}.$$

Now $$\frac{dy}{dx} = \lim_{\Delta x \to 0} \cos\left(x + \frac{\Delta x}{2}\right) \cdot \lim_{\Delta x \to 0} \frac{\sin \frac{\Delta x}{2}}{\frac{\Delta x}{2}} = \cos x \cdot \lim_{\Delta x \to 0} \frac{\sin \frac{\Delta x}{2}}{\frac{\Delta x}{2}},$$

if the last limit exists, since $\cos\left(x + \frac{\Delta x}{2}\right)$ is a continuous function of Δx. We are thus led to examine the limit

$$\lim_{\Delta x \to 0} \frac{\sin \frac{\Delta x}{2}}{\frac{\Delta x}{2}} = \lim_{\theta \to 0} \frac{\sin \theta}{\theta},$$

where the substitution $\theta = \frac{\Delta x}{2}$ has been made for convenience.

To evaluate this limit, we turn to a geometrical argument. The figure

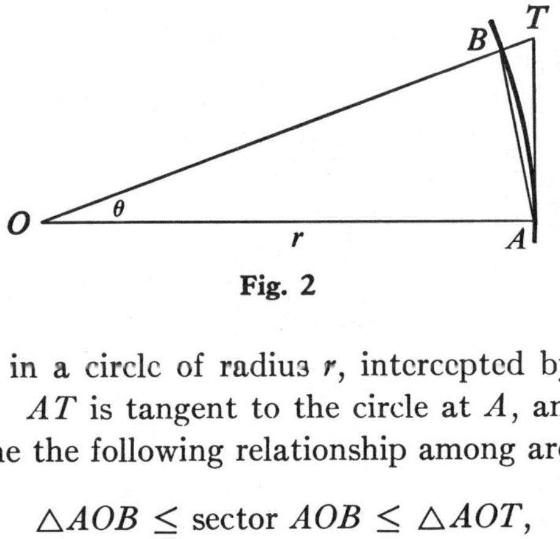

Fig. 2

shows an arc AB, in a circle of radius r, intercepted by an acute central angle of θ *radians*. AT is tangent to the circle at A, and therefore $AT = r \tan \theta$. We assume the following relationship among areas:

$$\triangle AOB \leq \text{sector } AOB \leq \triangle AOT,$$

or, introducing the appropriate area formulas,

$$\tfrac{1}{2}r^2 \sin \theta \leq \tfrac{1}{2}r^2\theta \leq \tfrac{1}{2}r^2 \tan \theta,$$

and dividing this inequality by the positive quantity $\frac{1}{2}r^2$, we have $\sin \theta \leq \theta \leq \tan \theta$. Dividing by $\sin \theta$, we then have

$$1 \leq \frac{\theta}{\sin \theta} \leq \frac{1}{\cos \theta}, \quad \text{or} \quad 1 \geq \frac{\sin \theta}{\theta} \geq \cos \theta,$$

a relationship which, though derived for $\pi/2 > \theta > 0$, is also valid for

$-\pi/2 < \theta < 0$. As $\theta \to 0$, the right and left members of this relationship have the common limit 1; it therefore follows by Theorem 5.3-5 that the middle element has the same limit, that

$$\lim_{\theta \to 0} \frac{\sin \theta}{\theta} = 1.$$

If we introduce this important result in tne expression for $\frac{dy}{dx}$, we have

▶ **Rule 9.** $\frac{d}{dx}(\sin x) = \cos x,$

or, introducing differentials,

▶ **Rule 9a.** $d(\sin x) = \cos x \, dx.$

Note that the formula for the area of sector AOB, and therefore this result, depend on our measuring all angles in radians. If we measure x in degrees, we have

$$\frac{d}{dx}(\sin x^\circ) = \frac{\pi}{180} \cos x^\circ.$$

It is the awkwardness of this form, as compared to Rule 9, that leads to the practice, universal in advanced mathematics, of always measuring angles in radians.

If $y = \sin u$, where u is a function of x, then the chain rule and Rule 9 give $\frac{dy}{dx} = \cos u \frac{du}{dx}$, a result which can be used to find the derivative of $y = \cos x$, for if we set $u = \frac{\pi}{2} - x$, we have

$$y = \sin u = \sin\left(\frac{\pi}{2} - x\right) = \cos x,$$

and therefore $\frac{dy}{dx} = \cos u \frac{du}{dx} = \cos\left(\frac{\pi}{2} - x\right) \cdot (-1) = -\sin x,$

which gives us

▶ **Rule 10.** $\frac{d}{dx}(\cos x) = -\sin x.$ **Rule 10a.** $d(\cos x) = -\sin x \, dx.$

Since the other circular functions are defined in terms of sin and cos, the student should have no difficulty in establishing the rules for differentiating these functions:

▶ **Rule 11.** $\frac{d}{dx}(\tan x) = \sec^2 x.$ **Rule 11a.** $d(\tan x) = \sec^2 x\, dx.$

Rule 12. $\frac{d}{dx}(\cot x) = -\csc^2 x.$ **Rule 12a.** $d(\cot x) = -\csc^2 x\, dx.$

Rule 13. $\frac{d}{dx}(\sec x) = \sec x \tan x.$ **Rule 13a.** $d(\sec x) = \sec x \tan x\, dx.$

Rule 14. $\frac{d}{dx}(\csc x) = -\csc x \cot x.$ **Rule 14a.** $d(\csc x) = -\csc x \cot x\, dx.$

Since integration is the inverse of differentiation, each of these rules for differentiation implies a rule for integration. Thus, for example, $\int \sec^2 x\, dx = \tan x + C$. We shall not, however, make a separate list of these integration rules; the student who is thoroughly familiar with Rules 9–14 should be able to apply them backward as well as forward, in integration as well as in differentiation.

Example 1. Differentiate $y = \sin x \cos x$.

Solution:

METHOD 1. Using the product rule and Rules 9 and 10, we have

$$\frac{dy}{dx} = \sin x \frac{d(\cos x)}{dx} + \cos x \frac{d(\sin x)}{dx}$$

$$= \sin x(-\sin x) + \cos x(\cos x) = \cos^2 x - \sin^2 x = \cos 2x.$$

METHOD 2. $y = \sin x \cos x = \frac{1}{2} \sin 2x$, and therefore, by Rule 9 and the chain rule,

$$\frac{dy}{dx} = \frac{1}{2} \cos 2x \frac{d(2x)}{dx} = \frac{1}{2} \cos 2x \cdot 2 = \cos 2x.$$

In this section, we meet for the first time applications of the chain rule (Rule 7) other than the power formula, Rule 8. Thus, in deriving Rule 10, we used the substitution $u = (\pi/2) - x$, and in Example 1 we made implicit use of the substitution $u = 2x$; neither of these is of the form $u = x^n$. The chain rule is quite general, however, and can be combined with Rule 9, for example, to give us

$$\frac{d}{dx}(\sin u) = \cos u \frac{du}{dx}$$

if u is any differentiable function of x whatever.

Example 2. Differentiate $\sec(x^2 + 5)$.

Solution: This is of the form $\sec u$, where

$$u = x^2 + 5,$$

and therefore

$$du = 2x\,dx.$$

Combining Rule 13a with the chain rule, we have, in general,

$$d(\sec u) = \sec u \tan u\,du,$$

and in this particular instance

$$d(\sec(x^2 + 5)) = \sec(x^2 + 5)\tan(x^2 + 5)\cdot 2x\,dx.$$

You should not lean too heavily on the explicit substitution of an auxiliary variable, such as u in this case. You should recognize that

$$d(\sec(\quad)) = \sec(\quad)\tan(\quad)\,d(\quad)$$

is valid if the formula for any differentiable function whatever is used to fill all four empty parentheses. You should learn to think in terms of this pattern, not in terms of any particular letters.

Example 3. Differentiate $\sin\sqrt{3x + 1}$.

Solution: From Rule 9a and the chain rule, we get

$$d(\sin\sqrt{3x + 1}) = \cos\sqrt{3x + 1}\,d(\sqrt{3x + 1}), \tag{1}$$

but this is only the first step, the first link in the chain. Next, we find

$$d(\sqrt{3x + 1}) = \tfrac{1}{2}(3x + 1)^{-1/2}\,d(3x + 1), \tag{2}$$

the second link in the chain. Finally, we determine

$$d(3x + 1) = 3\,dx, \tag{3}$$

the third and last link in the chain. In actual practice, the student should not set down in (1) the factor $d(\sqrt{3x + 1})$, but should write, in its stead, the equivalent found in (2)—except that, in writing (2), he should write, not $d(3x + 1)$, but its equivalent $3\,dx$ from (3). The actual working of the problem should then look, *at the very first writing*, like this:

$$d(\sin\sqrt{3x + 1}) = \cos\sqrt{3x + 1}\cdot\tfrac{1}{2}(3x + 1)^{-1/2}\cdot 3\,dx.$$

This can now be simplified in a second step:

$$d(\sin\sqrt{3x + 1}) = \frac{3\cos\sqrt{3x + 1}}{2\sqrt{3x + 1}}\,dx.$$

The process is, in effect, one of peeling away successive layers, differentiating until we get to the innermost layer, x itself, and not until this point is reached does the process come to a stop.

Example 4. Differentiate $y = \tan \dfrac{a-x}{a+x}$.

Solution: $$\frac{dy}{dx} = \sec^2 \frac{a-x}{a+x} \cdot \frac{(a+x)(-1)-(a-x)(1)}{(a+x)^2}$$
$$= \frac{-2a}{(a+x)^2} \sec^2 \frac{a-x}{a+x}.$$

The methods of this section yield two more important differentiation formulas:

▶ **Rule 15.** $\dfrac{d}{dx}(\sin^{-1} x) = \dfrac{1}{\sqrt{1-x^2}}$.

▶ **Rule 15a.** $d(\sin^{-1} x) = \dfrac{dx}{\sqrt{1-x^2}}$.

▶ **Rule 16.** $\dfrac{d}{dx}(\tan^{-1} x) = \dfrac{1}{1+x^2}$.

▶ **Rule 16a.** $d(\tan^{-1} x) = \dfrac{dx}{1+x^2}$.

Proofs of these rules are left to the student as exercises.

Exercises 6.5 [A]

1. Derive Rules 11–14 by expressing the function involved in terms of sines, cosines, or both, and then differentiating by the preceding rules.

2. Find dy/dx if

(a) $y = \cos 3x$.
(b) $y = sec\ (x - s)$.
(c) $y = \tan kx$.
(d) $y = \sin \dfrac{a}{x}$.
(e) $y = \cot (3x + 10)$.
(f) $y = \cos (7 - 4x)$.
(g) $y = 3 \sin^2 x$.
(h) $y = \sqrt{\tan x}$.
(i) $y = \tan \sqrt{x}$.
(j) $y = \tan^2 \sqrt{x}$.
(k) $y = x \sin x$.
(l) $y = \cos x \cos 2x$.
(m) $y = \dfrac{x}{\cos x}$.
(n) $y = x^2 \cos (4x - 3)$.
(o) $y = \sqrt{1 - \sin 5x}$.
(p) $\cos (x + y) = \frac{1}{2}\sqrt{3}$.
(q) $\tan xy = x^2$.
(r) $x \sin y = y \sin x + 1$.

3. Find the limit, as $x \to 0$, of

(a) $\dfrac{\tan x}{x}$. (b) $\dfrac{\sin 2x}{x}$. (c) $\dfrac{1 - \cos x}{x}$.

4. Find the abscissas of all maxima, minima, and inflection points, $0 \le x < 2\pi$:

(a) $y = \sin x - \cos x$. (b) $y = \sin^2 x - 2 \cos x$.

5. Prove Rule 15. (*Hint:* Let $y = \sin^{-1}x$; then $x = \sin y$. Differentiate implicitly, and eliminate y from the result.)

6. Find the area between the x-axis and one arch of the curve $y = \sin x$.

7. Show that an isosceles triangle with fixed legs has maximum area when it is a right triangle.

8. Find the angle between $y = \tan x$ and $y = \cot x$.

9. Use L'Hôpital's Rule to find the limit.

(a) $\lim\limits_{x\to 0} \dfrac{\tan^{-1} x}{x}$ (b) $\lim\limits_{x\to 0} \dfrac{\sin ax}{\sin bx}$ (c) $\lim\limits_{x\to 0} \dfrac{x - \sin x}{x^3}$

Exercises 6.5 [B]

1. Find dy/dx if

(a) $y = 3 \tan 2x$.
(b) $y = \cos(\pi - x)$.
(c) $y = \csc(5x - 3)$.
(d) $y = 4 \cot \dfrac{x}{2}$.
(e) $y = \sec^2 \dfrac{x-3}{4}$.
(f) $y = \sin(x + x^{-1})$.
(g) $y = \cos\sqrt{a - x}$.
(h) $y = \sqrt{1 + \cos 2x}$.
(i) $y = \dfrac{\tan x}{x}$.
(j) $y = \sin(x - k)\cos(x + k)$.
(k) $\sin 2x = 2 \sin y$.
(l) $\dfrac{\tan x + \tan y}{1 - \tan x \tan y} = 1$.

2. Find the second derivative of

(a) $\sec\theta$. (b) $\sin 3x$. (c) $t \cos t$. (d) $\tan 2\phi$.

3. Find the limit, as $x \to 0$, of

(a) $\dfrac{\sin 3x}{x}$. (b) $\dfrac{\sin^2 x}{x}$. (c) $\cot x \sin x$. (d) $\dfrac{1 - \cos x}{x^2}$.

4. Find the abscissas of all maxima, minima, and inflection points, $0 \le x < 2\pi$:

(a) $y = 3 \sin x + 2 \cos x$. (b) $y = 2 \sec x - \tan x$.

5. Find the area enclosed by the curves $y = \sin x$ and $y = \cos x$, between two intersections of those curves.

6. Show that $f(x) = x - \sin x$ is an increasing function for all values of x.

7. Prove Rule 16. (See the *Hint* for Ex. 5 of Exercises 6.5 [A].)

8. Express the volume of a cone in terms of the slant height e and the semivertical angle θ, and find the value of θ for which this volume is a maximum, if e is constant (Fig. 3 shows an axial section).

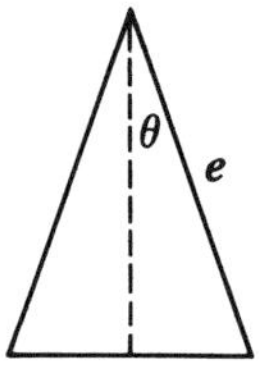

Fig. 3

9. A particle moves along a line in such a way that its distance s from a fixed pont O of the line is $s = A \sin kt$ at time t. Show that the acceleration is proportional to the distance of the point from O and is directed toward O.

10. Use L'Hôpital's Rule to find the limit.

(a) $\lim_{x \to 0} \frac{\sin^{-1} x}{x}$ (b) $\lim_{x \to 0} \frac{\sec x - 1}{x^2}$ (c) $\lim_{x \to 0} \frac{\tan x - x}{x^3}$

Chapter Review

1. If $y = x^2 + 3$, find dy, Δy, and $\Delta y/dy$ when $x = 1$ and $dx =$ (a) 10; (b) -1; (c) 0.1.

2. Find dy if

(a) $y = (x - 1)^2(x + 2)^3$.

(b) $y = x\sqrt{x^2 + 9}$.

(c) $y = \csc(3x - 4)$.

(d) $y = x \cot 2x$.

(e) $y = \frac{x^2}{x - 3}$.

(f) $y = \sin^3 x^2$.

(g) $y = \frac{x^2}{\sqrt{x^2 - x + 1}}$.

(h) $y = \frac{\sin^2 x}{1 - \cos x}$.

3. Find an equation of the tangent to $2x^2 + y^2 = 12$ at $(2, 2)$.

4. If $ay^3 = x^4$, show that $\frac{d^2y}{dx^2} = \frac{4y}{9x^2}$.

5. The region bounded by $y = \sec x$, the x-axis, the y-axis, and the line $x = \pi/4$ is rotated about the x-axis. Find the volume of the solid thus generated.

6. Evaluate: (a) $\lim_{x \to 0} \frac{\sec x - 1}{\sin x}$ (b) $\lim_{x \to 0} \frac{\sin x}{1 - \cos x}$

7

Applications of the Derivative and the Differential

7.1 Related rates

When the surface of a pond is disturbed by the fall of a stone, a circular pattern of ripples is formed, a circle which grows, ever widening, until it strikes the shore or dies out. As the ripples move over the surface of the pond, the radius R of the circle, the diameter D, the circumference C, and the area A all increase; each is a function of the time t. Not only are these quantities functions of the time, but their values are also interconnected, at each instant, by familiar relationships: $D = 2R$, $C = 2\pi R$, $A = \pi R^2$, and so on. Hence their rates of change are also related to one another, and through these relationships we can find the rate of change of each from the rate of any one.

Suppose, for example, that the ripples move at the rate of 10 in./sec.; this means that the radius of the circle is increasing at the rate of 10 in./sec., or, in the notation of calculus,

$$\frac{dR}{dt} = 10.$$

We can then use the chain rule to find the rate dC/dt at which the circumference is changing. We have

$$\frac{dC}{dt} = \frac{dC}{dR} \cdot \frac{dR}{dt},$$

and since
$$\frac{dC}{dR} = 2\pi \quad \text{and} \quad \frac{dR}{dt} = 10,$$

this gives
$$\frac{dC}{dt} = (2\pi)(10) = 20\pi.$$

Similarly, to find the rate dA/dt at which the area increases, we have

$$\frac{dA}{dt} = \frac{dA}{dR} \cdot \frac{dR}{dt} = (2\pi R)(10) = 20\pi R.$$

Note that, because the circumference C is a linear function of the radius R, the derivative dC/dR is constant, and the rate at which the circumference changes, dC/dt, is therefore a constant multiple (2π) of the rate at which the radius changes. On the other hand, A is a nonlinear function of R, the derivative $dA/dR = 2\pi R$ has different values for different values of R, and the rate dA/dt is therefore a variable multiple of dR/dt. In our example, when the radius is 6 in., we have $dA/dt = 120\pi$ sq. in./sec., but when the radius is 10 ft., $dA/dt = 2400\pi$ sq. in./sec.; the bigger the circle, the faster its area increases.

There are many situations, like the preceding example, in which the rate at which one physical quantity changes is related, by some constraint, to the rates at which others change, and in which, knowing the rate of change of one, we can use the chain rule to determine the rates of change of the others. The following examples exhibit some of these.

Example 1. A 12-foot ladder stands upright against a vertical wall. If the lower end of the ladder is pulled away from the wall at the rate of 2 ft./sec., how fast is the top of the ladder coming down the wall at the instant it is 6 ft. above the ground?

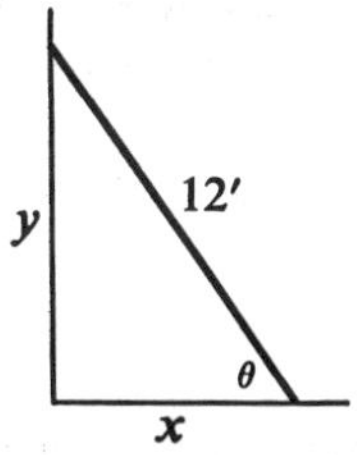

Fig. 1

Solution: If we denote by x feet and y feet, respectively, the distances of the foot and the top of the ladder from the bottom of the wall, then, for any values of x and y, we must have

$$x^2 + y^2 = 144. \tag{1}$$

Let us measure time from the beginning of the motion; it then follows that, at any time t,

$$x = 2t. \tag{2}$$

Using (1) and (2), we find

$$y = \sqrt{144 - 4t^2}. \tag{3}$$

Hence
$$\frac{dy}{dt} = \frac{1}{2}(144 - 4t^2)^{-1/2}(-8t),$$

or
$$\frac{dy}{dt} = \frac{-2t}{\sqrt{36 - t^2}}. \tag{4}$$

Setting $y = 6$ in (3) gives $t = 3\sqrt{3}$. Substituting this result in (4), we find

$$\frac{dy}{dt} = -2\sqrt{3}.$$

Hence, at the specified instant, the top of the ladder is sliding down the wall with a velocity of $2\sqrt{3}$ ft./sec.

The technique of implicit differentiation makes possible a simpler and more elegant solution of this problem. We know from the physics of the problem that x and y must be functions of the time t. (As a matter of fact, we have exhibited possible forms of these functional relationships in (2) and (3), but we do not need all this detail.) We can therefore differentiate (1) implicitly with respect to t, obtaining

$$2x\frac{dx}{dt} + 2y\frac{dy}{dt} = 0, \tag{5}$$

a relationship which is valid at any time during the course of the motion. We also know that $dx/dt = 2$, and when $y = 6$ we find $x = 6\sqrt{3}$ from (1). Substituting all this in (5), we find

$$\frac{dy}{dt} = -2\sqrt{3}$$

as before.

Example 2. In the preceding problem, find how fast the angle θ between the ladder and the ground is changing when $y = 6$.

Solution: We need a general relationship, valid at any point of the motion, between θ and x or y. There are several, but perhaps the simplest and most useful is

$$x = 12 \cos \theta. \tag{6}$$

Since x and θ are functions of t, it makes sense to differentiate (6) implicitly with respect to t, obtaining

$$\frac{dx}{dt} = -12 \sin \theta \frac{d\theta}{dt}.$$

We know $dx/dt = 2$ at any point of the motion, and when $y = 6$ we find $\sin \theta = \frac{1}{2}$ and therefore

$$\frac{d\theta}{dt} = -\frac{1}{3};$$

at this instant, the angle is decreasing at $\frac{1}{3}$ rad./sec.

Example 3. A railroad which runs north and south is crossed at right angles by a highway. At noon, a car is 20 mi. west of the crossing, heading east at 40 mi./hr., and a train is 10 mi. north of the crossing, heading south at 60 mi./hr. Find **(a)** when the car and train are closest to one another; **(b)** how far apart they are at that time; and **(c)** how fast the distance between them is increasing at 1 o'clock.

Solution: Let s be the distance in miles between the car and the train; then the first thing we seek to do is to find a general relationship describing s as a function of the time. To do this, we note that at t hours past noon the car will be $(40t - 20)$ miles east of the crossing and the train will be $(60t - 10)$ miles south of the crossing. (Figure 2 illustrates this relationship.)

Hence

$$s^2 = (40t - 20)^2 + (60t - 10)^2 = 5200t^2 - 2800t + 500 \tag{7}$$

and

$$s = 10\sqrt{52t^2 - 28t + 5}. \tag{8}$$

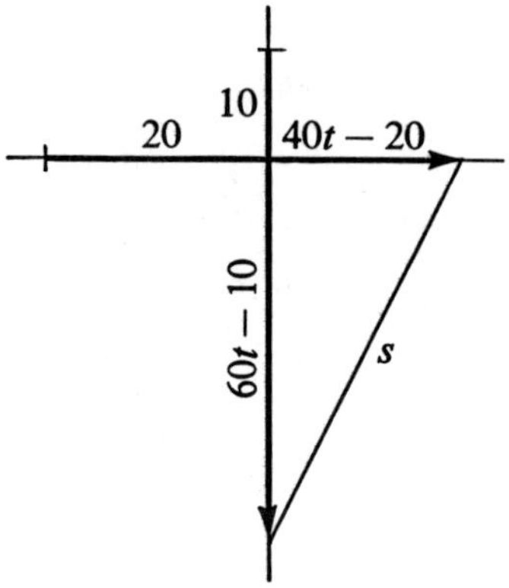

Fig. 2

We must now find the derivative ds/dt. This can be done by differentiating (8), but it is easier to avoid the radical by differentiating (7) implicitly:

$$2s\,ds = (10400t - 2800)dt;$$

$$\frac{ds}{dt} = \frac{200(26t - 7)}{s}. \tag{9}$$

We note that the only zero of ds/dt is $t = 7/26$, and any of the tests of Chapter 3 makes it clear that this corresponds to a minimum. Hence the answer to (**a**) is 7/26 hr. past noon, or approximately 12:16 P.M. For (**b**), we substitute $t = 7/26$ in (8), obtaining $40\sqrt{13}/13$, or approximately 11.1 miles, and for (**c**) we put $t = 1$ in (8) and (9), getting $ds/dt = 3800/10\sqrt{29}$, or approximately 70.6 mi./hr.

Example 4. A conical reservoir (shown in axial section in Fig. 3), 12 ft.

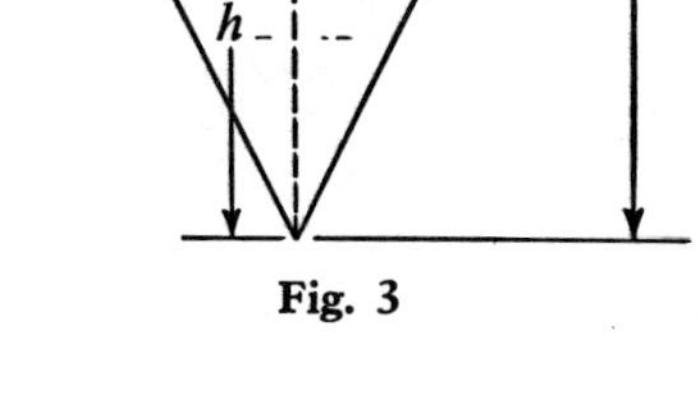

Fig. 3

deep and also 12 ft. across the top, is being filled with water at the rate of 5 cu. ft./min. How fast is the water rising when it is 4 ft. deep?

Solution: If the volume of water in the cone is V cu. ft. and the depth is h ft., we are told that $dV/dt = 5$ and are asked to find dh/dt when $h = 4$. We must find some relationship, valid at any time, between h and V. Such a relationship is implied by the formula for the conical volume of the water,

$$V = \tfrac{1}{3}\pi r^2 h. \tag{10}$$

By similar triangles, $r/h = 6/12$, or $r = h/2$; this permits us to eliminate r, in which we are not interested, in favor of h:

$$V = \tfrac{1}{12}\pi h^3.$$

Hence
$$\frac{dV}{dt} = \frac{1}{4}\pi h^2 \frac{dh}{dt},$$

and substituting 5 for dV/dt and 4 for h, we find $dh/dt = 5/4\pi$, or approximately 0.39 ft./min.

The procedure used in the preceding examples can be summarized thus:

▶ **1. State as an equation a relationship, *valid at any time,* among the variables involved.**

2. Differentiate with respect to time, thus obtaining a relationship among the *rates* which is valid at any time.

3. Now, and not before this, introduce the specific values which the variables take at the instant in question.

To introduce specific values of the variables before differentiating will produce a static situation, one in which no change is possible, and in which, therefore, all derivatives are identically zero.

If y is given as a function of x, and x as a function of t, the chain rule can be used, in the form

▶
$$\frac{dy}{dt} = \frac{dy}{dx} \cdot \frac{dx}{dt}, \tag{11}$$

to find the derivative dy/dt without expressing y as a function of t. This principle was illustrated in Example 5 of § 6.3. On the other hand, it may be that x and y are known as functions of t: in such a case, (11) can be used in the form

▶
$$\frac{dy}{dx} = \frac{dy/dt}{dx/dt} \tag{12}$$

to find the derivative dy/dx (generally in terms of t) without first going to the trouble of expressing y in terms of x. This procedure is illustrated in part (a) of the following example.

Example 5. If a gun, elevated at an angle E, fires a shell with a muzzle velocity of k ft./sec., then t seconds later the vertical distance of the shell from the gun, in feet, is given by

$$y = kt \sin E - 16t^2,$$

and the horizontal distance by

$$x = kt \cos E.$$

Find (**a**) dy/dx in terms of t; (**b**) the maximum height of the shell; (**c**) the range of the shell; and (**d**) the value of E for maximum range.

Solution: (**a**) We find $\frac{dy}{dt} = k \sin E - 32t$ and $\frac{dx}{dt} = k \cos E$; hence, by (12),

$$\frac{dy}{dx} = \frac{k \sin E - 32t}{k \cos E} = \tan E - \frac{32 \sec E}{k} t.$$

(**b**) To find the maximum height, we may solve either $\frac{dy}{dt} = 0$ or $\frac{dy}{dx} = 0$. In either case

$$t = \frac{k \sin E}{32},$$

and therefore the maximum value of y is given by

$$y = \frac{k^2 \sin^2 E}{64}.$$

(**c**) To find the range, we must determine where the shell returns to earth, that is, where $y = 0$.

Using $$kt \sin E - 16t^2 = 0,$$

we find $$= 0 \quad \text{or} \quad t = \frac{k \sin E}{16},$$

and the latter gives as the range $x = k\left(\frac{k \sin E}{16}\right) \cos E = \frac{k^2}{32} \sin 2E.$

(**d**) Since $\sin 2E$ has its maximum value of 1 when $2E = \pi/2$, or $E = \pi/4$, the maximum range will be $k^2/32$, when the elevation of the gun is 45°.

We see from (11) that if x and y are both changing with time, then the rate at which y changes, dy/dt, is dy/dx times the rate at which x changes, dx/dt: if $dy/dx = 10$, then y changes 10 times as fast as x; if $dy/dx = \frac{1}{2}$, then y changes half as fast as x; if $dy/dx = -3$, then y changes 3 times as fast as x and in the opposite sense, decreasing when x increases, increasing when x decreases. It is perhaps for this reason that dy/dx is sometimes called the *rate at which y changes with x* (see also § 2.8).

Example 6. If $y = x^3 - 3x$, find the rate at which y changes with x when $x = -1$ and when $x = 2$.

Solution: We find $\dfrac{dy}{dx} = 3x^2 - 3$. When $x = -1$, $\dfrac{dy}{dx} = 0$, and when $x = 2$, $\dfrac{dy}{dx} = 9$.

Example 7. Show that the rate at which the volume of a sphere changes with the radius is equal to the area of the sphere.

Solution: Since the volume V is given in terms of the radius R by the formula $V = \frac{4}{3}\pi R^3$, the required rate is, by definition, $\dfrac{dV}{dR} = 4\pi R^2$. Since the area $A = 4\pi R^2$, this is the required result. The student may be interested in giving a geometrical explanation of this result, and in comparing this with the relationship between the circumference of a circle and the rate at which the area changes with the radius, in the discussion at the beginning of this section.

Exercises 7.1 [A]

1. A rectangular cistern is 6 ft. long, 4 ft. wide, and 8 ft. deep.
   (a) When the water in the cistern is x ft. deep, find the volume V of water in the cistern.
   (b) If water is running into or out of the cistern, what mathematical symbols represent the rate at which the volume is changing and the rate at which the depth is changing?
   (c) Write an equation relating the two symbols found in (b).
   (d) If water is running in at 3 cu. ft./sec., how fast is the level of the water rising?

2. (a) If $A = 4\pi r^2$, write (i) the symbol that designates the rate at which A changes with the time t; (ii) the symbol that designates the rate at which r changes with t; and (iii) an equation relating these two symbols.
   (b) If A is increasing at the rate of 2 sq. ft./min., how fast is r changing when (i) $r = 5$? (ii) $r = 10$?

3. (a) If $y = f(x)$, and if x and y vary with the time t, write an equation relating dy/dt and dx/dt.
   (b) If $y = x^2 + 5x$ and $dx/dt = 3$, find dy/dt when $x = -1$.

4. Find the derivatives dx/dt and dy/dt from the given equations, and then use them to find dy/dx in terms of t:

   (a) $x = t^2,\ y = t - t^3$ (b) $x = a \cos t,\ y = b \sin t$

5. A highway crosses a railroad at right angles. A car going 45 mi./hr. crosses the track 10 min. before a train comes through at 60 mi./hr.
   (a) How far apart are the car and the train t min. after the car crosses the tracks?
   (b) When is this distance a minimum?
   (c) What is the minimum distance?

6. A ship is anchored 2 mi. off a straight shore, and its searchlight is following a car that is going along the shore at 40 mi./hr. How fast is the searchlight turning when the car is 4 mi. from the ship? (See Fig. 4.)

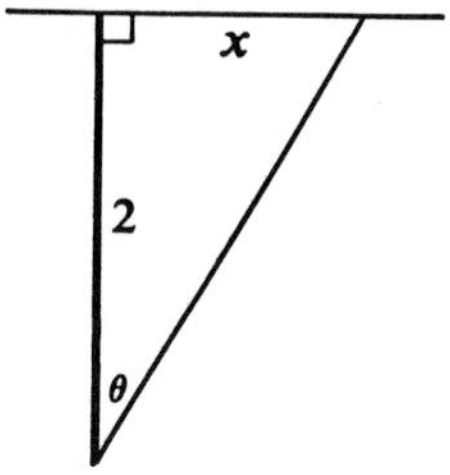

Fig. 4

Exercises 7.1 [B]

1. Find dy/dx in terms of t if
   (a) $x = a \sec (kt - n),\ y = b \tan (kt - n)$.
   (b) $x = \dfrac{1 - t^2}{1 + t^2},\ y = \dfrac{2t}{1 + t^2}$.

2. (a) Write a formula for the volume V of a right circular cylinder, in terms of its radius r and its altitude h.
   (b) If h is constant but r varies, find the rate at which V changes with r, and show that this rate is equal to the area of the curved surface of the cylinder.

3. A spherical balloon is being inflated.
   (a) If the volume of the balloon is V, the surface area is A, and the radius is r, explain the physical significance of the derivatives dV/dt, dA/dt, dr/dt.
   (b) What is the relationship between V and r? between A and r?
   (c) What is the relationship between dV/dt and dr/dt? between dA/dt and dr/dt?
   (d) If air is being blown into the balloon at the rate of 2 cu. in./sec., how fast is the radius increasing at the instant when it is 3 in.?
   (e) How fast is the area increasing at the same instant?

4. A conical container 5 cm. deep and 8 cm. across the top is being filled with water.

   (a) If the volume of water in the container is V, the depth of the water is h, and the radius of the surface of the water is r, write an equation relating V, h, and r.

   (b) What is the physical significance of the derivatives dV/dt, dr/dt, dh/dt?

   (c) Use the geometry of the configuration to obtain a relationship between r and h, and use this to eliminate r from the equation found in (a).

   (d) What is the relationship between dV/dt and dh/dt?

   (e) If the water is running in at the rate of 3 c.c./min., at what rate is the water rising at the instant when it is 4 cm. deep?

5. A boat is being pulled in to a wharf by a line that passes over the edge of the wharf and is made fast to the boat at a point that is 8 ft. below the edge of the wharf.

   (a) If there are z ft. of rope between the wharf and the boat at the time t when the horizontal distance between the boat and the wharf is x ft., explain the physical significance of the derivatives dz/dt and dx/dt.

   (b) Obtain an equation relating z and x, and use it to obtain an equation relating dz/dt and dx/dt.

   (c) If the rope is being pulled in at the rate of 24 ft./min., how fast is the boat moving when there are 17 ft. of rope still out?

   (d) At the same instant, how fast is the angle between the rope and the horizontal changing?

6. A bridge crosses a canal at right angles, 60 ft. above the surface of the water. A woman walking across the bridge at 5 ft./sec. passes directly over a boat passing through the canal at 12 ft./sec. How fast are the woman and the boat separating one minute later?

Miscellaneous Exercises 7.1

1. The altitude of a cone is equal to the radius. Find the rate at which the volume changes with the radius.

2. A light 2 miles off a straight shore rotates 4 times a minute. How fast is the beam of light moving along the shore when it makes an angle of $\pi/6$ *with the shore?*

3. Sand pours out of a spout at the rate of 2 cu. ft./sec. and forms a conical pile in which the height is always equal to the radius of the base. When the pile is 20 ft. high, (**a**) how fast is its height increasing? (**b**) how fast is the area which it covers increasing?

4. A kite 100 ft. high is being blown horizontally at 8 ft./sec. When there are 300 ft. of string out, (**a**) how fast is the string running out? (**b**) how fast is the angle between the string and the horizontal changing?

5. In triangle ABC, $AB = 8$ ft., $AC = 12$ ft., and $\angle A$ is increasing at the rate of 0.1 rad./sec. How fast is (**a**) the area of triangle ABC, and (**b**) the length of BC, changing when $\angle A = \pi/3$?

6. A lamp 15 ft. off the ground casts the shadow of a 6-foot man who is walking away from the lamp at 6 ft./sec. When the man is 30 ft. from the lamppost, (**a**) how fast is his shadow lengthening? (**b**) how fast is the tip of his shadow moving?

7. An airplane 10 mi. high passes overhead at 600 mi./hr. How fast is its angle of elevation changing 1 min. later?

8. Two roads meet at an angle of 60°. A girl starts from the intersection and walks down one at 3 mi./hr., and one hour later a second girl starts down the other at 4 mi./hr. How fast are they separating 2 hours after the second girl starts?

9. A light is 50 ft. above the ground. A stone is released at the same level as the light and 20 ft. away from it. How fast is the shadow of the stone moving along the ground 1 sec. later? (Take 32 ft./sec.² as the acceleration of gravity.)

10. Show that, when $x = \pi/3$, the rate at which $\tan x$ changes with x is 8 times the rate at which $\sin x$ changes.

11. The point $P(x, y)$ is tracing the curve $x^2 = 4y$. If $dx/dt = 2$ when P is at (4, 4), find, at that instant, (**a**) dy/dt; (**b**) the rate of turning of the line OP which joins P to the origin.

12. If $y = x^2 - 2x$, $z = x^3y$, and $dy/dt = -2$, find dz/dt when $x = 2$.

13. If x and y are both functions of t, and if $x^2 + y^2 = a^2$ and $V = \pi x^2 y$, find dV/dt when $y = 2a/3$ and $dy/dt = -2a$.

14. P is a point on the circle with center at the origin O and radius a. Q is the projection of P on the x-axis. If the radius OP is rotating about O at 2 rad./min., how fast is the area of triangle PQO changing when $\angle POQ = \pi/6$?

15. If $x^2 + y^2 = 25$ and x is increasing at 2 units/sec., how fast is dy/dx changing when $x = 3$, $y = -4$?

16. A ship with a constant speed of 12 knots sails for one hour on a heading of 000° and for a second hour on a heading of 060°. At the end of the second hour, how fast is the distance between the ship and its starting point increasing?

17. A light is on the ground, 25 ft. from a wall. A woman 5 ft. tall walks from the light toward the wall at 4 ft./sec. Describe the motion of her shadow on the wall when she is 15 ft. from the light.

7.2 The differential as an approximation

As was noted in § 6.1, the tangent to a curve may often be used, over a short interval, as an approximation to the curve itself. As the figure indi-

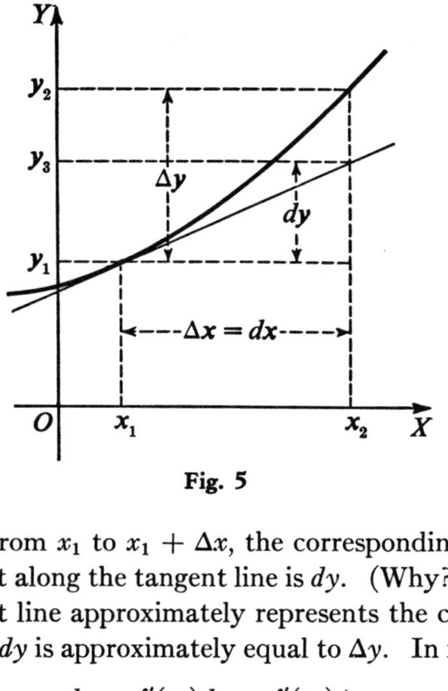

Fig. 5

cates, if x changes from x_1 to $x_1 + \Delta x$, the corresponding change along the curve is Δy, and that along the tangent line is dy. (Why? What is dy?) To say that the tangent line approximately represents the curve is then equivalent to saying that dy is approximately equal to Δy. In more precise terms, we have

$$dy = f'(x_1)dx = f'(x_1)\Delta x,$$

and if $dy \neq 0$, we may then write

$$\frac{\Delta y}{dy} = \frac{\Delta y}{f'(x_1)\Delta x} = \frac{1}{f'(x_1)} \cdot \frac{\Delta y}{\Delta x},$$

and therefore

$$\lim_{\Delta x \to 0} \frac{\Delta y}{dy} = \frac{1}{f'(x_1)} \cdot \lim_{\Delta x \to 0} \frac{\Delta y}{\Delta x} = \frac{1}{f'(x_1)} \cdot f'(x_1) = 1.$$

Hence, by making Δx small enough, we can make the ratio $\Delta y/dy$ arbitrarily close to 1, or dy arbitrarily close to Δy. This is the implication of the statement that dy is a "good" approximation to Δy if Δx is "small." In many cases, Δy is awkward to compute, dy is relatively easy, and the latter may then serve as a convenient approximation to the former. How small Δx (or dx) must be to make the approximation satisfactory depends on the circumstances, as will be noted in the following examples. Procedures for measuring the error are developed in more advanced courses.

Example 1. The radius of a sphere is 1 ft. If the radius is increased by 1 in., by about how much will the volume be increased?

Solution: The volume is given by the formula $V = \frac{4}{3}\pi r^3$. We then have $dV = 4\pi r^2\,dr$, and since we are given $r = 1$ ft., $dr = \frac{1}{12}$ ft., we find $dV = 4\pi(1)^2 \cdot \frac{1}{12} = \frac{1}{3}\pi$; the increase in volume is about $\frac{1}{3}\pi$ cu. ft.

The exact increase, ΔV, is $\Delta V = \frac{4}{3}\pi\left(\frac{13}{12}\right)^3 - \frac{4}{3}\pi(1)^3 = \frac{469}{432}\cdot\frac{\pi}{3}$, and the error involved in using dV as an approximation to ΔV is about 7.9%.

Example 2. Find $\sqrt{65}$, approximately.

Solution: If we take $y = f(x) = \sqrt{x}$, and introduce $x = 64$, $\Delta x = 1$, we have, exactly,

$$\Delta y = f(x + \Delta x) - f(x) = \sqrt{65} - \sqrt{64} = \sqrt{65} - 8,$$

and therefore

$$\sqrt{65} = 8 + \Delta y.$$

We now use dy as an approximation to the unknown Δy. Since $y = x^{1/2}$,

$$dy = \tfrac{1}{2}x^{-1/2}\,dx = \tfrac{1}{2}(64)^{-1/2}(1) = \tfrac{1}{16},$$

and therefore

$$\sqrt{65} = 8\tfrac{1}{16}, \text{ approximately.}$$

By tables, to seven significant figures, $\sqrt{65} = 8.062258$; our approximation, 8.062500, is in error by about 0.003% of the correct value.

In problems of this sort, it is important to choose suitable values of x and dx. In this case, for example, one might have taken $x = 63$, $dx = 2$, or $x = 81$, $dx = -16$. You should be sure you understand the disadvantages of each of these choices.

Example 3. If the measurement of an edge of a cube may be in error by 1%, what is the consequent relative error in the computed volume of the cube?

Solution: If the true length of the edge is e units, and the measurement is in error by Δe units, then the *relative* error is defined to be $\frac{\Delta e}{e}$; in this case, we are told that $\dfrac{\Delta e}{e} = 0.01$ and are asked to find $\dfrac{\Delta V}{V}$, where V represents the volume. We shall use dV as an approximation to ΔV. We have

$$V = e^3; \tag{1}$$

hence

$$dV = 3e^2\,de, \tag{2}$$

and dividing (2) by (1), we have $\dfrac{dV}{V} = \dfrac{3e^2\,de}{e^3} = 3\dfrac{de}{e}$.

The relative error in V is approximately 3 times the relative error in e; in this case, therefore, it is 3%.

An important application of the differential is to the approximation of irrational roots of equations. This process, which is known as *Newton's method*, is best explained through a specific example.

Example 4. Find a rational approximation, correct to four significant figures, to the real root r of

$$f(x) = x^3 + 3x^2 + 4x - 3 = 0.$$

Solution: We find $f(0) = -3$, $f(1) = 5$, and conclude that there is a zero of f between 0 and 1. We take $x_1 = 0$ as a first approximation to r. Now, as x changes from 0 to r by an unknown amount Δx, y changes from $f(0) = -3$ to $f(r) = 0$ by a known amount $\Delta y = 3$. Insofar as we may use $dy\ (= f'(x)dx)$ as an approximation to Δy, it is then true that

$$3 = f'(0)\Delta x, \quad \text{or} \quad \Delta x = \frac{3}{f'(0)}.$$

Since $f'(x) = 3x^2 + 6x + 4$, it then follows that $f'(0) = 4$ and therefore $\Delta x = \frac{3}{4}$. Hence we take, as a second approximation to r, $x_2 = x_1 + \Delta x = 0.75$.

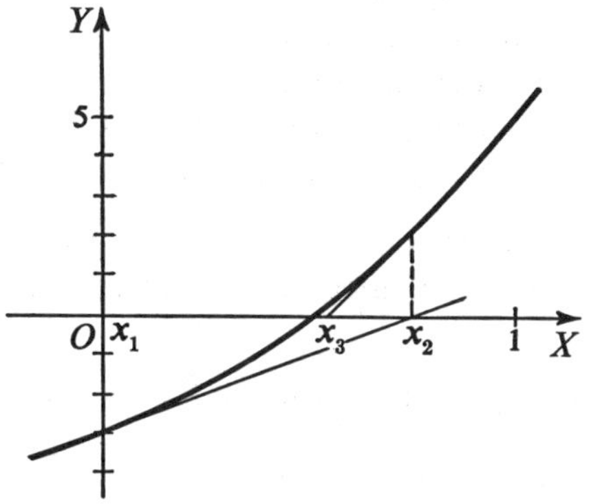

Fig. 6

We now repeat the process. We first compute $f(0.75) = 2.11$, approximately. As x changes from 0.75 to r by an unknown amount Δx, y changes from 2.11 to $f(r) = 0$ by an amount $\Delta y = -2.11$. Hence, approximately,

$$-2.11 = f'(0.75)\Delta x = 10.2 \cdot \Delta x, \quad \text{or} \quad \Delta x = -0.21,$$

and we therefore take $x_3 = x_2 + \Delta x = 0.75 - 0.21 = 0.54$ as a third approximation to r. Repeating the process yet again, we compute $f(0.54) = 0.192, f'(0.54) = 8.11$, $\Delta x = -0.192/8.11 = -0.024$, and hence $x_4 = 0.516$. Still another repetition yields $x_5 = 0.5160$, which is correct to the required four significant figures.

Let us now consider the general case. If x_n is an approximation to a zero r of a function f, we wish to estimate the error $\Delta x = r - x_n$ in this approximation. We know that the corresponding error in the value of f is $\Delta y = f(r) - f(x_n)$, and using $dy = f'(x_n)\Delta x$ as an approximation to Δy, we find

$$\Delta x = -f(x_n)/f'(x_n),$$

and we therefore have

$$x_{n+1} = x_n - [f(x_n)/f'(x_n)]$$

as our new (and generally improved) estimate of r.

Exercises 7.2 [A]

In the following exercises, use differentials to obtain approximate answers:

1. Find the approximate change in x^2 when x is changed (**a**) from 1 to 1.1; (**b**) from 10 to 10.1.
2. Estimate $\sqrt{27}$. (*Hint:* Let $y = \sqrt{x}$, find an expression for dy in terms of x and dx, and substitute $x = 25$, $dx = 2$.)
3. Estimate (**a**) $\sqrt{34}$; (**b**) $\sqrt[3]{65}$; (**c**) $(28)^{2/3}$.
4. Find the approximate increase in $\sin x$ per 0.01 increase in x, when (**a**) $x = 0$; (**b**) $x = \pi/4$; (**c**) $x = \pi/2$.
5. For what $x \in [0, 2\pi)$ does a given error in x produce the greatest error in $\cos x$?
6. The outside dimensions of a cubical box are $20 \times 20 \times 20$ cm., and the thickness of the walls of the box is 0.15 cm. What is the approximate inner volume of the box?
7. Use Newton's method to find the positive root of $x^3 + 6x - 13 = 0$ correct to the nearest 0.01. How many negative roots has this equation?

Exercises 7.2 [B]

In the following exercises, use differentials to obtain approximate answers:

1. Estimate
   (a) $(12.1)^3$ (b) $\sqrt[4]{79}$ (c) $\sqrt[3]{123}$
2. Express in terms of the inner radius r and the width Δr the approximate area of an annulus, or circular ring.
3. Find the approximate increase in $\tan x$ per 1° increase in x when
   (a) $x = 0$; (b) $x = \pi/4$; (c) $x = 3\pi/4$.
4. If x is determined with an error Δx, what is the resulting error in $1/x$?
5. If a brass sphere 1 in. in diameter is given a chrome plating 0.005 in. thick, by how much is the volume of the sphere increased?
6. The area of a triangle is to be determined from the measured values of two sides and the included angle. What relative error in the computed area will result from an error of 1% in the measurement of the angle, if the angle is about 30°?
7. Use Newton's method to find each of the positive roots of $x^3 + 3x^2 - 7x + 1 = 0$, correct to the nearest 0.01. Use these values and the sum of the roots to infer the value of the negative root. Using a calculator, find the product of the roots and use it as a check. Is this a complete check?

Miscellaneous Exercises 7.2

1. Use differentials to approximate
   (a) $\sqrt{50}$ (e) $1/99.9$ (i) $\cos 0.01$
   (b) $\sqrt[3]{25}$ (f) $30^{3/5}$ (j) $(1 + h)^n$ ($|nh|$ small)
   (c) $15^{1/4}$ (g) $1/10.1$ (k) $\tan\left(\frac{\pi}{4} + 0.03\right)$
   (d) $1000^{1/10}$ (h) 9.7^{-2}
2. A thin cylindrical shell is h in. high, has an inner radius of r in., and is Δr in. thick. What is its approximate volume?
3. If the volume of a cube must be within 0.2 cu. in. of 27 cu. in., how accurately must the edge be determined?
4. Show that the relative error in the product of two numbers may be as great as the sum of the absolute values of the relative errors in the factors. (*Hint:* If $y = uv$, then the relative errors in y, u, and v are dy/y, du/u, and dv/v. Use Rule 5a, § 6.2.)

5. Show that the relative error in the quotient of two numbers may be as great as the sum of the absolute values of the relative errors in the dividend and divisor.

6. The area of an isosceles triangle is determined by measuring the equal sides and the included angle. **(a)** For what value of the angle does a given error in the measurement of the sides produce the greatest error in the area? **(b)** Answer part (a) with reference to the *relative* error in the area.

7. For what values of x can $\sqrt{x}$ be used as an approximation to $\sqrt{x+1}$ with **(a)** an error no greater than 0.01? **(b)** a relative error no greater than 1%?

8. **(a)** For what x does an error in the determination of x produce the greatest error in $\sin x$?

(b) Answer part (a) with reference to $\tan x$.

9. Triangle ABC has a right angle at C, and $AC = 8$ yd. Find the error in computing BC caused by an error of $6''$ in measuring $\angle A$, if $\angle A = 85°$, approximately.

10. For what x is it possible to have x in error by as much as 0.01 but still have $\sin x$ correct to within 0.001?

11. Use Newton's method to find all real roots correct to 0.01:

(a) $x^3 + 3x - 2 = 0$

(b) $x^3 - x^2 + x - 2 = 0$

(c) $x^3 + 7x + 1 = 0$

(d) $2x^3 - 3x^2 + 6x + 2 = 0$

(e) $x^5 - 3 = 0$

(f) $x^3 - 7x + 7 = 0$

12. **(a)** Show that if x_n is an approximation to $\sqrt[3]{a}$, then Newton's method gives $x_{n+1} = \frac{1}{3}[2x_n + (a/x_n{}^2)]$ as a next approximation.

(b) Use this result to approximate $\sqrt[3]{8}$ to the nearest hundredth, using $x_1 = 1$.

(c) Guess a similar formula for $\sqrt[r]{a}$, and describe your formula in terms of *averages*.

13. Use Newton's method to find a root between 0 and $\pi/2$, correct to 0.01:

(a) $\sin x + x - 1 = 0$

(b) $\tan x = 2x$

(c) $x + 1 = 2\cos x$

(d) $x^2 + \sin x - 2 = 0$

7.3 Extreme values: further considerations

As we have seen (§ 3.2), a function has a (relative) maximum value at a point where it changes from an increasing to a decreasing function, a (relative) minimum at a point where it changes from a decreasing to an increasing function. Such changes generally occur at zeros of the first derivative (stationary points), but, as the following example shows, this is not always the case.

Example 1. Find and identify any extreme values of $f = \{(x, y): y = x^{2/3}\}$.

Solution: Our standard procedure in such problems has been to find the zeros of the first derivative, but

$$\frac{dy}{dx} = \frac{2}{3}x^{-1/3}$$

is different from zero for all $x \neq 0$ and is not defined for $x = 0$. But $y = (x^{1/3})^2$ is defined for all x and is the square of a real number, and therefore $y = 0$ when $x = 0$, $y > 0$ when $x \neq 0$; hence $f(x)$ has a minimum value (of 0) when $x = 0$.

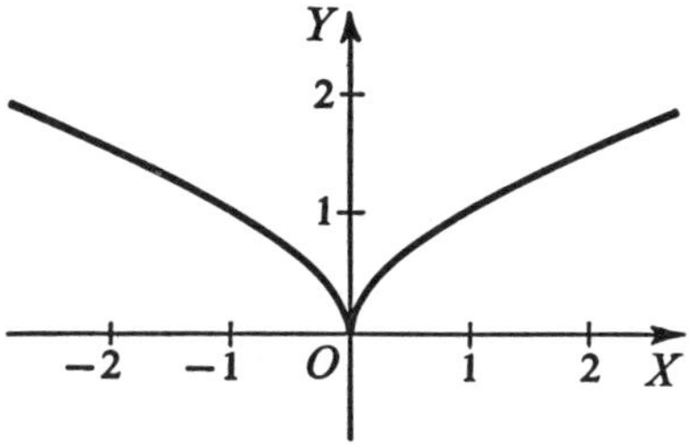

Fig. 7

The same conclusion can be reached by an examination of the derivative (§ 3.2, test 2). Since $dy/dx < 0$ when $x < 0$, and $dy/dx > 0$ when $x > 0$, we see that f changes at 0 from a decreasing to an increasing function, and the value of f at this point is therefore a minimum. This extreme value is thus seen to occur, not at a zero of the derivative, but at a point in the domain of f at which the derivative dy/dx does not exist. Such points must then be examined in any search for extreme values; that they do not always correspond to such values is shown by the next example.

Example 2. Find and identify any extreme values of $g = \{(x, y): y = x^{1/3}\}$.

Solution: We find $g'(x) = \frac{1}{3}x^{-2/3}$. Although g' has no zeros, it fails to exist at $x = 0$, and Example 1 suggests that this may therefore give an extreme value. But

$$g'(x) = \tfrac{1}{3}(x^{-1/3})^2$$

is positive for all $x \neq 0$, and g is therefore an increasing function for all such x. Moreover, $y < 0$ when $x < 0$, $y = 0$ when $x = 0$, $y > 0$ when $x > 0$, so g is therefore increasing for all x. Hence there are no extreme values of g. As the graph (Fig. 8) suggests, $x = 0$ corresponds to an inflection point with vertical tangent.

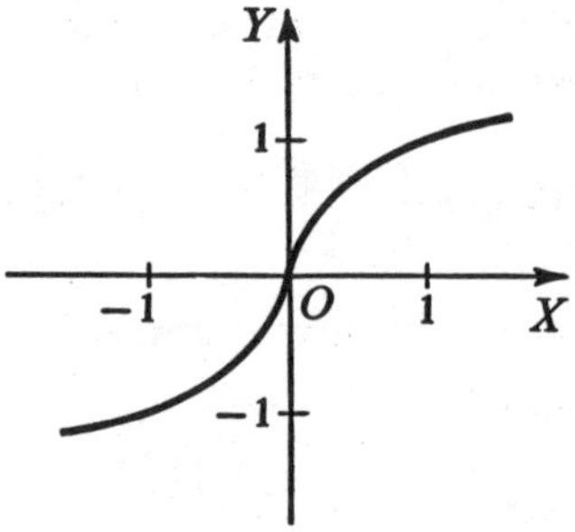

Fig. 8

Example 3. Find two non-negative numbers whose sum is 10 and the sum of whose squares is a maximum.

Solution: If we call the numbers x and $10 - x$, then the sum of their squares is

$$y = x^2 + (10 - x)^2 = 2x^2 - 20x + 100,$$

and

$$\frac{dy}{dx} = 4x - 20;$$

this derivative is defined for all x and is zero if $x = 5$. But, by any test, $y = 50$, the value which corresponds to $x = 5$, is a *minimum*, and any value of $x \neq 5$ will make $y > 50$. When $x = 0$ or when $x = 10$, $y = 100$, and 100 is therefore the largest value of y if $x \in [0, 10]$.

The considerations raised in the preceding examples lead us to enlarge the notion of a critical point introduced in § 3.1; we state this expanded idea as a formal definition:

Definition. The critical points of a function are those points of its domain

(a) at which the derivative is zero (stationary points), or
(b) at which the derivative does not exist, or
(c) that are endpoints of intervals of the domain.

Any (relative) maxima and minima of a function will occur at critical points.

The two examples that follow are not intended to show new sources of extreme values but rather to exhibit certain details of technique which will be convenient in some of the subsequent exercises.

Example 4. Find the shortest distance between the parabola $y^2 = 8x$ and **(a)** the point $(4, 0)$; **(b)** the point $(1, 0)$.

Solution: **(a)** The distance between $(4, 0)$ and an arbitrary point (x, y) is

$$s = \sqrt{(x - 4)^2 + y^2}. \tag{1}$$

If the point (x, y) lies on the given parabola, we must have $y^2 = 8x$, a condition on x and y which we can use to simplify (1) by eliminating from it the variable y:

$$s = \sqrt{(x-4)^2 + 8x} = \sqrt{x^2 + 16}. \tag{2}$$

To find the minimum value of s we must examine the derivative $\frac{ds}{dx}$; it is easier to differentiate if we first square (2):

$$s^2 = x^2 + 16.$$

Then $$2s\frac{ds}{dx} = 2x,$$

$$\frac{ds}{dx} = \frac{x}{s}.$$

Hence $$\frac{ds}{dx} = 0 \text{ if and only if } x = 0.$$

If $x = 0$, $s = 4$, and an examination of (2) makes it clear that $s > 4$ for any $x \neq 0$. Hence 4 is the required minimum distance; the point on the parabola nearest to $(4, 0)$ is the vertex $(0, 0)$.

(b) Proceeding as in part (a), we find

$$s = \sqrt{(x-1)^2 + y^2} = \sqrt{(x-1)^2 + 8x};$$

$$s^2 = x^2 + 6x + 1;$$

$$\frac{ds}{dx} = \frac{x+3}{s}.$$

Hence $$\frac{ds}{dx} = 0 \text{ if and only if } x = -3.$$

But the equation $y^2 = 8x$ has a meaning only for $x \geq 0$, the domain of $\{(x, s)\}$ is therefore $[0, \infty)$, and -3 (which gives us the absurd equation $s^2 = -8$) is not in this domain. Since ds/dx is positive for all admissible values of x (why?), s is everywhere increasing, and the required minimum is therefore at the left-hand boundary of its domain, at $x = 0$. Here $s = 1$, and, as in (a), the vertex is the point of the parabola nearest the given point.

Example 5. Point P is a units above the line AB, and Q is b units below AB (Fig. 9). The velocity of light is u units/sec. above AB and v units/sec. below, where $u > v$. Find the path which will take a ray of light from P to Q in the shortest possible time.

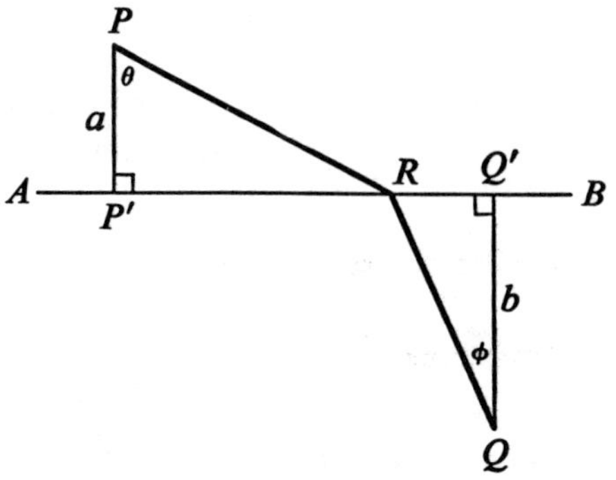

Fig. 9

Solution: From the figure, $PR = a \sec \theta$, $RQ = b \sec \phi$, and the total time T is therefore

$$T = \frac{a}{u} \sec \theta + \frac{b}{v} \sec \phi. \tag{3}$$

In (3), T depends on two variables, θ and ϕ, and the procedure followed in all such problems to this point requires us to find a condition on these variables which will permit the elimination of one of them. We can find such a condition in the fact that the distance $P'Q'$ has a fixed value, say c, and therefore

$$a \tan \theta + b \tan \phi = c. \tag{4}$$

However, we shall not try to carry out the actual elimination of either θ or ϕ, an elimination which would be awkward and would result in an expression difficult to differentiate. Instead, we note that (4) defines ϕ implicitly as a function of θ, we consider θ to be the independent variable, and we differentiate both (3) and (4), as they stand, with respect to θ.

$$\frac{dT}{d\theta} = \frac{a}{u} \sec \theta \tan \theta + \frac{b}{v} \sec \phi \tan \phi \frac{d\phi}{d\theta}; \tag{5}$$

$$a \sec^2 \theta + b \sec^2 \phi \frac{d\phi}{d\theta} = 0. \tag{6}$$

If we solve (6) for $\frac{d\phi}{d\theta}$ and substitute the resulting expression in (5), we get $\frac{dT}{d\theta} = \frac{a}{u} \sec \theta \tan \theta - \frac{a}{v} \sin \phi \sec^2 \theta$, and if $\frac{dT}{d\theta} = 0$, this reduces to

$$\frac{\sin \theta}{\sin \phi} = \frac{u}{v}. \tag{7}$$

The physical circumstances of the problem make it clear that the condition (7) corresponds to a minimum. This result is known as Snell's law of refraction. We have, of course, not explicitly found the location of the point R, but that location is fully determined by (4) and (7), two equations in two unknowns, once a, b, c, u, and v are known.

The important technical feature of this example is that no attempt was made to express T in terms of only one variable before differentiating; this same device may prove useful in some of the problems that follow.

Exercises 7.3 [A]

1. Find the largest and the smallest values of $x^3 - 4x$ if $x \in [-2, 3]$.
2. A rectangle is to be inscribed in a semicircle of radius a, with one side on the bounding diameter and two vertices on the semicircle. What is the maximum area of the rectangle?
3. One side of a triangle is 10 ft., and the opposite angle is 40°. Find the other two angles of the triangle if the area is a maximum.
4. (a) A silo, to hold a given volume, is in the shape of a right circular cylinder topped by a hemisphere. What proportions will use the least metal? (Neglect the floor of the silo.)

   (b) If the hemispherical top costs twice as much per square foot as the cylindrical portion, what then are the most economical proportions?
5. Find the least and the greatest straight-line distance between (0, 2) and the ellipse $x^2 + 9y^2 = 9$.

Exercises 7.3 [B]

1. Find all maxima and minima of $2 \sin x + \cos 2x$, $0 \leq x \leq \pi/2$.
2. Find the area of the largest rectangle that can be inscribed in the ellipse $4x^2 + 9y^2 = 36$. (*Hint:* Let (x, y) be the first-quadrant vertex of the rectangle, express the area in terms of x and y, and find a condition that x and y must satisfy.)
3. Find the dimensions of the largest cylinder that can be turned from a wooden cone of radius r and altitude h.
4. A log is 20 ft. long and tapers from a diameter of 4 ft. at one end to 3 ft. at the other. Find the diameter of the largest cylinder that can be cut from the log.
5. A messenger is to go ashore from a ship 8 miles offshore and deliver a message to a camp which is 20 miles up the beach from the point nearest the ship. The messenger can go 15 mi./hr. by boat and will be met by a jeep which can go 25 mi./hr. over the beach. Where should the landing be to complete the trip in as short a time as possible?

Miscellaneous Exercises 7.3

1. A line through (3, 5) cuts a triangle from the first quadrant. Find the minimum area of this triangle.

2. A piece of wire a in. long is to be used to make a circle, a square, or both. Find a side of the square if the total area enclosed is (**a**) a maximum; (**b**) a minimum.

3. The equation $y^2 = x^3 - 9x$ defines implicitly two functions of x, one of which takes no negative values and the other of which takes no positive values. Obtain explicit formulas for each function, state the domains, find any extreme values, and sketch.

4. Find the least and the greatest straight-line distance between the ellipse $x^2 + 4y^2 = 4$ and the point (**a**) (1, 0); (**b**) (3, 0).

5. Find the maximum area of a rectangle inscribed in the segment bounded by the parabola $y^2 = 6x$ and the line $x - 8 = 0$.

6. The sides of a triangle are 10 in., 10 in., and 16 in. A rectangle has one side on the 16-inch side and its other two vertices on the 10-inch sides. Find the maximum area of the rectangle.

7. The configuration of Ex. 6 is rotated about an altitude of the triangle to generate a cylinder inscribed in a cone. Find the maximum volume of the cylinder.

8. (**a**) A messenger leaves a ship which is 10 miles offshore with a message for a camp which is 4 mi. up the beach from the point nearest the ship. The messenger can go 12 mi./hr. by boat and will be met by a jeep which can go 30 mi./hr. along the beach. Where should the landing be to complete the trip as quickly as possible?

(**b**) Suppose the camp is 20 mi. up the beach. Where then should the landing be?

9. Two vertices of a rectangle are on the x-axis, and the other two are on the curve $y = 8/(x^2 + 4)$ (see Ex. 7, page 197). Find the maximum area of the rectangle.

10. One base of a trapezoid is a diameter of a circle of radius a, and the other base is a chord of that circle. Find the maximum area of the trapezoid.

11. A strip of metal 24 in. wide is to be made into a gutter by bending up an 8-inch strip along each side. What should be the angle between sides and bottom for maximum capacity?

12. A rectangular solid with a square base is inscribed in a sphere of radius a. What is the maximum volume of the solid?

Exercises 7.3 [C]

1. Find the minimum volume of a cone circumscribed about a sphere of radius a.
2. Find the minimum curved surface of a cone circumscribed about a sphere of radius a.
3. Find the extreme values of $x^2 + xy + y^2$ if (x, y) is a point on the circle $x^2 + y^2 = a^2$.
4. Find the length of the shortest segment cut by the coordinate axes on a line through (a, b). (Take $a > 0$, $b > 0$.)
5. Analyze the critical points of $y = (x + 1)^{1/3}(x - 2)^{2/3}$, and sketch the curve.
6. Find the points nearest the origin on the curve $y = 8/(x^2 + 4)$.
7. A 27-foot wall is 8 ft. from the side of a building. Find the length of the shortest ladder that will reach the building from the ground outside the wall.
8. A painting 10 ft. tall hangs on a wall with its bottom edge 8 ft. above the level of a woman's eyes. How far from the wall should she stand so that the picture will subtend the greatest possible angle at her eye?
9. A 27-foot girder has to be moved down an 8-foot corridor and around a right-angle turn into a connecting corridor. If the width of the girder is neglected, how wide must the second corridor be?
10. A strip of metal 10 in. wide is to be bent into a gutter. If the cross section of the gutter is a circular arc, what radius (to the nearest 0.1 in.) will give the maximum capacity?

7.4 Integration by substitution

As we have seen (§ 4.1), if

$$y = \int f(x)dx, \tag{1}$$

then

$$\frac{dy}{dx} = f(x)$$

and therefore

$$dy = f(x)dx. \tag{2}$$

A comparison of (1) and (2) suggests that the dx in $\int f(x)dx$, formerly interpreted (§ 4.1) as simply a part of the symbol $\int \cdots dx$ for the integral, might now be interpreted as the differential dx, and, since from (1) and (2) we get

$y = \int dy$, the integral sign $\int$ might then be interpreted as being, in some sense, an inverse of the d symbol for the differential. (The constant of integration causes trouble if we try to think of integration as being strictly the inverse of differentiation.) Under this interpretation, which we now adopt, the y of (1) is the same as the y of (2), and this new interpretation of our symbols will therefore give us results that will be consistent with those obtained earlier. The principal reason for this identification of the dx in the integral with the differential dx is that it permits the easy introduction of an auxiliary variable, thus enabling us to use the chain rule to evaluate integrals which could not be evaluated by our earlier methods.

Virtually all the integrals which we have so far considered have been of the form

$$\int kx^n\,dx = \frac{k}{n+1}x^{n+1} + C \quad (n \neq -1);$$

it will be our concern, in the rest of this section, to see how a variety of integrals not already of this form can be reduced to this form by appropriate substitutions. The form of $\int f(x)dx$ can be changed by expressing x as a function of an auxiliary variable u, $x = g(u)$. Then $dx = g'(u)du$, and therefore

$$\int f(x)dx = \int f[g(u)]g'(u)du.$$

If an immediate evaluation of the original integral is not possible, then a skillful choice of the variable u may produce a result which can be integrated by known formulas. The method will be illustrated in the following examples.

Example 1. Integrate: $\int \sqrt{a - x}\,dx.$

Solution: If we were differentiating $\sqrt{a - x}$, we would introduce (mentally, it is to be hoped) an auxiliary variable $u = a - x$, because it is easy to differentiate $u^{1/2} = \sqrt{a - x}$ *with respect to* u, and the chain rule then gives us the required derivative *with respect to* x. It is also easy to integrate $u^{1/2}$ with respect to u, and the same substitution is therefore useful in integration (the student will soon learn to make this substitution mentally, too). Since $x = a - u$, we find $dx = -du$, and therefore

$$\int \sqrt{a - x}\,dx = \int u^{1/2}(-du) = -\int u^{1/2}\,du = -\tfrac{2}{3}u^{3/2} + C = -\tfrac{2}{3}(a - x)^{3/2} + C.$$

If we were to write $\int \sqrt{a - x}\,dx = \int u^{1/2}\,dx$, the result, though correct, would be useless; an essential feature of this process is that the differential dx can be replaced by its equivalent $-du$, so that the integrand (the expression affected by the integral sign, $\int$) can be expressed in terms of u *as the only variable.*

The substitution used is not the only possible one. We might, for example, let $v = \sqrt{a - x}$; then $x = a - v^2$, $dx = -2v\,dv$, and

$$\int \sqrt{a - x}\,dx = \int v(-2v\,dv) = -2\int v^2\,dv = -\tfrac{2}{3}v^3 + C = -\tfrac{2}{3}(a - x)^{3/2} + C.$$

The best substitution to use in this problem is, however, no substitution at all, or, to be a little more accurate, an *implicit* substitution which is never made explicit. You should recognize that

$$\int (\quad)^{1/2}\,d(\quad) = \tfrac{2}{3}(\quad)^{3/2} + C \tag{3}$$

is a valid integration provided any expression for which the statement has meaning is used to fill all three parentheses. The trick is to tinker with the given integration so as to produce something in the form (3). We have $(a - x)^{1/2}$, and nothing much can be done to change the form of this, so we turn to the dx and inquire what we can do to turn this into $d(a - x)$; since $d(a - x) = -dx$, all we need is to supply a minus sign, and the integration using an implicit substitution then looks like this:

$$\int \sqrt{a - x}\,dx = -\int (a - x)^{1/2}\,d(a - x) = -\tfrac{2}{3}(a - x)^{3/2} + C.$$

Although you should not hesitate to use an explicit substitution, as we used u and v above, you should also try to practice the skill of using implicit substitutions; in the long run this skill will save you a great deal of time and effort.

Example 2. Integrate: $\int x\sqrt{a^2 - x^2}\,dx.$

Solution: The situation here somewhat resembles that in Example 1 and therefore suggests a similar substitution, $u = a^2 - x^2$. This gives $du = -2x\,dx$, and, *thanks to the factor x in the original integrand*, we have

$$\int x\sqrt{a^2 - x^2}\,dx = \int \sqrt{a^2 - x^2}\,(x\,dx) = \int u^{1/2}(-\tfrac{1}{2}\,du) = -\tfrac{1}{2}\int u^{1/2}\,du$$
$$= -\tfrac{1}{2}\cdot\tfrac{2}{3}u^{3/2} + C = -\tfrac{1}{3}(a^2 - x^2)^{3/2} + C.$$

It must be emphasized that the success of this substitution hinges on the presence of the factor x in the integrand, and that without this factor the problem would have been entirely different and much more difficult (see Example 6, p. 248). You may therefore conclude that there is a considerable element of luck involved here. The conclusion is valid. On the other hand, you must learn to *observe*, and the combination $x\,dx$ should suggest to you the differential of x^2 or even of $a^2 - x^2$, and might, therefore, suggest the substitution actually used.

The substitution $v = \sqrt{a^2 - x^2}$ also works.

The implicit technique is a bit more difficult here, but it is still possible and may serve to emphasize the part played by the factor x:

$$\int x\sqrt{a^2 - x^2}\,dx = \int \sqrt{a^2 - x^2}\,(x\,dx) = -\tfrac{1}{2}\int \sqrt{a^2 - x^2}\,(-2x\,dx)$$

$$= -\tfrac{1}{2}\int \sqrt{a^2 - x^2}\,d(a^2 - x^2) = -\tfrac{1}{3}(a^2 - x^2)^{3/2} + C.$$

Exercises 7.4 (1) [A]

Algebraic Substitutions

Perform the indicated integrations.

1. $\int \sqrt{x - 2}\,dx$
2. $\int (2x + 3)^{12}\,dx$
3. $\int \sqrt{3t + 2}\,dt$
4. $\int (2x - 5)^{43}\,dx$
5. $\int 3(4x - 5)^{2/3}\,dx$
6. $\int 2(3 - 5x)^{3/5}\,dx$
7. $\int \sqrt[3]{4x + 7}\,dx$
8. $\int \frac{dx}{(2x - 3)^4}$
9. $\int y(y^2 + 4)^{2/3}\,dy$
10. $\int 5x\sqrt{1 - x^2}\,dx$
11. $\int x\sqrt{2x^2 + 7}\,dx$
12. $\int x^2\sqrt{2x^3 - 7}\,dx$
13. $\int 5x^2(3 - 2x^3)^{1/3}\,dx$
14. $\int x^3\sqrt{x^4 - 16}\,dx$
15. $\int \frac{t\,dt}{\sqrt{3t^2 + 1}}$
16. $\int x^{1/3}(x^{4/3} + 81)^6\,dx$
17. $\int \sqrt{1 + \sqrt{x}} \cdot \frac{dx}{\sqrt{x}}$
18. $\int x^{-1/2}(x^{1/2} + 2)^{1/5}\,dx$
19. $\int (2x + 1)\sqrt{x^2 + x - 3}\,dx$
20. $\int (x + 5)\sqrt{x^2 + 10x + 24}\,dx$
21. $\int \frac{x - 2}{\sqrt{x + 1}}\,dx$
22. $\int x^3\sqrt{3x^2 + 1}\,dx$
23. $\int (x + 1)\sqrt{x + 2}\,dx$
24. $\int \frac{x^2\,dx}{\sqrt{x - 3}}$

Example 3. Integrate: $\int \sin 3\theta \, d\theta$.

Solution: It is clear from the rule for differentiation of the cosine (Rule 10, § 6.5) that

$$\int \sin u \, du = -\cos u + C.$$

We can obtain the same form here by the substitution $u = 3\theta$, which gives $\theta = u/3$, $d\theta = du/3$, and therefore

$$\int \sin 3\theta \, d\theta = \int \sin u \left(\tfrac{1}{3} du\right) = \tfrac{1}{3}\int \sin u \, du = -\tfrac{1}{3}\cos u + C = -\tfrac{1}{3}\cos 3\theta + C.$$

In this case, an implicit substitution is particularly easy. The integrals $\int \sin u \, du$, $\int \sin y \, dy$, $\int \sin f(x) d(f(x))$ all have the same form and can therefore be integrated in exactly the same manner. The essential feature of each of these, missing from the integral $\int \sin 3\theta \, d\theta$, is that we have the sine and the differential of exactly the same expression. We can correct this, in the given integral, without an explicit substitution: we must obtain $d(3\theta)$ in place of $d\theta$, and since $d(3\theta) = 3 \, d\theta$, we can write

$$\int \sin 3\theta \, d\theta = \tfrac{1}{3}\int \sin 3\theta \, d(3\theta) = -\tfrac{1}{3}\cos 3\theta + C.$$

Example 4. Integrate: $\int \sin^2 x \cos x \, dx$.

Solution: An observation of the combination $\cos x \, dx$ suggests $d(\sin x)$; hence we write

$$\int \sin^2 x \cos x \, dx = \int \sin^2 x \, d(\sin x) = \tfrac{1}{3}\sin^3 x + C$$

This is, of course, equivalent to making the substitution $u = \sin x$, which gives $du = \cos x \, dx$ from Rule 9a, and therefore

$$\int \sin^2 x \cos x \, dx = \int u^2 \, du = \tfrac{1}{3}u^3 + C = \tfrac{1}{3}\sin^3 x + C.$$

Example 5. Integrate: $\int \cos^2 y \, dy$.

Solution: The exponent 2 is annoying here. The standard way to eliminate it is to use equation (13) of § 2.3:

$$\int \cos^2 y \, dy = \tfrac{1}{2}\int (1 + \cos 2y) dy = \tfrac{1}{2}\int dy + \tfrac{1}{2}\int \cos 2y \, dy$$

$$= \tfrac{1}{2}y + \tfrac{1}{4}\sin 2y + C.$$

Equation (12) of § 2.3 is similarly convenient if it is desirable to eliminate the square of a sine from an integrand. Note, however, that such an elimination is not always desirable, as Example 4 shows.

Exercises 7.4 (2) [A]

Trigonometric Integrands

Perform the indicated integrations.

1. $\int \sin u \, du$
2. $\int \cos u \, du$
3. $\int \sin 5x \, dx$
4. $\int \sec^2 z \, dz$
5. $\int \csc^2 \theta \, d\theta$
6. $\int \sec 2x \tan 2x \, dx$
7. $\int \cot 3t \csc 3t \, dt$
8. $\int \tan x \sec^2 x \, dx$
9. $\int \cos^3 x \sin x \, dx$
10. $\int \sin 5\theta \cos 5\theta \, d\theta$
11. $\int \sin^2 2x \cos 2x \, dx$
12. $\int \cos (ax + b) dx$
13. $\int \sqrt{\sin x} \cos x \, dx$
14. $\int (2 + \cos x)^{2/3} \sin x \, dx$
15. $\int \frac{\sin x \, dx}{(1 - \cos x)^2}$
16. $\int \sqrt{3 - 2 \sin 5\theta} \cos 5\theta \, d\theta$
17. $\int \tan^2 x \sec^2 x \, dx$
18. $\int \tan^2 x \, dx$

    (*Hint:* Use equation (4), § 2.2.)
19. $\int \tan^4 x \, dx$
20. $\int \cot^2 x \, dx$
21. $\int \cot^4 t \, dt$
22. $\int \sec^4 x \, dx$
23. $\int \csc^4 x \, dx$
24. $\int \sin^2 x \, dx$
25. $\int \cos^2 4\theta \, d\theta$
26. $\int \sin^n x \cos x \, dx \quad (n \neq -1)$

27. $\int \cos^n x \sin x\, dx \quad (n \neq -1)$

28. $\int \sin^3 x \cos^2 x\, dx$

29. $\int \sin^2 x \cos^3 x\, dx$

30. $\int \tan^n x \sec^2 x\, dx \quad (n \neq -1)$

31. $\int \sec^n x \tan x\, dx \quad (n \neq 0)$

32. $\int \sin^3 x\, dx$

33. $\int \cos^3 x\, dx$

34. $\int \sin^4 x\, dx$

35. $\int \sec^6 x\, dx$

36. $\int \sin 2x \cos 3x\, dx$

(*Hint:* Use equation (19), § 2.3.)

37. $\int \sin 5x \cos x\, dx$

38. $\int \sin ax \cos bx\, dx \quad (a^2 \neq b^2)$

39. $\int \sin ax \sin bx\, dx \quad (a^2 \neq b^2)$

40. $\int \cos ax \cos bx\, dx \quad (a^2 \neq b^2)$

Example 6. Integrate: $\int \sqrt{a^2 - x^2}\, dx$.

Solution: The resemblance to Example 2 suggests $u = a^2 - x^2$. If we try this, we have $du = -2x\, dx$, and since $x^2 = a^2 - u$, $x = \sqrt{a^2 - u}$, and $dx = -\dfrac{du}{2x} = -\dfrac{1}{2}\dfrac{du}{\sqrt{a^2 - u}}$, we get

$$\int \sqrt{a^2 - x^2}\, dx = \int u^{1/2}\left(-\frac{1}{2}\frac{du}{\sqrt{a^2 - u}}\right) = -\frac{1}{2}\int \frac{u^{1/2}\, du}{\sqrt{a^2 - u}};$$

the substitution has produced a result even more complicated than was the original integral. Another plausible substitution might be $v = \sqrt{a^2 - x^2}$, but this is equally futile. It has, however, been observed that the substitution $\theta = \sin^{-1}(x/a)$ is useful in this case. We find $x = a \sin \theta$, $dx = a \cos \theta\, d\theta$, and

$$\begin{aligned}
\int \sqrt{a^2 - x^2}\, dx &= \int \sqrt{a^2 - a^2 \sin^2 \theta}\, (a \cos \theta\, d\theta) \\
&= \int \sqrt{a^2 \cos^2 \theta}\, (a \cos \theta\, d\theta) \\
&= a^2 \int \cos^2 \theta\, d\theta = a^2(\tfrac{1}{2}\theta + \tfrac{1}{4}\sin 2\theta) + C,
\end{aligned}$$

where we have used the result of Example 5 in the last step. It is still necessary for us to express this result in terms of x. Since $\theta = \sin^{-1}(x/a)$, we get from § 2.3

$$\sin 2\theta = 2 \sin\theta \cos\theta = 2\frac{x}{a}\sqrt{1-(x^2/a^2)} = \frac{2x}{a^2}\sqrt{a^2-x^2}.$$

Hence
$$\int \sqrt{a^2-x^2}\,dx = \frac{1}{2}a^2 \sin^{-1}\frac{x}{a} + \frac{1}{2}x\sqrt{a^2-x^2} + C.$$

Exercises 7.4 (3) [A]

Trigonometric Substitutions

Perform the indicated integrations.

1. $\displaystyle\int \frac{dx}{\sqrt{1-x^2}}$
2. $\displaystyle\int \sqrt{1-x^2}\,dx$
3. $\displaystyle\int \frac{dx}{\sqrt{4-x^2}}$
4. $\displaystyle\int (1-x^2)dx$
5. $\displaystyle\int \frac{dx}{\sqrt{9-4x^2}}$
6. $\displaystyle\int \frac{x\,dx}{\sqrt{4-x^2}}$
7. $\displaystyle\int \frac{x^2\,dx}{\sqrt{1-x^2}}$
8. $\displaystyle\int \frac{dx}{x^2+1}$

The technique of integration is one which can demand great ingenuity, and there are expressions which cannot be integrated, no matter how much ingenuity is exerted on them. The integrals in the following exercises can, however, all be integrated with the aid of a fairly limited variety of substitutions. The following suggestions may be helpful:

1. If an integrand contains a form like $\sqrt{ax+b}$, then the substitution $u = ax + b$ or $u = \sqrt{ax+b}$ may be helpful. Similar substitutions may be useful in problems involving powers of expressions like $a^2 \pm x^2$ or $ax^2 + bx + c$.

2. You should review the formulas for the differentiation of circular functions, in § 6.5, so that you will recognize these derivatives when you see them in an integrand. Thus, for example, $\int f(\tan\theta)\sec^2\theta\,d\theta$ has in it $\sec^2\theta\,d\theta = d(\tan\theta)$ and can therefore be put in the form $\int f(u)du$ by the substitution $u = \tan\theta$.

3. The identities $\sin^2 x = \frac{1}{2}(1 - \cos 2x)$ and $\cos^2 x = \frac{1}{2}(1 + \cos 2x)$ are often useful, as Example 5 showed, and should be remembered.

4. If the substitutions suggested under 1, above, fail, it may be helpful to introduce a trigonometric substitution in an algebraic integrand. The two following substitutions are particularly useful:

If the integrand contains an expression of the form $a^2 - x^2$, then the substitution $\theta = \sin^{-1}(x/a)$ $(a > 0)$, gives $a^2 - x^2 = a^2(1 - \sin^2\theta) = a^2\cos^2\theta$. Furthermore, since $-\pi/2 \leq \theta \leq \pi/2$ by definition of the inverse sine function, and since $\cos\theta \geq 0$ for all such θ, we then have $\sqrt{a^2 - x^2} = \sqrt{a^2\cos^2\theta} = a\cos\theta$, without ambiguity of sign.

If the integrand contains an expression of the form $a^2 + x^2$, then the substitution $\theta = \tan^{-1}(x/a)$ $(a > 0)$ gives $a^2 + x^2 = a^2(1 + \tan^2\theta) = a^2\sec^2\theta$, and since the definition of the inverse tangent implies $-\pi/2 < \theta < \pi/2$, which in turn implies $\sec\theta > 0$, we have $\sqrt{a^2 + x^2} = \sqrt{a^2\sec^2\theta} = a\sec\theta$, again without ambiguity of sign.

Exercises 7.4 (4) [A]

Perform the following integrations:

1. $\int (2x - 5)^{10}\,dx$
2. $\int (3 - 4x)^{5/2}\,dx$
3. $\int \sin 4\theta\,d\theta$
4. $\int x\sqrt{x^2 + 3}\,dx$
5. $\int \sec^2 5\theta\,d\theta$
6. $\int x^{-1/3}\sqrt{3x^{2/3} - 5}\,dx$
7. $\int \frac{dx}{\sin^2 x}$
8. $\int (1 - \cos x)^2\,dx$
9. $\int x^3\sqrt{2x^2 + 1}\,dx$
10. $\int x\cos x^2\,dx$
11. $\int \sin^2 3y\,dy$
12. $\int \frac{dx}{\sqrt{1 - x^2}}$
13. $\int \frac{x + 3}{\sqrt{x - 5}}\,dx$
14. $\int \frac{dx}{x^2 + 4}$

Exercises 7.4 (4) [B]

Perform the following integrations:

1. $\int \frac{dx}{\sqrt{3x - 7}}$

2. $\int x^2(x^3 - 5)^{2/3}\, dx$

3. $\int \frac{x\, dx}{\sqrt{3x^2 - 7}}$

4. $\int \cos \tfrac{1}{2}v\, dv$

5. $\int \sin (ax + b)dx$

6. $\int \sec \frac{x + 1}{3} \tan \frac{x + 1}{3}\, dx$

7. $\int \frac{dx}{x^2 + 2x + 1}$

8. $\int \frac{(2x + 3)dx}{(x^2 + 3x - 5)^3}$

9. $\int \cos^3 z \sin z\, dz$

10. $\int \cot^2 \theta \csc^2 \theta\, d\theta$

11. $\int (\sqrt{x} + \sqrt{a})^{2/3} x^{-1/2}\, dx$

12. $\int \sqrt{1 + \tan x} \sec^2 x\, dx$

13. $\int \frac{\sin \sqrt{x}}{\sqrt{x}}\, dx$

14. $\int \frac{x\, dx}{\sqrt{a^2 - x^2}}$

15. $\int \frac{dx}{\sqrt{a^2 - x^2}}$

16. $\int (x + 2)\sqrt{x - 1}\, dx$

Miscellaneous Exercises 7.4

Perform the indicated integrations.

1. $\int \cos 7\theta\, d\theta$

2. $\int \sqrt{14 - 11x}\, dx$

3. $\int \cos \left(x - \frac{\pi}{3}\right) dx$

4. $\int (2x - 1)^{3/2}\, dx$

5. $\int \frac{dx}{\sqrt{x + 2}}$

6. $\int (ax + b)^n\, dx \quad (n \neq -1)$

7. $\int \sin \theta \cos \theta\, d\theta$

8. $\int 3x^2\sqrt{4x^3 - 1}\, dx$

9. $\int \frac{x\, dx}{(x^2 - 1)^7}$

10. $\int \frac{(a^2 - x^2)^2}{x^2}\, dx$

11. $\int \frac{x^3\,dx}{(x^4-5)^3}$

12. $\int \tan x \sec^2 x\,dx$

13. $\int \frac{\cos x\,dx}{\sin^2 x}$

14. $\int \frac{\sin 2z}{\sqrt{\cos 2z}}\,dz$

15. $\int \frac{x^2+3x-1}{\sqrt{4-x}}\,dx$

16. $\int \frac{\csc y \cot y}{(2+\csc y)^2}\,dy$

17. $\int \frac{dx}{\sqrt{3-x^2}}$

18. $\int \frac{dx}{\cos^2 (x/3)}$

19. $\int \frac{dx}{4x^2+9}$

20. $\int (2x+7)\sqrt{x-1}\,dx$

21. $\int \cot^3 \theta \csc^2 \theta\,d\theta$

22. $\int \frac{dx}{\sqrt{1-9x^2}}$

23. $\int \tan^2 3z\,dz$

24. $\int \frac{x^2\,dx}{\sqrt{2x-3}}$

25. $\int \frac{dx}{x^2-2x+1}$

26. $\int \frac{dx}{x^2\sqrt{a^2-x^2}}$

27. $\int \sin^2 4\theta\,d\theta$

28. $\int \sqrt{1-\cos x}\,dx$

29. $\int \sin x \sin\left(x+\frac{\pi}{2}\right)dx$

30. $\int \frac{dx}{\sqrt{2-x^2}}$

31. $\int 2 \sin\left(\frac{\pi}{4}-\frac{x}{2}\right)\cos\left(\frac{\pi}{4}-\frac{x}{2}\right)dx$

32. $\int \frac{x+5}{\sqrt{2x+1}}\,dx$

33. $\int \sin(\pi-x)\sin(\pi+x)\,dx$

34. $\int \frac{x^2\,dx}{\sqrt{9-x^2}}$

35. $\int (3-\sin 2\theta)^2\,d\theta$

36. $\int \cos^4 z\,dz$

37. $\int \sin^5 t\,dt$

38. $\int \sin\theta \cos 2\theta\,d\theta$

7.5 Definite integrals

When there is a change of variable in a definite integral, it is important to bear in mind that the limits of integration are values of the original variable and not, in general, appropriate values of the new variable. It is then essential either to

(1) change the limits to the proper values of the new variable, or to
(2) restore the original variable before introducing the limits.

The first method is usually the easier, but there are many exceptions.

Example 1. Evaluate $\int_2^3 x\sqrt{x-1}\,dx.$

Solution: We set $u = \sqrt{x-1}$; then $u^2 = x - 1$, $dx = 2u\,du$, and

$$\int x\sqrt{x-1}\,dx = \int u(u^2+1)\cdot 2u\,du = 2\int (u^4+u^2)du.$$

Now, when $x = 2$, $u = 1$, and when $x = 3$, $u = \sqrt{2}$, so

$$\int_2^3 x\sqrt{x-1}\,dx = 2\int_1^{\sqrt{2}} (u^4+u^2)du = 2\left[\frac{1}{5}u^5 + \frac{1}{3}u^3\right]_1^{\sqrt{2}}$$

$$= 2\left[\left(\frac{4\sqrt{2}}{5} + \frac{2\sqrt{2}}{3}\right) - \left(\frac{1}{5} + \frac{1}{3}\right)\right] = \frac{44\sqrt{2}-16}{15}.$$

This illustrates the first of the two possible approaches listed above. In the second approach, we eliminate u from the result before introducing the limits:

Since $\int x\sqrt{x-1}\,dx = \frac{2}{5}u^5 + \frac{2}{3}u^3 + C = \frac{2}{5}(x-1)^{5/2} + \frac{2}{3}(x-1)^{3/2} + C,$

we have $\int_2^3 x\sqrt{x-1}\,dx = [\frac{2}{5}(x-1)^{5/2} + \frac{2}{3}(x-1)^{3/2}]_2^3 = \dfrac{44\sqrt{2}-16}{15}.$

If an algebraic integrand is transformed by a trigonometric substitution, as in Example 6 of § 7.4, an inverse circular function may appear in the result. The student should recall, from § 2.5, that

$$-\pi/2 \le \sin^{-1} x \le \pi/2; \quad 0 \le \cos^{-1} x \le \pi; \quad -\pi/2 < \tan^{-1} x < \pi/2.$$

Example 2. Evaluate $\int_0^a \sqrt{a^2-x^2}\,dx.$

Solution: From § 7.4, Example 6, we have, if $x = a\sin\theta$,

$$\int \sqrt{a^2-x^2}\,dx = a^2(\tfrac{1}{2}\theta + \tfrac{1}{4}\sin 2\theta) = C.$$

When $x = 0$, $\theta = \sin^{-1} 0 = 0$, and when $x = a$, $\theta = \sin^{-1} 1 = \pi/2$. Therefore

$$\int_0^a \sqrt{a^2-x^2}\,dx = a^2\left[\frac{1}{2}\theta + \frac{1}{4}\sin 2\theta\right]_0^{\pi/2}$$

$$= a^2\left[\left(\frac{\pi}{4} + \frac{1}{4}\cdot 0\right) - \left(\frac{1}{2}\cdot 0 + \frac{1}{4}\cdot 0\right)\right] = \frac{\pi a^2}{4}.$$

If we restore the original variables before evaluating, we have, again from § 7.4, Example 6,

$$\int_0^a \sqrt{a^2 - x^2}\, dx = \frac{1}{2}\left[a^2 \sin^{-1}\frac{x}{a} + x\sqrt{a^2 - x^2}\right]_0^a$$

$$= \frac{1}{2}\left[a^2 \sin^{-1} 1 - a^2 \sin^{-1} 0\right] = \frac{1}{2}\left[\frac{\pi a^2}{2} - 0\right] = \frac{\pi a^2}{4}.$$

Exercises 7.5 [A]

Evaluate each of the integrals in Exs. 1–8.

1. $\int_{-1}^{3} (x^2 + 2x - 7)dx$
2. $\int_{-1}^{2} \sqrt{x + 2}\, dx$
3. $\int_{1}^{2} \frac{t\, dt}{(5 - t^2)^4}$
4. $\int_{0}^{2} y\sqrt{y + 1}\, dy$
5. $\int_{-1}^{1} \frac{dx}{x^2 + 1}$
6. $\int_{0}^{7} z(z + 1)^{1/3}\, dz$
7. $\int_{0}^{1} \frac{dx}{\sqrt{4 - x^2}}$
8. $\int_{0}^{1} \frac{x^2\, dx}{\sqrt{9 - x^2}}$
9. Find the area in the first quadrant bounded by $y = \sqrt{4 - x}$ and the coordinate axes.
10. The region bounded by the x-axis and one arch of $y = \sin x$ is rotated about the x-axis. Find the volume generated.
11. Find, by integration, the area of the circle $x^2 + y^2 = a^2$.
12. Find the area of the ellipse $x^2 + 4y^2 = 16$.

Exercises 7.5 [B]

Evaluate each of the integrals in Exs. 1–8.

1. $\int_{1}^{2} x\sqrt{2x^2 + 1}\, dx$
2. $\int_{-4}^{-1} \sqrt{x + 5}\, dx$
3. $\int_{0}^{3} \sqrt{25 - 3s}\, ds$
4. $\int_{1}^{4} \frac{dz}{\sqrt{5 - z}}$
5. $\int_{0}^{2} \frac{dx}{x^2 + 4}$
6. $\int_{0}^{a} (\sqrt{a} - \sqrt{y})^2\, dy$
7. $\int_{0}^{2} \frac{dx}{\sqrt{8 - x^2}}$
8. $\int_{1}^{2} \frac{dx}{x^2\sqrt{9 - x^2}}$

9. Find the area bounded by $y\sqrt{x+2} = 1$, the x-axis, and the lines $x = -1$ and $x = 7$.

10. Find the volume of the solid generated by rotating about the x-axis the region bounded by $y = \tan x \sec x$, the x-axis, and the line $x = \pi/4$.

11. Find the area of the ellipse $9x^2 + y^2 = 36$.

12. Find, by integration, the area of the minor segment cut from the circle $x^2 + y^2 = a^2$ by the line $x = a/2$.

7.6 Differential equations

The differential equation

$$F'(x) = f(x) \tag{1a}$$

was considered in § 4.4. If $y = F(x)$, (1a) can be written in an equivalent differential form as

$$dy = f(x)dx; \tag{1b}$$

this form, involving differentials rather than derivatives, is also known as a differential equation.

In this section we shall consider a somewhat more general form of differential equation, one which can be written in the form

$$f(x)dx + g(y)dy = 0, \tag{2}$$

in which the variables x and y are separated and collected each in a single term. Thus, the variables are separable in $dy/dx = \sqrt{xy}$, for this can be written $x^{1/2}\,dx - y^{-1/2}\,dy = 0$, but they are not separable in $dy/dx = x + y$, and we shall therefore not concern ourselves with the latter. Note that (1b) can be considered a special case of the type (2), with $g(y) = -1$.

If there exists a function $h = \{(x, y)\}$ that satisfies (2), then $g(y) = g(h(x))$, $dy = h'(x)dx$. Equation (2) can then be written

$$[f(x) + g(h(x))h'(x)]dx = 0.$$

Hence $\int[f(x) + g(h(x))h'(x)]dx = \int f(x)dx + \int g(h(x))h'(x)dx = C$, or

$$\int f(x)dx + \int g(y)dy = C. \tag{3}$$

If F and G are any functions such that $F' = f$, $G' = g$, then (3) is equivalent to

$$F(x) + G(y) = C, \tag{4}$$

and implicit differentiation of (4) shows that it is a solution to (2).

Example 1. Solve $dy/dx = \sqrt{xy}$.

Solution: As we have seen, this can be written in the form $x^{1/2}\,dx - y^{-1/2}\,dy = 0$, and the solution is therefore

$$\int x^{1/2}\,dx - \int y^{-1/2}\,dy = C,$$

$$\tfrac{2}{3}x^{3/2} - 2y^{1/2} = C,$$

or, somewhat more neatly,

$$x^{3/2} - 3y^{1/2} = C',$$

where $C' = \frac{3}{2}C$ is also an arbitrary constant.

Example 2. The slope of a curve at the point (x, y) is given by the value of $-\dfrac{x}{y}$. Find the equation of the family of curves which satisfy this condition and of the particular solution that passes through $(1, -3)$.

Solution: The given condition is equivalent to the differential equation $\dfrac{dy}{dx} = -\dfrac{x}{y}$; separating the variables and integrating, we have

$$x\,dx + y\,dy = 0,$$

$$\int x\,dx + \int y\,dy = 0,$$

$$\tfrac{1}{2}x^2 + \tfrac{1}{2}y^2 = C,$$

or

$$x^2 + y^2 = C'.$$

For $C' > 0$, this represents the family of circles with center at the origin and radius $\sqrt{C'}$; for $C' < 0$, there is no corresponding curve (what does the case $C' = 0$ represent?). If a member of the family passes through $(1, -3)$, we have $1^2 + (-3)^2 = C'$ and the particular solution required is therefore $x^2 + y^2 = 10$.

Exercises 7.6 [A]

1. If $dy/dx = xy^2$ and $y = 1$ when $x = 2$, find y when $x = -3$.
2. Find the solution of $dy = x\sqrt{y}\,dx$ which passes through $(2, 9)$.
3. Find the solution of $dx = x^2y\,dy + 3x^2\,dy$ which goes through $(-\frac{1}{2}, -1)$.
4. Solve: (a) $\sin x \sec y\,dx + \sin y \sec x\,dy = 0$

   (b) $\sqrt{xy}\,dy = dx$

   (c) $\sqrt{1 - x^2}\,dy + \sqrt{1 - y^2}\,dx = 0$

   (d) $(1 + x^2)dy - (1 + y^2)dx = 0$

   (e) $dy = \sqrt{xy - x + y - 1}\,dx$

5. Describe the curves in the family of solutions of $x\,dx = y\,dy$.
6. Find the solution of $2y \cos^2 x\,dy = dx$ that goes through $(\pi/4, \sqrt{2})$.
7. Find the family of curves for which the slope at any point is equal to the square of the ordinate.
8. Find the solution of $\sqrt{1 - y^2}\,dx = dy$ that passes through $(\pi/3, 1)$.
9. A particle moving in a straight line is opposed by a frictional force which is proportional to the square of the velocity v. Show that this implies the differential equation of motion $dv/dt = -kv^2$. If $v = 1$ when $t = 0$ and $v = \frac{1}{2}$ when $t = 1$, find v when $t = 9$.
10. Water is leaking from a cylindrical tank at a rate which is proportional to the square root of the depth of water remaining. If the depth drops from 36 ft. to 25 ft. in one hour, how much longer will it take the tank to empty?

Chapter Review

1. A ship going due south at 16 knots is 10 nautical miles north of a second ship going due west at 12 knots. Find the minimum distance between the two ships.
2. Evaluate:

   (a) $\displaystyle\int_2^7 \frac{x - 1}{\sqrt{x + 2}}\,dx$ (b) $\displaystyle\int_0^1 \frac{\tan x}{\cos^2 x}\,dx$
3. Use differentials to approximate:

   (a) $(1.01)^{10}$ (b) $\sqrt[3]{126}$
4. Integrate:

   (a) $\displaystyle\int \frac{dx}{(3x - 5)^2}$ (b) $\displaystyle\int \frac{dx}{\sqrt{1 + \sqrt{x}}}$
5. A point P is tracing the circle $x^2 + y^2 = 100$ at a rate of 3 units/sec. How fast is the projection of P on the x-axis moving when P is 5 units above that axis?
6. The motion of a particle P with coordinates (x, y) is governed by the differential equations $\dfrac{dx}{dt} = 2t$, $\dfrac{dy}{dt} = 2$. If P is at the origin when $t = 0$, find its coordinates when $t = 2$.
7. The hourly fuel consumption of a vessel is proportional to the cube of its speed through the water. If it operates on a river which has a steady 6-knot current, what speed through the water uses the least fuel per mile (a) going upstream, and (b) going downstream?

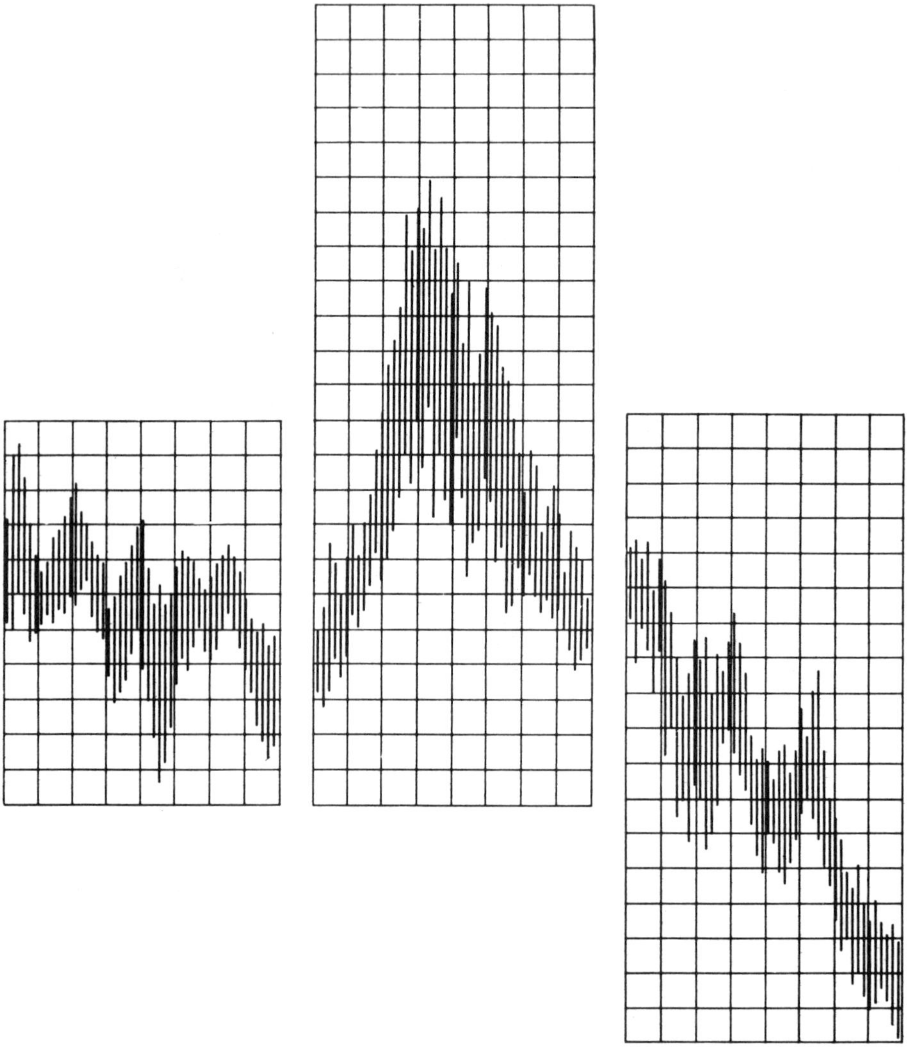

8

Graphs

8.1 Curve tracing

If a graph is drawn for the purpose of providing accurate readings of coordinates, there is no substitute for plotting a large number of points. This is a long and painstaking process. There are, however, many important mathematical questions which can be answered with the aid of a sketch of a graph, drawn, perhaps, without actually plotting any points whatsoever. Such a sketch may, for example, show clearly that a critical value of a function is a maximum or is a minimum. We shall consider in this section various bits of information, usually easy to obtain, which reveal important characteristics of a curve and permit the drawing of such a sketch.

a. Extent. The coordinates (x, y) of a point are necessarily real numbers. If an equation in x and y is solved, in turn, for each of these variables, it may become clear that certain values of one variable do not yield real values for the other, and this will then limit those values which need be considered. In particular, values will be limited by the fact that an even power of a real number is never negative and by the fact that division by zero is impossible.

Example 1. The graph of $y^2 = 4x$ does not extend to the left of the y-axis, for negative values of x would yield negative values of y^2.

Example 2. There are no points on the graph of $x^2 + 4y^2 = 16$ outside the rectangle bounded by the lines $x = \pm 4$, $y = \pm 2$, for $x^2 = 4(4 - y^2)$ will be negative if $|y| > 2$, and $y^2 = (16 - x^2)/4$ will be negative if $|x| > 4$; neither is possible.

Example 3. The cubic function of x defined by the formula $x^3 - 9x = x(x + 3)(x - 3)$ is negative if $x < -3$ or if $0 < x < 3$. Hence the graph of $y^2 = x^3 - 9x$ has no points corresponding to these values of x; it is confined to regions of the plane in which $-3 \leq x \leq 0$ or $x \geq 3$.

b. Intercepts. The intercepts on the coordinate axes are usually easy to compute and often give important information about the location and shape of the graph.

c. Derivatives. The first derivative gives useful information about critical points (see § 7.3) and regions in which the graph is ascending or descending. Points of the domain of $y = f(x)$ at which the derivative dy/dx fails to exist are commonly points at which the tangent line is vertical (parallel to the y-axis); this can be checked by examining the derivative dx/dy (see Exercises 6.4 [A], Prob. 9, and also § 7.3, Examples 1 and 2).

The second derivative gives information about the concavity of a curve and about inflection points (§ 3.3). The student should bear in mind that an inflection point is a point at which the second derivative d^2y/dx^2 changes sign. This is *usually* at a point where $d^2y/dx^2 = 0$, but there are exceptions. It is possible to have $d^2y/dx^2 = 0$ at a point which is not an inflection point (see § 3.3, Example 3), and an inflection point (with vertical tangent) may occur at a point at which d^2y/dx^2 does not exist (see § 7.3, Example 2).

d. Symmetry. If a graph can be folded along some line in such a way that the two parts exactly coincide, then that line is an *axis of symmetry* of the graph. More formally:

► **Definition.** A line L is an axis of symmetry of a graph G if the points of G can be paired in such a way that L is the perpendicular bisector of the line segment determined by each pair.

For example, every parabola has one such axis, an ellipse or a hyperbola has two, and every diameter is an axis of symmetry of a circle. There are three special cases in which the existence of such an axis is readily determined by an examination of the equation of the curve.

(i) *Symmetry in the y-axis:* If the point $(-a, b)$ is on a curve whenever the point (a, b) is on the curve, then the y-axis is an axis of symmetry. Because $(-a)^2 = (+a)^2$, an equation in which x occurs only with even exponents has a graph which is symmetric in the y-axis. The curve of Example 2 is one such graph, and because $\cos(-a) = \cos a$, the curve $y = \cos x$ is another.

(ii) *Symmetry in the x-axis:* A similar argument will show that a curve defined by an equation in which y occurs only with even exponents is symmetric in the x-axis. The curves of Examples 1, 2, and 3 are of this type.

(iii) *Symmetry in the line $y = x$:* If the point (b, a) is on a curve whenever the point (a, b) is on the curve, then the curve is symmetric in the line $y = x$; a demonstration of this is left to the student as an exercise. An equation which remains essentially unchanged (which is changed, that is, into an equivalent equation) when x and y are interchanged defines such a curve. Some examples are:

$$x + y = 0, \qquad x^2 + y^2 = 25, \qquad xy = 12, \qquad x^3 + y^3 = 3axy.$$

Another type of symmetry is symmetry in a *point*, called a *center of symmetry:*

▶ **Definition.** A point P is a center of symmetry of a graph G if the points of G can be paired in such a way that P is the midpoint of the line segment determined by each pair.

If, in particular, the point $(-a, -b)$ is on a curve whenever the point (a, b) is on the curve, then the *origin* is a center of symmetry; the demonstration is left as an exercise. If, therefore, an equation remains essentially unchanged when $-x$ is substituted for x and also $-y$ is substituted for y, its graph is symmetric in the origin. Some examples are:

$$y = x^3, \qquad y = \sin x, \qquad x^2 - xy + y^2 = 10, \qquad xy = 12.$$

If a curve is symmetric in both coordinate axes, it is also symmetric in the origin; the converse is not valid, as is shown by the example $y = x^3$.

e. Asymptotes. If, as one traces a curve in a direction of increasing $|x|$ or $|y|$ (or both), the distance between the curve and a line has zero as a limit, then that line is said to be an asymptote of the curve. It is entirely possible to consider asymptotes which are curved, but we shall generally concern ourselves only with straight-line asymptotes and especially with horizontal or vertical asymptotes.

Evidently the line $x = c$ is an asymptote of the curve $y = f(x)$ if either $\lim_{y\to\infty} x = c$ or $\lim_{y\to-\infty} x = c$. It is also possible to attack the problem from an opposite point of view and note that $x = c$ is an asymptote if $|y|$ increases without bound as x approaches c. This is often the more convenient point of view, for if f can be expressed as the quotient of functions n and d, so that $y = n(x)/d(x)$, and if c is a zero of the denominator d but not of the numerator n, then evidently $|y|$ increases without bound as x approaches c, and $x = c$ is therefore an asymptote.

Similarly, the curve $x = g(y)$ has the asymptote $y = c$ if either $\lim_{x\to\infty} y = c$ or $\lim_{x\to-\infty} y = c$, or if $|x|$ increases without bound as y approaches c.

Example 4. The equation $xy = 12$ can be written $y = 12/x$; therefore the line $x = 0$ (the y-axis) is an asymptote, and since the equation can also be written $x = 12/y$, the x-axis is also an asymptote.

Example 5. Since $|\tan x|$ increases without bound as x approaches $\pi/2$, the line $x = \pi/2$ is an asymptote of $y = \tan x$.

f. Check points. It may also be desirable to compute the coordinates of a few points.

We have listed above six possible sources of information about a curve. In a given instance, one or more of these may prove to involve difficult algebraic manipulation; if so, an effort should be made to infer the necessary information from other sources. The list is merely a guide, and you should not feel required to check off each of the six items in full detail each time a sketch is made. As you analyze the equation, you should indicate on the graph the information you obtain, continuing your analysis only until you are able to complete the sketch.

Example 6. Analyze and sketch: $y^2 = x^3$.

Solution:

a. The curve is not defined for negative x, but all non-negative values of x are admissible.

b. When $x = 0$, $y = 0$, and vice versa; the curve makes its only contact with the axes at the origin.

c. $\dfrac{dy}{dx} = \dfrac{3x^2}{2y}$; this has the same sign as y and is indeterminate at $(0, 0)$. If, however, we make the substitution $y = \pm x^{3/2}$, we get $\dfrac{dy}{dx} = \dfrac{3x^2}{\pm 2x^{3/2}} = \pm \dfrac{3}{2} x^{1/2}$, and this equals 0 when $x = 0$; the curve has zero slope at the origin.

We have determined that the origin is the only critical point; that it is neither a maximum nor a minimum can be inferred from the next item in the analysis.

d. The curve is symmetric in the x-axis.

e. There are no asymptotes.

f. It is worth noting that $(1, 1)$ and $(1, -1)$ are on the curve.

For the graph see Fig. 1.

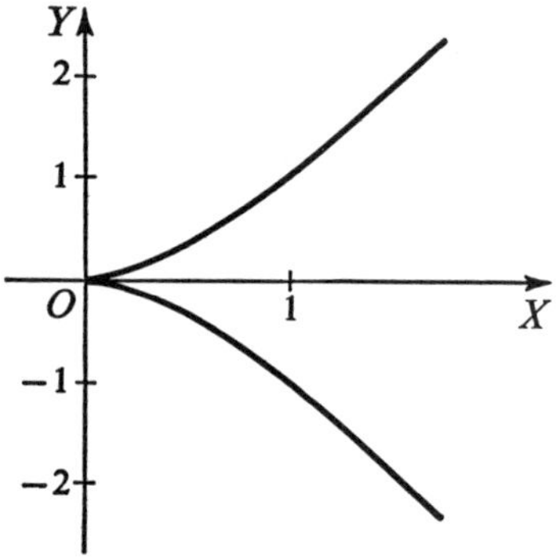

Fig. 1

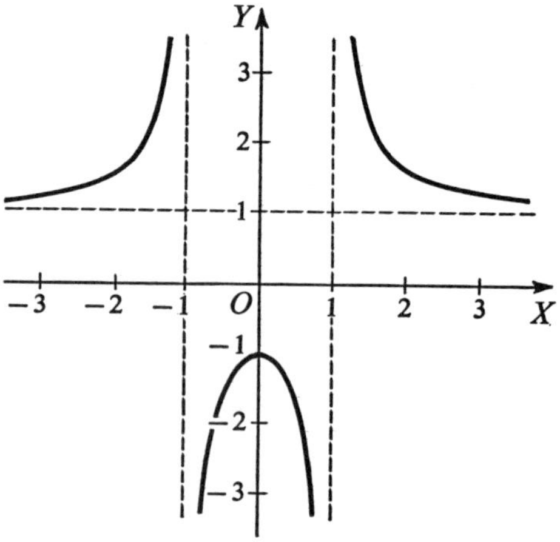

Fig. 2

Example 7. Analyze and sketch: $y = \dfrac{x^2 + 1}{x^2 - 1}$. (1)

Solution:

a. No limitations on the extent of the curve are immediately apparent, except that the values $x = \pm 1$ involve division by zero and are therefore impossible. If, however, we solve for x^2,

$$x^2 = \frac{y + 1}{y - 1}, \tag{2}$$

we find a fraction that will be negative if numerator and denominator have opposite signs, that is, if $-1 < y < 1$; this would imply a negative square, and there are, therefore, no points of the curve in this region.

b. We see from the preceding paragraph that the curve does not cross the x-axis. If $x = 0$, $y = -1$: the only y-intercept is -1.

c. Since $\dfrac{dy}{dx} = -\dfrac{4x}{(x^2 - 1)^2}$, the slope is of opposite sign to x; the curve is ascending on the left of the y-axis, descending on the right. The only critical point, $(0, -1)$, is therefore a maximum.

d. The curve is symmetric in the y-axis.

e. From (1), it is evident that the lines $x = \pm 1$ are asymptotes, since these are zeros of the denominator; the same information can be obtained by finding the limit of (2) as $y \to \infty$. By examining the denominator of (2), or by letting $x \to \infty$ in (1), we discover that $y = 1$ is also an asymptote.

f. No special need for further points is apparent.

It may be useful to note, from (1), that y is negative for $-1 < x < 1$ but otherwise positive (and therefore > 1; see part **a**). The graph is shown in Fig. 2 at the foot of the opposite page.

Exercises 8.1 [A]

Analyze and sketch:

1. $y = x - \dfrac{1}{x}$
2. $x^2 y = 1$
3. (a) $y = x^2 + 1$

   (b) $y = \dfrac{1}{x^2 + 1}$
4. $x^4 + y^4 = a^4$
5. $y = \dfrac{x + 1}{x - 1}$
6. (a) $y = x^3 - 4x$

   (b) $y^2 = x^3 - 4x$
7. $y = \dfrac{1}{x^2 - 1}$
8. $\sqrt{x} + \sqrt{y} = \sqrt{a}$

Exercises 8.1 [B]

Analyze and sketch:

1. $y^3 = x^2$
2. $x^2y^2 = 16$
3. $y = \dfrac{x}{x^2 + 1}$
4. $y = \dfrac{x^2}{x^2 + 4}$
5. $y^2 = x^3 - x^2 - 2x$
6. $x^3 + y^3 = a^3$
7. $y = \dfrac{x}{x^2 - 1}$
8. $x^{2/3} + y^{2/3} = a^{2/3}$

Miscellaneous Exercises 8.1

1. Prove that the segment joining (a, b) and $(-a, -b)$ is bisected by the origin.
2. Prove that the line $y = x$ is the perpendicular bisector of the segment joining (a, b) and (b, a).
3. State and prove a test for symmetry in the line **(a)** $y = -x$; **(b)** $x = k$; **(c)** $y = k$.

Analyze and sketch:

4. **(a)** $y = x(x - 2)^2$
   (b) $y^2 = x(x - 2)^2$
5. $y = \dfrac{x}{x^2 + x + 2}$
6. $y = \dfrac{x}{x^2 + x - 2}$
7. $y = \dfrac{x - 1}{x(x - 2)}$
8. $y^2 = x^2 - x^4$
9. $y = \dfrac{x^2 - 1}{x^2 + 1}$
10. $y^2 = \dfrac{x^3}{2a - x}$ (cissoid)
11. $y^2 = \dfrac{x^2(a + x)}{a - x}$ (strophoid)
12. $y^2 = x^3(1 - x)$
13. $y^2 = \dfrac{x^2 + 1}{x^2 - 1}$
14. $y^2 = x^4(1 + x)$
15. $(y - x^2)^2 = x^5$

8.2 Parametric equations

There are many ways of giving a mathematical description of a curve. We have seen that one coordinate may be given as a function of the other, as $y = f(x)$ or $x = g(y)$. Because of the special conditions that the graph of a function must meet (§ 2.1), this method is rather severely limited in its application; we cannot use it, for example, to describe a curve that has a loop. We gain a little generality if we state the conditions on the coordinates x and y implicitly, as in the equation $x^2 + y^2 = 25$, but this, too, has many limitations; in particular, it is inconvenient to have to specify both x and y to determine a unique point or to evaluate a derivative. In this section, we shall consider still another way of describing a curve, a way more general and often more convenient than either of the foregoing.

Suppose a curve, drawn on a piece of paper, is traced by the point of a pencil. The curve may loop, zigzag, cross itself many times, but throughout the motion of the pencil the coordinates x and y of its point are determined as functions of the time t:

$$x = f_1(t), \quad y = f_2(t). \tag{1}$$

We may, in fact, take the equations (1) as a description of the curve, and, abandoning the physical picture of a moving pencil, think of t as a simple numerical variable, having no necessary connection with time. Under these circumstances, the variable t is called a parameter, and the equations (1) are a parametric representation of the curve. It may be possible to eliminate t between the equations (1) and thus obtain an equation for the curve in a more familiar form (called a *rectangular*, or *Cartesian*, form), but for many purposes the parametric form is the more convenient.

Example 1. Sketch the curve represented by $x = t$, $y = 1 - 2t^2$, and obtain a Cartesian equation.

Solution: The table on the following page gives the computed values of x and y used in plotting the curve, which is shown below.

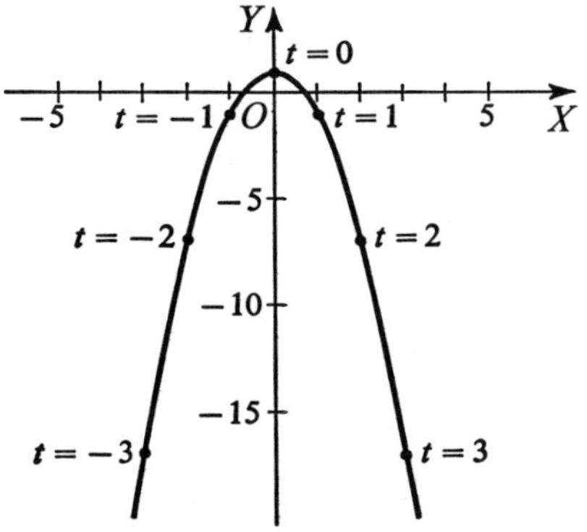

Fig. 3

Using the equations $x = t$ and $y = 1 - 2t^2$, and substituting integral values of t on the interval $[-3, 3]$, we have:

$t =$	-3	-2	-1	0	1	2	3
$x =$	-3	-2	-1	0	1	2	3
$y =$	-17	-7	-1	1	-1	-7	-17

As Fig. 3 suggests, the curve is a parabola.

To obtain the required Cartesian equation, we substitute in the second of the parametric equations the value of t in terms of x that is given by the first: $y = 1 - 2x^2$.

Example 2. Sketch, and derive a Cartesian equation: $x = \sin t$, $y = \sin 2t$.

Solution: It is clear from the equations that we need consider only $t \in [0, 2\pi)$. We find

$t = 0$	$\frac{\pi}{4}$	$\frac{\pi}{2}$	$\frac{3\pi}{4}$	π	$\frac{5\pi}{4}$	$\frac{3\pi}{2}$	$\frac{7\pi}{4}$
$x = 0$	0.7	1	0.7	0	-0.7	-1	-0.7
$y = 0$	1	0	-1	0	1	0	-1

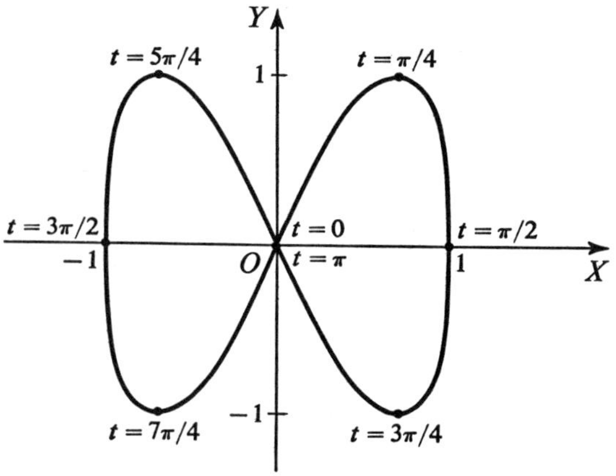

Fig. 4

(Some of the values of x are approximations.) A more finished result could be obtained by using more values of t, but these will serve for a sketch.

Since
$$y = \sin 2t = 2 \sin t \cos t,$$
$$\sin t = x,$$
and
$$\cos t = \pm\sqrt{1 - \sin^2 t} = \pm\sqrt{1 - x^2},$$
we have
$$y = \pm 2x\sqrt{1 - x^2},$$
or
$$y^2 = 4x^2(1 - x^2),$$
as the required Cartesian equation.

In many cases the Cartesian equation obtained by elimination of a parameter represents more than do the original parametric equations. It is, therefore, essential to state explicitly any limitations of domain which may be inherent in the parametric form but which may not appear in the Cartesian form.

Example 3. Convert to Cartesian form: $x = \sin\theta$, $y = \cos 2\theta$.

Solution: The parameter θ is easily eliminated:

$$y = \cos 2\theta = 1 - 2\sin^2\theta = 1 - 2x^2.$$

This is the same as the Cartesian equation of Example 1. In Example 1, however, there is no restriction on the domain of x, whereas in this case the parametric definition clearly restricts x to $[-1, 1]$. As θ takes all possible values, the point (x, y) merely traces and retraces the arc bounded by the points $(-1, -1)$ and $(1, -1)$. The Cartesian equation of this curve should therefore be stated

$$y = 1 - 2x^2, \quad -1 \leq x \leq 1.$$

As was observed in § 7.1, the derivative dy/dx can be obtained directly from parametric equations, as a function of the parameter, without changing to the Cartesian form. If parametric equations define y as a differentiable function of x, and this is usually the case, then the derivative dy/dx exists, but since x and y are given as functions of the parameter t, we apply the chain rule:

$$\frac{dy}{dt} = \frac{dy}{dx} \cdot \frac{dx}{dt}$$

and therefore

$$\frac{dy}{dx} = \frac{dy/dt}{dx/dt}.$$

In terms of differentials, this relation, like any application of the chain rule, reduces to an algebraic identity.

The computation of second derivatives from parametric equations holds a trap for the unwary. One must carefully bear in mind that the second derivative d^2y/dx^2 is to be found by differentiating the first derivative dy/dx *with respect to* x, but since the first derivative is given as a *function of* t, it is therefore necessary to use the chain rule:

$$\frac{d^2y}{dx^2} = \frac{d}{dx}\left(\frac{dy}{dx}\right) = \frac{d}{dt}\left(\frac{dy}{dx}\right) \cdot \frac{dt}{dx} = \frac{\dfrac{d}{dt}\left(\dfrac{dy}{dx}\right)}{dx/dt},$$

since dx/dt is the reciprocal of dt/dx.

Example 4. Find the derivatives dy/dx and d^2y/dx^2 from the parametric equations of Examples 1–3.

Solution:

(1) We have $x = t$, $y = 1 - 2t^2$; hence $dx = dt$, $dy = -4t\,dt$, and $dy/dx = -4t$. Therefore

$$\frac{d^2y}{dx^2} = \frac{\dfrac{d}{dt}(-4t)}{1} = -4.$$

(2) We have $x = \sin t$, $y = \sin 2t$; hence $dx = \cos t\,dt$, $dy = 2\cos 2t\,dt$, and $\dfrac{dy}{dx} = \dfrac{2\cos 2t}{\cos t}$.

$$\frac{d^2y}{dx^2} = \frac{\dfrac{d}{dt}\left(\dfrac{2\cos 2t}{\cos t}\right)}{\cos t} = \frac{\cos t(-4\sin 2t) - 2\cos 2t(-\sin t)}{\cos^2 t} \cdot \frac{1}{\cos t}$$

$$= \frac{-2\sin t(2\cos^2 t + 1)}{\cos^3 t}.$$

The student should carry through the final reduction.

(3) Since $x = \sin\theta$, $y = \cos 2\theta$, we find $dx = \cos\theta\,d\theta$, $dy = -2\sin 2\theta\,d\theta$, and therefore

$$\frac{dy}{dx} = -\frac{2\sin 2\theta}{\cos\theta} = -4\sin\theta \quad \text{(why?)}.$$

$$\frac{d^2y}{dx^2} = -\frac{4\cos\theta}{\cos\theta} = -4.$$

Compare this result with that in part (1) of this example.

In using integration to find areas or to find volumes of revolution, we have usually encountered integrals of the form

$$\int_a^b y\,dx \quad \text{or} \quad \pi\int_a^b y^2\,dx.$$

Since y is generally a known function of x, $y = f(x)$, we have customarily replaced y in these integrals by its equal $f(x)$, thus obtaining an integrand involving only the one variable x. An alternative procedure is to replace dx by its equivalent in terms of y and dy, and another is to replace both y and dx by their equivalents in terms of a third variable. This is what we do when we introduce an auxiliary variable (often called u in earlier chapters), and this is what we usually do when x and y are given in terms of a parameter.

Example 5. Find the area of the region enclosed by the curve of Example 2, and also the volume obtained by rotating that region about the x-axis.

Solution: To take advantage of the symmetry of the curve, we shall find the total area as four times the area in the first quadrant. Then

$$A = 4\int_0^1 y\,dx.$$

But

$$y = \sin 2t, \quad x = \sin t;$$

therefore

$$dx = \cos t\,dt,$$

and x goes from 0 to 1 as t goes from 0 to $\pi/2$. Hence

$$A = 4\int_0^{\pi/2} \sin 2t \cos t\,dt = 8\int_0^{\pi/2} \sin t \cos^2 t\,dt = -8\int_0^{\pi/2} \cos^2 t\,d(\cos t)$$

$$= -\tfrac{8}{3}\cos^3 t\Big|_0^{\pi/2} = \tfrac{8}{3}.$$

In finding the volume, symmetry permits us to take twice the right-hand half:

$$V = 2\pi\int_0^1 y^2\,dx$$

(see § 4.6). But we have values of y and dx in terms of t, so

$$V = 2\pi\int_0^{\pi/2} \sin^2 2t \cos t\,dt = 8\pi\int_0^{\pi/2} \sin^2 t \cos^3 t\,dt$$

$$= 8\pi\int_0^{\pi/2} \sin^2 t(1 - \sin^2 t)\cos t\,dt$$

$$= 8\pi\int_0^{\pi/2} (\sin^2 t - \sin^4 t)d(\sin t) = 8\pi\left[\frac{1}{3}\sin^3 t - \frac{1}{5}\sin^5 t\right]_0^{\pi/2}$$

$$= 8\pi\left(\frac{1}{3} - \frac{1}{5}\right) = \frac{16\pi}{15}.$$

We could, of course, have obtained the same results without using the parameter t.

Exercises 8.2 [A]

1. Derive a Cartesian equation, and sketch:

(a) $x = a \sin t,\ y = b \sin t$ (b) $x = a \sec t,\ y = b \sec t$

(c) $x = a \tan t,\ y = b \tan t$

2. Sketch, find the derivatives dy/dx and d^2y/dx^2 in parametric form, and derive a Cartesian equation:

(a) $x = a \cos \theta,\ y = a \sin \theta$ (b) $x = pt^2,\ y = 2pt$

(c) $x = 1 + \cos 2\phi,\ y = \sin 2\phi$

3. Derive parametric equations for $y^2 = x^2 - x^3$ by putting $y = tx$ and solving for x and y in terms of t.

4. Find, by integrating with respect to the parameter t, the area of the region enclosed by the parabola $x = t + 2,\ y = 9 - t^2$ and the x-axis, and also the volume of the solid generated by rotating this region about the x-axis.

5. A curve is defined in terms of the parameter t by the differential equations

$$\frac{dx}{dt} = 3 + 2t, \quad \frac{dy}{dt} = 5.$$

If the value $t = 0$ of the parameter corresponds to the point (0, 0) on the curve, find parametric equations of the curve.

Exercises 8.2 [B]

1. Express in Cartesian form, and sketch:

(a) $x = t^2,\ y = 2t^2 + 1$ (b) $x = \cos 2\theta,\ y = 3 - 4 \sin^2 \theta$

(c) $x = \frac{1}{t},\ y = \frac{t+2}{t}$

2. Sketch, find the derivatives dy/dx and d^2y/dx^2 in parametric form, and derive a Cartesian equation:

(a) $x = a \sec t,\ y = a \tan t$ (b) $x = \frac{4p}{t^2},\ y = \frac{4p}{t}$

(c) $x = t - t^2,\ y = t^2 - 1$

3. In the equation of the straight line through (x_1, y_1) and (x_2, y_2), set $y = y_1 + t(y_2 - y_1)$ and show that the corresponding expression for x is then $x = x_1 + t(x_2 - x_1)$. Explain geometrically the distinctions among $t < 0$, $0 \leq t \leq 1$, and $t > 1$.

4. Find, by integrating with respect to the parameter θ, the area of the circle $x = a \cos \theta,\ y = a \sin \theta$, and the volume of the sphere generated by rotating this circle about the x-axis.

5. A curve is defined in terms of the parameter t by the differential equations $dx/dt = 2t - 1$, $dy/dt = x$, and the conditions that $x = 2$ and $y = 5$ when $t = 0$. Find parametric equations of the curve.

8.3 The circle

A circle is defined to be the locus (set) of points in a plane at a given distance from a given fixed point. If the given distance (the radius) is r, and the given point (the center) is (h, k), then the distance formula (equation (1) of § 1.3) tells us that $P(x, y)$ is on the circle if and only if

$$(x - h)^2 + (y - k)^2 = r^2. \tag{1}$$

This is the standard Cartesian equation of the circle in *center-radius form.* If we perform the indicated multiplications and rearrange the terms, we obtain

$$x^2 + y^2 - 2hx - 2ky + h^2 + k^2 - r^2 = 0;$$

if we now set $a = -2h$, $b = -2k$, and $c = h^2 + k^2 - r^2$, we have

$$x^2 + y^2 + ax + by + c = 0, \tag{2}$$

which is called the *general form* of the equation. Conversely, we can transform (2) into (1) by completing the squares of the terms in x and the terms in y, as illustrated in Example 1, page 272.

In many applications it is possible to choose coordinates in such a way that the origin is at the center of the circle; in this case, of course, (1) takes the particularly simple form

$$x^2 + y^2 = r^2. \tag{3}$$

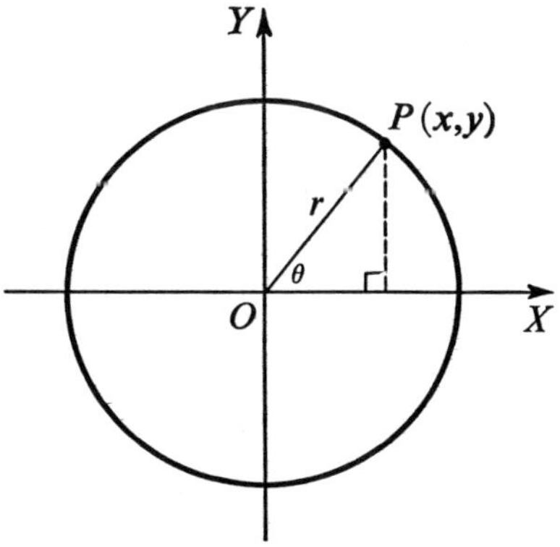

Fig. 5

If a circle has radius r and center at the origin, and the radius to $P(x, y)$ makes a positive angle θ with the positive x-axis, as shown in Fig. 5, we find

$$x = r\cos\theta, \quad y = r\sin\theta. \tag{4}$$

Conversely, if the coordinates (x, y) of a point satisfy (4), then

$$x^2 + y^2 = r^2(\cos^2\theta + \sin^2\theta) = r^2,$$

and the point lies on the circle (3). Equations (4) are therefore a set of parametric equations for (3).

Example 1. Find the center and radius of $2x^2 + 2y^2 - 6x + 18y - 5 = 0$.

Solution: If we divide the given equation by 2, we have

$$x^2 + y^2 - 3x + 9y - \tfrac{5}{2} = 0,$$

which is in the general form (2). If we rewrite this in the center-radius form (1), we can then read the coordinates of the center, and the radius, directly from the equation. To make this transformation, we complete the squares of the terms in x and the terms in y:

$$(x^2 - 3x + \quad) + (y^2 + 9y + \quad) = \tfrac{5}{2};$$

$$(x^2 - 3x + \tfrac{9}{4}) + (y^2 + 9y + \tfrac{81}{4}) = \tfrac{5}{2} + \tfrac{9}{4} + \tfrac{81}{4} = 25;$$

$$(x - \tfrac{3}{2})^2 + (y + \tfrac{9}{2})^2 = 5^2.$$

By comparison with (1), we see that the center is at $(\tfrac{3}{2}, -\tfrac{9}{2})$ and that the radius is 5.

Example 2. Find an equation of the tangent to the circle (3) in terms of

(**a**) the coordinates (x_1, y_1) of the point of tangency;
(**b**) the corresponding value θ_1 of the parameter θ.

Solution: (**a**) From (3), we find $\dfrac{dy}{dx} = -\dfrac{x}{y}$; the equation of the tangent can therefore be written

$$y - y_1 = -\frac{x_1}{y_1}(x - x_1),$$

or

$$x_1x + y_1y = x_1^2 + y_1^2.$$

Since (x_1, y_1) lies on the circle, its coordinates satisfy (3):

$$x_1^2 + y_1^2 = r^2.$$

The equation of the tangent can therefore be written in the simpler form

$$x_1x + y_1y = r^2.$$

The device used to simplify this equation will be of use in many exercises: if a point lies on a curve, its coordinates must satisfy the equation of the curve, and that equation may often be used to simplify expressions containing those coordinates.

(**b**) From (4), $dx = -r \sin\theta\, d\theta$, $dy = r\cos\theta\, d\theta$, and the equation of the tangent can therefore be written

$$y - r\sin\theta_1 = \frac{r\cos\theta_1}{-r\sin\theta_1}(x - r\cos\theta_1),$$

or

$$x\cos\theta_1 + y\sin\theta_1 = r(\cos^2\theta_1 + \sin^2\theta_1) = r.$$

Exercises 8.3 [A]

1. Find the center and radius of the following:

   (a) $x^2 + y^2 - 2x - 10y + 10 = 0$

   (b) $x^2 + y^2 + x - 6y + 9 = 0$

   (c) $x^2 + y^2 + 2x - 2y + 6 = 0$

   (d) $12x^2 + 12y^2 - 36x + 8y - 21 = 0$

2. Derive a set of parametric equations, similar to (4), for the circle with radius r and center at (h, k).

3. Show that an equation of a line tangent at (x_1, y_1) to the circle with radius r and center at (h, k) can be written

$$(x_1 - h)(x - h) + (y_1 - k)(y - k) = r^2.$$

4. Prove analytically that the angle inscribed in a semicircle is a right angle.

5. Find equations of the tangents from the origin to $x^2 + y^2 - 6x - 10y + 32 = 0$. (*Hint:* It is easier *not* to use calculus in this problem; represent the required line by $y = mx$ and use the discriminant of a quadratic equation to find the values of m for which the line meets the circle in two coincident points.)

6. Show that the line $y = mx \pm r\sqrt{m^2 + 1}$ is tangent to the circle $x^2 + y^2 = r^2$ for every real number m. (See hint for Ex. 5.)

7. A rectangle is inscribed in a semicircle. Express its area in terms of the radius r and the value at one of its vertices of the parameter θ, and use this expression to find the maximum value of that area.

8. Find an equation of the family of curves defined by the differential equation $\dfrac{dy}{dx} = -\dfrac{x - h}{y - k}$, and find an equation of the member of the family that goes through the origin.

9. Find an equation of the circle that goes through $(3, -4)$ and is tangent at $(3, 2)$ to the line $x + y - 5 = 0$.

10. Show that the circles $x^2 + y^2 - 10y = 0$ and $x^2 + y^2 + 10x = 0$ are orthogonal.

Exercises 8.3 [B]

1. Find the center and radius of the circle $x^2 + y^2 + ax + by + c = 0$. Under what conditions does this equation not define a real locus?

2. Show that an equation of the line tangent to $x^2 + y^2 + ax + by + c = 0$ at (x_1, y_1) is

$$x_1x + y_1y + \frac{a}{2}(x + x_1) + \frac{b}{2}(y + y_1) + c = 0.$$

3. Find an equation of the circle through (0, 1), (3, 0), and (0, 4).
4. Find the slopes of the lines through (0, 10) tangent to $x^2 + y^2 = 25$.
5. Find equations of the circles that are tangent to the coordinate axes and have their centers on the line $x + 2y - 12 = 0$.
6. Show that the points (4, 0), (0, 0), (0, 6), and (5, 5) are concyclic.
7. The point $P(x, y)$ is traveling around a circle of radius r in the counterclockwise direction at a constant speed of v units/sec. Taking $t = 0$ when the parameter $\theta = 0$, show that then $\theta = vt/r$, and that the abscissa x of P therefore satisfies the differential equation

$$\frac{d^2x}{dt^2} = -\left(\frac{v}{r}\right)^2 x.$$

This differential equation describes the motion of the projection of P on the x-axis, called simple harmonic motion. What is the general nature of this motion?

8. Find the angles of intersection of the circles $x^2 + y^2 = 100$ and $x^2 + y^2 - 28x + 124 = 0$.
9. Prove analytically that the perpendicular to a diameter from a point on a circle is the mean proportional between the segments into which it divides the diameter.
10. Show that the line $x_1x + y_1y = r^2$ meets the circle $x^2 + y^2 = r^2$ in two distinct points if and only if ${x_1}^2 + {y_1}^2 > r^2$. What does this inequality imply about the point (x_1, y_1)?

8.4 The ellipse

An ellipse is defined to be the locus (set) of points in a plane the sum of whose distances from two fixed points is constant. The fixed points are called the foci (singular, focus); we shall designate them F_1 and F_2. It is convenient to represent the constant sum of the definition by $2a$; the point P is then a point of the locus if and only if

$$PF_1 + PF_2 = 2a. \tag{1}$$

If we take the x-axis through the foci, with the origin O midway between them, and let the distance $F_1F_2 = 2c$, then F_1 and F_2 have coordinates

$(c, 0)$ and $(-c, 0)$, respectively. A segment that connects a point of the curve with a focus is a focal radius, and a segment that has both ends on the curve and passes through a focus is a focal chord. If the curve crosses the coordinate axes at A_1 and A_2, B_1 and B_2, as shown in Fig. 6, then the

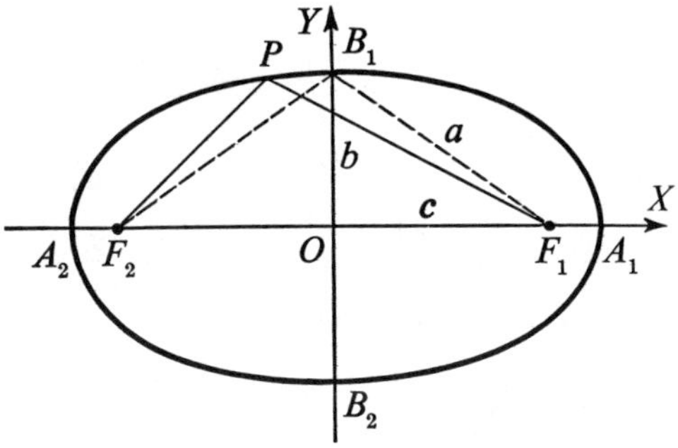

Fig. 6

segments A_1A_2 and B_1B_2 are called, respectively, the major axis and the minor axis of the curve, and the points A_1 and A_2, the ends of the major axis, are the vertices.

Since A_1 is on the curve,

$$A_1F_1 + A_1F_2 = 2a,$$

or

$$(OA_1 - OF_1) + (F_2O + OA_1) = 2a,$$

and therefore, since $OF_1 = F_2O$,

$$OA_1 = a = OA_2.$$

We let $B_1B_2 = 2b$; then, since $B_1F_1 = B_1F_2 = a$, we have, in the right triangle B_1F_1O,

$$a^2 = b^2 + c^2. \tag{2}$$

If $c = 0$, F_1 and F_2 are both at O, and (1) reduces to $PO = a$; the ellipse reduces to a circle of radius a. As the foci are moved away from the origin, the curve becomes longer and narrower, less like a circle. The quantity

▶ $$\mathbf{e = \frac{c}{a},} \tag{3}$$

called the eccentricity, provides a measure of the departure from circularity. Since $c < a$, we have $e < 1$. Any two ellipses with the same eccentricity are similar.

If $P(x, y)$ is any point on the ellipse, $PF_1 = \sqrt{(x-c)^2 + y^2}$, $PF_2 = \sqrt{(x+c)^2 + y^2}$, and (1) can therefore be stated

$$\sqrt{(x-c)^2 + y^2} + \sqrt{(x+c)^2 + y^2} = 2a.$$

This can be considerably simplified:

$$\sqrt{(x-c)^2 + y^2} = 2a - \sqrt{(x+c)^2 + y^2},$$

$$x^2 - 2cx + c^2 + y^2 = 4a^2 - 4a\sqrt{(x+c)^2 + y^2} + x^2 + 2cx + c^2 + y^2,$$

$$a\sqrt{(x+c)^2 + y^2} = a^2 + cx,$$

$$a^2x^2 + 2a^2cx + a^2c^2 + a^2y^2 = a^4 + 2a^2cx + c^2x^2,$$

$$(a^2 - c^2)x^2 + a^2y^2 = a^2(a^2 - c^2).$$

From (2),

$$a^2 - c^2 = b^2,$$

so we have

$$b^2x^2 + a^2y^2 = a^2b^2,$$

or, dividing by a^2b^2,

$$\blacktriangleright \qquad \frac{\mathbf{x}^2}{\mathbf{a}^2} + \frac{\mathbf{y}^2}{\mathbf{b}^2} = \mathbf{1}. \tag{4}$$

We have thus proved that if a point P satisfies (1) then its coordinates (x, y) satisfy (4). The derivation can be retraced; it is possible to start with (4) and, paying careful attention to the signs of various square roots, work back to (1). Therefore we see that a point satisfies (1) if its coordinates satisfy (4). Hence a point lies on the ellipse if and only if its coordinates satisfy equation (4), the standard Cartesian equation of this curve.

We note from (4) that the ellipse is symmetric in both axes and hence in the origin; this, then, permits us to call O the center of the curve. The extent of the curve (see § 8.1) is limited by the conditions

$$x^2 \leq a^2, \quad y^2 \leq b^2.$$

The curve is therefore entirely contained within the rectangle bounded by the four lines

$$x = \pm a, \quad y = \pm b.$$

This rectangle, which is tangent to the curve at the ends of its axes, is a great convenience in sketching the curve.

If $b > a$, (4) is still the equation of an ellipse, but the foci, vertices, and major axis all lie on the y-axis, and the preceding geometrical discussion must be modified in certain rather obvious ways. In particular, (2) and (3) become

$$b^2 = a^2 + c^2 \quad \text{and} \quad e = \frac{c}{b},$$

respectively.

If we consider the ellipse with major axis horizontal and center at (h, k), it will evidently have foci at $(h \pm c, k)$ and vertices at $(h \pm a, k)$, and therefore

$$PF_1 = \sqrt{[x - (h + c)]^2 + (y - k)^2} = \sqrt{[(x - h) - c]^2 + (y - k)^2},$$

$$PF_2 = \sqrt{[x - (h - c)]^2 + (y - k)^2} = \sqrt{[(x - h) + c]^2 + (y - k)^2}.$$

These expressions for PF_1 and PF_2 differ from the earlier ones only in that x has been replaced by $(x - h)$ and y by $(y - k)$. The derivation which led to (4) can then be carried through with these replacements and will evidently lead to

▶ $$\frac{(x - h)^2}{a^2} + \frac{(y - k)^2}{b^2} = 1, \tag{5}$$

a Cartesian equation of the ellipse with center at (h, k) and axes parallel to the coordinate axes.

In the circle $x^2 + y^2 = a^2$, we draw a radius OQ making an angle θ with the positive x-axis, draw QM perpendicular to the x-axis at M, and mark a point P on QM such that $PM = (b/a)QM$, where b is any positive number whatever. (See Fig. 7.) Then the coordinates (x, y) of P will be such that

▶ $$x = a \cos \theta, \quad y = b \sin \theta. \tag{6}$$

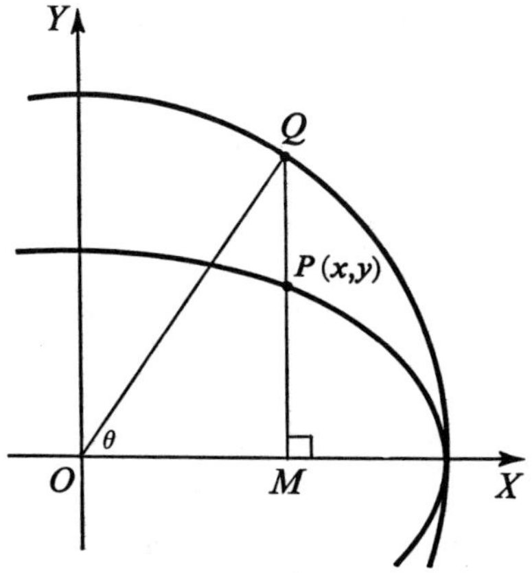

Fig. 7

But then
$$\frac{x^2}{a^2} + \frac{y^2}{b^2} = \cos^2\theta + \sin^2\theta = 1,$$

and the equations (6) are therefore equivalent to (4) and are, then, a set of parametric equations for the ellipse (4).

Example 1. Locate the foci and vertices and determine the eccentricity of $x^2 + 16y^2 = 25$.

Solution: We can write this equation in the form
$$\frac{x^2}{5^2} + \frac{y^2}{(\frac{5}{4})^2} = 1.$$

This is the form (4), and we can therefore read directly from it $a = 5$, $b = \frac{5}{4}$. Then from (2),
$$c^2 = 25 - \frac{25}{16} = \frac{375}{16},$$

and therefore
$$c = \tfrac{5}{4}\sqrt{15}.$$

Hence the foci are at $(\pm\frac{5}{4}\sqrt{15}, 0)$, the vertices are at $(\pm 5, 0)$, and, from (3), the eccentricity $e = \frac{1}{4}\sqrt{15}$.

Example 2. Locate the foci and vertices and determine the eccentricity of the ellipse $x^2 + 16y^2 - 4x + 96y + 123 = 0$.

Solution: We first locate the center, as in the case of the circle, by completing the squares of the terms in x and in y:
$$(x^2 - 4x + \quad) + 16(y^2 + 6y + \quad) = -123;$$
$$(x^2 - 4x + 4) + 16(y^2 + 6y + 9) = -123 + 4 + 144;$$
$$(x - 2)^2 + 16(y + 3)^2 = 25;$$
$$\frac{(x - 2)^2}{5^2} + \frac{(y + 3)^2}{(\frac{5}{4})^2} = 1.$$

This is the equation of an ellipse with center at $(2, -3)$, and because the values of a and b are the same as those found in Example 1, we conclude that the foci are at $(2 \pm \frac{5}{4}\sqrt{15}, -3)$, that the vertices are at $(7, -3)$ and $(-3, -3)$, and that the eccentricity is $\frac{1}{4}\sqrt{15}$.

Example 3. Find the length of a focal chord perpendicular to the major axis in the ellipse (4) (this chord is known as the latus rectum of the ellipse).

Solution: The latus rectum through F_1 is a segment of the line $x = c$. If we substitute this value of x in (4), we have

$$\frac{c^2}{a^2} + \frac{y^2}{b^2} = 1,$$

or

$$a^2y^2 = a^2b^2 - b^2c^2 = b^2(a^2 - c^2) = b^4,$$

since, from (2),

$$a^2 - c^2 = b^2.$$

Hence, $y = \pm \frac{b^2}{a}$, and the required length is $\frac{2b^2}{a}$.

Example 4. Show that the focal radii to a point on an ellipse make equal angles with the tangent at that point.

Solution: From (4), we find $\frac{dy}{dx} = -\frac{b^2x}{a^2y}$, so if the given point is $P(x_1, y_1)$, then the slope of the tangent at P is $-\frac{b^2x_1}{a^2y_1}$. The slope of PF_1 is $\frac{y_1}{x_1 - c}$, and the angle ϕ_1 between the tangent and PF_1 is then given by (§ 1.3)

$$\tan \phi_1 = \frac{-\frac{b^2x_1}{a^2y_1} - \frac{y_1}{x_1 - c}}{1 - \frac{b^2x_1y_1}{a^2y_1(x_1 - c)}} = \frac{-b^2x_1^2 + b^2cx_1 - a^2y_1^2}{a^2x_1y_1 - a^2cy_1 - b^2x_1y_1}$$

$$= \frac{b^2cx_1 - (b^2x_1^2 + a^2y_1^2)}{y_1[(a^2 - b^2)x_1 - a^2c]}.$$

But, from (4), $b^2x_1^2 + a^2y_1^2 = a^2b^2$, and, from (2), $a^2 - b^2 = c^2$, so

$$\tan \phi_1 = \frac{b^2cx_1 - a^2b^2}{y_1(c^2x_1 - a^2c)} = \frac{b^2(cx_1 - a^2)}{cy_1(cx_1 - a^2)} = \frac{b^2}{cy_1}.$$

Similarly, the slope of PF_2 is $\frac{y_1}{x_1 + c}$ and the angle ϕ_2 between PF_2 and the tangent is then determined by

$$\tan \phi_2 = \frac{\frac{y_1}{x_1 + c} + \frac{b^2x_1}{a^2y_1}}{1 - \frac{b^2x_1y_1}{a^2y_1(x_1 + c)}} = \frac{a^2y_1^2 + b^2x_1^2 + b^2cx_1}{a^2x_1y_1 + a^2cy_1 - b^2x_1y_1}$$

$$= \frac{(a^2y_1^2 + b^2x_1^2) + b^2cx_1}{y_1[(a^2 - b^2)x_1 + a^2c]} = \frac{a^2b^2 + b^2cx_1}{y_1(c^2x_1 + a^2c)}$$

$$= \frac{b^2}{cy_1} = \tan \phi_1.$$

A sketch will show that the angles ϕ_1 and ϕ_2 are therefore equal and not supplementary (for a discussion of this possible ambiguity, see § 1.3).

Example 4 proves what is known as the "reflection property" of the ellipse, for if a reflecting surface is generated by rotating an ellipse about its major axis, and a source of light is placed at one focus, then a consequence of this property and the law of reflection is that all the light, in whatever direction it may leave the source, will be reflected to the other focus.

Example 5. Show that the midpoints of a family of parallel chords of the ellipse (4) lie on a line through the center. (The segment of this line intercepted by the ellipse is a diameter.)

Solution: Each chord of slope m is, for some k, a segment of the line $y = mx + k$. If we use this equation to eliminate y from (4), we find, on clearing of fractions, that $b^2x^2 + a^2(mx + k)^2 = a^2b^2$, or $(a^2m^2 + b^2)x^2 + 2a^2kmx + a^2k^2 - a^2b^2 = 0$; the roots of this equation are the abscissas of the endpoints of the chord. The abscissa of the midpoint is then half the sum of these roots, by equation (2) of § 1.3, but we need not solve our equation to find this abscissa. It is shown in algebra that the sum of the roots of $px^2 + qx + r = 0$ is $-q/p$, and the abscissa, x, that we seek is therefore half the corresponding quantity in our equation; that is, $x = -a^2km/(a^2m^2 + b^2)$. To find the corresponding ordinate, we substitute this value of x in the equation of the chord:

$$y = m\left(\frac{-a^2km}{a^2m^2 + b^2}\right) + k = \frac{-a^2km^2 + a^2km^2 + b^2k}{a^2m^2 + b^2} = \frac{b^2k}{a^2m^2 + b^2}.$$

Since these coordinates satisfy the equation $b^2x + a^2my = 0$ for any k (check this!), it follows that all chords of slope m have their midpoints on the line $b^2x + a^2my = 0$; moreover, this line passes through the origin, which is the center of the ellipse (4).

Exercises 8.4 [A]

1. Find the coordinates of foci and vertices and the eccentricity, and sketch:

(a) $\dfrac{x^2}{5} + \dfrac{y^2}{4} = 1$

(b) $25x^2 + 9y^2 = 225$

(c) $x = 2\cos\theta$, $y = \sin\theta$

(d) $9x^2 + 25y^2 + 18x - 100y - 116 = 0$

2. Find an equation of an ellipse if

(a) its foci are at $(\pm 3, 0)$ and its minor axis is 4 units long.

(b) its center is at the origin, one focus is $(0, 2)$, and one vertex is $(0, 3)$.

(c) its vertices are at $(-2, 3)$ and $(6, 3)$ and one focus is on the y-axis.

3. Find the eccentricity of an ellipse in which the foci are midway between the center and the vertices.

4. Find, in simplest form, the slope of the tangents to the ellipse (4) at the ends of a latus rectum.

5. Show that an equation of the line tangent to the ellipse (4) at (x_1, y_1) is $x_1x/a^2 + y_1y/b^2 = 1$.

6. Find each point on the ellipse (4) such that the tangent at that point makes equal angles with the coordinate axes.

7. Find the area of an ellipse with axes $2a$ and $2b$. (*Hint:* The parametric form is particularly convenient.)

8. Find the maximum area of the rectangle inscribed in an ellipse.

9. $P(x, y)$ is tracing the first-quadrant arc of $x^2 + 4y^2 = 4$ at such a rate that $dx/dt = 2$ units/min. when $x = \sqrt{3}$. Find the values of dy/dt and $d\theta/dt$ at that same instant.

Exercises 8.4 [B]

1. Find the coordinates of foci and vertices and the eccentricity, and sketch:
   (a) $x^2 + 2y^2 = 9$
   (b) $9x^2 + 4y^2 - 54x + 45 = 0$
   (c) $16x^2 + 25y^2 - 64x + 200y + 464 = 0$
   (d) $x = \cos\theta$, $y = 3\sin\theta$

2. Find an equation of an ellipse if
   (a) its vertices are at $(1, 7)$ and $(1, -3)$ and it passes through $(5, 2)$.
   (b) one vertex is at $(-6, -2)$ and one end of the minor axis is at $(-2, 0)$.
   (c) its axes lie on the coordinate axes and its slope at $(3, 2)$ is -1.

3. Find the points on the ellipse (4) at which the tangent is parallel to the chord connecting the points A_1 and B_1 (Fig. 6).

4. Show that an equation of the normal to (4) at (x_1, y_1) is $\dfrac{a^2x}{x_1} - \dfrac{b^2y}{y_1} = c^2$.

5. Find the volume generated by revolving the ellipse (4) about (a) the x-axis; (b) the y-axis.

6. Show that the lines $y = mx \pm\sqrt{a^2m^2 + b^2}$ are tangent to (4) for all values of m.

7. Show that the lengths of the focal radii to the point (x_1, y_1) on the ellipse (4) are given by $a \pm ex_1$.

8. Find the point on the ellipse (4) at which the tangent makes, with the coordinate axes, a triangle of minimum area.

8.5 The hyperbola

A hyperbola is defined to be the locus of points in a plane the difference of whose distances to two fixed points is constant. The geometry of the hyperbola contains many striking analogies to that of the ellipse, and much of the terminology of § 8.4 will apply as well here. The fixed points, for instance, are called the foci, and if we place the hyperbola as shown in Fig. 8, we can designate them $F_1(c, 0)$ and $F_2(-c, 0)$. The constant difference is $2a$, and P is therefore on the curve if and only if

$$|PF_1 - PF_2| = 2a. \tag{1}$$

The curve crosses the x-axis at the vertices, A_1 and A_2, and the distance between these is readily shown to be $2a$, as in the ellipse; the distance from

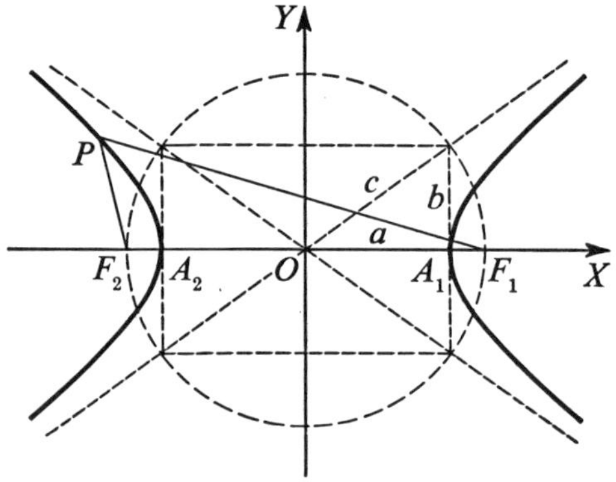

Fig. 8

the center O to a vertex is therefore a. From triangle PF_1F_2, we observe that $|PF_1 - PF_2| < F_1F_2$, from which we conclude $a < c$; as a consequence, the eccentricity,

$$\mathbf{e} = \frac{\mathbf{c}}{\mathbf{a}}, \tag{2}$$

is always greater than 1 in the hyperbola.

Since $PF_1 = \sqrt{(x - c)^2 + y^2}$ and $PF_2 = \sqrt{(x + c)^2 + y^2}$, we can write (1) in the form

$$\left|\sqrt{(x - c)^2 + y^2} - \sqrt{(x + c)^2 + y^2}\right| = 2a,$$

and this can be reduced to $(c^2 - a^2)x^2 - a^2y^2 = a^2(c^2 - a^2)$ by steps much like those used in simplifying the equation of the ellipse. Since $c > a$, the quantity $c^2 - a^2$ which appears in this equation is necessarily positive and is, therefore, the square of some real number b: $c^2 - a^2 = b^2$, or

$$a^2 + b^2 = c^2. \tag{3}$$

We therefore obtain $b^2x^2 - a^2y^2 = a^2b^2$, or

$$\blacktriangleright \qquad \frac{x^2}{a^2} - \frac{y^2}{b^2} = 1. \tag{4}$$

If a point satisfies (1), its coordinates satisfy (4), and conversely, since the steps leading from (1) to (4) can be retraced, a point P satisfies (1) if its coordinates satisfy (4). We therefore have in (4) the equation of the locus, the standard Cartesian equation of the hyperbola.

From (4), we note that the hyperbola is symmetric in both axes and therefore in the origin. We also note that there is no point on the curve in the region $-a < x < a$, since these values of x yield imaginary values of y; there are, however, real values of x for every real y.

Solving (4) for y, we find

$$b^2x^2 - a^2y^2 = a^2b^2;$$

$$a^2y^2 = b^2(x^2 - a^2) = b^2x^2\left(1 - \frac{a^2}{x^2}\right);$$

$$y = \pm \frac{b}{a} x \sqrt{1 - \frac{a^2}{x^2}}.$$

Since the term a^2/x^2 can be made arbitrarily small by making $|x|$ large enough, we infer that the lines

$$\blacktriangleright \qquad y = \pm \frac{b}{a} x \tag{5}$$

are asymptotes of (4); a proof of this is left to the student as an exercise. These lines are convenient in sketching the curve.

We have seen that the orientation of the ellipse is determined by the relative sizes of the constants a and b which occur in the equation of the curve (§ 8.4, equation (4)): if $a > b$, the foci and vertices are on the x-axis, and if $b > a$, they are on the y-axis. The analogous change in the hyperbola is brought about not by a change in the relative sizes of a and b but by a change of sign. The equation

$$\blacktriangleright \qquad \frac{y^2}{b^2} - \frac{x^2}{a^2} = 1 \tag{6}$$

represents a hyperbola with the same asymptotes as (4), but its foci and vertices are on the y-axis and its eccentricity is $e = \dfrac{c}{b}$. The hyperbolas (4) and (6) are said to be conjugate to each other.

As in the case of the ellipse, the steps which lead from (1) to (4) can easily be modified to give

$$\frac{(x - h)^2}{a^2} - \frac{(y - k)^2}{b^2} = 1 \tag{7}$$

as an equation of a hyperbola identical to (4) but with its center at (h, k). It can also be shown that

$$(x - a)(y - b) = k \tag{8}$$

represents a hyperbola with the lines $x = a$ and $y = b$ as asymptotes.

If, in (4), we set

$$x = a \sec \theta,$$

we obtain (9)

$$y = b \tan \theta;$$

these, then, are parametric equations for the hyperbola (4). As with the ellipse, the parameter θ can be interpreted as an angle having a specific geometric meaning in connection with the curve; this construction is left as an exercise (see Exercises 8.8 [C]).

Exercises 8.5 [A]

1. Find the coordinates of foci and vertices, the eccentricity, and the equations of the asymptotes, and sketch:
   (a) $4x^2 - y^2 = 20$
   (b) $x^2 - y^2 = -1$
   (c) $x = 5 \sec \theta,\ y = 12 \tan \theta$
   (d) $4x^2 - y^2 - 16x - 6y + 3 = 0$
2. Find an equation for each of the hyperbolas described below:
   (a) vertices at $(\pm 2, 0)$, one focus at $(3, 0)$.
   (b) center on the x-axis, one vertex at $(3, 4)$, one focus at $(3, -5)$.
   (c) vertices at $(\pm 3, 0)$, curve passes through $(5, 4)$.
3. The vertices of a hyperbola (4) are midway between the foci and the center. Find the eccentricity and the slopes of the asymptotes.
4. Find the length of a latus rectum (a focal chord perpendicular to the x-axis) and the slopes of the tangents at the ends of a latus rectum of the hyperbola (4).
5. Show that, in terms of the parameter θ, we have as the slope of the ellipse $-\frac{b}{a} \cot \theta$ and as the slope of the hyperbola $\frac{b}{a} \csc \theta$.

6. Show that an equation of the tangent to (4) at (x_1, y_1) is $\frac{x_1 x}{a^2} - \frac{y_1 y}{b^2} = 1$.

7. Use the following procedure to prove that the lines (5) are asymptotes of the hyperbola (4):

   (a) Show that the vertical distance between the line $y = bx/a$ and the first-quadrant branch of (4) is $z = b\left(x - \sqrt{x^2 - a^2}\right)/a$.

   (b) By rationalizing the numerator, show that $z = ab/\left(x + \sqrt{x^2 - a^2}\right)$.

   (c) Infer that $\lim_{x \to \infty} z = 0$, thus establishing the desired result in the first quadrant.

   (d) Show how this result can be immediately extended to the other three quadrants without further computation.

Exercises 8.5 [B]

1. Find the coordinates of foci and vertices, the eccentricity, and the equations of the asymptotes, and sketch:

   (a) $x^2 - 3y^2 = 4$

   (b) $9x^2 - 4y^2 = k^2 \quad (k > 0)$

   (c) $9x^2 - 4y^2 - 18x - 24y - 36 = 0$

   (d) $x^2 - 4y^2 - 4x + 40 = 0$

2. Find the slopes of the asymptotes of (4) in terms of the eccentricity.

3. Show that the normal to (4) at (x_1, y_1) is $\frac{a^2 x}{x_1} + \frac{b^2 y}{y_1} = c^2$.

4. Find the slopes of the lines through (0, 4) tangent to $x^2 - y^2 = 9$.

5. Find the volume of the solid of revolution formed by rotating about the x-axis the region enclosed by $x^2 - 2y^2 = 18$ and $x = 6$.

6. A tangent to the hyperbola $xy = k^2$ makes a triangle with the coordinate axes. Show that the area of this triangle does not depend on the point of tangency.

7. Show that the lines $y = mx \pm \sqrt{a^2 m^2 - b^2}$ are tangent to the hyperbola (4) for all values of m.

8. Show that the lengths of the focal radii to the point (x_1, y_1) of (4) are given by $|ex_1 \pm a|$.

8.6 The parabola

A parabola is defined to be the locus of points in a plane equidistant from a given fixed line and a given fixed point not on that line. The given point is called the focus, and the given line the directrix, of the curve. If we call the focus F and the directrix D, and if PN is a segment perpendicular to D at N, then P is on the curve if and only if

$$PF = PN. \tag{1}$$

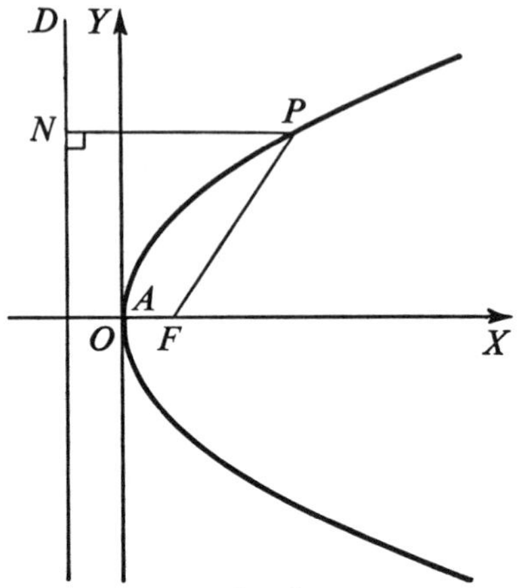

Fig. 9

The line through the focus and perpendicular to the directrix is the axis of the curve, and the point A, where the curve crosses its axis, is the vertex. The terms focal chord, focal radius, and latus rectum are defined as for the ellipse and the hyperbola. The parabola has no center.

We shall choose coordinates so that the focus F has coordinates $(p, 0)$, where $p > 0$, and the directrix D has the equation $x = -p$. The axis of the curve then falls on the x-axis, with the vertex at the origin. Then for any point $P(x, y)$, $PF = \sqrt{(x - p)^2 + y^2}$ and $PN = |x + p|$. Condition (1) therefore becomes

$$\sqrt{(x - p)^2 + y^2} = |x + p|,$$

or

$$x^2 - 2px + p^2 + y^2 = x^2 + 2px + p^2,$$

or, finally,

▶ $$\mathbf{y^2 = 4px.} \tag{2}$$

This is the standard Cartesian equation of the parabola: if a point satisfies (1), its coordinates satisfy (2), and conversely. Evidently, a different choice of axes could have led to one of the forms

$$x^2 = 4py, \quad y^2 = -4px, \quad x^2 = -4py;$$

the student should sketch each. Note that, in each case, the line we have called the axis of the curve is, indeed, an axis of symmetry of the curve.

If the vertex of the parabola is at (h, k) and the focus at $(h + p, k)$, it is easily verified that the equation of the curve is

$$(y - k)^2 = 4p(x - h), \tag{3}$$

with three other forms, analogous to those listed above, in which the vertex is at (h, k).

If, in (2), we set

$$y = 2pt,$$

we find

$$x = pt^2; \tag{4}$$

these are parametric equations of the parabola (2).

Example 1. Show that the tangents to a parabola at the ends of the latus rectum intersect each other orthogonally.

Solution: The equation of the latus rectum is $x = p$; to find the coordinates of its endpoints we substitute this value of x in (2): $y^2 = 4px = 4p(p) = 4p^2$, and therefore $y = \pm 2p$. Differentiating (2), we find

$$2y\,dy = 4p\,dx, \quad \text{and therefore} \quad \frac{dy}{dx} = \frac{2p}{y};$$

substituting $\pm 2p$ for y, we find $\dfrac{dy}{dx} = \pm 1$, which establishes the required fact.

It is instructive to work the same problem using the parametric form (4). Since this form gives $x = pt^2$, when $x = p$ we have $pt^2 = p$ and $t = \pm 1$ as the values of the parameter t at the ends of the latus rectum. Moreover,

$$dx = 2pt\,dt, \quad dy = 2p\,dt, \quad \text{and therefore} \quad \frac{dy}{dx} = \frac{1}{t}$$

at any point on the parabola. Substituting ± 1 for t in this expression gives the required result.

Note that this example provides a geometric interpretation of the parameter t; since $1/t$ is the slope of the tangent, its negative reciprocal, $-t$, is the slope of the normal.

Example 2. If $t = t_1$ at one end of a focal chord, find the value of t at the other end of the chord.

Solution: Corresponding to $t = t_1$, we have the point $(pt_1^2, 2pt_1)$. The chord also passes through $F(p, 0)$, and its slope is therefore $\dfrac{2pt_1 - 0}{pt_1^2 - p} = \dfrac{2t_1}{t_1^2 - 1}$. In terms of the unknown value of t at the other end of the chord, the slope must also be $\dfrac{2t}{t^2 - 1}$. Since these two expressions for the slope must be equal, we have the equation $\dfrac{2t}{t^2 - 1} = \dfrac{2t_1}{t_1^{\,2} - 1}$, which has roots $t = t_1, \; -\dfrac{1}{t_1}$. The first of these is the value with which we started; the second must then be the required value.

Example 3. Show that the focal radius PF to the point P on a parabola, and the line through P parallel to the axis of the curve, make equal angles with the tangent at P.

Solution: From Example 1, we know that the slope of the tangent is $1/t$, and from Example 2 we know that the slope of the focal chord is $2t/(t^2 - 1)$; the angle between these two lines is then given by

$$\tan\phi = \frac{\dfrac{2t}{t^2 - 1} - \dfrac{1}{t}}{1 + \dfrac{2t}{t^2 - 1}\cdot\dfrac{1}{t}} = \frac{2t^2 - t^2 + 1}{t^3 - t + 2t} = \frac{t^2 + 1}{t^3 + t} = \frac{1}{t}.$$

But this quantity is equal to the slope of the curve, the tangent of the inclination, and therefore to the tangent of the angle which the tangent line makes with any line parallel to the x-axis; the demonstration is therefore complete.

The preceding example proves the "reflecting property" of the parabola. If a parabolic arc is rotated about the axis of the curve, and the concave side of the surface thus generated is made reflecting, then any ray of light parallel to the axis will be reflected to the focus. This therefore is the ideal shape for the reflector of an astronomical telescope.

Exercises 8.6 [A]

1. State the coordinates of focus and vertex and the equation of the directrix, and sketch:

   (a) $x^2 = 8y$ (c) $3y^2 = 10x$ (e) $y^2 = 8x + 8y + 8$

   (b) $y^2 + 12x = 0$ (d) $x = 2t^2, y = 4t$ (f) $x(x + 4) = -2(y + 1)$

2. Write a Cartesian equation of the parabola
   (a) with vertex at $(0, 0)$ and focus $(0, -2)$.
   (b) with focus $(3, 0)$ and directrix $x = -3$.
   (c) with focus $(0, 5)$ and directrix $x = 2$.

3. Show that the tangent to the parabola (2) at (x_1, y_1), $y_1 \neq 0$, is $y_1 y = 2p(x + x_1)$.

4. Show that the length of the focal radius of (2) to the point (x_1, y_1) is $x_1 + p$.
5. The tangent to a parabola at P meets the axis of the curve at Q. Prove $PF = QF$.
6. Find the angles of intersection of $y^2 = 4px$ and $x^2 = 4py$.
7. Find the area of the region enclosed by the parabolas of Ex. 6.
8. The segment cut from the parabola $y^2 = 4px$ by the line $x = a$ is rotated about the x-axis. Find the volume of the solid of revolution thus generated, and compare this volume with the volumes of the circumscribed cylinder and the inscribed right circular cone.

Exercises 8.6 [B]

1. Find the vertex and focus of $y = ax^2 + bx + c$.
2. The equation $\sqrt{x} + \sqrt{y} = \sqrt{a}$ represents an arc of a parabola. Sketch the curve and determine an equation of its axis and the coordinates of its vertex.
3. Show that the line $x - t_1 y + p{t_1}^2 = 0$ is tangent to the parabola (4) at the point $t = t_1$.
4. Prove that the tangents to a parabola at the ends of a focal chord meet on the directrix.
5. The normal to a parabola at P meets the axis of the curve at Q. Prove that the length of the projection of PQ on the axis is independent of the choice of P. (This projection is called the subnormal.)
6. A point P is moving along the first-quadrant arc of $y^2 = 4x$. If $dx/dt = 2$ when $x = 1$, find, at that same instant, (**a**) dy/dt; (**b**) the rate at which the focal radius PF is turning.
7. The latus rectum of (2) forms the longer base of an isosceles trapezoid inscribed in the parabola. Find the maximum area of this trapezoid.
8. Show that two parabolas which have the same axis and focus but which open in opposite directions are orthogonal.

8.7 Equations of the second degree*

In previous sections, we have seen that a variety of geometric loci can be described by equations of the second degree. In this section, we consider a converse question: Given an equation of the second degree in two variables, what locus does it describe? We shall see that, with the ex-

* None of the material of this section will be required in subsequent sections.

ception of certain special cases, the locus is one of the curves that we have studied in §§ 8.3–8.6, and we shall derive methods for reducing each such equation to a familiar, standard form.

The general equation of the second degree in two variables can be written

$$\mathbf{Ax^2 + Bxy + Cy^2 + Dx + Ey + F = 0,} \tag{1}$$

where A, B, C are not all zero. It is convenient to investigate separately the two cases $B = 0$ and $B \neq 0$.

The case $B = 0$ has been examined in preceding sections. Thus, for example, in Example 1 of § 8.3 we saw that

$$2x^2 + 2y^2 - 6x + 18y - 5 = 0, \tag{2}$$

which corresponds to (1) with $A = C = 2$, $B = 0$, $D = -6$, $E = 18$, $F = -5$, can be put in the form

$$(x - \tfrac{3}{2})^2 + (y + \tfrac{9}{2})^2 = 5^2 \tag{3}$$

by completing the squares of the terms in x and in y. In this form, we readily recognize the equation of a circle with center at $(\frac{3}{2}, -\frac{9}{2})$ and radius 5.

Let us carry the simplification of (2) a step farther. The geometric nature of this circle is independent of any coordinate system, and we have already seen (especially in § 1.3) that a good choice of axes can greatly simplify a problem. We have also seen that the equation of a circle takes its simplest form when the center of the circle is at the origin. Let us, then, introduce a new set of coordinates, with origin at the center of our circle, the point that has coordinates $(\frac{3}{2}, -\frac{9}{2})$ in the original coordinate system. Each point in the plane will then have two sets of coordinates, a set (x, y) which refers to the original axes and a set, say (x', y'), which refers to the new axes. Figure 10 illustrates the fact that, for each point in the plane, the two sets

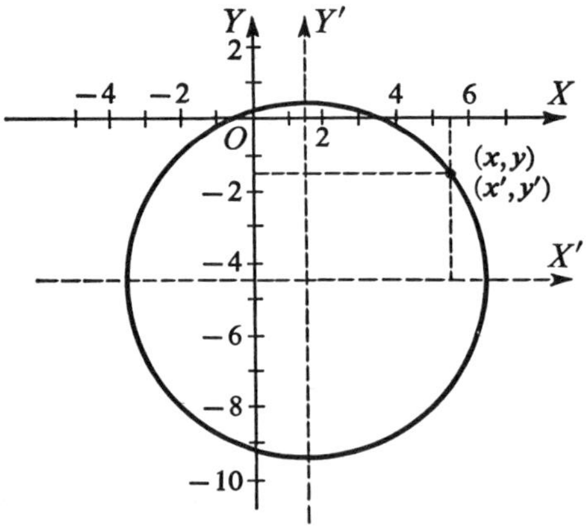

Fig. 10

of coordinates are related by the equations

$$x' = x - \tfrac{3}{2}, \quad y' = y + \tfrac{9}{2}.$$

In terms of the new coordinates, the equation of the circle (3) becomes

$$x'^2 + y'^2 = 5^2. \tag{4}$$

The transition from (3) to (4) may be interpreted in two ways. We may regard it as moving the entire coordinate plane, axes and all, $\frac{3}{2}$ units to the right and $\frac{9}{2}$ units down; this is sometimes picturesquely called the "alibi" interpretation of the transformation. Such a motion, in which the axes remain parallel to their original positions, is called a translation of axes. Alternatively, we may regard the transformation as leaving everything fixed but giving the circle a new name, the equation (4); this interpretation as a change of coordinates may be called the "alias" interpretation.

In like manner, we saw in Example 2 of § 8.4 that

$$x^2 + 16y^2 - 4x + 96y + 123 = 0$$

can be written as
$$\frac{(x-2)^2}{5^2} + \frac{(y+3)^2}{(\frac{5}{4})^2} = 1,$$

the equation of an ellipse with center at $(2, -3)$; the substitution

$$x' = x - 2, \quad y' = y + 3,$$

which is equivalent to translating the origin to the point $(2, -3)$, then gives

$$\frac{x'^2}{5^2} + \frac{y'^2}{(\frac{5}{4})^2} = 1.$$

Similarly, we can rewrite $\quad y^2 - 4x - 4y = 0$

as $\quad (y - 2)^2 = 4(x + 1),$

and the translation $\quad x' = x + 1, \quad y' = y - 2$

transforms this into $\quad y'^2 = 4x',$

a standard form of the equation of a parabola.

This same method of completing squares can be used to reduce the general equation (1) to standard form, given $B = 0$, but the resulting formulas are rather formidable, and we shall therefore treat each problem from the beginning, in what follows, rather than develop a general solution.

The case $B \neq 0$ is less familiar, although it was noted in § 8.5 that the equation $(x - a)(y - b) = k$ represents a hyperbola with the lines $x = a$, $y = b$ as asymptotes. Let us consider a particular case, the equation

$$xy = k, \tag{5}$$

in which the coordinate axes are the asymptotes; this corresponds to (1) with $A = C = D = E = 0$, $B = 1$, $F = -k$. We shall transform this equation into one of the standard forms of the equation of the hyperbola, equation (4) of § 8.5.

A study of the graph of (5) (the diagram assumes $k > 0$), and comparison with the graph of a hyperbola in standard position, suggest that if we use

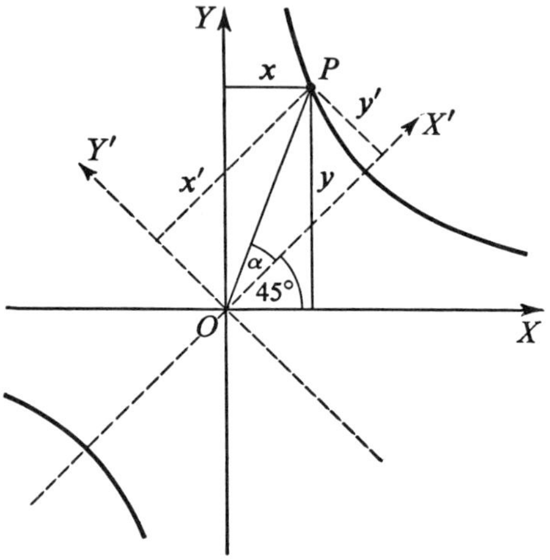

Fig. 11

the line $y = x$ as an x'-axis and the line $y = -x$ as a y'-axis, we may get the standard form we seek. Because this transformation can be thought of as keeping the origin fixed and turning the axes through an angle of 45°, it is called a rotation of axes. To see what effect this rotation has on (5), we must consider an arbitrary point P and find the relationship between its coordinates (x, y) in the original coordinate system and its coordinates (x', y') in the rotated system. This relationship will pemit us to eliminate x and y from (5), in favor of x' and y'.

It is convenient to make use of the angle α from the positive x'-axis to the ray OP. In terms of this angle and the distance OP, we find that

$$x' = OP \cos \alpha, \quad y' = OP \sin \alpha;$$

hence

$$x = OP \cos (\alpha + 45°) = OP \cos \alpha \cos 45° - OP \sin \alpha \sin 45°,$$

or $$x = \frac{\sqrt{2}}{2}(OP \cos \alpha - OP \sin \alpha) = \frac{\sqrt{2}}{2}(x' - y'),$$

and $$y = OP \sin (\alpha + 45°) = OP \sin \alpha \cos 45° + OP \cos \alpha \sin 45°,$$

or $$y = \frac{\sqrt{2}}{2}(OP \cos \alpha + OP \sin \alpha) = \frac{\sqrt{2}}{2}(x' + y').$$

Substituting these results in (5), we obtain

$$\frac{\sqrt{2}}{2}(x' - y') \cdot \frac{\sqrt{2}}{2}(x' + y') = k, \quad \text{or} \quad \frac{x'^2}{2k} - \frac{y'^2}{2k} = 1,$$

the desired standard form.

In this particular instance, then, we were able to eliminate the xy term from an equation of the second degree by a rotation of the coordinate axes. Were we just lucky or can we always find a rotation to accomplish this result? To answer this question, let us consider the effect, on each pair of coordinates (x, y), of an arbitrary rotation θ.

The development directly follows the one we have just seen. If P is any point other than the origin O, with coordinates (x, y) in the original system and (x', y') in the rotated system, if θ is the angle from the positive x-axis to the positive x'-axis, and if α is the angle from the positive x'-axis to the ray OP, we then have

$$x' = OP \cos \alpha, \quad y' = OP \sin \alpha,$$

$$x = OP \cos(\theta + \alpha) = OP \cos\theta \cos\alpha - OP \sin\theta \sin\alpha$$

$$= (OP \cos\alpha)\cos\theta - (OP \sin\alpha)\sin\theta,$$

or
$$\mathbf{x = x' \cos\theta - y' \sin\theta,} \qquad \textbf{(6a)}$$

and
$$y = OP \sin(\theta + \alpha) = OP \sin\theta \cos\alpha + OP \cos\theta \sin\alpha$$

$$= (OP \cos\alpha)\sin\theta + (OP \sin\alpha)\cos\theta,$$

or
$$\mathbf{y = x' \sin\theta + y' \cos\theta.} \qquad \textbf{(6b)}$$

Substituting these results in (1), we obtain an equation of the form

$$A'x'^2 + B'x'y' + C'y'^2 + D'x' + E'y' + F' = 0;$$

it is left to the student as an exercise to show that expressions for the new coefficients in terms of the old are

$$\begin{aligned} A' &= A\cos^2\theta + B\sin\theta\cos\theta + C\sin^2\theta, \\ B' &= 2(C - A)\sin\theta\cos\theta + B(\cos^2\theta - \sin^2\theta), \\ C' &= A\sin^2\theta - B\sin\theta\cos\theta + C\cos^2\theta, \\ D' &= D\cos\theta + E\sin\theta, \\ E' &= -D\sin\theta + E\cos\theta, \\ F' &= F. \end{aligned} \qquad (7)$$

Our primary concern in deriving these formulas was to see whether we could eliminate the xy term from (1); this will be accomplished if we can find a value of θ which will make $B' = 0$ in (7). Setting $B' = 0$, we find

$$2(C - A) \sin \theta \cos \theta + B(\cos^2 \theta - \sin^2 \theta) = 0;$$

$$(C - A) \sin 2\theta + B \cos 2\theta = 0;$$

$$\cot 2\theta = \frac{A - C}{B}. \tag{8}$$

(We are free to divide by B since we are here concerned only with the case $B \neq 0$; it is left to the student to determine why we use the cotangent rather than the tangent.) Since the range of the cotangent function contains all real numbers, we can always find values of θ that satisfy (8). Thus, for example, in the case of the hyperbola $xy = k$ that we considered above, we have $A = C = 0$, $B = 1$, $\cot 2\theta = 0$, $2\theta = 90°$, and $\theta = 45°$.

Example 1. Express in a standard form and identify:

$$13x^2 - 6\sqrt{3}xy + 7y^2 = 144.$$

Solution: From (8), we have $\cot 2\theta = (13 - 7)/(-6\sqrt{3}) = -\sqrt{3}/3$. A value of 2θ that satisfies this is 120°, and we therefore take $\theta = 60°$, $\sin \theta = \sqrt{3}/2$, $\cos \theta = 1/2$; (6a) and (6b) then become $x = (x' - y'\sqrt{3})/2$, $y = (x'\sqrt{3} + y')/2$. Substituting these results in the given equation, we find

$$\frac{13}{4}(x' - y'\sqrt{3})^2 - \frac{6\sqrt{3}}{4}(x' - y'\sqrt{3})(x'\sqrt{3} + y') + \frac{7}{4}(x'\sqrt{3} + y')^2 = 144,$$

$$4x'^2 + 16y'^2 = 144,$$

or

$$\frac{x'^2}{36} + \frac{y'^2}{9} = 1,$$

an ellipse.

Example 2. Reduce to standard form:

$$9x^2 - 24xy + 16y^2 - 50x - 100y - 175 = 0. \tag{9}$$

Solution: From (8), we have $\cot 2\theta = (9 - 16)/(-24) = 7/24$. If we take the acute value of 2θ that satisfies this equation, we find $\cos 2\theta = 7/25$, and hence $\sin \theta = \sqrt{(1 - 7/25)/2} = 3/5$, $\cos \theta = 4/5$; (6a) and (6b) then become

$$x = \tfrac{1}{5}(4x' - 3y'), \quad y = \tfrac{1}{5}(3x' + 4y'). \tag{10}$$

Substituting these results in (9) and simplifying, we have

$$y'^2 - 4x' - 2y' - 7 = 0,$$

or, completing the square of the terms in y,

$$(y' - 1)^2 = 4(x' + 2). \tag{11}$$

If we now translate to a set of $x''y''$-axes, with origin at the point whose $x'y'$-coordinates are $(-2, 1)$, the equations of the transformations are

$$x'' = x' + 2, \quad y'' = y' - 1, \tag{12}$$

and (11) becomes

$$y''^2 = 4x'', \tag{13}$$

a standard form of the equation of a parabola.

We have made two transformations: a rotation, described by equations (10), and a translation, described by equations (12). It is possible to combine these two transformations into one. If we solve (12) for x' and y' in terms of x'' and y'', and then substitute the resulting expressions in (10), we obtain

$$x = \tfrac{1}{5}(4x'' - 3y'' - 11), \quad y = \tfrac{1}{5}(3x'' + 4y'' - 2)$$

as the equations of the combined transformation. If we now substitute these values of x and y in (9), we can get (13) directly.

The techniques used in the preceding examples are entirely general; they suffice, in every case, to transform (1) into a standard form which the student, at this stage, should be able to identify by inspection. In most cases, it will be found that (1) represents a circle, an ellipse, a hyperbola, or a parabola. The exceptions, so-called "degenerate" cases, lead to one or to two straight lines, to a point, or to no locus. These conclusions are further developed and illustrated in the exercises that follow.

Exercises 8.7 [C]

1. Reduce to a standard form, identify, and sketch, showing both sets of axes:

(a) $5x^2 + 8xy + 5y^2 = 36$
(b) $5x^2 + 8xy + 5y^2 = 0$
(c) $5x^2 + 8xy + 5y^2 = -36$
(d) $2x^2 + 5xy + 2y^2 = 18$
(e) $2x^2 + 5xy + 2y^2 = 0$
(f) $2x^2 + 5xy + 2y^2 = -18$
(g) $x^2 + 2xy + y^2 = 18$
(h) $x^2 + 2xy + y^2 = 0$

2. Transform to a standard form by rotating the axes, and discuss the various forms which the locus can take:

$$ax^2 + bxy + ay^2 = k.$$

(*Hint:* Assume $a \geq 0$ (why can this be done without loss of generality?), and consider all possible combinations of the alternatives (i) $b^2 = 4a^2$, $b^2 \neq 4a^2$; (ii) $k > 0$, $k = 0$, $k < 0$.)

3. Reduce to a standard form, identify, and sketch:
   (a) $3x^2 + \sqrt{3}xy + 2y^2 = 42$
   (b) $4x^2 + 3\sqrt{3}xy + y^2 = 22$
   (c) $3x^2 + 10xy + 3y^2 + 8\sqrt{2}x - 8\sqrt{2}y = 0$
   (d) $13x^2 - 10xy + 13y^2 - 82x + 98y + 157 = 0$
   (e) $16x^2 + 24xy + 9y^2 + 4x + 3y - 6 = 0$
4. Verify that the rotation (6a) and (6b) produces the relationships (7).
5. (a) Show that the coefficients A, B, C of (1) are unchanged under any translation.

   (b) Show that if (1) is transformed by a rotation, the new coefficients A', B', C' will not in general be equal to the old ones, but that

$$B'^2 - 4A'C' = B^2 - 4AC \quad \text{and} \quad A' + C' = A + C.$$

   (*Hint:* Use (7).)

6. The previous exercise shows that the quantity $B^2 - 4AC$ is invariant under any combination of translations and rotations. Use this fact, and refer to the standard forms of the corresponding equations, to show that (1) represents

$$\begin{array}{lll} \text{an ellipse} & \text{if} & B^2 - 4AC < 0; \\ \text{a parabola} & \text{if} & B^2 - 4AC = 0; \\ \text{a hyperbola} & \text{if} & B^2 - 4AC > 0. \end{array}$$

   (The "degenerate" cases are exceptions.)

7. Show that the line tangent to (1) at (x_1, y_1) is

$$Ax_1x + \frac{B}{2}(x_1y + y_1x) + Cy_1y + \frac{D}{2}(x_1 + x) + \frac{E}{2}(y_1 + y) + F = 0.$$

8.8 Selected loci

In §§ 8.3–8.6 we considered in some detail four special loci. The reasons for paying particular attention to these four are partly historical, partly that they have many important applications, and partly that their equations are relatively simple. There are, however, many other important loci, and there are other definitions that lead us back to the very loci we have been studying. The following examples and exercises illustrate these points.

Example 1. Find the locus of points equidistant from $A(2, -1)$ and $B(3, 0)$.

Solution: A point P will be on the required locus if and only if $PA = PB$. If the coordinates of P are (x, y), then this condition can be expressed as

$$\sqrt{(x-2)^2 + (y+1)^2} = \sqrt{(x-3)^2 + y^2}.$$

Squaring and simplifying, we have $x + y - 2 = 0$, the equation of a straight line. A result of this nature was, of course, to be expected; a theorem of elementary geometry states that the locus thus defined is a straight line, the perpendicular bisector of AB.

Example 2. Find the locus of a point P whose distance from a fixed point F is always a constant e times its distance from a fixed line D. (F is a *focus* and D is a *directrix* of the curve.)

Solution: If $e = 1$, the point P is equally distant from F and D; this case is then the parabola. We shall therefore consider only $e \neq 1$.

We choose coordinates so that the y-axis falls on D and the x-axis passes through F, to which we shall assign coordinates $(k, 0)$. Then $P(x, y)$ is a point of the locus if and only if

$$\sqrt{(x-k)^2 + y^2} = e|x|.$$

If we square both sides of this equation and then complete the square of the terms in x, we get

$$\left(x - \frac{k}{1-e^2}\right)^2 + \frac{y^2}{1-e^2} = \frac{e^2k^2}{(1-e^2)^2}; \tag{1}$$

the details are left to the student. We now consider two cases:

Case 1. $e < 1$: In this case, $1 - e^2 > 0$. We set $a = \dfrac{ek}{1-e^2}$ and get, in (1),

$$\left(x - \frac{k}{1-e^2}\right)^2 + \frac{y^2}{1-e^2} = a^2,$$

or

$$\frac{\left(x - \dfrac{k}{1-e^2}\right)^2}{a^2} + \frac{y^2}{a^2(1-e^2)} = 1;$$

if we also set $b = a\sqrt{1-e^2}$, this becomes the standard form of the equation of an ellipse with center at $\left(\dfrac{k}{1-e^2}, 0\right)$. Since $c^2 = a^2 - b^2$ in an ellipse, we have here $c^2 = a^2 - a^2(1-e^2) = a^2e^2$; hence $e = \dfrac{c}{a}$, and the constant e used in defining this locus is therefore precisely the eccentricity.

CASE 2. $e > 1$: In this case, $1 - e^2 < 0$. We set $-a = \dfrac{ek}{1 - e^2}$ and get, in (1),

$$\left(x - \frac{k}{1 - e^2}\right)^2 + \frac{y^2}{1 - e^2} = a^2;$$

dividing by a^2 and setting $b = a\sqrt{e^2 - 1}$, we get

$$\frac{\left(x - \dfrac{k}{1 - e^2}\right)^2}{a^2} - \frac{y^2}{b^2} = 1,$$

a hyperbola with center at $\left(\dfrac{k}{1 - e^2}, 0\right)$. In this case, as well, the constant e can be shown to be the eccentricity of the curve.

We have thus shown, in Example 2 and in § 8.6, that a single definition gives us the ellipse, the parabola, or the hyperbola, according as $e < 1$, $e = 1$, or $e > 1$; it can also be shown, though not by direct application of this definition, that the circle corresponds to the case $e = 0$. These four curves, then, form a family, and because each of them can be obtained as the intersection of a plane with a conical surface, they are collectively known as the conic sections. They occur frequently in nature and in human achievements. We see circles everywhere. The orbits of planets and of recurring comets, like Halley's, are approximately ellipses, with the sun at one focus, and comets that come but once have nearly parabolic, or occasionally hyperbolic, orbits. Most of the earth satellites travel in elliptical orbits (see the illustration facing page 1). Near-circular orbits can be achieved by repositioning satellites that have been placed in elliptical orbits. The hyperbola is involved in the paths of space vehicles intended for landing on other planets. The curve outlined by the main cables of a suspension bridge is, very nearly, a parabola. Modern architects have used this curve in the design of several outstanding buildings, among them the Ingalls Hockey Rink at Yale and the Kresge Auditorium at MIT. The applications of the parabola to optics are suggested on page 288.

Some of the similarities and differences among curves of this family can be shown by writing an equation of each with a vertex at the origin, as we usually do with the parabola. We can do this for the standard ellipse by placing the center of the curve at $(a, 0)$; an equation of the ellipse is then $(x - a)^2/a^2 + y^2/b^2 = 1$, and this reduces to $y^2 = 2b^2x/a - b^2x^2/a^2$ (for the circle, set $b = a$). Similarly, an equation of a hyperbola with center at $(-a, 0)$ is $y^2 = 2b^2x/a + b^2x^2/a^2$. For each curve, $2b^2/a$ is the length of a latus rectum (see Example 3, page 278, and Exercise 4, page 284); the corresponding length for the parabola $y^2 = 4px$ is $4p$. If we put k for this length in the equations of the ellipse, the parabola, and the hyperbola, these equations respectively become

$$y_e^2 = kx - \frac{b^2x^2}{a^2}, \quad y_p^2 = kx, \quad y_h^2 = kx + \frac{b^2x^2}{a^2};$$

hence $y_e^2 \leq y_p^2 \leq y_h^2$. If, like the Greek geometer Apollonius of Perga (3d century B.C.), you accept the idea that the parabola is in some sense "just right," then clearly the ellipse lacks something ($y_e^2 \leq y_p^2$), whereas the hyperbola has too much ($y_p^2 \leq y_h^2$). If you will now look up the words "ellipsis" and "hyperbole" in a dictionary, noting especially their derivations, you should get a fair idea of why Apollonius gave the ellipse and the hyperbola the names they still bear.

Example 3. The cycloid. A circle rolls, without slipping, on a straight line. Find the locus of a point on the circle.

Solution: We take the given line as the x-axis and start the point, $P(x, y)$, at the origin. When the circle has rolled through an angle θ, we have, if the radius of the circle is a,

$$\text{arc } AP = a\theta = OA$$

(this expresses the condition that there is no slipping),

$$PQ = a \sin \theta,$$

and therefore $\quad x = OA - PQ = a(\theta - \sin \theta),$

and similarly $\quad y = PB = QA = CA - CQ = a(1 - \cos \theta).$

These are parametric equations of the locus.

The preceding derivation tacitly assumes $0 < \theta < \frac{\pi}{2}$. It will be found, however, that the result is valid for all θ.

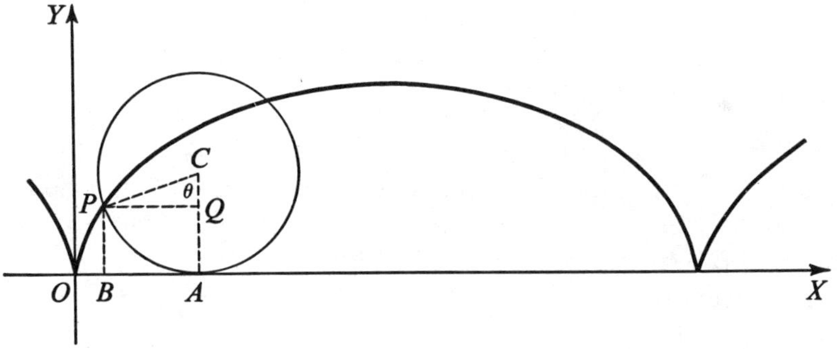

Fig. 12

Exercises 8.8 [A]

In Exs. 1–5, find the equation of the locus of $P(x, y)$ under the stated conditions, and, if the locus is a conic section, identify and describe it.

1. A and B are $(3, 4)$ and $(-3, -4)$, respectively, and $PA^2 + PB^2 = 100$.
2. A and B are $(3, 0)$ and $(-3, 0)$, respectively, and $PA = 2 \cdot PB$.
3. P is twice as far from the y-axis as from $(3, 0)$.
4. A and B are $(2, 0)$ and $(2, 6)$, and $PA + PB = 10$.
5. A and B are $(4, 0)$ and $(-4, 0)$, and the slope of PA is always 2 more than the slope of PB.

6. If A and B are fixed points, and $\angle APB$ is always a right angle, what is the locus of P? Obtain parametric equations of this locus as follows: Let A, B, and P be $(-a, 0)$, $(a, 0)$, and (x, y), respectively, and set the slope of PA equal to t. Then the slope of PB will be $-1/t$ (why?). Solve these slope equations for x and y in terms of the constant a and the parameter t.
7. Find the area included between the x-axis and one arch of the cycloid (Example 3).

Exercises 8.8 [B]

In Exs. 1–5, find the equation of the locus of $P(x, y)$ under the stated conditions, and, if the locus is a conic section, identify and describe it.

1. A and B are $(-a, 0)$ and $(a, 0)$, and the slope of PA is 1 more than the slope of PB.
2. A and B are the points described in Ex. 1, and the product of the slopes of PA and PB is 2.
3. P is 2 units farther from $(-3, 0)$ than from $(3, 0)$.
4. The sum of the distances of P from $(0, 2)$ and $(2, 0)$ is constant, and the origin is a point of the locus.
5. P is half as far from $x + 12 = 0$ as it is from $(-3, 0)$.

6. A and B are fixed points, and the product of the slopes of PA and PB is always 1. Find parametric equations of the locus of P as follows: Let A, B, and P be $(-a, 0)$, $(a, 0)$, and (x, y), respectively, set the slopes of PA and PB respectively equal to t and $1/t$, and solve for x and y in terms of the constant a and the parameter t. Check by eliminating t to form a Cartesian equation.

   Interpret geometrically the cases $t = \pm 1$.

Exercises 8.8 [C]

1. The prolate and curtate cycloids: The definition of the cycloid (Example 3) is modified so that the point $P(x, y)$ is not on the rolling circle but on a fixed radius of that circle, at a distance b from the center. Obtain parametric equations, using the parameter θ of the example. Sketch the curve for $b > a$ (the prolate cycloid) and for $b < a$ (the curtate cycloid).

2. The Witch of Agnesi: A circle of radius a has its center at $(0, a)$. A line is drawn tangent to this circle at $T(0, 2a)$, and a variable secant through the origin O meets the circle at A and the tangent line at B, as shown in Fig. 13. $P(x, y)$ has the same ordinate as A and the same abscissa as B. Obtain parametric equations of the locus of P, using $\angle TOA = \theta$ as parameter. Also obtain a Cartesian equation of the locus.

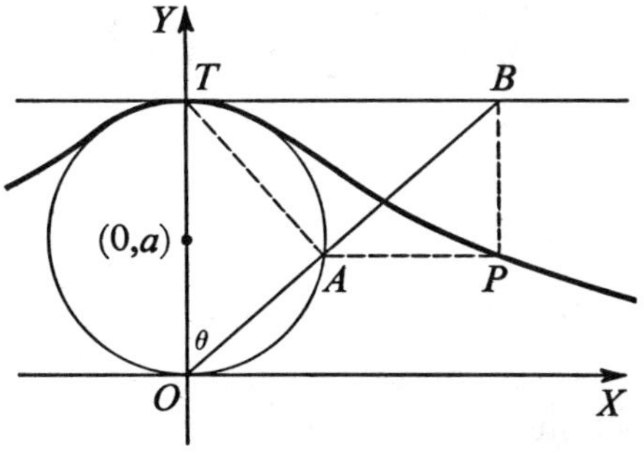

Fig. 13

3. The four-cusped hypocycloid, or astroid: This curve is the locus of a point on a circle which rolls on the inside of a fixed circle, the radius of the fixed circle being four times that of the moving circle. Take axes as indicated (Fig. 14), let the radii be a and $4a$, use $\angle BOD = \theta$ as parameter, and derive parametric equations of the locus. (*Hint:* Show that $\angle PAC = \pi/2 - 3\theta$.)

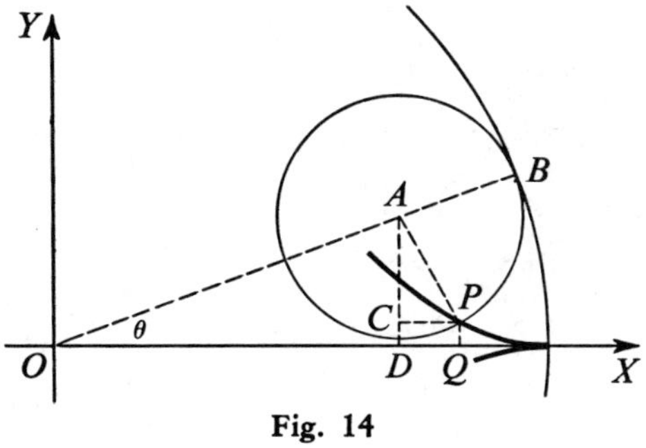

Fig. 14

4. The circles with centers at the origin and radii a and b meet the x-axis at A and B, as shown in Fig. 15. A radius OC is drawn, making an angle θ with the x-axis. A tangent to one circle at B meets OC at D, and a tangent to the other circle at C meets the x-axis at E. A point P has the same ordinate as D and the same abscissa as E. Find equations of the locus of P in terms of the parameter θ, and also obtain a Cartesian equation.

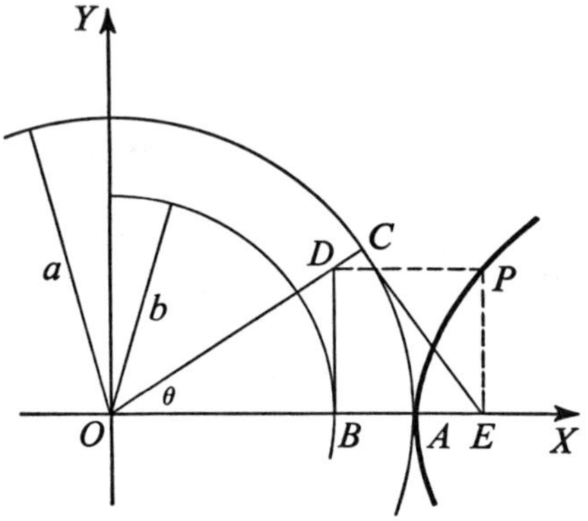

Fig. 15

5. The cissoid of Diocles: MN is tangent at one end of diameter OA to a circle of radius a. A variable secant through O meets the circle at B and MN at C, as shown in Fig. 16. On OC take $OP = BC$. Taking axes as indicated, and using $\angle AOC = \theta$ as parameter, find parametric equations of the locus of P, and also derive a Cartesian equation.

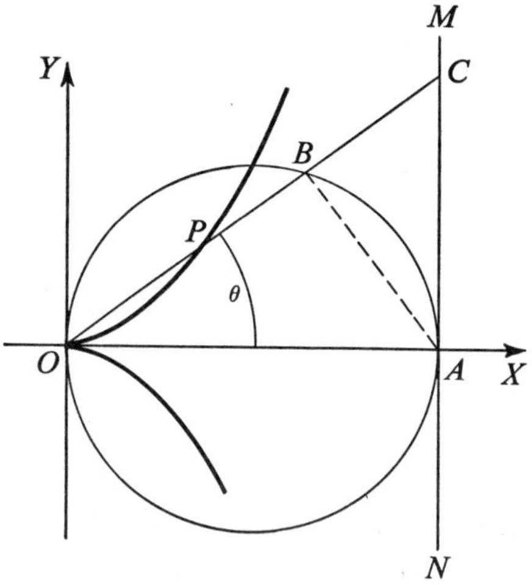

Fig. 16

Chapter Review

1. Analyze and sketch: $y = \dfrac{x + 2}{x^2 - 1}$.
2. Find the area of the region enclosed by $y^2 = x^2(1 - x^2)$ and the volume of the solid generated when this region is rotated about the x-axis.
3. Reduce to standard form and sketch:

   (a) $16x^2 - 9y^2 - 64x - 18y - 89 = 0$

   (b) $9x^2 + 16y^2 + 12x - 24y - 131 = 0$
4. If $x = 2(1 - \cos\theta)$, $y = 4\sin^2\theta$, (a) find dy/dx in terms of θ; (b) obtain a Cartesian equation.
5. Find an equation of (a) an ellipse with foci at $(\pm 2, 0)$ and vertices at $(\pm 3, 0)$; (b) a hyperbola with vertices at $(\pm 3, 0)$ and a focus at $(6, 0)$.
6. Find an equation of a circle which is tangent to the coordinate axes and passes through $(-4, 2)$.
7. Show that $y = mx + 2c\sqrt{-m}$ is tangent to $xy = c^2$, and find the coordinates of the point of tangency.
8. Given $A(a, 0)$ and $B(-a, 0)$, find the locus of P if the slope of PA is always twice the slope of PB.
9. Show that the chord that joins the points $t = t_1$ and $t = t_2$ on the parabola $x = pt^2$, $y = 2pt$ has the equation

$$2x - (t_1 + t_2)y + 2pt_1t_2 = 0.$$

10. The tangent to a parabola at P meets the directrix at Q. Prove that PQ subtends a right angle at the focus.

The photograph shows Hoover Dam, on the border of Nevada and Arizona. An essential step in the design of a dam is the determination of the total force of water that the dam will have to withstand. This is a task for which calculus is a highly valuable tool. (See § 9.8 for a discussion of fluid pressure.)

9

The Definite Integral

9.1 Sigma notation

In this chapter we shall return to the discussion of the definite integral begun in Chapter 4. Recall, then, that the integral of a function f over an interval $[a, b]$ is obtained by subdividing $[a, b]$ into n subintervals, each of length $\Delta x = (b - a)/n$, picking a point in each subinterval, say x_i^* in the ith subinterval, and forming the sum

$$S = f(x_1^*)\Delta x + f(x_2^*)\Delta x + \cdots + f(x_n^*)\Delta x. \tag{1}$$

Then the integral is defined as

$$\int_a^b f(x)dx = \lim_{\Delta x \to 0} S, \tag{2}$$

provided the limit exists.

The way the sum S is written in (1) is often criticized on at least two counts:

(1) It takes a lot of space.

(2) It does not make absolutely clear what the missing terms are, the terms represented by "$\cdots$". A clear and efficient notation that meets both objections is, for the sum S,

$$S = \sum_{i=1}^{n} f(x_i^*)\Delta x.$$

This is read "The sum, as i goes from 1 to n, of $f(x_i^*)\Delta x$." The Greek letter Σ, capital sigma, is used as an abbreviation of the word "sum." The equation $i = 1$, and the numeral n, indicate the *variable of summation*, i, and the *limits of summation*, 1 and n. These show that the letter i is to be replaced, successively, by the integers from 1 to n. Finally, the Σ is an instruction to add the expressions so obtained. Thus, for instance,

$$\sum_{i=1}^{4} \frac{2i}{i+1} = \frac{2}{2} + \frac{4}{3} + \frac{6}{4} + \frac{8}{5} = \frac{163}{30},$$

$$\sum_{i=0}^{2} (i^3 + 2) = (0 + 2) + (1 + 2) + (8 + 2) = 15.$$

The variable of summation i is a dummy variable and can be replaced by another letter, but only i, j, and k seem to be widely used for this purpose. Thus we may represent the sum of the squares of the first 100 positive integers by

$$\sum_{j=1}^{100} j^2,$$

while the sum of the sines in Table 2 (not counting repetitions) is

$$\sum_{k=1}^{160} \sin 0.01k.$$

This sigma notation permits us to combine (1) and (2) into a single concise and unambiguous statement:

$$\int_a^b f(x)dx = \lim_{\Delta x \to 0} \sum_{i=1}^{n} f(x_i^*)\Delta x. \tag{3}$$

Observe, moreover, that since $\Delta x = (b - a)/n$, the condition $\Delta x \to 0$ is equivalent to $n \to \infty$. Thus we may also write (3) as

$$\int_a^b f(x)dx = \lim_{n \to \infty} \sum_{i=1}^{n} f(x_i^*)\Delta x. \tag{4}$$

Because (4) is more widely used than (3), we shall generally prefer it in what follows.

Example 1: Find $\lim_{n \to \infty} \frac{1}{n} \sum_{i=1}^{n} \left(1 + \frac{i}{n}\right)^2$.

Solution: The strategy here is to express this limit in the form $\lim_{n \to \infty} \sum_{i=1}^{n} f(x_i^*)\Delta x$, in which x ranges over some appropriate interval $[a, b]$; the limit is then, by definition, a definite integral, $\int_a^b f(x)dx$. The form of the given sum suggests that we might profitably take $f(x) = x^2$ and set $x_i = 1 + (i/n)$. The points x_i, $0 \leq i \leq n$, then divide $[1, 2]$ into n equal subintervals of width

$$\begin{aligned} \Delta x &= x_{i+1} - x_i \\ &= \left(1 + \frac{i+1}{n}\right) - \left(1 + \frac{i}{n}\right) \\ &= \frac{1}{n}, \end{aligned}$$

and therefore

$$\begin{aligned} \frac{1}{n} \sum_{i=1}^{n} \left(1 + \frac{i}{n}\right)^2 &= \sum_{i=1}^{n} \left(1 + \frac{i}{n}\right)^2 \cdot \frac{1}{n} \\ &= \sum_{i=1}^{n} f(x_i)\Delta x, \end{aligned}$$

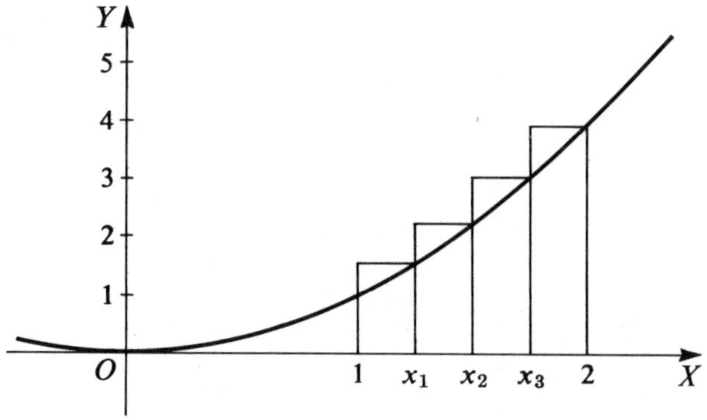

Fig. 1

which is an upper sum S_U approximating $\int_1^2 f(x)dx$. (Figure 1 illustrates the case $n = 4$.) Hence we have

$$\lim_{n\to\infty} \frac{1}{n}\sum_{i=1}^{n}\left(1+\frac{i}{n}\right)^2 = \lim_{n\to\infty}\sum_{i=1}^{n} x_i^2\,\Delta x$$
$$= \int_1^2 x^2\,dx$$
$$= \frac{7}{3}.$$

We can also take $f(x) = (1 + x)^2$ and $x_i = i/n$; this then leads to

$$\int_0^1 (1+x)^2\,dx = \tfrac{7}{3}.$$

Example 2: Find $\lim_{n\to\infty} \frac{1}{n}\sum_{k=1}^{n} \sin\frac{k}{n}$.

Solution: Pursuing the same strategy as in Example 1, we take $f(x) = \sin x$ and let $x_k = k/n$. It follows that

$$\Delta x = x_{k+1} - x_k$$
$$= \frac{k+1}{n} - \frac{k}{n}$$
$$= \frac{1}{n}$$

as in Example 1, and therefore

$$\frac{1}{n}\sum_{k=1}^{n} \sin\frac{k}{n} = \sum_{k=1}^{n} f(x_k)\Delta x$$

is an upper sum approximating $\int_0^1 f(x)dx$. Hence

$$\lim_{n\to\infty} \frac{1}{n}\sum_{k=1}^{n} \sin\frac{k}{n} = \int_0^1 \sin x\,dx$$

$$= -\cos x \Big|_0^1$$
$$= \cos 0 - \cos 1$$
$$= 1 - \cos 1 = 0.4597,$$

approximately.

Miscellaneous Exercises 9.1

1. Evaluate:

(a) $\sum_{k=1}^{4} k^{-1}$

(b) $\sum_{i=3}^{5} \frac{i}{i-2}$

(c) $\sum_{i=1}^{5} (0 \cdot i + 1)$

(d) $\sum_{i=1}^{5} 1$

(e) $\sum_{i=1}^{3} i^3$

(f) $\left(\sum_{i=1}^{3} i\right)^3$

(g) $\sum_{j=0}^{5} (2j+1)$

(h) $\sum_{k=1}^{6} (2k-1)$

2. Let $x_1 = 2$, $x_2 = 5$, $x_3 = 1$, $x_4 = -2$, $x_5 = 0$.

(a) Evaluate (i) $\sum_{i=1}^{5} x_i$; (ii) $\sum_{i=1}^{5} x_i^2$; (iii) $\sum_{j=2}^{4} 1/x_j$; (iv) $\sum_{k=1}^{3} x_k/(x_k + 1)$.

(b) Show that (i) $\sum_{i=1}^{5} 2x_i = 2 \sum_{i=1}^{5} x_i$; (ii) $\sum_{i=1}^{5} (x_i + 2) = \sum_{i=1}^{5} x_i + 10$.

3. Given numbers $x_1, x_2, \ldots, x_n$ and $y_1, y_2, \ldots, y_n$, explain why

$$\sum_{i=1}^{n} (ax_i + by_i) = a \sum_{i=1}^{n} x_i + b \sum_{i=1}^{n} y_i$$

for any constants a and b.

4. Write in sigma notation:
(a) $\sqrt{2} + \sqrt{3} + \sqrt{4} + \cdots + \sqrt{17}$.
(b) $\sqrt{1} + \sqrt{2} + \sqrt{3} + \cdots + \sqrt{n}$.
(c) $\sqrt{1} + \sqrt{2} + \sqrt{3} + \cdots + \sqrt{2n}$.
(d) $\sqrt{3} + \sqrt{4} + \sqrt{5} + \cdots + \sqrt{n+1}$.
(e) $\sqrt{n} + \sqrt{n+1} + \sqrt{n+2} + \cdots + \sqrt{2n}$.
(f) The sum in (c) minus the sum in (b).

5. Write in sigma notation the sum of
(a) the first n positive integers.
(b) the cubes of the integers from 7 to 43, inclusive.
(c) the first n positive even integers.

(**d**) the first n positive odd integers.
(**e**) n terms of $2 + 4 + 8 + 16 + \cdots$.
(**f**) n terms of $-1 + 2 - 3 + 4 - 5 + \cdots$. (*Hint*: $(-1)^i$ is equal to 1 if i is even, and to -1 if i is odd.)
(**g**) n terms of $1 - \frac{1}{2} + \frac{1}{3} - \frac{1}{4} + \cdots$.

6. Write in sigma notation, not including the repetition of the values for 5.5, (**a**) the sum of the squares, and (**b**) the sum of the reciprocals, in Table 1, p. 436.

7. Find $\lim_{n\to\infty} \frac{1}{n} \sum_{k=1}^{n} \sqrt{\frac{k}{n}}$. (*Hint:* See Example 1. Use the function $f(x) = \sqrt{x}$, divide the interval [0, 1] into equal parts by the points $x_k = k/n, 0 \le k \le n$, and show that the given sum then takes the form $\sum_{k=1}^{n} f(x_k)\Delta x$. What dictates the choice of the interval [0, 1]?)

8. Evaluate the following limits:

(**a**) $\lim_{n\to\infty} \frac{1}{n} \sum_{i=1}^{n} \left(2 + \frac{i}{n}\right)$

(**b**) $\lim_{n\to\infty} \frac{1}{n} \sum_{k=1}^{n} \sin \frac{k\pi}{2n}$

(**c**) $\lim_{n\to\infty} n^{-3} \sum_{j=1}^{n} (n + j)^2$

(**d**) $\lim_{n\to\infty} \frac{1}{n} \sum_{k=1}^{n} \sec^2 \frac{k\pi}{4n}$

(**e**) $\lim_{n\to\infty} n^{-4} \sum_{j=1}^{n} j^3$

(**f**) $\lim_{n\to\infty} n^{-3} \sum_{i=1}^{n} (2n + i)^2$

Exercises 9.1 [C]

1. Evaluate:

(**a**) $\lim_{n\to\infty} \sum_{j=1}^{n} n/(j + n)^2$

(**b**) $\lim_{n\to\infty} n^{-3/2} \sum_{k=1}^{n} \sqrt{k + n}$

(**c**) $\lim_{n\to\infty} n^{-2} \sum_{k=1}^{n} \sqrt{n^2 + 3kn}$

2. Use the following procedure to approximate the sum

$$S = (1.01)^2 + (1.02)^2 + \cdots + (1.99)^2 + (2.00)^2 = \sum_{k=1}^{100} (1 + 0.01k)^2.$$

(**a**) Show, as in Example 2, that $0.01S = \sum_{k=1}^{100} (1 + 0.01k)^2(0.01)$ is an upper sum S_U approximating the definite integral $I = \int_1^2 x^2\, dx$.

(b) Show that $S_L = \sum_{k=0}^{99} (1 + 0.01k)^2(0.01)$ is a lower sum approximating I.

(c) Show that $S_L - 0.01 = S_U - 0.04$, and that therefore $S_L = 0.01(S - 3)$ and $0.01(S - 3) < I < 0.01S$.

(d) Conclude that $100I < S < 100I + 3$, or, since $I = \frac{7}{3}$, $233\frac{1}{3} < S < 236\frac{1}{3}$.

(e) With the aid of a diagram, show that S is very nearly midway between $233\frac{1}{3}$ and $236\frac{1}{3}$, and that a good estimate is therefore $S = 234\frac{5}{6}$. (The exact value is $S = 234.835$.)

3. Show that $83\frac{1}{3} < \sum_{i=1}^{25} \sqrt{i} < 88\frac{1}{3}$.

4. In Table 1 estimate the sum of the entries in the column headed (a) n^2; (b) $\sqrt{n}$; (c) $\sqrt{10n}$; (d) $\sqrt[3]{n}$. (Note that the entries for $n = 5.5$ are repeated at the top of the second page of the table; omit this second occurrence.)

5. Find an approximation to the sum of the sines in Table 2, pages 438 and 439.

6. A table lists the sines of angles from $0°$ to $89°54'$, at intervals of $6'$. Estimate the sum of the sines in this table. (*Hint*: $\Delta x = 6' = \pi/1800$ radians.)

9.2 Rational approximations

There is a tendency to think of approximations as inferior substitutes. This is unfortunate, because a good approximation may be more useful and more informative than an exact expression. Most of our intuitive understanding of irrational numbers, for instance, is associated with rational approximations. If somebody asks you what is the (positive) square root of 2, there is essentially only one correct answer: $\sqrt{2}$. This answer will get you no thanks, however, because it carries no information; almost certainly the question seeks a rational approximation, such as 1.414. Not only is the approximation more useful than the exact expression in many computations and applications, it also conveys important information that the exact form conceals: it tells where to locate the number on the real number line. Similarly, π and tan 2 are numerals; each is a symbol representing a unique real number, but these symbols tell us less about the corresponding numbers than do the rational approximations 3.14 and -2.18. A happy feature of rational approximations, moreover, is that they converge: there exist arbitrarily good rational approximations to every real number.

Like $\sqrt{2}$, a definite integral is a numeral. For instance, $\int_0^2 x^2\,dx$ represents the number more familiarly known as 8/3. Since $\sqrt{1+x^4}$ is integrable on [0, 2], it follows that $\int_0^2 \sqrt{1+x^4}\,dx$ also represents a real number. In this case, however, there is no more familiar numeral; the integral itself is the only symbol that exactly represents the number. Hence the question "What is the value of this integral?" like our earlier question about $\sqrt{2}$, has no answer. But, again like $\sqrt{2}$, we can obtain excellent rational approximations to this. In this section we shall investigate some ways of computing such approximations. This problem has been a matter of mathematical interest for many centuries, but its study has taken on new life and vigor since the development of high-speed computers has reduced to seconds computations that in former years would have been measured in lifetimes.

Rectangular approximations. A definite integral is the limit, as $n \to \infty$, of a certain type of sum of n terms. This suggests that the value of the sum for large but finite n might be a useful approximation to the limit. Such approximations are often called *rectangular approximations* because for a positive function each term can be interpreted as the area of a rectangle.

A few special rectangular approximations are particularly interesting and useful. If the integrand is either increasing or decreasing throughout the interval of integration, then it is relatively easy to form an upper sum S_U and a lower sum S_L, as in § 4.1. This procedure has the advantage that the difference $S_U - S_L$ then sets an upper bound to the error incurred in using either S_U or S_L as an approximation to the integral.

Example 1. To approximate $\int_0^2 f(x)dx$, where $f(x) = \sqrt{1+x^4}$, it is convenient to divide [0, 2] into $2n$ subintervals, each of length $1/n$, by the points $0, 1/n, 2/n, \ldots, (2n-1)/n, 2$. The ith subinterval is then $[(i-1)/n, i/n]$, and since f is increasing on [0, 2], the smallest value of f on the ith subinterval is $f[(i-1)/n]$ and the largest is $f(i/n)$. Hence

$$S_L = \sum_{i=1}^{2n} \frac{1}{n}\sqrt{1+\left(\frac{i-1}{n}\right)^4},$$

$$S_U = \sum_{i=1}^{2n} \frac{1}{n}\sqrt{1+\left(\frac{i}{n}\right)^4}.$$

The following values of S_L and S_U were obtained by computer, using these formulas:

n	S_L	S_U
10	3.50056	3.81287
20	3.57621	3.73237
50	3.62238	3.68484
100	3.63790	3.66913
200	3.64568	3.66130
500	3.65036	3.65660

For $n = 500$ we have $S_U - S_L = 0.00624$, and we conclude that

$$\int_0^2 \sqrt{1 + x^4}\, dx = 3.65,$$

with perhaps some doubt about the second decimal place.

It is clear that S_U gives an overestimate and S_L an underestimate of the integral. Some intermediate sum should therefore give a better estimate, for a given subdivision of the interval of integration, than either S_U or S_L alone. One such rectangular approximation, a particularly efficient one, is formed by the *midpoint rule*: In each term of the sum whose limit is the integral, the function is evaluated at the midpoint of the corresponding subinterval.

Example 1 (continued). The midpoint of the ith subinterval is

$$\frac{1}{2}\left(\frac{i-1}{n} + \frac{i}{n}\right) = \frac{2i-1}{2n}.$$

Hence the midpoint rule gives an approximation

$$S_M = \sum_{i=1}^{2n} \frac{1}{n} \sqrt{1 + \left(\frac{2i-1}{2n}\right)^4}.$$

The following table shows not only S_M but also, for comparison, the previously obtained values of S_L and S_U.

n	S_L	S_M	S_U
10	3.50056	3.65187	3.81287
20	3.57621	3.65308	3.73237
50	3.62238	3.65342	3.68484
100	3.63790	3.65347	3.66913
200	3.64568	3.65348	3.66130
500	3.65036	3.65348	3.65660

Observe that the midpoint rule gives, for much smaller n, the same result previously obtained from S_L and S_U, and strongly suggests that

$$\int_0^2 \sqrt{1 + x^4}\, dx = 3.65348,$$

to five decimal places. The three-figure approximation obtained for $n = 10$ is well within the range of human calculation.

Trapezoidal rule. There is no necessity that approximating sums be rectangular, and many nonrectangular sums have proved useful in practice. The simplest of these is formed by what is called the *trapezoidal rule*. As the name suggests, this rule is motivated by geometric considerations, and we therefore assume, merely for the sake of explanation, that the integrand f is non-negative. If we divide the interval of integration $[a, b]$ into n subintervals of length $\Delta x = (b - a)/n$ by inserting the points $x_1, x_2, \ldots, x_{n-1}$, and if we then erect vertical lines at the points of division, we could, as in the previous approximations, use these verticals as sides of rectangles. Common sense suggests, however, that we might instead draw chords joining the points where consecutive verticals meet the graph of f, as shown in Fig. 2. Each such chord then completes a trapezoid whose bases lie on

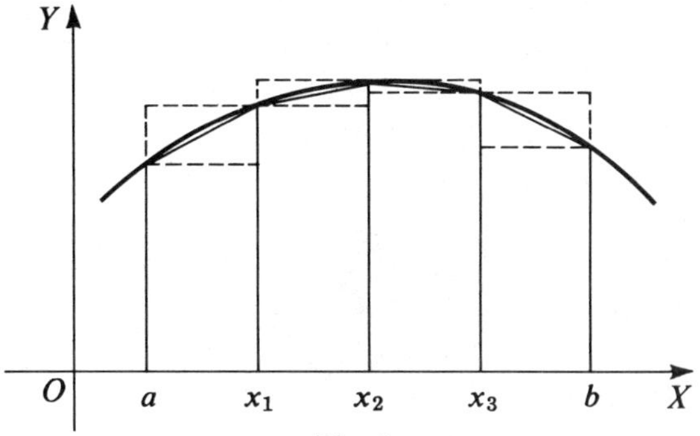

Fig. 2

two consecutive vertical lines and whose altitude is Δx, and the sum of the areas of these trapezoids then gives a *trapezoidal approximation* to $\int_a^b f(x)dx$.

Since the area of a trapezoid is one-half the sum of its bases times its height, the trapezoidal approximation is

$$\begin{aligned} S_T &= \tfrac{1}{2}[f(a) + f(x_1)]\Delta x + \tfrac{1}{2}[f(x_1) + f(x_2)]\Delta x + \cdots + \tfrac{1}{2}[f(x_{n-1}) + f(b)]\Delta x \\ &= \left[\tfrac{1}{2}f(a) + \sum_{i=1}^{n-1} f(x_i) + \tfrac{1}{2}f(b)\right]\Delta x. \end{aligned}$$

Example 1 (continued). We show values obtained by the trapezoidal rule approximating $\int_0^2 \sqrt{1 + x^4}\, dx$ and, for comparison, the values previously obtained by the midpoint rule:

n	S_M	S_T
10	3.65187	3.65672
20	3.65308	3.65429
50	3.65342	3.65361
100	3.65347	3.65352
200	3.65348	3.65349
500	3.65348	3.65348

Observe that the two rules eventually give the same approximation, but that the midpoint rule is more efficient, in that it gives this approximation for a smaller value of n.

Simpson's rule. The trapezoidal rule uses chords to approximate the graph of the given integrand f. A look at Fig. 2 suggests that if we could somehow bend these chords towards the graph of f we would obtain a better approximation to $\int_a^b f(x)\,dx$. One way of "bending" the chords is to replace them by arcs of parabolas. Thus instead of approximating f by a set of linear (or constant) functions, as in the trapezoidal rule, we would be approximating it by a set of quadratic functions. One way of doing this is called *Simpson's rule.*

The problem, then, is to fit a set of parabolic arcs to the graph of f. As the three coefficients in the standard form $y = ax^2 + bx + c$ suggest, it is possible to make each parabolic arc meet the given graph in three points, and these points can be chosen in a variety of ways.* The procedure in Simpson's rule is to divide the interval of integration $[a, b]$ into an *even* number of subintervals, say $2n$, by the points $a, x_1, x_2, \ldots, x_{2n-1}, b$. The length of each subinterval is then $\Delta x = (b - a)/2n$. If the first parabolic arc meets the graph of f at the points with abscissas a, x_1, and x_2, then the area under this arc is

$$\tfrac{1}{3}\Delta x[f(a) + 4f(x_1) + f(x_2)];$$

a demonstration of this fact is left to the student as an exercise. Similarly, the second parabola, meeting the curve at points with abscissas x_2, x_3, and x_4, determines an area

$$\tfrac{1}{3}\Delta x[f(x_2) + 4f(x_3) + f(x_4)],$$

and so on. We thus obtain the Simpson's-rule approximation

$$\begin{aligned} S_S = {} & \tfrac{1}{3}\Delta x[f(a) + 4f(x_1) + f(x_2)] \\ & + \tfrac{1}{3}\Delta x[f(x_2) + 4f(x_3) + f(x_4)] \end{aligned}$$

* If the three points happen to be collinear, then $a = 0$ and the "parabola" is a line.

$$\begin{aligned} &\cdot \\ &\cdot \\ &\cdot \\ &+ \tfrac{1}{3}\Delta x[f(x_{2n-2}) + 4f(x_{2n-1}) + f(b)] \\ &= \tfrac{1}{3}\Delta x[f(a) + 4\sum_{i=1}^{n} f(x_{2i-1}) + 2\sum_{i=1}^{n-1} f(x_{2i}) + f(b)]. \end{aligned}$$

Example 1 (concluded). The efficiency of Simpson's rule as an approximation to $\int_0^2 \sqrt{1 + x^4}\, dx$ is striking:

n	S_S
1	3.59332
2	3.65392
3	3.65354
4	3.65347
5	3.65348

Polynomial approximations. The problem of finding rational approximations is by no means disposed of when we have solved it for definite integrals. Our knowledge of a function is meager indeed if we have no way of evaluating it at an arbitrary point of its domain, and by "evaluating" we usually mean obtaining rational approximations such as are listed in tables. For polynomial functions this presents no great difficulty, since only the operations of addition and multiplication are needed to evaluate a polynomial at any real number. But what about the circular functions? What, for instance, is sin 0.5? A rational approximation can easily be obtained from a table, but this merely evades the mathematical question: How are such tables computed in the first place? In the great majority of instances, the answer to this question is to be found in a polynomial approximation to the given function. We shall show here how definite integrals can be used to obtain polynomial approximations to the sine and cosine functions. Since the other circular functions are defined in terms of these two, we shall thus solve the problem of computing tables of all the circular functions.

We start with the inequality, true for all x,

$$\cos x \leq 1. \tag{1}$$

It then follows, for $t \geq 0$, that

$$\int_0^t \cos x\, dx \leq \int_0^t 1\, dx,$$

$$\sin x\Big|_0^t \leq x\Big|_0^t,$$

or

$$\sin t \leq t.$$

We now make a simple change of variable to obtain

$$\sin x \leq x,$$

true by the preceding derivation for all $x \geq 0$, and repeat the maneuver:

$$\int_0^t \sin x \, dx \leq \int_0^t x \, dx,$$

$$-\cos x \Big|_0^t \leq \tfrac{1}{2}x^2 \Big|_0^t,$$

$$-\cos t + 1 \leq \tfrac{1}{2}t^2,$$

$$1 - \tfrac{1}{2}t^2 \leq \cos t,$$

or, finally,

$$1 - \tfrac{1}{2}x^2 \leq \cos x.$$

Combining this with (1) then yields

$$1 - \tfrac{1}{2}x^2 \leq \cos x \leq 1. \tag{2}$$

If we integrate this double inequality in the pattern of the preceding steps (the details are left to the student), we obtain

$$x - \frac{x^3}{3!} \leq \sin x \leq x.^* \tag{3}$$

Repeating the same maneuver, we obtain in turn

$$1 - \frac{x^2}{2!} \leq \cos x \leq 1 - \frac{x^2}{2!} + \frac{x^4}{4!}, \tag{4}$$

$$x - \frac{x^3}{3!} \leq \sin x \leq x - \frac{x^3}{3!} + \frac{x^5}{5!}, \tag{5}$$

$$1 - \frac{x^2}{2!} + \frac{x^4}{4!} - \frac{x^6}{6!} \leq \cos x \leq 1 - \frac{x^2}{2!} + \frac{x^4}{4!}, \tag{6}$$

$$x - \frac{x^3}{3!} + \frac{x^5}{5!} - \frac{x^7}{7!} \leq \sin x \leq x - \frac{x^3}{3!} + \frac{x^5}{5!}, \tag{7}$$

and so on. Thus the process generates a sequence of polynomial bounds for, alternately, the sine and the cosine. Moreover, these polynomials follow

* For each non-negative integer n, $n!$ (read "n factorial") is defined by the two-part rule

$$0! = 1, \quad (n+1)! = (n+1) \cdot n!.$$

Thus $1! = 1 \cdot 0! = 1$, $2! = 2 \cdot 1! = 2$, $3! = 3 \cdot 2! = 6$, and so on. One important observation is that $n!$ increases very rapidly with increasing n. Thus, $10! = 3{,}628{,}800$, and $25! > 10^{25}$.

a pattern which is easy to generalize. If we look first at the bounds for the cosine, we see that each term is of even degree (including the term 1, which has degree 0). Hence the exponent in each term is $2k$, for some non-negative integer k. The denominator of each term is just the factorial of the exponent. Finally, the signs alternate, thus following the pattern of $(-1)^k$, where k takes successive integer values starting with 0. Hence a typical term in each polynomial bound for the cosine is $(-1)^k x^{2k}/(2k)!$, and if we represent by $C_{2n}(x)$ the polynomial of degree $2n$ in this sequence, we have

$$C_{2n}(x) = 1 - \frac{x^2}{2!} + \frac{x^4}{4!} - \cdots + (-1)^n \frac{x^{2n}}{(2n)!}$$

$$= \sum_{k=0}^{n} (-1)^k \frac{x^{2k}}{(2k)!}. \tag{8}$$

A similar analysis shows that each polynomial bound for the sine has the form

$$S_{2n+1}(x) = x - \frac{x^3}{3!} + \frac{x^5}{5!} - \cdots + (-1)^n \frac{x^{2n+1}}{(2n+1)!}$$

$$= \sum_{k=0}^{n} (-1)^k \frac{x^{2k+1}}{(2k+1)!}. \tag{9}$$

Now, -1 and 1 are polynomial bounds for both the sine and the cosine, but this information, while important, is not of much value in computing a table of sines. The important feature of the polynomials in (8) and (9) is that for given x they *converge*, with increasing n, to $\cos x$ and $\sin x$, respectively:

$$\lim_{n \to \infty} C_{2n}(x) = \cos x, \qquad \lim_{n \to \infty} S_{2n+1}(x) = \sin x.$$

To see why this is, observe that for the cosine we have

$$C_{2n+2}(x) \leq \cos x \leq C_{2n}(x)$$

when n is even, with inequalities reversed when n is odd. In either case, the error incurred in using either $C_{2n}(x)$ or $C_{2n+2}(x)$ as an approximation to $\cos x$ does not exceed their difference,

$$E_n = |C_{2n+2}(x) - C_{2n}(x)| = \left|(-1)^{n+1} \frac{x^{2n+2}}{(2n+2)!}\right| = \frac{|x|^{2n+2}}{(2n+2)!},$$

and for any fixed x, $\lim_{n \to \infty} E_n = 0$. A similar result holds for our approximations to the sine.

We have one small bit of unfinished business to attend to. The preceding derivation has assumed $x \geq 0$. What happens if $x < 0$? To see what happens, look first at, say, (6). If $x < 0$, then $-x > 0$, and substituting $(-x)$ for x in (6) makes no change whatever because $\cos(-x) = \cos x$ and even powers of $-x$ are the same as the corresponding powers of x. Quite generally, $C_{2n}(-x) = C_{2n}(x)$, and so our analysis of the cosine holds, unchanged, for negative x.

The situation with respect to the sine is scarcely more complicated. We saw in (7), for instance, that

$$x - \frac{x^3}{3!} + \frac{x^5}{5!} - \frac{x^7}{7!} \leq \sin x \leq x - \frac{x^3}{3!} + \frac{x^5}{5!}$$

for $x \geq 0$. If $x < 0$, we may then substitute $-x$ for x in this inequality. Since $\sin(-x) = -\sin x$, and since $(-x)^n = -x^n$ for odd n, we thus obtain

$$-x + \frac{x^3}{3!} - \frac{x^5}{5!} + \frac{x^7}{7!} \leq -\sin x \leq -x + \frac{x^3}{3!} - \frac{x^5}{5!}.$$

If we now multiply through by -1, the inequalities reverse, and we have

$$x - \frac{x^3}{3!} + \frac{x^5}{5!} - \frac{x^7}{7!} \geq \sin x \geq x - \frac{x^3}{3!} + \frac{x^5}{5!}$$

for $x < 0$. The convergence to $\sin x$ of the polynomials $S_{2n+1}(x)$ is not affected by this reversal of inequalities, so our analysis still holds. We conclude that for any real x, the polynomials $C_{2n}(x)$ and $S_{2n+1}(x)$ give useful approximations to $\cos x$ and $\sin x$, respectively, and that the accuracy of these approximations increases with increasing n. This is illustrated in Figure 3, which shows graphs of the sine function and the approximating polynomials S_1, S_3, S_5, and S_7.

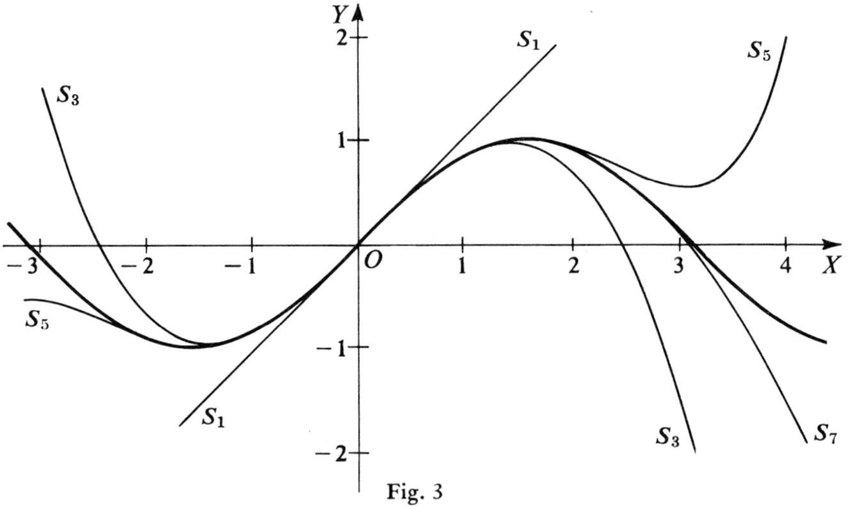

Fig. 3

Miscellaneous Exercises 9.2

1. Show that for an increasing function or for a decreasing function the trapezoidal approximation S_T is the average of the rectangular approximations S_L and S_U.

2. Show that the midpoint rule is equivalent to using as approximating regions a set of trapezoids each of which is tangent to the graph of f at the midpoint of its interval (Fig. 4). For this reason, the midpoint rule is sometimes called the *tangent rule*.

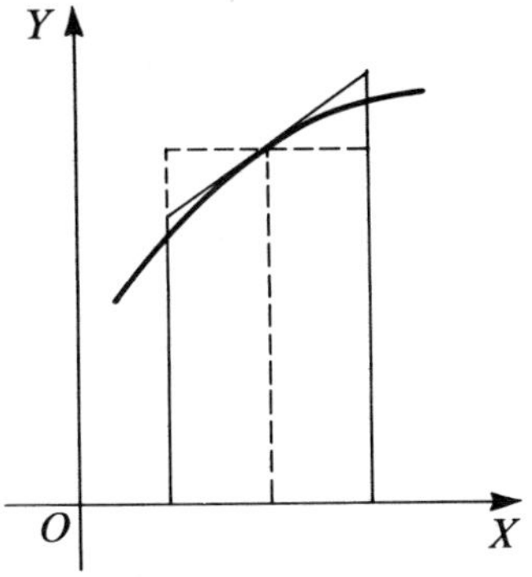

Fig. 4

3. If the graph of the integrand is concave down throughout the interval of integration, the trapezoidal rule gives a result that is too small and the midpoint rule a result that is too large. Explain. (Use Exercise 2.) What if the graph is concave up? What if the direction of concavity changes?

4. (a) If $f(x) = ax^2 + bx + c$, show that

$$\int_{-h}^{h} f(x)dx = \tfrac{1}{3}h(2ah^2 + 6c).$$

(b) Show that

$$f(-h) + 4f(0) + f(h) = 2ah^2 + 6c,$$

and thus conclude that

$$\int_{-h}^{h} f(x)dx = \tfrac{1}{3}h[f(-h) + 4f(0) + f(h)].$$

Show the relationship between this result and Simpson's rule.

5. Explain why Simpson's rule gives exact results when the integrand is a quadratic function. Use this rule to evaluate

(a) $\int_{-1}^{3}(2x^2 - x + 7)dx$ (b) $\int_{0}^{6}(x^2 - 6x + 9)dx.$

6. (a) Let
$$I = \int_{-h}^{h}(ax^2 + bx + c)\,dx$$
$$= \tfrac{2}{3}ah^3 + 2ch.$$
(See Exercise 4(a).)

(b) Show that if the trapezoidal rule is used, with $n = 1$, to approximate the integral in (a), the approximation obtained is
$$T_1 = 2ah^3 + 2ch.$$

(c) Show that if the midpoint rule is used, with $n = 1$, to approximate the integral in (a), the approximation obtained is
$$M_1 = 2ch.$$

(d) Show that
$$|I - T_1| = 2|I - M_1|$$
Conclude that in this case the error in the trapezoidal approximation is twice the error in the midpoint approximation.

(e) Repeat (b) and (c) with $n = 2$ to obtain the approximations
$$T_2 = ah^3 + 2ch,$$
$$M_2 = \tfrac{1}{2}ah^3 + 2ch.$$
Show that these approximations are four times as accurate as T_1 and M_1, in that
$$|I - T_2| = \tfrac{1}{4}|I - T_1|$$
and
$$|I - M_2| = \tfrac{1}{4}|I - M_1|,$$
but that the error in the trapezoidal approximation T_2 is still twice the error in the midpoint approximation M_2.

7. (a) (See Exercises 4 and 5) If $f(x) = kx^3 + ax^2 + bx + c$, show that
$$\int_{-h}^{h} f(x)\,dx = \tfrac{1}{3}h(2ah^2 + 6c)$$
$$= \tfrac{1}{3}h[f(-h) + 4f(0) + f(h)].$$
Conclude that Simpson's rule gives exact results when the integrand is a cubic function.

(**b**) Use Simpson's rule to evaluate

(**i**) $\int_{-2}^{2} (143x^3 + 4x^2 + 517x - 1)\, dx$;

(**ii**) $\int_{0}^{6} (x^3 - x)\, dx$;

(**iii**) $\int_{-1}^{3} (3x^3 - 7x^2 + 2x - 5)\, dx$.

8. A barrel-shaped solid is contained between two parallel plane faces, its bases, and the area of each section of the barrel by a plane x units from the lower base is $f(x)$ square units, for some given function f. It can be shown by methods to be developed in § 9.3, that if the height of the barrel is H units, then its volume is $V = \int_0^H f(x)\, dx$ cubic units.

(**a**) Show that if Simpson's rule is used with $n = 2$ to approximate this volume, the approximation obtained is

$$V = \tfrac{1}{6}H(B_1 + 4M + B_2),$$

where $B_1 = f(0)$ and $B_2 = f(H)$ are the areas of the bases and $M = f(\tfrac{1}{2}H)$ is the area of the midsection. This result is known as the *prismatoid* (or prismoid) *formula*. (A prismatoid is a polyhedron all of whose vertices lie in two parallel planes; the formula is exact for such a solid.)

(**b**) Show that the prismatoid formula gives the exact volumes of the following solids: (**i**) a cylinder; (**ii**) a pyramid; (**iii**) a cone; (**iv**) a sphere.

9. (See Exercise 8.) An arc of an ellipse lies above the x-axis, with its midpoint at (0, 3) and its endpoints at (−3, 2) and (3, 2). This arc is rotated about the x-axis to form the curved surface of a barrel-shaped solid. Show that the prismatoid formula gives the exact volume for this solid.

10. Use $n = 4$ subintervals to approximate $\int_0^{\pi} \sin x\, dx$ by (**a**) the trapezoidal rule, (**b**) the midpoint rule, and (**c**) Simpson's rule.

11. Show that $\int_0^1 dx/(1 + x^2) = \pi/4$. Obtain a rational approximation to

π by approximating this integral. Use $n = 4$ subintervals and (a) the trapezoidal rule, (b) the midpoint rule, and (c) Simpson's rule.

12. Approximate $\int_1^2 x^{-1}\,dx$ by Simpson's rule, using 8 subintervals.

13. A motorist on a two-hour trip noted the speed at 10-minute intervals as 0, 57, 51, 55, 0, 62, 60, 58, 35, 60, 33, 35, and 0 mi./hr. Estimate the distance traveled, using (a) the trapezoidal rule, and (b) Simpson's rule.

14. (For students with access to a computer.) Find rational approximations to each of the following integrals, correct to 4 significant figures.

(a) $\int_2^4 x^{-1}\,dx$ (d) $\int_0^3 \sqrt{1+x^4}\,dx$

(b) $\int_4^8 x^{-1}\,dx$ (e) $\int_0^\pi \sqrt{2-\sin^2 x}\,dx$

(c) $\int_0^4 x\,dx/(x^2+1)$ (f) $\int_0^2 dx/\sqrt{1+x^3}$

15. Show how inequalities (3) through (7) can be obtained, sequentially, from (2).

16. Use a polynomial $S_{2n+1}(x)$ to obtain a rational approximation, correct to 6 places of decimals, to sin 0.5. Compare with the value given by a calculator or Table 2.

17. Compute sin 0.1 and cos 0.1, each to 6 places.

18. Show that the error incurred in using either $S_{2n+1}(x)$ or $S_{2n+3}(x)$ as an approximation to $\sin x$ does not exceed $E_n = |S_{2n+3}(x) - S_{2n+1}(x)|$. Show also that, for fixed x, $\lim_{n\to\infty} E_n = 0$.

9.3 Volumes

In the rest of this chapter we shall consider applications of the definite integral to a variety of problems from geometry and physics. We begin by returning to the question of volume, already studied in § 4.5. In that earlier treatment we restricted our attention to volumes of solids generated by rotating about the x-axis regions that had one boundary on that axis. The principles used in § 4.5 apply, however, to rotation of a plane region about any line in its plane, as the examples that follow will show.

Example 1. The region bounded by the parabola $y = x^2$, the x-axis, and the line $x = 2$ is rotated about the x-axis. Find the volume of the solid thus generated.

Solution: A typical rectangular element, indicated in Fig. 5, will generate a cylinder of volume

$$\Delta V_i = \pi y_i^2\, \Delta x;$$

hence

$$S = \sum_{i=1}^{n} \pi y_i^2\, \Delta x,$$

and

$$V = \lim_{n\to\infty} S = \pi \int_0^2 y^2\, dx = \pi \int_0^2 x^4\, dx = \pi \left.\frac{x^5}{5}\right|_0^2 = \frac{32\pi}{5}.$$

The required volume is, then, $32\pi/5$ cubic units.

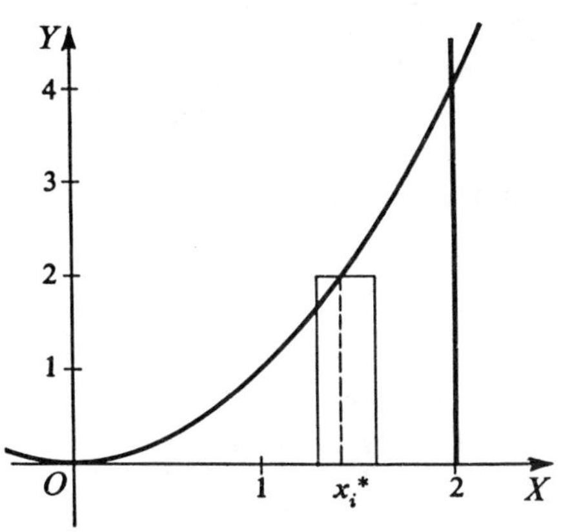

Fig. 5

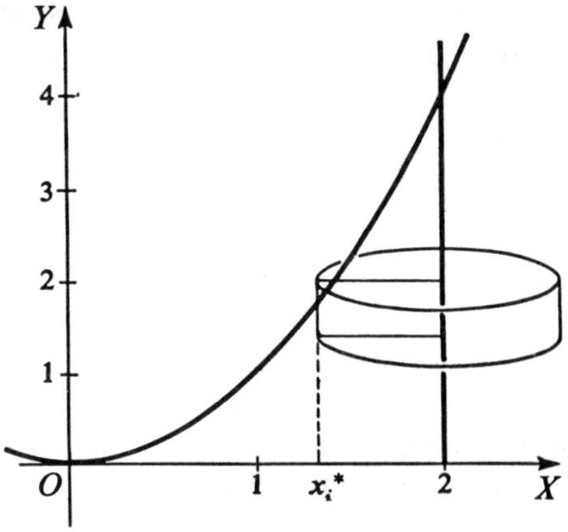

Fig. 6

Example 2. Find the volume of the solid generated by rotating the region of Example 1 about the line $x = 2$.

Solution: In this case, a typical rectangular element generates a cylinder of radius $2 - x_i^*$ and height Δy (see Fig. 6). Hence

$$\Delta V_i = \pi(2 - x_i^*)^2\, \Delta y,$$

and summing and passing to the limit gives

$$V = \pi \int_0^4 (2 - x)^2\, dy = \pi \int_0^4 (2 - \sqrt{y}\,)^2\, dy$$

$$= \pi \int_0^4 (4 - 4y^{1/2} + y)dy$$

$$= \pi \left[4y - \frac{8}{3} y^{3/2} + \frac{1}{2} y^2\right]_0^4 = \frac{8\pi}{3}.$$

The required volume is $8\pi/3$ cubic units.

Exercises [A]

Cylindrical Disks

Find the volume of the solid generated by rotating the given region about the given line. For each solid:

(**a**) sketch the generating region, showing a typical generating rectangle;
(**b**) write an expression for the volume generated by this rectangle;
(**c**) express the exact volume of the solid as a definite integral;
(**d**) evaluate the integral.

1. The region in the first quadrant bounded by $y = 4 - x^2$ and the coordinate axes, about (**i**) the x-axis; (**ii**) the y-axis.

2. The region in the first quadrant bounded by $x = y^3$, the x-axis, and $x = 8$, about (**i**) the x-axis; (**ii**) $x = 8$.

3. The region between $y = x^2 - 4x + 5$ and $y = 5$, about $y = 5$.

If a solid of revolution has a hole in it, the device of finding its volume as the limit of a sum of cylindrical disks must be modified. In many cases, it is possible to use as an approximating element a cylindrical disk with a cylindrical hole through the center, a washer. If such an element has an outer radius R, an inner radius r, and height Δh, then its volume can be found by subtracting the volume of the hole from that of the disk:

$$\Delta V = \pi R^2\,\Delta h - \pi r^2\,\Delta h = \pi(R^2 - r^2)\Delta h.$$

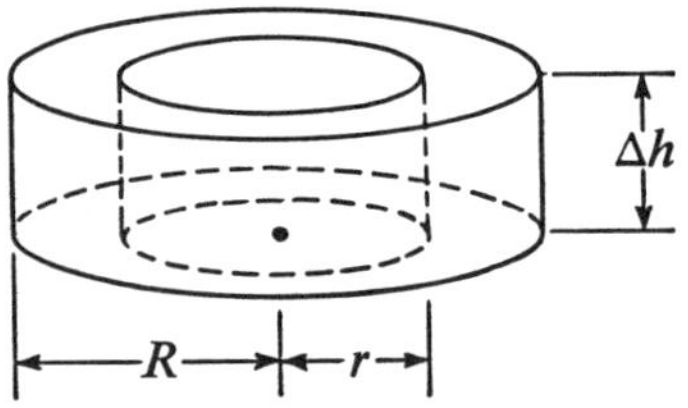

Fig. 7

If we divide our solid into n parallel slices of height Δh, then the sum $S = \sum_{i=1}^{n} \pi(R_i^2 - r_i^2)\Delta h$ is an approximation to its volume, and we can define that volume as the limit $V = \lim_{n\to\infty} S$, a definition entirely consistent with the one stated on page 150. If, now, we can express R and r in terms of the variable h, it follows that the sum S is of the form to which the definition of definite integral applies, and we may determine V by integrating some such form as $\pi\int(R^2 - r^2)dh$ between appropriate limits.

Example 3. Find the volume of the solid generated by rotating the region of Example 1 about the y-axis.

Solution: A typical element, indicated in Fig. 8, will generate a washer-shaped volume with inner radius x_i^* and outer radius 2, so

$$\Delta V_i = \pi(2^2 - x_i^{*2})\Delta y,$$

and

$$V = \pi\int_0^4 (4 - x^2)dy = \pi\int_0^4 (4 - y)dy$$
$$= \pi[4y - \tfrac{1}{2}y^2]_0^4 = 8\pi.$$

The volume is 8π cubic units.

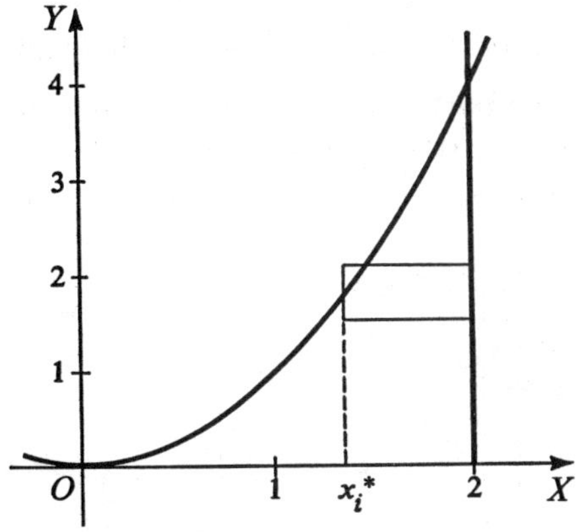

Fig. 8

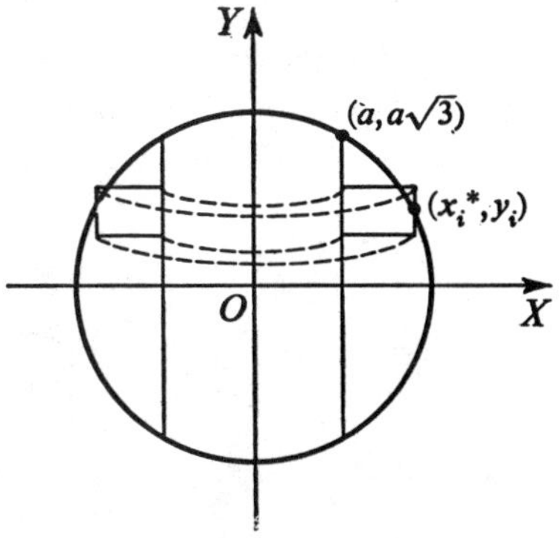

Fig. 9

Example 4. A hole of radius a is bored symmetrically through a sphere of radius $2a$. Find the volume remaining.

Solution: The volume required can be generated by rotating about the y-axis an appropriate segment of the circle $x^2 + y^2 = 4a^2$, as indicated in Fig. 9. If we approximate this volume by washers, we have

$$\Delta V_i = \pi(x_i^{*2} - a^2)\Delta y,$$

and therefore

$$V = \pi\int_{-a\sqrt{3}}^{a\sqrt{3}} (x^2 - a^2)dy$$

$$= 2\pi\int_0^{a\sqrt{3}} (3a^2 - y^2)dy = 2\pi[3a^2y - \tfrac{1}{3}y^3]_0^{a\sqrt{3}} = 4\pi a^3\sqrt{3}.$$

The volume is $4\pi a^3\sqrt{3}$ cubic units.

Exercises [A]

Cylindrical Washers

Find the volume of the solid generated by rotating the given region about the given line. For each solid:

(**a**) sketch the generating region, showing a typical generating rectangle;
(**b**) write an expression for the volume generated by this rectangle;
(**c**) express the exact volume of the solid as a definite integral;
(**d**) evaluate the integral.

1. The region bounded by $y = 2 - x^2$ and $y = 1$, about (**i**) the x-axis; (**ii**) $y = -1$.

2. The region bounded by $x = y^2 - 6y$ and $x = -5$, about the y-axis.

3. The region enclosed by $y = h^2 - x^2$ and $2y = h^2 - x^2$, about the x-axis.

In some cases, the method of approximating a volume by cylindrical washers is less convenient than an approximation by cylindrical shells. We found, in Ex. 2 of Miscellaneous Exercises 7.2, that if a cylinder of fixed

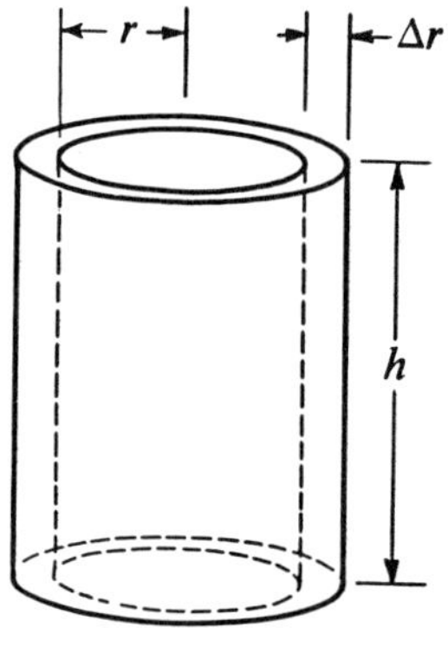

Fig. 10

height h has its radius r increased by an amount $\Delta r = dr$, the increase ΔV in the volume V is approximately equal to the differential dV when Δr is small. Since $V = \pi r^2 h$, this gives $dV = 2\pi rh\,\Delta r$ as an approximation to the volume of a cylindrical shell of height h and thickness Δr. Because the shell is "thin," because, that is, we shall be concerned with the limit as Δr approaches zero, it makes no difference whether we take r to be the inside radius, the outside radius, or some intermediate radius.

Note that if the cylindrical shell is cut along a line parallel to its axis and then flattened out, the result will approximate a rectangular solid with length $2\pi r$, height h, and thickness Δr, and therefore with a volume which is approximately $2\pi rh\,\Delta r$, as before.

Example 5. Use cylindrical shells to find the volume of Example 4.

Solution: When the rectangle outlined in Fig. 11 is rotated about the y-axis, it will generate a cylindrical shell of height $2y_i$, radius approximately x_i^*, and thickness Δx. Hence $\Delta V_i = 2\pi x_i^*(2y_i)\Delta x$, and the volume is

$$V = 4\pi \int_a^{2a} xy\,dx = 4\pi \int_a^{2a} x\sqrt{4a^2 - x^2}\,dx$$

$$= -2\pi[\tfrac{2}{3}(4a^2 - x^2)^{3/2}]_a^{2a} = -2\pi[0 - \tfrac{2}{3}(3a^2)^{3/2}] = 4\pi a^3\sqrt{3}.$$

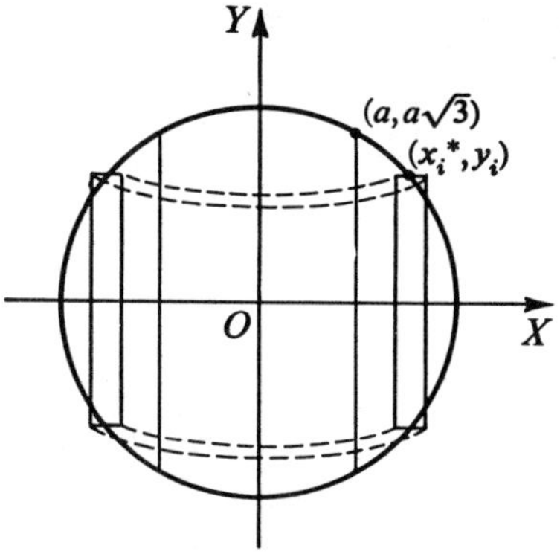

Fig. 11

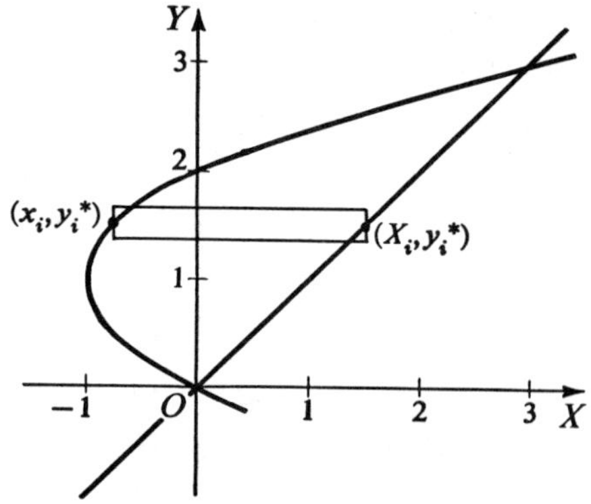

Fig. 12

Example 6. Find the volume of the solid generated by rotating about the x-axis the region enclosed by $x = y^2 - 2y$ and $y = x$.

Solution: We find that these curves (Fig. 12) intersect at (0, 0) and (3, 3). The line $y = y_i^*$ meets the parabola at (x_i, y_i^*) and the given line at (X_i, y_i^*), and a rectangular element of width Δy extending from one of these intersections to the other will generate a cylindrical shell of height $X_i - x_i$, radius y_i^*, and thickness Δy. Hence $\Delta V_i = 2\pi y_i^*(X_i - x_i)\Delta y$, and

$$V = 2\pi \int_0^3 y(X - x)dy = 2\pi \int_0^3 y[y - (y^2 - 2y)]dy$$

$$= 2\pi \int_0^3 (3y^2 - y^3)dy = 2\pi[y^3 - \tfrac{1}{4}y^4]_0^3 = \frac{27\pi}{2}.$$

Hence the required volume is $27\pi/2$ cubic units.

Note that it would be awkward to use washers as approximations to this volume. To the right of the y-axis, the inner radius of a washer would be the ordinate of a point on the line; to the left of the y-axis, the situation is otherwise and more complex. The cylindrical shell offers by far the simpler approach.

Exercises [A]

Cylindrical Shells

Find the volume of the solid generated by rotating the given region about the given line. For each solid:

(a) sketch the generating region, showing a typical generating rectangle;
(b) write an approximate expression for the volume generated by this rectangle;
(c) express the exact volume of the solid as a definite integral;
(d) evaluate the integral.

1. The region enclosed by $y = 4x - x^2$ and the x-axis, about (i) the y-axis; (ii) $x = -1$.
2. The region in the first quadrant enclosed by $y = x^3$ and $y = 4x$, about the x-axis.
3. The region enclosed by $y = x^2 - 2x - 6$ and $2x - y - 6 = 0$, about the y-axis.
4. The loop of $y^2 = 2x^2 - x^3$, about the y-axis.

The volumes of certain solids not generated by rotation can be determined by the methods of this section. If, in particular, the areas of a family of parallel plane sections of a solid are known, the problem of determining the volume can be reduced to the problem of evaluating an integral.

Example 7. Using the fact that the volume of a prism is found by multiplying the area of the base by the height, derive the formula for the volume of a triangular pyramid (Fig. 13).

Solution: Consider a pyramid of basal area B and altitude h. From a theorem of solid geometry, any section of this pyramid by a plane parallel to the base will be a triangle similar to the base, and if the section is x units from the apex, the ratio of similitude will be x/h; the area of such a section will therefore be $(x/h)^2B$. We now cut the pyramid into n slices by a system of planes parallel to the base and $\Delta x = h/n$ units apart. If x_i^* is the distance from the apex of a point in the ith slice, then the volume of that slice is approximately that of a prism of base $(x_i^*/h)^2B$ and height Δx: that is, $\Delta V_i = (x_i^*/h)^2B\,\Delta x$. The volume of the pyramid is therefore approximately

$$S = \sum_{i=1}^{n} \Delta V_i,$$

and the exact value is

$$V = \lim_{n\to\infty} S = \int_0^h \left(\frac{x}{h}\right)^2 B\,dx = \frac{B}{h^2}\int_0^h x^2\,dx = \frac{B}{h^2}\left[\frac{1}{3}x^3\right]_0^h = \frac{1}{3}Bh.$$

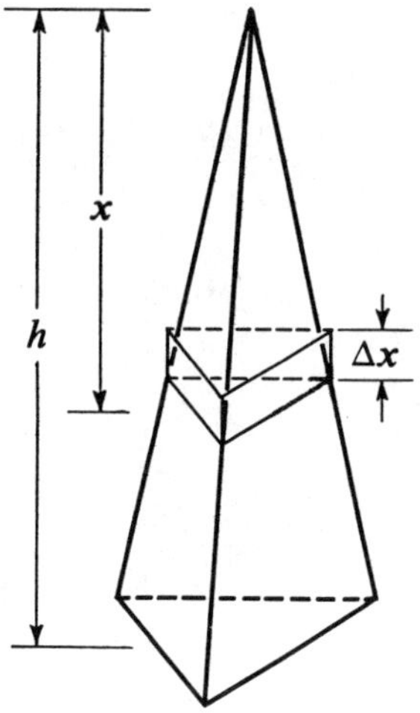

Fig. 13

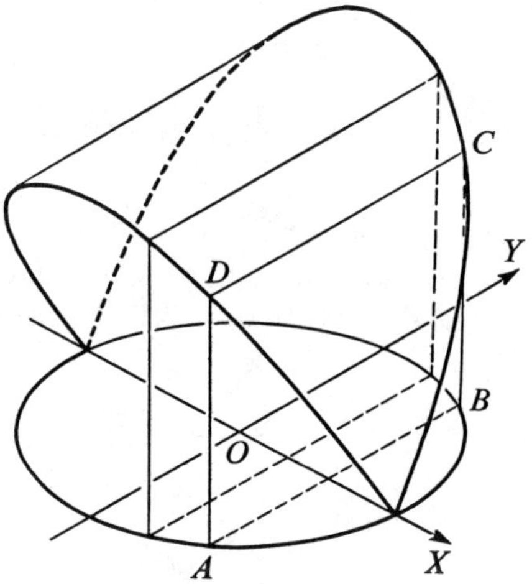

Fig. 14

Example 8. The base of a solid is the circle $x^2 + y^2 = 9$, and every section of the solid by a plane perpendicular to the x-axis is a square. Find the volume.

Solution: If the vertex B of the section $ABCD$ shown in Fig. 14 is at (x_i, y_i), then the length of AB is $2y_i$ and the area of $ABCD$ is $4y_i^2$. The volume enclosed between this section and one Δx units away is therefore approximately $\Delta V_i = 4y_i^2 \, \Delta x$, and the volume of the solid is therefore

$$V = 4\int_{-3}^{3} y^2\,dx = 8\int_{0}^{3} (9 - x^2)dx$$

$$= 8[9x - \tfrac{1}{3}x^3]_0^3 = 144 \text{ (cubic units).}$$

Exercises [A]

Known Cross Section

Find the volume of the following solids. For each solid:

(a) sketch the solid, showing a typical cross section;

(b) write an approximate expression for the volume included between two neighboring cross sections;

(c) express the volume of the solid as a definite integral;

(d) evaluate the integral.

1. The base of the solid is the circle $x^2 + y^2 = 4$, and every section of the solid by a plane perpendicular to the x-axis is **(i)** a square; **(ii)** an equilateral triangle; (**iii**) an isosceles right triangle with its hypotenuse in the plane of the circle.

2. Every section of the solid by a plane perpendicular to the x-axis is a square with the ends of one diagonal on the ellipse $x^2 + 4y^2 = 16$.
3. The solid is such that every section by a plane perpendicular to the y-axis is a circle with one end of a diameter on the parabola $x = y^2 - 4$ and the other end of the same diameter on the parabola $x = 2y^2 - 8$. The solid lies entirely to the left of the y-axis.

Exercises 9.3 [A]

1. Find the volume of the solid generated by rotating the region in the first quadrant bounded by the y-axis, $y = x^2$, and $y = 4$ about (**a**) the y-axis; (**b**) $y = 4$; (**c**) the x-axis; (**d**) $x = 2$.
2. The region enclosed by $y = x$ and $y^2 = 4x$ is rotated about the x-axis. Find the volume generated, using (**a**) washers; (**b**) shells.
3. Find the volume generated by rotating about the y-axis the region in the first quadrant bounded by $y = 2x$ and $y = 6x - x^3$.
4. Find the volume of the solid generated by rotating about the x-axis the region bounded by $y = \sec x$, the y-axis, the x-axis, and $x = \pi/4$.
5. The base of a solid is the circle $x^2 + y^2 = r^2$, and every section of the solid by a plane perpendicular to the x-axis is an isosceles triangle of altitude b, with its base in the plane of the circle. Find the volume.

Exercises 9.3 [B]

1. Find the volume of the solid generated by rotating the region in the first quadrant bounded by $x^2 = 4y$, the x-axis, and $x = 2$ about (**a**) the x-axis; (**b**) the y-axis; (**c**) $x = 2$; (**d**) $y = -1$; (**e**) $x = -1$.
2. Find the volume of the solid generated by rotating about the y-axis the region bounded by $y^2 = 4px$ and $y = x$, using (**a**) washers; (**b**) shells.
3. Find the volume of the solid generated by rotating about the x-axis the loop of $y^2 = x^3 - 4x$.
4. Find the volume of a spherical segment of one base if the radius of the sphere is r and the height of the segment is h. (This is the solid cut from the sphere by a plane which is $r - h$ units from the center.) Check by setting $h = 2r$.
5. The base of a solid is the circle $x^2 + y^2 = a^2$, and every section by a plane perpendicular to the y-axis is an isosceles right triangle with one leg in the plane of the circle. Find the volume.

Miscellaneous Exercises 9.3

1. Derive the formula for the volume of a right circular cone.
2. The region included between the parabolic arc $x^{1/2} + y^{1/2} = a^{1/2}$ and the coordinate axes is rotated about the x-axis. Find the volume of the solid generated.
3. The triangular region bounded by the x-axis and the lines $y = x - 2$ and $x = 5$ is rotated about the y-axis. Find the volume generated.
4. The radius of a right circular cylinder is a. A wedge is cut out of the bottom of the cylinder by a plane which passes through a diameter of the base and makes a 45° angle with the plane of the base. Find the volume of the wedge.
5. A conical surface has its vertex at the center of a sphere of radius a. If the semivertical angle of the cone is θ ($\theta < \pi/2$), find the volume of that part of the sphere which is inside the cone.
6. Find the volume of the solid generated by rotating the four-cusped hypocycloid $x^{2/3} + y^{2/3} = a^{2/3}$ about the y-axis.
7. The region bounded by $xy = 1$, the x-axis, and the lines $x = 1$ and $x = b$ ($b > 1$) is rotated about the x-axis. Find the limit of the volume thus generated as $b \to \infty$.
8. The base of a solid is the region enclosed by $y = x^2 - 4x + 5$ and $y = 11 - x^2$. Find the volume of the solid if each section by a plane perpendicular to the x-axis is **(a)** a square; **(b)** an equilateral triangle.
9. Find the volume of the solid generated by rotating about the y-axis the region enclosed by $y = x^2 - 4x + 4$ and $y = x$.

Exercises 9.3 [C]

1. Find the volume of the solid generated by rotating about the x-axis the region enclosed by the x-axis and one arch of the cycloid $x = a(\theta - \sin\theta)$, $y = a(1 - \cos\theta)$.
2. Find the volume of the solid generated by rotating about the t-axis the smallest region enclosed by $y = \sin t$ and $y = \sin 2t$.
3. A cylindrical hole through the center of a sphere is 6 in. long. Find the remaining volume of the sphere.
4. The interior of the circle $x^2 + y^2 = 25$ is rotated about the line $x = 10$. Find the volume of the torus thus generated.
5. The region bounded by the witch $y = 8a^3/(x^2 + 4a^2)$, the x-axis, and the lines $x = \pm b$ is rotated about the x-axis. Find the volume of the solid thus generated, and the limit of this volume as $b \to \infty$.

6. Find the general formula for the volume of a torus: the circle $x^2 + y^2 = a^2$ is rotated about the line $x = b$ $(b > a)$.

7. The axes of two right circular cylinders meet at right angles. If the radius of each cylinder is r, find the volume common to both. (*Hint:* What is the shape of a section by a plane parallel to the axes of the cylinders?)

8. Work Ex. 7 for the case of three equal cylinders, the axis of each being perpendicular to the axes of the other two.

9.4 Work

The work done by a force is defined, in physics, to be the product of the force and the distance through which it acts: Work = Force × Distance, $W = Fs$. Thus, for example, if gravity exerts a force of 15 lb. on an object (the force which gravity exerts on an object is called the weight of the object) and moves it a distance of 10 ft., the work done is $15 \cdot 10 = 150$ foot-pounds (abbreviated ft.-lb.). If the force is not constant, or if the parts of the object moved do not all travel the same distance, it is generally necessary to use integration to measure the work done.

Example 1. When a spring is stretched, the force used is, within certain limits, proportional to the increase in length. If a spring which is normally 10 in. long is extended to 11 in. by a force of 5 lb., find the work done in stretching it from a length of 11 in. to a length of 14 in.

Solution: If the extension of the spring is x and the force is F, we have the general rule $F = kx$. For this particular spring, we have $5 = k(1)$, and therefore $F = 5x$. To extend the spring to 14 in. corresponds to $x = 4$ and therefore requires 5(4), or 20, lb., but the work done will not be (20 lb.)(3 in.) = 60 in.-lb., for the force will be less than 20 lb. until the full 4-inch extension is reached and will vary continuously with x. Let us divide the interval [1, 4] into n subintervals of length Δx. Then, if x_i is any value of x in the ith subinterval, and if Δx is small, the value of F will be nearly $5x_i$ throughout that subinterval, and the work done in stretching the spring over this subinterval is approximately $\Delta W_i = 5x_i\,\Delta x$. Hence the total work done in stretching the spring is

$$W = \lim_{n\to\infty} \sum_{i=1}^{n} 5x_i\,\Delta x = 5\int_1^4 x\,dx = \tfrac{5}{2}x^2 \Big|_1^4 = 37\tfrac{1}{2} \text{ (in.-lb.)}.$$

Example 2. A rectangular tank, 6 ft. by 6 ft. by 4 ft. high, is full of water. Find the work done in pumping all the water to a height 8 ft. above the top of the tank. (The gravitational force on a cubic foot of water, which we represent by the Greek letter ρ (rho), does not vary during the pumping operation. It remains constant at 62.5 lb.)

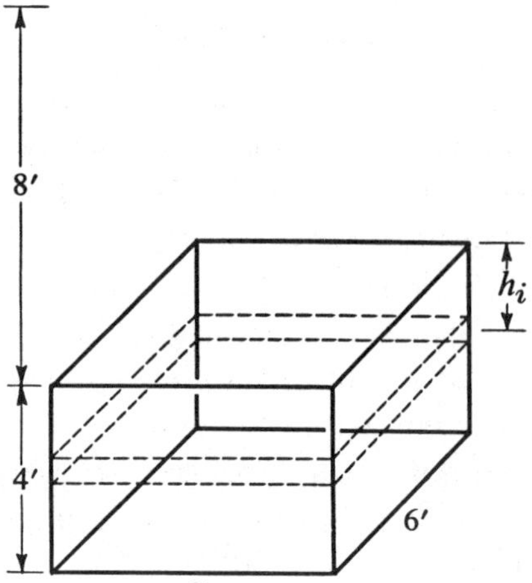

Fig. 15

Solution: Although $\rho = 62.5$ lb. throughout, we see that a cubic foot of water at the top of the tank must be lifted 8 ft., at a cost of (8)(62.5), or 500, ft.-lb., and a cubic foot at the bottom of the tank must be lifted 12 ft., at a cost of (12)(62.5), or 750, ft.-lb. Letting h measure the distance in feet down from the top of the tank, we divide the tank into horizontal layers of thickness Δh. Then, if h_i is a value of h in the ith layer down, that layer, which weighs $(6)(6)(\rho)\Delta h$ pounds, must be lifted approximately $8 + h_i$ feet, at an approximate expenditure, in foot-pounds, of $\Delta W_i = (8 + h_i) \cdot 36\rho\, \Delta h$. The total work done in pumping the water out of the tank is therefore

$$W = 36\rho \int_0^4 (8 + h)dh = 36\rho[8h + \tfrac{1}{2}h^2]_0^4$$
$$= 1440\rho \text{ ft.-lb.} = 45 \text{ ft.-tons.}$$

Example 3. The gravitational force which the earth exerts on an object is inversely proportional to the square of the distance of the object from the center of the earth. Using 4000 miles as the radius of the earth, find the work done in lifting to an altitude of 1000 miles an object that weighs 500 lb. at the surface of the earth.

Solution: The gravitational force at a distance s from the center of the earth is $F = k/s^2$. For the object of this example, we have $500 = k/(4000)^2$; hence $k = 500(4000)^2 = 8 \cdot 10^9$, and $F = 8 \cdot 10^9/s^2$. If the 1000 miles up from the surface of the earth is divided into n subintervals of length Δs, and if s_i is some value of s in the ith subinterval, then the work done in traversing that interval will be approximately $\Delta W_i = (8 \cdot 10^9/s_i^2)\Delta s$, and the total amount of work done will be

$$W = k\int_{4\cdot 10^3}^{5\cdot 10^3} \frac{ds}{s^2} = k\left[-\frac{1}{s}\right]_{4\cdot 10^3}^{5\cdot 10^3} = 4\cdot 10^5 \text{ mi.-lb.} = 200 \text{ mi.-tons.}$$

Exercises 9.4 [A]

1. A water tank is a right circular cylinder, axis vertical, 20 ft. high and 8 ft. in diameter. (**a**) If the tank is full, how much work will be required to pump all the water to the top of the tank? (**b**) If the tank stands with its bottom 25 ft. above the surface of a lake, how much work will be required to fill it through a pipe which leads out of the bottom of the lake?
2. A water tank has the shape of a hemisphere, with its curved surface down; its radius is 6 ft. How much work will be required to pump its contents to a level 20 ft. above the top of the tank?
3. A spring that is naturally 24 in. long can be compressed to a length of 20 in. by a force of 16 lb. How much work will be required to compress it from a length of 22 in. to a length of 16 in.?
4. Two like electrical charges repel each other with a force that is inversely proportional to the square of the distance between them. If two charges repel each other with a force of 2 dynes when they are 10 cm. apart, how many dyne-centimeters (ergs) of work will be required to bring them (**a**) from a separation of 50 cm. to a separation of 1 cm.? (**b**) from 10^6 cm. to 1 cm.? (**c**) from outer space to 1 cm.?
5. A 100-foot chain, which weighs 10 lb./ft., hangs vertically with its upper end at the top of a cliff. How much work will be required to raise all the chain to the top of the cliff?

Exercises 9.4 [B]

1. An underground gasoline storage tank is a right circular cylinder, 16 ft. long and 6 ft. in diameter, with its axis horizontal. If gasoline weighs 50 lb./cu. ft., how much work will be required to pump the contents of the tank to a height 8 ft. above the top of the tank? What part of this work will have been done when the tank is half empty?
2. A mine elevator weighs 2 tons, and the cable by which it is raised weighs 12 lb./ft. How much work is done in raising the elevator to the surface from the 500-foot level?
3. How much work is required to move a 1-pound mass from the surface of the earth to a height of h miles? (See Example 3.) What is the limit of this quantity as $h \to \infty$?
4. A spring is normally 12 in. long. A force of 24 lb. will stretch it 2 in. How much work will be done in stretching it one foot?
5. A water tank is an inverted square pyramid 8 ft. high, with its top 5 ft. on an edge and with its lowest point 5 ft. above the ground. Find how

much work is needed to fill the tank with water pumped from ground level.

6. A water tank is shaped like an ellipsoid of rotation, formed by rotating about its minor (vertical) axis an ellipse with an 8-foot minor axis and a 16-foot major axis. The bottom of the tank is 16 ft. above the ground. How much work is required to fill the tank with water pumped from 80 ft. below ground level?

9.5 Length of arc; motion on a curve

Length of arc. To determine the length of the curve $y = f(x)$, $a \leq x \leq b$, we divide the interval $[a, b]$ of the x-axis into n subintervals of length Δx, and, at the points of division, erect vertical lines which meet the curve at $P_0, P_1, \cdots, P_n$ (Fig. 16 shows the case $n = 4$), and join these points, in order, by straight-line segments of length $P_0P_1 = \Delta s_1$, $P_1P_2 = \Delta s_2, \cdots, P_{n-1}P_n = \Delta s_n$. The sum of these segments is an approximation to the length of the curve, and if this sum has a limit as $\Delta x \to 0$ (and $n \to \infty$), we define that limit to be the length of the arc P_0P_n. If such a limit exists, the curve is said to be rectifiable. A smooth curve (i.e., one that has a parametric representation in which the parametric functions have continuous first derivatives) is rectifiable. Most familiar curves are smooth, but this does not necessarily mean that their lengths can easily be computed. Even for a circular arc we must use the inverse circular functions,* for a parabolic arc we need a function that we shall meet for the first time in Chapter 10, and for the ellipse we need a whole new class of functions, the elliptic functions, which are studied in more advanced courses. Hence the

Fig. 16

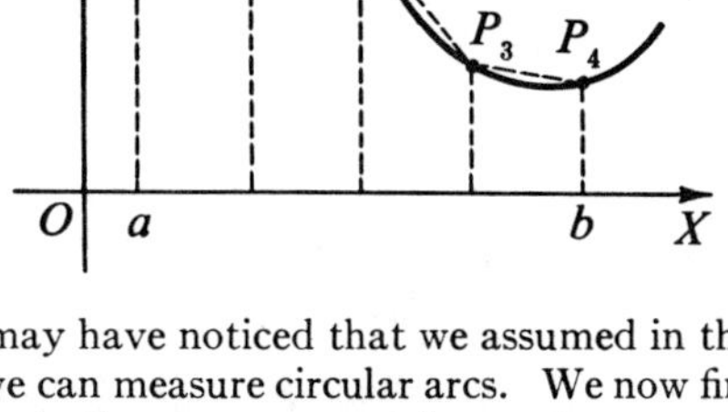

* The alert student may have noticed that we assumed in the definition of the circular functions (§ 2.2) that we can measure circular arcs. We now find that we must use the inverses of those functions in the measurement; hence the argument is circular, too. There are ways out of this logical impasse, but they are best postponed to a more advanced course.

Fig. 17

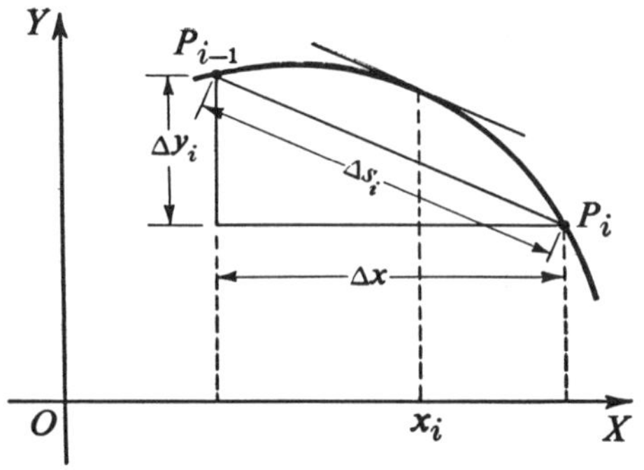

actual computation of arc lengths must, at this stage, be confined to the few cases that lead to limits we can evaluate.

We must now find a way to evaluate these limits. Let us consider, then, the ith subinterval. From Fig. 17, it is clear that

$$(\Delta s_i)^2 = (\Delta x)^2 + (\Delta y_i)^2; \tag{1}$$

hence $$\Delta s_i = \sqrt{(\Delta x)^2 + (\Delta y_i)^2} = \sqrt{1 + (\Delta y_i/\Delta x)^2}\,\Delta x.$$

(We justify the choice of the positive square root by agreeing to measure s in such a way that it increases as x increases.) We now assume that f satisfies the conditions of the mean value theorem, Theorem 5.4-3: f is continuous on $[a, b]$ and differentiable on (a, b). It then follows from that theorem that there is, in the ith subinterval, a number x_i such that $f'(x_i) = \Delta y_i/\Delta x$. Hence

$$\Delta s_i = \sqrt{1 + (f'(x_i))^2}\,\Delta x.$$

The radical is a function of x, and therefore

$$s = \lim_{\Delta x \to 0} \sum_{i=1}^{n} \sqrt{1 + (f'(x_i))^2}\,\Delta x = \int_a^b \sqrt{1 + (f'(x))^2}\,dx. \tag{2}$$

Example 1. Find the length of $y = x^{2/3}$ from $(1, 1)$ to $(8, 4)$.

Solution: Since $dy/dx = \frac{2}{3}x^{-1/3}$, we have

$$\sqrt{1 + \left(\frac{dy}{dx}\right)^2} = \sqrt{1 + \tfrac{4}{9}x^{-2/3}} = \tfrac{1}{3}x^{-1/3}\sqrt{9x^{2/3} + 4},$$

and therefore

$$\begin{aligned} s &= \int_1^8 \tfrac{1}{3}x^{-1/3}\sqrt{9x^{2/3} + 4}\,dx = \tfrac{1}{18}\int_1^8 (9x^{2/3} + 4)^{1/2}(6x^{-1/3}\,dx) \\ &= \tfrac{1}{18}\int_1^8 (9x^{2/3} + 4)^{1/2}\,d(9x^{2/3} + 4) = \tfrac{1}{18}\left[\tfrac{2}{3}(9x^{2/3} + 4)^{3/2}\right]_1^8 \\ &= \tfrac{1}{27}[(40)^{3/2} - (13)^{3/2}]. \end{aligned}$$

It is now convenient to change the dummy variable of integration in (2) from x to t—this has no effect on the value of the integral—and to take as the upper limit of integration an arbitrary $x \in [a, b]$. This juggling of symbols gives us

$$s = \int_a^x \sqrt{1 + (f'(t))^2}\, dt,$$

which defines s as a function of x. The first fundamental theorem of calculus (Theorem 4.2-1) tells us that, on $[a, b]$, the derivative of this function is precisely the integrand, or

$$\frac{ds}{dx} = \sqrt{1 + (f'(x))^2},$$

and since $f'(x) = dy/dx$, we can rewrite this in the form

$$(ds)^2 = (dx)^2 + (dy)^2, \tag{3}$$

which is probably the easiest form in which to remember (2). Note especially the similarity between (1) and (3).

Equation (3) is not only easy to remember, it is also highly versatile. If y is defined as a function of x, we can retrace our steps via

$$\frac{ds}{dx} = \sqrt{1 + \left(\frac{dy}{dx}\right)^2} \tag{3a}$$

to

$$s = \int_a^b \sqrt{1 + \left(\frac{dy}{dx}\right)^2}\, dx, \tag{2a}$$

which is precisely equivalent to (2). It may, however, be convenient or even essential to regard x as a function of y; if the function involved meets the conditions of the mean value theorem, (3) is then valid and can be written

$$\frac{ds}{dy} = \sqrt{\left(\frac{dx}{dy}\right)^2 + 1}, \tag{3b}$$

which leads to

$$s = \int_\alpha^\beta \sqrt{\left(\frac{dx}{dy}\right)^2 + 1}\, dy, \tag{2b}$$

where α and β are the values of y that determine the endpoints of the curve. Finally, it may be that x and y are both determined as functions of a parameter t. If both these functions meet the conditions of the mean value theorem, it can be shown that (3) is then valid; it can then be written

$$\frac{ds}{dt} = \sqrt{\left(\frac{dx}{dt}\right)^2 + \left(\frac{dy}{dt}\right)^2}, \tag{3c}$$

whence

$$s = \int_{t_1}^{t_2} \sqrt{\left(\frac{dx}{dt}\right)^2 + \left(\frac{dy}{dt}\right)^2}\, dt. \tag{2c}$$

In this case, it is not even necessary that the curve be the graph of a function (see § 8.2).

Example 2. Find the circumference of the circle $x^2 + y^2 = r^2$.

Solution:

METHOD 1. We have $2x\,dx + 2y\,dy = 0$, $dy = -\frac{x}{y}\,dx$, and, from (3),

$$(ds)^2 = (dx)^2 + \frac{x^2}{y^2}(dx)^2 = \frac{x^2 + y^2}{y^2}(dx)^2 = \frac{r^2}{y^2}(dx)^2,$$

and, on the upper semicircle, $ds = \frac{r}{y}\,dx$. Hence the length of that semicircle is $s = \int_{-r}^{r} \frac{r}{y}\,dx$. To evaluate this integral, it is convenient to introduce the parametric representation $x = r\cos\theta$, $y = r\sin\theta$, whence $dx = -r\sin\theta\,d\theta$ and

$$s = \int_{\pi}^{0} \frac{r(-r\sin\theta\,d\theta)}{r\sin\theta} = -r\int_{\pi}^{0} d\theta = -r\theta\Big|_{\pi}^{0} = \pi r.$$

The length of the entire circle is therefore $2\pi r$.

METHOD 2. If we use the parametric form from the beginning, with $t = \theta$ in (3c), we have

$$\left(\frac{ds}{d\theta}\right)^2 = \left(\frac{dx}{d\theta}\right)^2 + \left(\frac{dy}{d\theta}\right)^2 = (-r\sin\theta)^2 + (r\cos\theta)^2 = r^2$$

and therefore

$$s = \int_0^{2\pi} r\,d\theta = r\theta\Big|_0^{2\pi} = 2\pi r.$$

Example 3. Find the length of one arch of the cycloid $x = a(\theta - \sin\theta)$, $y = a(1 - \cos\theta)$. (See § 8.8, Example 3.)

Solution: We find $\frac{dx}{d\theta} = a(1 - \cos\theta)$, $\frac{dy}{d\theta} = a\sin\theta$; hence

$$\left(\frac{ds}{d\theta}\right)^2 = a^2(1 - 2\cos\theta + \cos^2\theta + \sin^2\theta) = 2a^2(1 - \cos\theta),$$

and $$s = a\sqrt{2}\int_0^{2\pi} \sqrt{1 - \cos\theta}\, d\theta = 2a\int_0^{2\pi} \sin\frac{\theta}{2}\, d\theta \qquad \text{(why?)}$$

$$= -4a\cos\frac{\theta}{2}\Big|_0^{2\pi} = 8a.$$

Motion on a curve.* Recall that if $O(0,0)$ is the origin and $P(x, y)$ is any point in the plane, then

$$\mathbf{r} = \mathbf{OP} = (x - 0, y - 0) = (x, y)$$

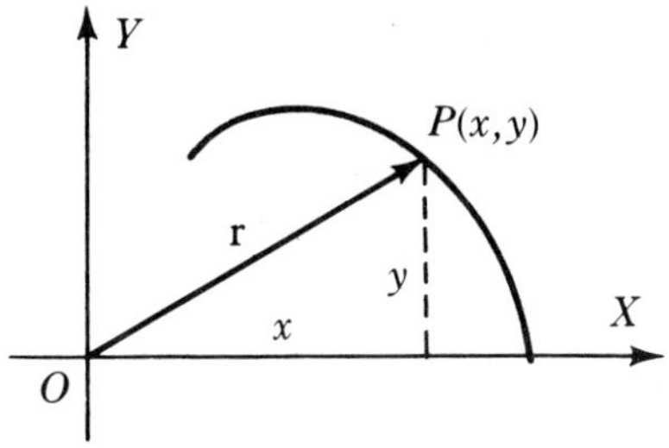

Fig. 18

is the position vector of P. If x and y are functions of a variable t, say $x = f_1(t)$, $y = f_2(t)$, then

$$\mathbf{r}(t) = (f_1(t), f_2(t))$$

is a vector-valued function of t. If we interpret t as time, we can then view P as moving, the motion being controlled by the function $\mathbf{r}$. In this case, the velocity of P is defined to be the derivative

$$\mathbf{v}(t) = \mathbf{r}'(t) = \lim_{\Delta t\to 0} \frac{\mathbf{r}(t + \Delta t) - \mathbf{r}(t)}{\Delta t}$$

$$= \lim_{\Delta t\to 0} \frac{(f_1(t + \Delta t), f_2(t + \Delta t)) - (f_1(t), f_2(t))}{\Delta t}$$

$$= \lim_{\Delta t\to 0} \frac{(f_1(t + \Delta t) - f_1(t), f_2(t + \Delta t) - f_2(t))}{\Delta t}$$

$$= \lim_{\Delta t\to 0} \left(\frac{f_1(t + \Delta t) - f_1(t)}{\Delta t}, \frac{f_2(t + \Delta t) - f_2(t)}{\Delta t}\right)$$

$$= (f_1'(t), f_2'(t))$$

$$= \left(\frac{dx}{dt}, \frac{dy}{dt}\right)$$

Thus the components of the derivative of $\mathbf{r}$ are the derivatives of its components; a vector-valued function is differentiated component by component.

* A concise review of the properties of vectors needed for this section will be found in the Appendix.

The speed, which is defined to be the length of the velocity vector, is then

$$|\mathbf{v}(t)| = \sqrt{\left(\frac{dx}{dt}\right)^2 + \left(\frac{dy}{dt}\right)^2}$$

$$= \frac{ds}{dt} \text{ by (3c).}$$

Example 4. The motion of a particle $P(x, y)$ is given by $\mathbf{r}(t) = (\cos t, \sin t)$. Find the velocity, the speed, and a Cartesian equation of the path.

Solution: We have $x = \cos t$, $y = \sin t$.

Hence $$\frac{dx}{dt} = -\sin t, \ \frac{dy}{dt} = \cos t,$$

$$\mathbf{v}(t) = (-\sin t, \cos t),$$

and $$|\mathbf{v}(t)| = \sqrt{(-\sin t)^2 + (\cos t)^2} = 1.$$

Since

$$x^2 + y^2 = \cos^2 t + \sin^2 t = 1,$$

the path is a circle. Thus the particle is going around the unit circle with a constant speed of 1 unit/unit of time. The velocity, however, is constantly changing because its direction changes. The figure shows P with its position vector **r** and its velocity vector **v**. Note that the direction of **v** is along the tangent to the circle.

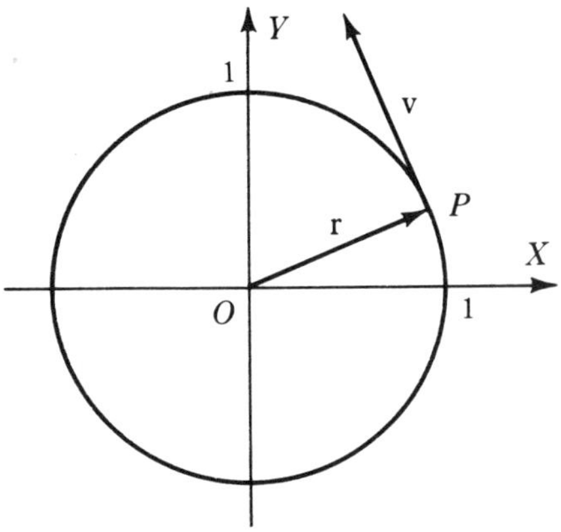

The acceleration of a moving particle is defined to be the derivative of the velocity,

$$\mathbf{a}(t) = \mathbf{v}'(t) = \lim_{\Delta t \to 0} \frac{\mathbf{v}(t + \Delta t) - \mathbf{v}(t)}{\Delta t}.$$

Since we have shown in the case of the velocity that a vector-valued function is differentiated component by component, and we already have $\mathbf{v}(t) = (dx/dt, dy/dt)$, it follows that

$$\mathbf{a}(t) = \left(\frac{d^2x}{dt^2}, \frac{d^2y}{dt^2}\right)$$

In Example 4, we find

$$\mathbf{a}(t) = (-\cos t, -\sin t) = -\mathbf{r}(t).$$

Thus in the figure on p. 340, $\mathbf{a}(t) = \mathbf{PO}$; the acceleration is always directed toward the center of the circle.

As Figure 18 suggests, the slope of a vector $\mathbf{r} = (x, y)$ is the ratio y/x of its components ($x \neq 0$). If x and y are functions of the time t, so that $\mathbf{r}(t)$ is the position vector of a moving particle P, the velocity is

$$\mathbf{v}(t) = (dx/dt, dy/dt).$$

Then the slope of $\mathbf{v}$ is the ratio of its components, but

$$\frac{dy/dt}{dx/dt} = \frac{dy}{dx}$$

by the Chain Rule. Thus the observation we made in Example 4 is a general property of the velocity vector: its direction is always along the line tangent to the path of P. This mathematical observation corresponds to the physically obvious fact that the velocity of a moving object is always in the direction of its motion.

Example 5. The motion of a particle is given by $\mathbf{r}(t) = (t/2, t^2/4)$. Find (i) a Cartesian equation of its path, and (ii) the velocity, speed, and acceleration when $t = 0$ and when $t = 1$. Draw a figure showing the velocity and acceleration vectors at both times.

Solution: (i) The given vector equation is equivalent to the parametric equations

$$x = t/2, \; y = t^2/4.$$

Eliminating t, we obtain the Cartesian equation

$$y = x^2.$$

Thus the particle moves on a parabola.

(ii) $\dfrac{dx}{dt} = \dfrac{1}{2}, \; \dfrac{d^2x}{dt^2} = 0; \; \dfrac{dy}{dt} = \dfrac{t}{2}, \; \dfrac{d^2y}{dt^2} = \dfrac{1}{2}.$

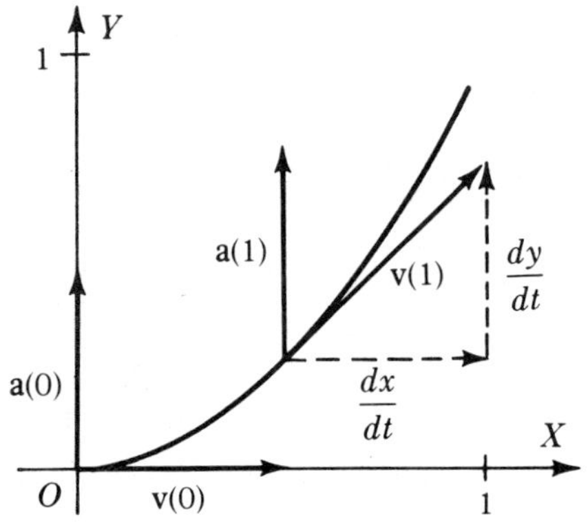

Fig. 19

Thus $\mathbf{v}(t) = \left(\frac{1}{2}, \frac{t}{2}\right)$, $\mathbf{a}(t) = \left(0, \frac{1}{2}\right)$. It then follows that

$$\mathbf{v}(0) = \left(\frac{1}{2}, 0\right); |\mathbf{v}(0)| = \sqrt{\left(\frac{1}{2}\right)^2 + 0^2} = \frac{1}{2}; \mathbf{a}(0) = \left(0, \frac{1}{2}\right).$$

$$\mathbf{v}(1) = \left(\frac{1}{2}, \frac{1}{2}\right); |\mathbf{v}(1)| = \sqrt{\left(\frac{1}{2}\right)^2 + \left(\frac{1}{2}\right)^2} = \frac{1}{2}\sqrt{2}; \mathbf{a}(1) = \left(0, \frac{1}{2}\right).$$

These vectors are shown in Figure 19.

Exercises 9.5 [A]

1. Find the length of $y = \frac{2}{3}x^{3/2}$ from (0, 0) to (9, 18).
2. Find the length of $y = \frac{1}{3}x^3 + \frac{1}{4}x^{-1}$ from $x = 1$ to $x = 3$.
3. Find the length of $x = \frac{1}{3}(y^2 + 2)^{3/2}$ from $y = 2$ to $y = 3$. (*Hint:* Use equation (2b).)
4. If the particle P has coordinates (x, y) given at time t by $x = t - t^2$, $y = 2t$, find the velocity, speed, and acceleration of P, **(a)** when $t = 2$; **(b)** when $t = 0$.
5. The coordinates (x, y) of a point P are $x = a \sec t$, $y = a \tan t$. **(a)** Find the velocity and the speed of P when $t = \pi/4$. **(b)** Find a Cartesian equation of the locus of P.
6. The four-cusped hypocycloid has the Cartesian equation $x^{2/3} + y^{2/3} = a^{2/3}$ and the parametric equations $x = a \cos^3 \theta$, $y = a \sin^3 \theta$. Find the entire length of the curve, using **(a)** the Cartesian equation; **(b)** the parametric equations.

Exercises 9.5 [B]

1. Find the length of $9x^2 = 4y^3$ from (0, 0) to $(\frac{2}{3}, 1)$.
2. Find the length of $y = \dfrac{x^3}{6} + \dfrac{1}{2x}$ from $(1, \frac{2}{3})$ to $(3, \frac{14}{3})$.
3. Find the circumference of the loop of $3y^2 = x(x - 1)^2$.
4. If $x = \dfrac{1 - t^2}{1 + t^2}$, $y = \dfrac{2t}{1 + t^2}$, **(a)** find the velocity and the speed of $P(x, y)$ when $t = 0$ and when $t = 1$; **(b)** find the distance traveled by $P(x, y)$ from $t = 0$ to $t = 1$.
5. If $x = 4p/t^2$, $y = 4p/t$, find ds/dt in terms of t.

6. The hypocycloid of three cusps has the parametric equations $x = 2\cos\theta + \cos 2\theta$, $y = 2\sin\theta - \sin 2\theta$. Show that

$$(ds/d\theta)^2 = 8(1 + \sin\theta\sin 2\theta - \cos\theta\cos 2\theta) = 8(1 - \cos 3\theta),$$

and hence find the length of the curve as θ goes from 0 to 2π.

Miscellaneous Exercises 9.5

1. (**a**) Find the length of $x = \cos 2\theta$, $y = 3 - 2\sin^2\theta$, $0 \le \theta \le \pi/2$.
(**b**) Find a Cartesian equation.

2. Find the length of $x = \cos\theta + \theta\sin\theta$, $y = \sin\theta - \theta\cos\theta$, $0 \le \theta \le 2\pi$.

3. A particle $P(x, y)$ is tracing the hypocycloid described in Prob. 6 of Exercises 9.5 [A] in such a way that $d\theta/dt = 2$. Find the velocity, speed, and acceleration of P when $\theta = \pi/6$.

4. A particle $P(x, y)$ moves subject to $x = r\cos t$, $y = r\sin t$.

(**a**) Obtain a Cartesian equation of the locus of P.

(**b**) Find **v** and **a** as functions of t.

(**c**) Show that P moves with constant speed.

(**d**) Show that **a** is orthogonal to **v**.

(**e**) Show that $|\mathbf{a}| = |\mathbf{v}|^2/r$.

5. Find the minimum speed of the particle $P(x, y)$ if

(**a**) $x = t - t^2$, $y = t^2 - 1$.

(**b**) $x = k\sec t$, $y = k\tan t$, $-\pi/2 < t < \pi/2$.

6. A particle moves with acceleration $\mathbf{a}(t) = (\cos t, -\sin t)$. It goes through the origin at time $t = 0$ with velocity $\mathbf{v}(0) = (0, 1)$. Find a vector equation of the path, and show that the speed is constant.

9.6 Mean value

The *mean*, or *average, value* of a finite set of numbers, sometimes also called the *arithmetic mean* to distinguish it from other averages, is commonly defined in elementary algebra. If the numbers are $x_1, x_2, \cdots, x_n$, then their mean, often denoted $\bar{x}$ (read "x-bar"), is $1/n$ of their sum:

$$\bar{\mathbf{x}} = \frac{1}{n}\sum_{i=1}^{n} \mathbf{x}_i. \qquad (1)$$

Consider now a function f defined on an interval $[a, b]$. Let us divide this interval into n subintervals, each of length $\Delta x = (b - a)/n$, by the points $x_0 = a$, x_1, x_2, $\cdots$, x_{n-1}, $x_n = b$, and let us compute the mean of the values that f takes at the points of division $x_1, \cdots, x_n$. This mean value is

$$\bar{y} = \frac{1}{n}\sum_{i=1}^{n} f(x_i)$$

by the definition (1), but since $b - a = n \cdot \Delta x$, we have $\frac{1}{n} = \Delta x/(b - a)$, and therefore

$$\bar{y} = \frac{\Delta x}{b - a}\sum_{i=1}^{n} f(x_i) = \frac{1}{b - a}\sum_{i=1}^{n} f(x_i)\Delta x.$$

We shall be concerned, in this section, with $M = \lim_{\Delta x \to 0} \bar{y}$. This limit has a familiar form, for

$$M = \lim_{\Delta x \to 0} \frac{1}{b - a}\sum_{i=1}^{n} f(x_i)\Delta x = \frac{1}{b - a}\lim_{\Delta x \to 0}\sum_{i=1}^{n} f(x_i)\Delta x = \frac{1}{b - a}\int_a^b f(x)dx$$

by definition of the definite integral. The limit M is defined to be the *mean value of f on* $[a, b]$:

Definition. The mean value of the function f on the interval $[a, b]$ is

$$M = \frac{1}{b - a}\int_a^b f(x)dx. \tag{2}$$

Example 1. If $k > 0$, then the mean value on $[0, 1]$ of the function $f = \{(x, x^k)\}$ is

$$M = \frac{1}{1 - 0}\int_0^1 x^k\, dx = \frac{1}{k + 1}x^{k+1}\Big|_0^1 = \frac{1}{k + 1}\cdot$$

The mean value of a function has a simple geometric interpretation. From (2), we have

$$(b - a)\cdot M = \int_a^b f(x)dx,$$

or, in other words, the area of the region bounded by the graph of f, the x-axis, and the lines $x = a$, $x = b$, is the same as the area of a rectangle with the same base and with height M, as Fig. 20 indicates.

A theorem asserts that a continuous function f actually takes the value M at some interior point of $[a, b]$, that there is a point $c \in (a, b)$ such that $f(c) = M$. (As the figure indicates, there may be more than one such point.) We have already met the mean value theorem (Theorem 5.4–3). It would be better to call that theorem the mean value theorem for deriva-

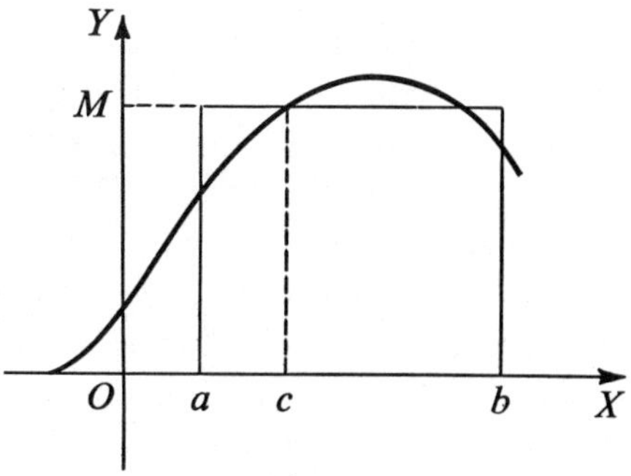

Fig. 20

tives, for the theorem we are concerned with here is also a mean value theorem:

Theorem 9.6-1 (Mean Value Theorem for Integrals). Let f be a function continuous on $[a, b]$. Then there exists a point $c \in (a, b)$ such that

$$\int_a^b f(x)dx = (b - a) \cdot f(c).$$

Proof: The theorem is an easy consequence of the first fundamental theorem of calculus (Theorem 4.2-1) and the mean value theorem for derivatives (Theorem 5.4-3). If we define a function F by the rule

$$F(t) = \int_a^t f(x)dx$$

for each $t \in [a, b]$, we know from Theorem 4.2-1 that $F' = f$ on $[a, b]$. From this we infer that F, being differentiable at all points of $[a, b]$, is also continuous on that interval (Theorem 5.3-3) and hence meets the conditions of the mean value theorem for derivatives (Theorem 5.4-3). Hence there exists a $c \in (a, b)$ such that

$$\frac{F(b) - F(a)}{b - a} = F'(c),$$

or

$$F(b) - F(a) = (b - a) \cdot F'(c). \tag{3}$$

But $F(a) = 0$ (why?), $F(b) = \int_a^b f(x)dx$ (why?), and $F' = f$, as has been noted. Hence (3) can be written

$$\int_a^b f(x)dx = (b - a) \cdot f(c),$$

which is the desired result.

The point c whose existence is thus demonstrated is, in a sense, a *typical* point of the interval for the given function. Theorem 9.6-1 says, in effect, that every interval of the domain of a continuous function contains an "average," or "typical," point, the point c. Note, however, that the theorem relates strictly to functions continuous on intervals; it is not, in general, possible to find a family with an average number of children.

The definition of the mean value of a function on a given interval is unambiguous, but it can lead to contradictions and confusion if it is carelessly applied, as the following example will show.

Example 2. Find the mean value of y on the semicircle $y = \sqrt{a^2 - x^2}$.

Solution: We shall show, by deriving two contradictory answers, that the problem as stated is ambiguous.

CASE 1. A direct application of the definition of the mean value of a function leads to

$$M_1 = \frac{1}{a - (-a)} \int_{-a}^{a} \sqrt{a^2 - x^2}\, dx.$$

Since the integral is precisely the area of the semicircle, and is therefore known to be $\pi a^2/2$, we then have $M_1 = \frac{1}{2a} \cdot \frac{\pi a^2}{2} = \frac{\pi}{4} a$. The mean value of y is then about 79% of the radius a.

CASE 2. It is often convenient to use a parametric representation of a curve, even if it is the graph of a function. For the semicircle in question, a common parametric representation is $x = a \cos \theta$, $y = a \sin \theta$, and the semicircle corresponds to the interval $[0, \pi]$ of values of θ. Hence the mean value of y is

$$M_2 = \frac{1}{\pi - 0} \int_0^{\pi} a \sin \theta \, d\theta = \frac{1}{\pi} [-a \cos \theta]_0^{\pi} = \frac{2}{\pi} a,$$

or about 64% of a.

The difficulty encountered in Example 2 can be analyzed as follows: We have defined the mean value of a *function*, but the quantity y whose value we sought in the example is not a function, it is a *variable*. If $y = f(x)$, it is a common practice to refer to the mean value of f as the mean value of y, and if the function f is unambiguously specified, no difficulty arises. A given variable may, however, be determined as a value of more than one function. This is precisely the situation in Example 2: the ordinate y of a point on the semicircle can be determined as a function f of the abscissa x, $y = f(x) = \sqrt{a^2 - x^2}$, $-a \leq x \leq a$, but it can also be determined as a function g of the parameter θ, $y = g(\theta) = a \sin \theta$, $0 \leq \theta \leq \pi$, and the mean value of f on $[-a, a]$ is not the same as the mean value of g on $[0, \pi]$. A conventional way of distinguishing between these two means is to speak of the mean M_1 of f as the mean value of y *with respect to* x, and of the mean M_2 of g as the mean value of y *with respect to* θ.

The two mean values found in Example 2 have fairly direct geometrical interpretations. In Case 1, suppose we divide the bounding diameter of the semicircle into n equal parts of length $\Delta x = 2a/n$ and erect verticals to the semicircle at the points of division (Fig 21). The limit of the mean

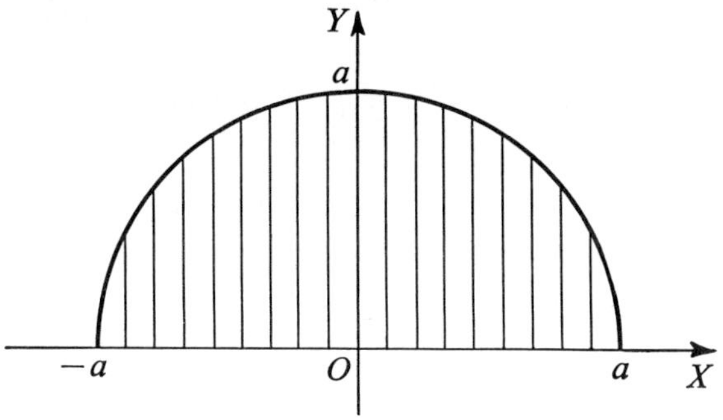

Fig. 21

length of these vertical segments, *spaced at equal intervals along the diameter*, is the mean $M_1 = \pi a/4$ found in Case 1. In Case 2, suppose we divide the central angle of the semicircle into n equal parts, of size $\Delta\theta = \pi/n$, by radii, and then drop perpendiculars to the bounding diameter from the outer ends of these radii. The limit of the mean length of these perpendiculars, *spaced at equal intervals around the arc*, is the mean $M_2 = 2a/\pi$ found in Case 2. As Fig. 22 suggests, an effect of the approach in Case 2 is to give greater weight to the smaller values of y that occur near the ends of the diameter, and thus to obtain a smaller mean than in Case 1.

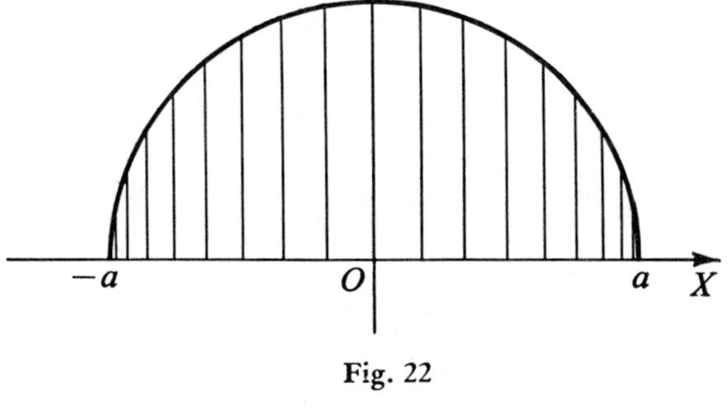

Fig. 22

Miscellaneous Exercises 9.6

1. If $f(x) = 4 - x^2$, find the average (mean) value of f on $[-2, 2]$. Use this result to find the area of the region enclosed between the graph of f and the x-axis.

2. (a) Find the mean value of y with respect to x on the arc of $y = x^3 - 9x$ that has endpoints $(0, 0)$ and $(3, 0)$.
   (b) If $x = \sqrt{t}$, find the mean value of y with respect to t on the same arc.

3. Find the mean value of y with respect to x on the line segment joining (x_1, y_1) and (x_2, y_2).

4. Find the average value of y on the parabola $y^2 = 4px$, from $(0, 0)$ to $(p, 2p)$, (a) with respect to x, (b) with respect to the parameter t, where $x = pt^2$, $y = 2pt$.

5. After t seconds of fall, a body falling from rest has velocity $v = 32t$ ft./sec. and has fallen $s = 16t^2$ ft. At the end of 3 sec. of fall, the velocity will then be 96 ft./sec. and the distance fallen will be 144 ft.

   (a) Find the mean velocity, with respect to time, over the first 3 sec. of fall.

   (b) Eliminate t from the given equations, thus expressing v as a function of s.

   (c) Find the mean velocity, with respect to distance, over the first 144 ft. of fall.

6. (a) Find the mean value of the function sin on $[0, \pi]$.

   (b) Find the mean value of the function $\sin^2$ on $[0, \pi]$.

   (c) Compare the mean found in (a) with the positive square root of the mean found in (b).

   (If $y = f(x)$, the positive square root of the mean of y^2 with respect to x is called the root-mean-square value of y. The root-mean-square value of $\sin x$ is important in the measurement of alternating current.)

7. A table gives values of $\sin x°$ at intervals of $6'$, $0 \leq x \leq 90$.

   (a) How many entries are there in the table?

   (b) What is the mean size of the entries?

   (c) Use the results of (a) and (b) to approximate the sum of the entries.

8. Rectangles are inscribed in the ellipse $x^2/a^2 + y^2/b^2 = 1$, with their sides parallel to the coordinate axes. Find the mean area of these rectangles if they are considered to have (**a**) their vertical sides equally spaced along the x-axis; (**b**) their horizontal sides equally spaced along the y-axis.
9. Find the mean height of one arch of the cycloid $x = a(\theta - \sin\theta)$, $y = a(1 - \cos\theta)$, (**a**) with respect to θ; (**b**) with respect to x; (**c**) with respect to the arc length s (see § 9.5, Example 3).
10. Find the average length of all chords of a circle of radius a, in two ways:
    (**a**) Express the length y of each chord as a function of the distance x between the chord and the center of the circle, and find the mean value of y with respect to x.
    (**b**) Express y as a function of the angle θ between the chord and the diameter through one of its endpoints, and find the mean value of y with respect to θ.

9.7 Surfaces of revolution

If an arc of the curve $y = f(x)$ is rotated about the x-axis, a surface of revolution will be generated. The area of such a surface can often be found by integration.

As in § 9.5, we divide the x-axis under the given arc into n subintervals of length Δx, we erect, at the points of division, vertical lines that meet the curve at $P_0, P_1, \cdots, P_n$, and we join these points of the curve by straight segments of length $P_0P_1 = \Delta s_1$, $P_1P_2 = \Delta s_2$, $\cdots$, $P_{n-1}P_n = \Delta s_n$. When this configuration is rotated about the x-axis, these chords will each generate the lateral surface of a conical frustum, and the sum of the areas of these conical surfaces is an approximation to the area of the surface generated by the curve. If this sum has a limit as $\Delta x \to 0$ (and $n \to \infty$), we define this limit to be the required area.

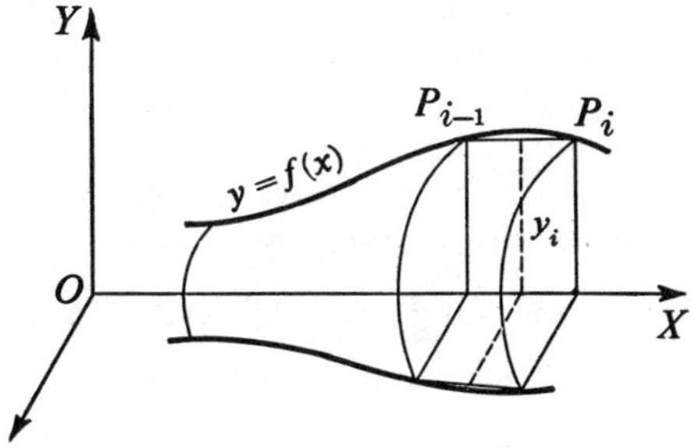

Fig. 23

Since the area of a conical frustum is the perimeter of the midsection times the slant height, the area generated by the ith chord $P_{i-1}P_i$ is then $2\pi y_i \cdot \Delta s_i$, where y_i is the ordinate of the midpoint of $P_{i-1}P_i$, and the area generated by the entire arc is therefore

$$S = \lim_{\Delta x \to 0} \sum_{i=1}^{n} 2\pi y_i \, \Delta s_i = 2\pi \int y \, ds$$

between appropriate limits, where the differential of arc ds is to be interpreted and determined as in § 9.5, and where $y \geq 0$.

If an arc is rotated about a line in its plane other than the x-axis, the preceding development can readily be modified to give, as the area of the surface thus generated, $S = 2\pi \int r \, ds$, where r is a variable that measures the distance from the points of the arc to the axis of rotation.

Example. Find the area of a sphere of radius a.

Solution: Using the circle $x = a \cos \theta$, $y = a \sin \theta$, we find (see § 9.5, Example 2) $ds = a \, d\theta$, and therefore

$$S = 2\pi \int y \, ds = 2\pi \int_0^{\pi} (a \sin \theta)(a \, d\theta)$$

$$= 2\pi a^2 \int_0^{\pi} \sin \theta \, d\theta = 2\pi a^2 [-\cos \theta]_0^{\pi} = 4\pi a^2.$$

Miscellaneous Exercises 9.7

1. Find the area of the surface generated by rotating about the x-axis the arc of $y = x^3$ from $(0, 0)$ to $(1, 1)$.

2. Find the area of the paraboloid of revolution generated by rotating about the x-axis the arc of $y^2 = 4px$ between $(0, 0)$ and $(p, 2p)$. Use (**a**) the Cartesian equation; (**b**) the parametric equations $x = pt^2$, $y = 2pt$.

3. The arc of $3y^2 = x(x - 1)^2$ between $(1, 0)$ and $(3, 2)$ is rotated about the x-axis. Find the area of the surface generated.

4. Find the area of the zone cut from a sphere of radius r by two parallel planes h units apart. Check by setting $h = 2r$.

5. Find the area of the surface generated by rotating about the x-axis the four-cusped hypocycloid $x^{2/3} + y^{2/3} = a^{2/3}$.

6. Find the area of the torus generated by rotating about the line $x = b$ the circle $x^2 + y^2 = a^2$ $(a < b)$.
7. Find the area of the surface generated by rotating about the x-axis the loop of $9y^2 = x(3 - x)^2$.
8. Find the area of the surface generated by rotating about the y-axis the arc of $y^2 = x^3$ from $(0, 0)$ to $(1, 1)$.
9. Find the area of the ellipsoid generated by rotating about the x-axis the ellipse $x^2/a^2 + y^2/b^2 = 1$ $(a > b)$. *Hint*: Show that

$$\left(\frac{ds}{dx}\right)^2 = \frac{b^2}{a^4y^2}[a^4 - (a^2 - b^2)x^2] = \frac{b^2c^2}{a^4y^2}\left[\left(\frac{a^2}{c}\right)^2 - x^2\right].$$

9.8 Hydrostatic force

The weight of a fluid exerts a force on any object immersed in the fluid. The force per unit of area exerted on the surface of such an object is called the pressure; it depends solely on the depth at which it is measured and on the density, or weight per unit volume, of the fluid. For example, the density of water is about 62.5 lb./cu. ft., and a horizontal plane region 1 sq. ft. in area is, in water at a depth of 10 ft., in effect supporting a column of water 10 ft. × 1 ft. × 1 ft., weighing 625 lb.; the pressure at a depth of 10 ft. in water is therefore 625 lb./sq. ft. In general, if the density of a fluid is ρ lb./cu. ft. (the Greek letter rho is commonly used to designate density), then the pressure at a depth of h feet is ρh lb./sq. ft., and a surface with an area of A sq. ft. at that depth is subject to a total force of ρhA lb.

If a submerged object has a plane surface that is all at one depth, the total force on that surface can be computed by finding its area and then multiplying by the appropriate pressure. There are, however, many situations in which a surface is not all at the same depth. The water pressure on the upstream face of a dam, for instance, varies from zero at the surface of the water to a large value at the bottom. To calculate the total force on a dam is a matter of great practical importance, and we shall see how definite integrals can be applied to this problem.

Two characteristics of fluid pressure sometimes cause confusion and should be clearly understood at the start. First, the pressure at a given depth is the same in all directions, up, down, and sideways, and the effective pressure on any surface is always normal to the surface. Second, the pressure does not depend on the shape or size of the container: the pressure on a dam at a depth of 50 feet is the same whether the lake behind the dam is 100 miles long or 1 mile long, and the pressure is also the same at a depth of 50 feet in a 1-inch pipe.

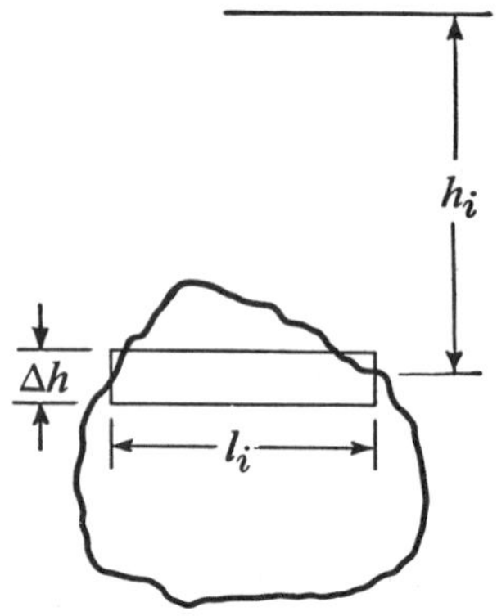

Fig. 24

Consider, now, an object having a vertical plane surface, submerged in a fluid of constant density ρ, and let h represent depth below the surface of the fluid. Various portions of the vertical surface are at various depths and are therefore subject to various pressures. If, however, we divide the surface into narrow horizontal strips of width Δh, then all of each strip will be at virtually one depth and therefore under virtually the same pressure at all its points. If, then, some point in the ith strip is at a depth h_i, the pressure on that strip will be everywhere approximately ρh_i. If the length of the strip is l_i, then its area will be $l_i\,\Delta h$; the total force of the fluid on this strip will then be approximately $\Delta F_i = \rho h_i l_i\,\Delta h$, and the total force on the entire surface will be approximately

$$S = \sum_{i=1}^{n} \rho h_i l_i\,\Delta h.$$

We could, of course, by appropriate choices of h_i and l_i, determine upper and lower sums S_U and S_L, but if the outline of our surface is a continuous curve, these sums will both have the same limit as $\Delta h \to 0$. Hence, where h_1 and h_2 are, respectively, the least and greatest depths to which the vertical surface is submerged, the total force on this surface is

$$F = \lim_{\Delta h \to 0} S = \int_{h_1}^{h_2} \rho h l\,dh.$$

Example 1. A steel plate is in the shape of an isosceles right triangle with legs 2 ft. long. This plate is submerged in water, as shown in Fig. 25, with its highest point 3 ft. below the surface of the water. Find the total force on one side. (Take $\rho = 62.5$ lb./cu. ft. for water.)

Solution: A typical element of area is indicated in the figure. If the depth of this element is approximately h_i, then, by similar triangles, its length is $l_i = 5 - h_i$, and therefore $\Delta F_i = \rho h_i(5 - h_i)\Delta h$, and the total force is

$$F = \rho \int_3^5 (5h - h^2)dh = \rho[\tfrac{5}{2}h^2 - \tfrac{1}{3}h^3]_3^5 = \frac{22\rho}{3} = 458\tfrac{1}{3} \text{ (lb.)}.$$

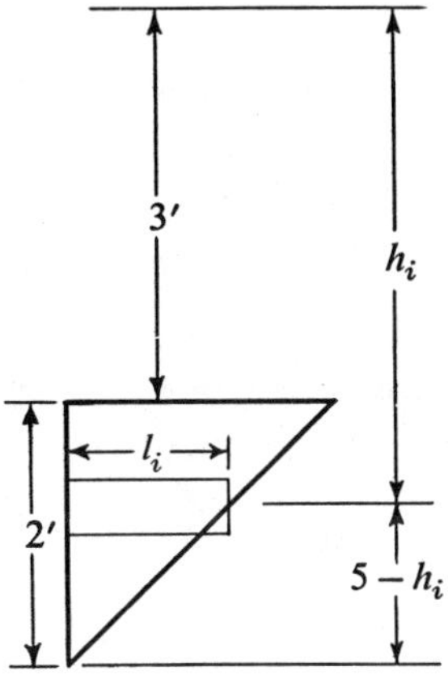

Fig. 25

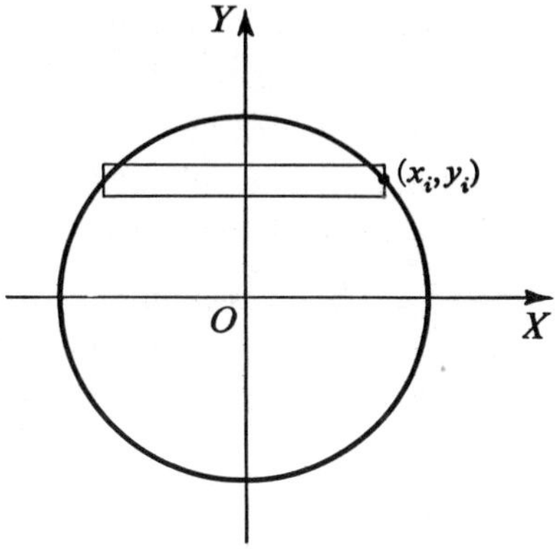

Fig. 26

Example 2. A circular valve in a water main has a one-foot radius. Find the total force on the valve if its center is 40 ft. below the surface of the water.

Solution: We introduce coordinate axes with the origin at the center of the valve and the y-axis directed up (Fig. 26). A typical horizontal element is then at a depth $(40 - y_i)$ ft. and has a length $2x_i$ ft., where (x_i, y_i) is on the circle $x^2 + y^2 = 1$. Then

$$F = 2\rho \int_{-1}^{1} x(40 - y)dy.$$

Introducing the parametric equations of the circle, $x = \cos\theta$, $y = \sin\theta$, we have

$$\begin{aligned} F &= 2\rho \int_{-\pi/2}^{\pi/2} \cos\theta\,(40 - \sin\theta)\cos\theta\,d\theta \\ &= 80\rho \int_{-\pi/2}^{\pi/2} \cos^2\theta\,d\theta - 2\rho \int_{-\pi/2}^{\pi/2} \cos^2\theta \sin\theta\,d\theta \\ &= 40\rho \int_{-\pi/2}^{\pi/2} (1 + \cos 2\theta)d\theta + 2\rho \int_{-\pi/2}^{\pi/2} \cos^2\theta\,d(\cos\theta) \\ &= 40\rho[\theta + \tfrac{1}{2}\sin 2\theta]_{-\pi/2}^{\pi/2} + 2\rho[\tfrac{1}{3}\cos^3\theta]_{-\pi/2}^{\pi/2} = 40\pi\rho; \end{aligned}$$

the total pressure is almost 4 tons.

Exercises 9.8 [A]

1. A rectangular plate 2 ft. by 4 ft. is vertically submerged in water, with its long dimension horizontal and its upper edge 6 ft. below the surface. Find the total force on one side.

2. A plate in the shape of an equilateral triangle 3 ft. on an edge is vertically immersed in water, with one edge in the surface of the water. Find the total force on one side of this plate.

3. Find the force on one side of the plate in Ex. 2 if the upper edge is 4 ft. below the surface of the water.

4. The end of a trough is bounded by that portion of the parabola $x^2 = 2y$ below the line $y = 2$, where the dimensions are in feet and the y-axis is directed vertically upward. Find the total force on one end when the trough is full of water.

5. The face of a vertical dam is in the form of an isosceles trapezoid, 50 ft. wide at the top, 30 ft. wide at the bottom, and 30 ft. high. Find the force on the dam when the water is 20 ft. deep.

6. A cylindrical oil tank, with its axis horizontal, is 18 ft. long and 4 ft. in diameter. It is filled to a depth of 3 ft. with oil weighing 50 lb./cu. ft. Find the force on one end.

7. The rudder of a sailboat has the shape of the segment cut from the parabola $x = y^2 + y$ by the line $y = -x$, where the dimensions are in feet and the y-axis is directed vertically upward. Find the force on one side of the rudder if its upper tip is at the surface of the water.

Exercises 9.8 [B]

1. A temporary bulkhead is made of vertical planks one foot wide. What is the force on a plank when the water is 12 ft. deep?

2. A plate in the shape of a right triangle is submerged in a tank, with a 4-foot leg resting horizontally on the bottom and a 2-foot leg vertical. Find the force on one side of this plate when the water is 4 ft. deep.

3. How deep must the water be to double the force on the plate in Ex. 2?

4. A semicircular plate, radius 2 ft., is vertically submerged with its bounding diameter in the surface of the water. Find the force on one side.

5. A gate in a dam is in the shape of an isosceles trapezoid 6 ft. wide at the bottom, 4 ft. wide at the top, and 3 ft. high. Find the force on the gate when the water level is 10 ft. above the top of the gate.

6. The end of an oil tank is in the form of an ellipse with its 6-foot major axis horizontal and its 4-foot minor axis vertical. The tank contains oil of density ρ. Find, in terms of ρ, the force on one end, if the oil is (**a**) 3 ft. deep; (**b**) 4 ft. deep.

9.9 Centroids

If a mass is concentrated at a single point, then the moment of that mass about an axis is defined to be the product of the mass by its distance from the axis. If there are several such masses in a plane, we take distances on opposite sides of the axis with opposite signs, and the combined moment of these masses is then just the algebraic sum of their individual moments. If, therefore, we have n masses $m_1, m_2, \cdots, m_n$ at points (x_1, y_1), (x_2, y_2), $\cdots$, (x_n, y_n), respectively, their moment M_y about the y-axis and their moment M_x about the x-axis are

$$M_y = \sum_{i=1}^{n} m_i x_i \quad \text{and} \quad M_x = \sum_{i=1}^{n} m_i y_i. \tag{1}$$

If all this mass were concentrated at a single point $P(x, y)$, these moments would then be

$$M_y = x \sum_{i=1}^{n} m_i \quad \text{and} \quad M_x = y \sum_{i=1}^{n} m_i. \tag{2}$$

If the moments given by (2) are the same as those in (1), the point P is called the center of mass of the system. We may therefore find the coordinates $(\bar{x}, \bar{y})$ of the center of mass by eliminating M_y and M_x from (1) and (2):

$$\bar{x} = \sum_{i=1}^{n} m_i x_i \Big/ \sum_{i=1}^{n} m_i, \quad \bar{y} = \sum_{i=1}^{n} m_i y_i \Big/ \sum_{i=1}^{n} m_i. \tag{3}$$

In many cases, a mass distribution responds to a gravitational force as though its entire mass were concentrated at its center of mass; in such cases, this point is usually called the center of gravity. If we think of the xy-plane as a rigid, weightless sheet, then the moment of a distribution of masses on that plane is a measure of the tendency of the weight of the distribution to cause the plane to tip; the moment of any distribution about a line through the center of gravity is zero. This, therefore, is the point at which the distribution can be balanced.

If each mass is a unit mass, $m_i = 1$ for all i, then the sum of the n masses is simply n, (3) becomes

$$\bar{x} = \frac{1}{n} \sum_{i=1}^{n} x_i, \quad \bar{y} = \frac{1}{n} \sum_{i=1}^{n} y_i,$$

and we see that $\bar{x}$ and $\bar{y}$ are then the simple averages of the abscissas and the ordinates, respectively, of the n points. The mass no longer appears, and the definition of $(\bar{x}, \bar{y})$ is now purely geometrical, independent of any physical concepts. In this case, $(\bar{x}, \bar{y})$ is called the centroid. It is with centroids that we shall be largely concerned in this section, but we shall continue to use the language of mass distributions as a descriptive convenience and an aid to intuition.

Example 1. Find the centroid of unit masses at $P_1(x_1, y_1)$ and $P_2(x_2, y_2)$.

Solution: Since the sum of the masses is 2, (3) gives us $\bar{x} = \frac{1}{2}(x_1 + x_2)$, $\bar{y} = \frac{1}{2}(y_1 + y_2)$, and we see that the centroid is the midpoint of P_1P_2.

Example 2. Similarly, the centroid of a distribution of unit masses at $P_1(x_1, y_1)$, $P_2(x_2, y_2)$, and $P_3(x_3, y_3)$ is, from (3), $\bar{x} = \frac{1}{3}(x_1 + x_2 + x_3)$, $\bar{y} = \frac{1}{3}(y_1 + y_2 + y_3)$, which is known to be the intersection of the medians of the triangle $P_1P_2P_3$.

If a mass is uniformly distributed along a curve in the xy-plane (the image of a thin, heavy wire may be helpful), we arrive at a definition of the centroid of this distribution by the following line of reasoning: Suppose the density of the curve is ρ units of mass per unit of length, here assumed constant. We divide the curve into n segments of length Δs. Then the mass of each segment will be $\rho\Delta s$, and if the abscissa of some point on the ith segment is x_i, the moment about the y-axis of that segment will be approximately $x_i \cdot \rho\, \Delta s$, and the moment about the y-axis of the whole curve will be approximately $\sum_{i=1}^{n} \rho x_i\, \Delta s$. If we increase n without bound, we can show, as in § 9.5, that the limit of this sum is $\int_0^L \rho x\, ds$, where L is the total length of the curve. Consideration of moments about the x-axis leads to a similar result. Since the total mass of the curve is ρL, we are thus led to *define* the centroid of this distribution as the point $(\bar{x}, \bar{y})$ determined by

$$\bar{x} = \frac{1}{\rho L}\int_0^L \rho x\, ds = \frac{1}{L}\int_0^L x\, ds, \quad \bar{y} = \frac{1}{L}\int_0^L y\, ds. \tag{4}$$

A comparison with § 9.6 will show that the quantities $\bar{x}$ and $\bar{y}$ thus defined are precisely the mean values of x and y, respectively, with respect to s.

Example 3. Find the centroid of a mass uniformly distributed along the semicircular arc $y = \sqrt{a^2 - x^2}$.

Solution: From the symmetry of the curve, $\bar{x} = 0$; $\bar{y} = \frac{1}{\pi a}\int_0^{\pi a} y\, ds = \frac{2}{\pi} a$, a result which has already been obtained in Case 2 of Example 2, § 9.6. This result puts the centroid about 64% of the way out on the central radius from the center of the circle, a position which seems intuitively plausible.

To find the centroid of a mass uniformly distributed over a region of the xy-plane, it is convenient to divide the region into narrow strips by lines parallel to a coordinate axis and treat each strip as though it were a rectangle, just as we do in finding plane areas. We assume that the centroid of a rectangle is at the intersection of its diagonals, and we then treat each rectangular approximation as if its mass were concentrated at that point.

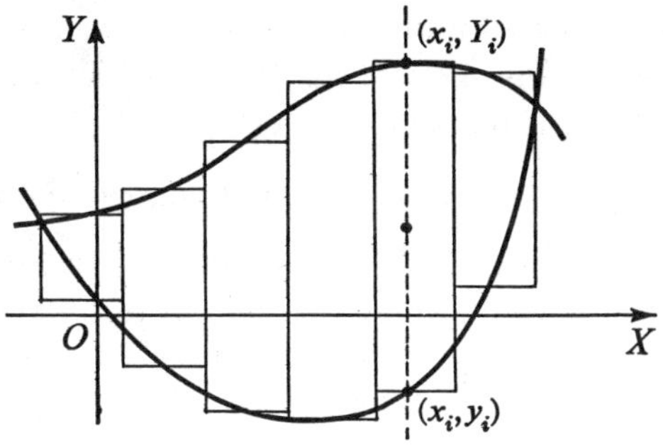

Fig. 27

If, in Fig. 27, we divide this region (of area A) into n rectangular strips of width Δx by lines parallel to the y-axis, and take x_i to be the abscissa of some point in the ith strip, then the line $x = x_i$ will meet the boundaries of the region in (x_i, Y_i) and (x_i, y_i). We then take the height of this ith rectangle to be $Y_i - y_i$, and if the density, or mass per unit area, is ρ, the mass of this rectangle is then $\rho(Y_i - y_i)\Delta x$. The intersection of the diagonals is approximately $\left(x_i, \dfrac{Y_i + y_i}{2}\right)$, and the moments of the rectangle about the y- and x-axes are then approximately

$$x_i \cdot \rho(Y_i - y_i)\Delta x \quad \text{and} \quad \tfrac{1}{2}(Y_i + y_i) \cdot \rho(Y_i - y_i)\Delta x,$$

respectively. Summing each of these, finding the limit as $\Delta x \to 0$, and dividing each moment thus obtained by the mass ρA of the entire region, we get

$$\bar{x} = \frac{1}{A}\int x(Y - y)dx,$$

$$\bar{y} = \frac{1}{A}\int \frac{Y + y}{2}(Y - y)dx = \frac{1}{2A}\int (Y^2 - y^2)dx, \tag{5}$$

between appropriate limits of integration, as formulas for the coordinates of the centroid. It should be understood that (5) is a definition, and that the preceding discussion is intended merely to show that this definition is a reasonable expression of the intuitive notions also involved in the definitions (3) and (4). This definition will, of course, take a somewhat simpler form if the x-axis is a boundary of the region; it will be modified in other obvious ways if it is more convenient to subdivide the region by lines parallel to the x-axis.

The concept of mass density, ρ, was borrowed from physics as an aid to intuition in the preceding discussion, but a uniform density drops out of the definitions (4) and (5), leaving integrals which depend solely on the geometry of the situation. This shows that the centroid is a purely geometric concept, as we pointed out on page 356.

Example 4. Find the centroid of a semicircular plane region of radius a.

Solution: Let us take the semicircle $x = \sqrt{a^2 - y^2}$. It is then clear from symmetry that $\bar{y} = 0$. If we divide the region into strips by lines parallel to the y-axis, and take an arbitrary line $x = x_i$ in the ith strip, this will meet the upper part of the curve at (x_i, y_i) and the lower part at $(x_i, -y_i)$, as shown in Fig. 28. We then have, from (5),

$$\bar{x} = \frac{2}{\pi a^2}\int_0^a x(2y)dx = \frac{4}{\pi a^2}\int_0^a x\sqrt{a^2 - x^2}\,dx = \frac{4a}{3\pi},$$

or about 42% of a. Compare this result with that of Example 3.

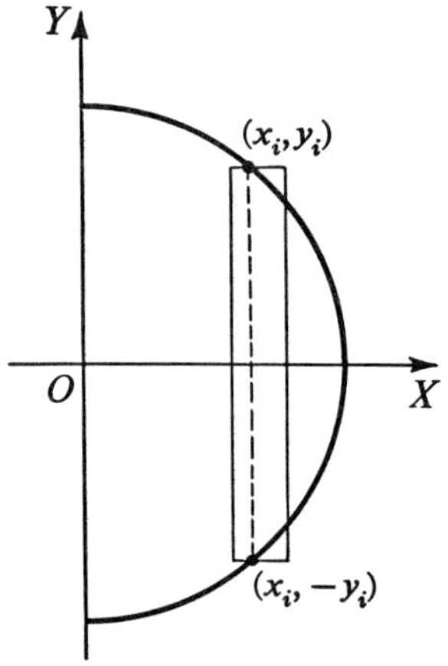

Fig. 28

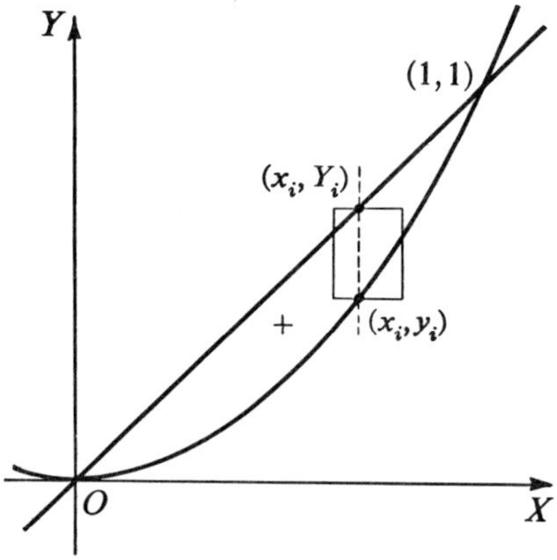

Fig. 29

Example 5. Find the centroid of the region enclosed by $y = x$ and $y = x^2$.

Solution: These curves intersect at (0, 0) and (1, 1). If we divide the region into strips by lines parallel to the y-axis, then a line $x = x_i$ in the ith strip will meet the given line at (x_i, Y_i) and the given parabola at (x_i, y_i), as shown in Fig. 29. We have the area

$$A = \int_0^1 (Y - y)dx = \int_0^1 (x - x^2)dx = \tfrac{1}{6},$$

and, from (5),

$$\bar{x} = 6\int_0^1 x(Y - y)dx = 6\int_0^1 (x^2 - x^3)dx = \tfrac{1}{2},$$

$$\bar{y} = 3\int_0^1 (Y^2 - y^2)dx = 3\int_0^1 (x^2 - x^4)dx = \tfrac{2}{5}.$$

The centroid, $(\frac{1}{2}, \frac{2}{5})$, is marked on the figure; the result seems reasonable.

Example 6. Find the centroid of a plane triangular region.

Solution: Let us call the altitude of the triangle h and the base b, and choose axes as indicated in Fig. 30. It is convenient to divide the figure by lines parallel

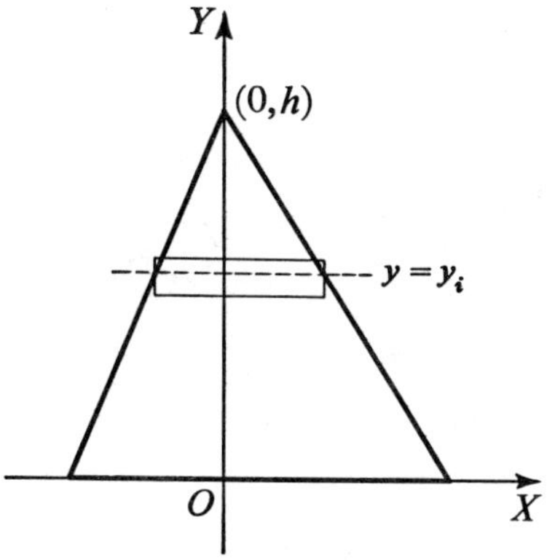

Fig. 30

to the x-axis. If y_i is the ordinate of a point in the ith strip up, and we call the length of this strip L_i, then we have, from similar triangles,

$$\frac{h - y_i}{L_i} = \frac{h}{b}, \quad \text{whence} \quad L_i = \frac{b(h - y_i)}{h},$$

and the area of the ith strip is approximately $\dfrac{b(h - y_i)}{h}\Delta y$. Since the center of this strip is at a height of about y_i, its moment about the x-axis (think of it, if you wish, as a mass distribution with $\rho = 1$) is then approximately $y_i \dfrac{b(h - y_i)}{h}\Delta y$.

Summing, passing to the limit, and dividing by the area $\frac{1}{2}bh$, we have

$$\bar{y} = \frac{2}{bh}\int_0^h y\frac{b(h-y)}{h}\,dy = \frac{2}{h^2}\int_0^h (hy - y^2)dy = \frac{1}{3}h;$$

the centroid of the region is $\frac{1}{3}$ of the way up from the base to the opposite vertex. Since any side of the triangle may be taken as base, the centroid has a similar location with respect to all three sides of the triangle and therefore lies at the intersection of the medians. Note that this is the same as the result obtained in Example 2, which was an entirely different problem.

Miscellaneous Exercises 9.9

1. Find the centroid of the first-quadrant arc of $x^2 + y^2 = a^2$.
2. Find the centroid of the first-quadrant arc of $x^{2/3} + y^{2/3} = a^{2/3}$.

Find the centroids of the following plane regions:

3. The first-quadrant sector of $x^2 + y^2 = a^2$
4. The region bounded on the left by the y-axis and on the right by the ellipse $\dfrac{x^2}{a^2} + \dfrac{y^2}{b^2} = 1$. Compare the result with that of Example 4, above.
5. The region enclosed by the x-axis and $y = \sin x$, $0 \le x \le \pi$
6. The region between the x-axis and $y = a^2 - x^2$
7. The region cut out of the first quadrant by $x = 3 - 2y - y^2$
8. The region cut out of the first quadrant by $y = 4x - x^3$
9. The region between $y = x^2$ and $x - y + 2 = 0$
10. The region in the first quadrant between $y = x^4$ and $y = 4x^2$

Exercises 9.9 [C]

The density ρ drops out of the formulas of the text only because it is constant. Find the centroids (more properly, centers of mass) of the distributions in Exs. 1–3, in which the density varies:

1. A thin rod, a units long, in which the density varies as the distance from one end.
2. An isosceles triangle of base b and altitude h, in which the density varies as **(a)** the distance from the base; **(b)** the square of the distance from the base; **(c)** the square root of the distance from the base.
3. A semicircular region in which the density is proportional to the distance from the bounding diameter.

The methods of the text can be used to develop formulas for the centroids of curved surfaces and of solids. The general problem is beyond the scope of this book, but a surface or solid of revolution will have its centroid on the axis of rotation, leaving only one coordinate to be found by integration.

4. Derive the formula $\bar{x} = \dfrac{2\pi}{S}\displaystyle\int xy\,ds$ for a mass uniformly distributed over the surface, of area S, formed by the rotation about the x-axis of an arc of the curve $y = f(x)$.

5. Find the centroid of a hemispherical surface of radius a.

6. Find the centroid of a conical surface of radius r and height h.

7. Derive the formula $\bar{x} = \dfrac{\pi}{V}\displaystyle\int xy^2\,dx$, and describe the geometrical configuration to which it applies.

8. Find the centroid of a hemispherical solid of radius a.

9. Find the centroid of a conical solid of radius r and height h.

10. Find the centroid of a regular square pyramid of height h and basal edge e.

Chapter Review

1. Find the area of the loop of $y^2 = x(x - 2)^2$.

2. Find the volume of the solid generated by rotating the region of Ex. 1 about (**a**) the x-axis; (**b**) the y-axis.

3. A cylindrical hole of radius 4 in. is bored through a sphere of radius 5 in. Find the remaining volume.

4. Every section of a solid by a plane perpendicular to the x-axis is a square with two diagonally opposite vertices on the ellipse $x^2 + 9y^2 = 36$. Find the volume of the solid.

5. Find the area of the surface generated by rotating about the x-axis the arc of $9y^2 = x(3 - x)^2$ from $x = 0$ to $x = 3$.

6. Show that the total force on a vertical surface of an object submerged in a fluid is equal to the area of the surface multiplied by the pressure at its centroid.

An artificial satellite must withstand extremes of heat and cold without damage to its delicate instruments. The equations governing the thermal balance of a particular satellite are determined initially from purely theoretical considerations. When the spacecraft is prepared for launch, extreme care is taken to protect the thermal control surfaces which provide the boundary conditions for the governing heat-transfer equations. Actual flight data compares favorably with the temperatures predicted by the mathematical model.

Item	Requirement (°C)	Prediction (°C)	Flight Data (°C)
ISP	10 to 30	15	17
Mission Adapter	–	12	13
Enclosure Panels	5 to 25	8 to 17	8 to 23
RIU's	5 to 45	29 to 33	28 to 31
FPSS			
Sensor	21 to 25	23	23
Electronics	5 to 45	37	39

10

The Logarithmic and Exponential Functions

10.1 A new function

We have seen that $\int x^n\,dx = \dfrac{1}{n+1}x^{n+1} + C$ for every rational exponent $n \neq -1$, but we have so far failed to discover an indefinite integral for the case $n = -1$, that is, a function L such that $L'(x) = x^{-1}$. Because many important problems lead to integrals of the form $\int x^{-1}\,dx$, we badly need to discover some such function L. It is our purpose in this section to define such a function and to examine its properties; when we have done so, we shall see that the function is, in fact, already well known from elementary algebra.

In general terms, our problem is to define a function with a given derivative. If the given derivative is continuous (as it is in the problem that concerns us here), then the first fundamental theorem of calculus (Theorem 4.2-1) points the way to a solution. Before we use this procedure to define a new function, however, we had better first see how it works in a familiar context, by using it to obtain new definitions for some familiar functions.

We have seen (§ 4.2) that if f is a function continuous on an interval $[a, b]$, and if c is a fixed point and t a variable point of that interval, then there is a function G defined by

$$G(t) = \int_c^t f(x)dx,$$

and the first fundamental theorem tells us that $G' = f$ on $[a, b]$. Let us consider a specific instance. Suppose $f(x) = x^2$; then f is continuous on the whole number line, and we can define a function F by

$$F(t) = \int_2^t x^2\,dx \tag{1}$$

for each real number t. There are, of course, other and more familiar ways of describing F. We can, for example, use the second fundamental theorem to obtain $\int_2^t x^2\,dx = \frac{1}{3}t^3 - \frac{8}{3}$, and we can therefore replace (1) by its equivalent,

$$F(t) = \tfrac{1}{3}t^3 - \tfrac{8}{3}, \tag{2}$$

a form which not only is more familiar than (1) but also lends itself more conveniently to evaluating F at a particular point: $F(5) = \frac{125}{3} - \frac{8}{3} = 39$, $F(-1) = -\frac{1}{3} - \frac{8}{3} = -3$, and so on. Note carefully, however, that we were able to obtain (2) from (1) *only because we already knew an indefinite integral of f.* Incidentally, we observe that differentiating (2) gives $F'(t) = t^2$, thus confirming that $F' = f$.

Let us consider another example. Let S be the function defined by

$$S(t) = \int_0^t \frac{dx}{\sqrt{1 - x^2}}, \quad -1 \le t \le 1. \tag{3}$$

The substitution $x = \sin\theta$ gives $dx = \cos\theta\,d\theta$; hence

$$\int \frac{dx}{\sqrt{1 - x^2}} = \int \frac{\cos\theta\,d\theta}{\sqrt{1 - \sin^2\theta}} = \int d\theta = \theta + C = \sin^{-1} x + C,$$

and

$$S(t) = [\sin^{-1} x]_0^t = \sin^{-1} t. \tag{4}$$

Thus, the function S defined by (3) turns out to be the familiar inverse-sine function, and because (3) is free from intuitive geometric arguments, it is often preferred, in more advanced work, to the definition of the inverse sine with which we are familiar.

The student should note that all (4) tells us about the function S is that we have met it before. It does not tell us any of the properties of S; it does not, for example, give us a formula for computing values of S. All the properties of S are already implicit in (3), although a good deal of work remains to be done to exhibit those properties.

Similarly, if $T(t) = \int_0^t dx/(1 + x^2)$, the student can use the substitution $x = \tan\theta$ to show that $T(t) = \tan^{-1} t$ (see Ex. 8, page 249). Once again, the whole story of this function is implicit in the definition we give here, though it may take skill and effort to read that story.

Our purpose in discussing this method of defining functions is not, however, to warm over old functions. We need it here, as noted above, to define a new function that will fill an annoying gap in our list of functions, a function L such that $L'(x) = 1/x$. The job is almost trivially easy. We set

$$L(t) = \int_1^t \frac{1}{x}\,dx \quad (t > 0), \tag{5}$$

and it is done, for, by the first fundamental theorem, we then have

$$L'(t) = \frac{1}{t}. \tag{6}$$

This is, however, a solution to our problem in only a formal sense. Until we know a little more about the function L—until, for example, we can find some way of determining or at least approximating $L(x)$ for any given $x > 0$—this solution is not of much practical use. Let us then see what we can discover about the properties of L.

Because L is defined as an integral, our investigation will have to use the properties of integrals. The most important of these properties are stated in equations (4)–(8) of § 4.1, and the student should review them before proceeding. Another property of the integral that we shall need is one to which we have so far paid but little attention: the value of an integral depends only on the limits of integration and on the *form* of the integrand. In particular, the variable of integration is a "dummy" variable which does not affect the value of the integral. Thus, for example,

$$\int_1^2 x^2\,dx = \tfrac{1}{3}x^3\Big|_1^2 = \tfrac{7}{3}, \quad \int_1^2 u^2\,du = \tfrac{1}{3}u^3\Big|_1^2 = \tfrac{7}{3}, \quad \int_1^2 \theta^2\,d\theta = \tfrac{1}{3}\theta^3\Big|_1^2 = \tfrac{7}{3},$$

and so on. Similarly, $L(t) = \int_1^t dx/x$, but also $L(t) = \int_1^t du/u$, no matter what relationship may exist between x and u.

Now to L. We note first that the integral in (5) can be interpreted as an area. If $t > 1$, then (5) is the area of the region (shaded in the figure) bounded by the x-axis, the hyperbola $y = 1/x$, and the lines $x = 1$, $x = t$. That (5) represents a positive number under these circumstances follows

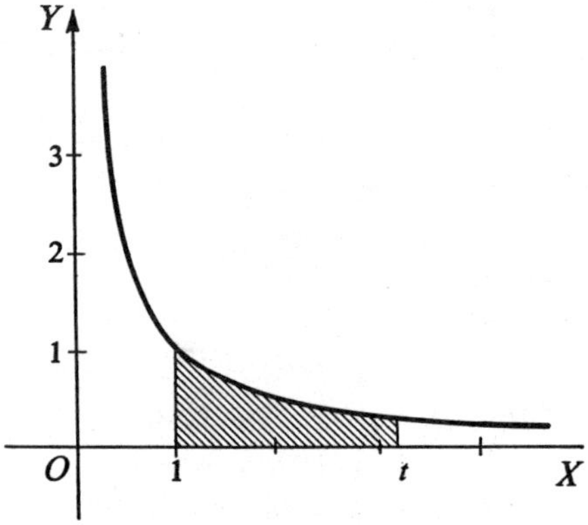

Fig. 1

directly from the definition of the definite integral, equation (2) of § 9.1. Hence $L(t) > 0$ if $t > 1$.

Next, if $t = 1$, we have

$$\blacktriangleright \qquad \mathbf{L(1) = \int_1^1 dx/x = 0} \tag{7}$$

by equation (5) of § 4.1. The situation when $0 < t < 1$ is slightly more complicated. Under these circumstances, the upper limit of integration in (5), namely t, is smaller than the lower limit, 1, and since the integrand is positive, the value of the integral is negative. More formally, $L(t) = \int_1^t dx/x = -\int_t^1 dx/x$, that is, $L(t)$ is the negative of the area of the region bounded by $y = 1/x$, the x-axis, and the lines $x = t$, $x = 1$, and is therefore negative. (We shall establish this same result independently below.) Hence, in summary, $L(t)$ is greater than, equal to, or less than 0 according as t is greater than, equal to, or less than 1.

If p and q are in the domain of L—if, that is, $p > 0$ and $q > 0$—then $pq > 0$ and pq is also in the domain of L. Then

$$L(pq) = \int_1^{pq} dx/x = \int_1^p dx/x + \int_p^{pq} dx/x$$

from equation (6) of § 4.1. We transform the last integral by introducing a variable of integration u defined by the equation $u = x/p$. This gives us $x = pu$, $dx = p\,du$, $dx/x = p\,du/pu = du/u$, and since $x = p$ implies $u = 1$, and $x = pq$ implies $u = q$, we have

$$\int_p^{pq} dx/x = \int_1^q du/u.$$

Hence

$$\blacktriangleright \qquad \mathbf{L(pq) = \int_1^p dx/x + \int_1^q du/u = L(p) + L(q),} \tag{8}$$

since, as noted above, the dummy variable of integration (x or u) does not affect the value of the integral.

If $q = 1/p$, then $pq = 1$, and (8) becomes $L(1) = L(p) + L(1/p)$. Since $L(1) = 0$ by (7), this is then equivalent to

$$\blacktriangleright \qquad \mathbf{L(1/p) = -L(p).} \tag{9}$$

This result has a simple geometric interpretation: If $0 < p < 1$, then $1/p > 1$ and $L(1/p) > 0$, as noted above. Hence (9) tells us that $0 < t < 1$ implies $L(t) < 0$, as we have already observed, but, more specifically, that the two regions shaded in Fig. 2 have equal areas.

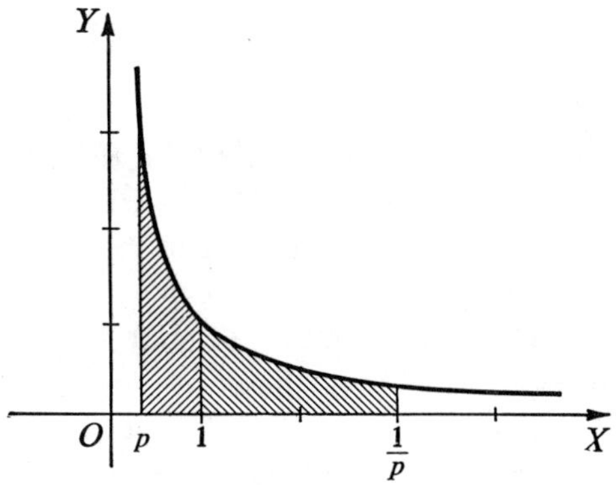

Fig. 2

We end this investigation of the properties of L by deriving two more formulas. First, using (8), we have

$$L(p/q) = L(p \cdot 1/q) = L(p) + L(1/q),$$

which, with the help of (9), becomes

$$\blacktriangleright \qquad \mathbf{L(p/q) = L(p) - L(q).} \qquad (10)$$

Second, we consider $L(p^n) = \int_1^{p^n} dx/x$. If we transform this by setting $u = x^{1/n}$, we have $x = u^n$, $dx = nu^{n-1}\,du$, $dx/x = nu^{n-1}\,du/u^n = n \cdot du/u$; also, $x = p^n$ implies $u = p$. Hence

$$\blacktriangleright \qquad \mathbf{L(p^n) = \int_1^{p^n} dx/x = n\int_1^{p} du/u = nL(p),} \qquad (11)$$

since it is still true that the dummy variable of integration does not affect the value of the integral.

Let us now take stock of what we have discovered about L. It is interesting to list the properties displayed in equations (7)–(11) in parallel with some properties of a function used in elementary algebra, the *logarithm* to an arbitrary positive base $b \neq 1$:

(7) $L(1) = 0$	$\log_b 1 = 0$
(8) $L(pq) = L(p) + L(q)$	$\log_b pq = \log_b p + \log_b q$
(9) $L\left(\frac{1}{p}\right) = -L(p)$	$\log_b\left(\frac{1}{p}\right) = -\log_b p$
(10) $L\left(\frac{p}{q}\right) = L(p) - L(q)$	$\log_b\left(\frac{p}{q}\right) = \log_b p - \log_b q$
(11) $L(p^n) = nL(p)$	$\log_b p^n = n \log_b p$

Except for notation, the two columns are identical; the function L has all the important properties of the logarithm, and we shall therefore call it a logarithmic function.

Actually, we have in L considerably more than the elementary logarithm. The latter is defined in terms of exponents: $\log_b x$ is the number n such that $b^n = x$, and b is therefore called the *base*. The exponential b^n, however, is defined only for rational values of the exponent n, and the logarithmic function thus defined therefore takes only rational values; it has neither a derivative nor an integral, and it is therefore almost useless to us in the context of calculus. We turn, therefore, to L as a function which (i) has all the important properties of the elementary logarithm, and (ii) is differentiable (and therefore continuous, and therefore integrable) throughout its domain, the interval $(0, \infty)$. In effect, we now abandon the elementary (rational) logarithm and use the word "logarithm" thus left unemployed to describe this new function L. In this, nothing is lost and much is gained.

We turn next to the problem of evaluating $L(x)$ for arbitrary $x > 0$. It is convenient to start by looking at $L(1 + t)$, for $t > -1$. We have

$$L(1 + t) = \int_1^{1+t} \frac{1}{x}\, dx$$

by definition. The change of variable $x = 1 + u$ then yields

$$L(1 + t) = \int_0^t \frac{du}{1 + u}.$$

As the student may easily show (multiply by $1 + u$), if $u > 0$ then

$$1 - u < \frac{1}{1 + u} < 1.$$

Hence $\displaystyle\int_0^t (1 - u)\, du < \int_0^t \frac{du}{1 + u} < \int_0^t 1\, du$, or $t - \frac{1}{2}t^2 < L(1 + t) < t.$

Similarly, $$1 - u + u^2 - u^3 < \frac{1}{1 + u} < 1 - u + u^2,$$

$$\int_0^t (1 - u + u^2 - u^3)\, du < \int_0^t \frac{du}{1 + u} < \int_0^t (1 - u + u^2)\, du,$$

and $$t - \tfrac{1}{2}t^2 + \tfrac{1}{3}t^3 - \tfrac{1}{4}t^4 < L(1 + t) < t - \tfrac{1}{2}t^2 + \tfrac{1}{3}t^3.$$

Further repetitions of this maneuver generate a sequence of polynomial bounds for $L(1 + t)$. If we define (with a change of variable to the more familiar x).

$$L_n(x) = x - \tfrac{1}{2}x^2 + \tfrac{1}{3}x^3 - \cdots + (-1)^{n-1}\frac{x^n}{n} = \sum_{k=1}^{n} (-1)^{k-1}\frac{x^k}{k},$$

we then have
$$L_n(x) \le L(1 + x) \le L_{n-1}(x) \tag{12}$$

for $x \ge 0$ and n even, with inequalities reversed if n is odd. If we use either $L_n(x)$ or $L_{n-1}(x)$ as an approximation to $L(1 + x)$, the error incurred does not exceed

$$E_n = |L_n(x) - L_{n-1}(x)| = \frac{1}{n}x^n,$$

and since $\lim_{n\to\infty} E_n = 0$ if $x < 1$, it follows that these approximations improve, with increasing n, for $0 \le x < 1$. We can now use these approximations to make a table of selected values of $L(1 + x)$ for $0 \le x < 1$, and equations (8) through (11) can then be used to extend the table.

For instance, since $\sqrt{2} = 1.414214$ to 7 significant figures, we can set $x = 0.414214$ in (12) to obtain, for $n = 10$,

$$0.34657 < L(1.414214) < 0.34658.$$

But by (11), $L(\sqrt{2}) = L(2^{1/2}) = \frac{1}{2}L(2)$, so $0.69314 < L(2) < 0.69316$. Similarly, $L(4) = L(2^2) = 2L(2)$, and therefore $1.38628 < L(4) < 1.38632$. We therefore have, to 3 places of decimals, $L(2) = 0.693$, $L(4) = 1.386$, and (9) then yields $L(0.5) = -0.693$, $L(0.25) = -1.386$.

In comparing L with the elementary logarithm $\log_b$, we have left unanswered one rather obvious question: What is the base of the logarithm L? Strictly, this question has no immediate answer, for if we do not define a logarithm as a power of some base, what then do we mean by the word "base"? One distinctive property of the base b in the elementary case is that it is the unique solution of the equation $\log_b x = 1$. We therefore ask whether there is, similarly, a unique root of the equation $L(x) = 1$. There is indeed, as we shall show, and we shall see in § 10.2 that there are strong reasons for calling it the base of the logarithmic function L. This root is universally designated e; hence e is the unique number such that

$$L(e) = \int_1^e \frac{1}{x}\,dx = 1.$$

The existence and uniqueness of e are easily demonstrated. It can also be shown that e is irrational, and that, to 4 significant figures,

$$e = 2.718.$$

We must now consider a question of notation. There is no logical reason why we should not continue to use the letter L to designate the function we have been considering, but there is a considerable weight of tradition in favor of another notation. Since we have identified this function as a logarithm, we could, of course, use conventional logarithmic notation, writing "$\log x$" in place of "$L(x)$"; this might, however, lead to confusion of this function with the logarithm to the base 10, or *common logarithm*. The logarithm to the base e, with which we deal here, is called the natural logarithm, and, to distinguish it from the common logarithm, it is often designated "$\ln x$" (read "natural logarithm of x," or often, for short, simply "$\log x$"). We shall adopt this notation; in what follows, we shall use "$\ln x$" in place of "$L(x)$."

Common logarithms remain the most practical variety for the computations of arithmetic, since it is only with these that there is a simple relationship between the characteristic and the position of the decimal point. Natural logarithms are not used in this sort of computation. They occur in the answers to certain problems which require calculus for their solution, and extensive tables of natural logarithms are printed for use in finding rational approximations to such answers; there is a short table on page 440 of this book. It is also possible to use a table of common logarithms to compute natural logarithms. If we start with the equation $10^n = x$ and take the common logarithms of both sides we get $n = \log_{10} x$, but if we take natural logarithms we get $n \ln 10 = \ln x$. Substituting $\log_{10} x$ for n in the second of these equations, and using the fact that $\ln 10 = 2.303$, we get

$$\ln x = (\log_{10} x)(\ln 10) = 2.303 \log_{10} x, \tag{13a}$$

or, alternatively,

$$\log_{10} x = \frac{\ln x}{\ln 10} = 0.4343 \ln x. \tag{13b}$$

Since we have discarded the elementary definition of the logarithm, we must, strictly speaking, regard (13a) and (13b) not as conversion formulas but as definitions of a function, the common logarithm, whose value at each positive number x we designate $\log_{10} x$. You need not fear, however, that all you have learned in your study of algebra about another function called the common logarithm is therefore obsolete; the properties you learned are also properties of this new function, and the tables you used can be used in just the same way with this new function.

The result (13b) can be generalized to give us a definition of logarithm to any positive base $b \neq 1$:

$$\blacktriangleright \qquad \log_b x = \frac{\ln x}{\ln b}. \tag{14}$$

The graph of $y = \ln x$ is a particularly simple one. We know from (7) that the x-intercept is 1, and since the logarithmic function has the interval $(0, \infty)$ for its domain, there is no y-intercept. Since the derivative $dy/dx = 1/x$ is always positive, the function is always increasing, and the

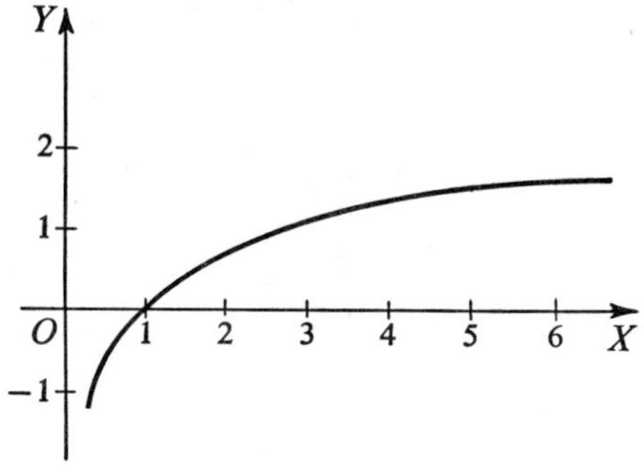

Fig. 3

graph rises steadily, though less and less steeply, toward the right. There is neither an upper nor a lower bound to the values of the function; to see this, note that, by (11) and (12), $\ln e^n = n \ln e = n$; given any y, however large it may be, there is a number $n > y$, and for this n, $\ln e^n = n > y$, and also $\ln e^{-n} = -n < -y$. Since $\lim\limits_{n \to \infty} e^{-n} = 0$, this last result also implies that $\ln x$ decreases without bound as $x \to 0$, and the y-axis is therefore a vertical asymptote.

We have, in (6), a new rule for differentiation:

▶ **Rule 15.** $\dfrac{d}{dx}(\ln x) = \dfrac{1}{x}\cdot$ **Rule 15a.** $d(\ln x) = \dfrac{1}{x}\,dx.$

The corresponding integral formula is not quite so simple as might appear, for if x is negative, $\ln x$ is not defined, but $\ln(-x)$ is. Differentiating $\ln(-x)$, we have

$$\frac{d}{dx}(\ln(-x)) = \frac{1}{-x}\cdot\frac{d(-x)}{dx} = \frac{1}{x},$$

which is the same as the derivative of $\ln x$. Hence we have

$$\int \frac{1}{x}\,dx = \ln x + C \quad \text{if} \quad x > 0,$$

and

$$\int \frac{1}{x}\,dx = \ln(-x) + C \quad \text{if} \quad x < 0.$$

Fortunately, there is a single formula which includes both cases, and this is our new rule for integration:

$$\blacktriangleright \qquad \int \frac{1}{x}\,dx = \ln|x| + C.$$

Example 1. Differentiate $\ln(x^2 - 2x + 3)$.

Solution: By Rule 15 and the chain rule, we have

$$\frac{d}{dx}\ln(x^2 - 2x + 3) = \frac{1}{x^2 - 2x + 3}\,\frac{d}{dx}(x^2 - 2x + 3)$$

$$= \frac{1}{x^2 - 2x + 3}(2x - 2) = \frac{2(x - 1)}{x^2 - 2x + 3}.$$

Example 2. Integrate $\int \tan x\,dx$.

Solution:

$$\int \tan x\,dx = \int \frac{\sin x\,dx}{\cos x} = -\int \frac{d(\cos x)}{\cos x} = -\ln|\cos x| + C = \ln|\sec x| + C.$$

Example 2 gives us an important new rule for integration:

$$\blacktriangleright \qquad \int \tan x\,dx = \ln|\sec x| + C.$$

Example 3. Integrate $\int \frac{x^2\,dx}{x - 1}$.

Solution: The integrand is an improper fraction, because its numerator is not of lower degree than its denominator. By long division, we find

$$\frac{x^2}{x - 1} = x + 1 + \frac{1}{x - 1},$$

and therefore

$$\int \frac{x^2\,dx}{x - 1} = \int\left(x + 1 + \frac{1}{x - 1}\right)dx = \tfrac{1}{2}x^2 + x + \ln|x - 1| + C.$$

Example 4. Differentiate $y = \sqrt{\frac{x^2 - 1}{x^2 + 1}}$.

Solution: The expression can, of course, be differentiated as it stands, but the task will be made much easier if we first take the logarithm of each side of the equation, use the properties of logarithms to simplify as much as possible, and then differentiate implicitly:

$$\ln y = \ln \sqrt{\frac{x^2 - 1}{x^2 + 1}} = \frac{1}{2}[\ln (x^2 - 1) - \ln (x^2 + 1)],$$

$$\frac{1}{y}\frac{dy}{dx} = \frac{1}{2}\left(\frac{2x}{x^2 - 1} - \frac{2x}{x^2 + 1}\right) = \frac{2x}{(x^2 - 1)(x^2 + 1)},$$

$$\frac{dy}{dx} = \frac{2xy}{(x^2 - 1)(x^2 + 1)} = \frac{2x}{(x^2 - 1)(x^2 + 1)}\sqrt{\frac{x^2 - 1}{x^2 + 1}} = \frac{2x}{(x^2 - 1)^{1/2}(x^2 + 1)^{3/2}}.$$

This technique is called *logarithmic differentiation.*

Exercises 10.1 [A]

1. **(a)** Using the given values and the basic properties of the natural logarithm given by equations (7)–(11), complete the following abbreviated table of natural logarithms:

x	$\ln x$
0.1	
0.2	
0.3	
0.4	
0.5	
0.8	

x	$\ln x$
1	
2	0.69
3	1.10
4	
5	1.61
8	

x	$\ln x$
10	
20	
30	
40	
50	
80	
100	

(b) Use a calculator to check the entries in your table.

(c) Sketch a graph of $y = \ln x$, $0.2 \leq x < 10$.

(d) On the same set of axes, without any further computation, sketch graphs of $y = \ln \sqrt{x}$ and $y = \ln x^2$.

2. Use the fact that $\ln x_1 = \ln x_2$ implies $x_1 = x_2$ (see Ex. 15, page 376), to find a number x such that

(a) $\ln x = \ln 2 + \ln 3$.

(b) $\ln x = \ln 2 - \ln 5$.

(c) $\ln x = 2 \ln 3 - 3 \ln 2$.

(d) $\ln x + \ln 2 = \frac{1}{3} \ln 8$.

(e) $\ln x = \frac{1}{2} \ln 9 + \frac{3}{2} \ln 2$.

(f) $\ln x = \ln e^2 - 1$.

(g) $\ln e^x = 5$.

(h) $\ln x = \ln 3 + \frac{1}{2}$.

3. Differentiate (wherever you can, use properties of logarithms to make the job easier):

(a) $\ln 7x$
(b) $\ln x^7$
(c) $\ln (x + 7)$
(d) $\ln (x^2 + 5)$
(e) $\ln (x^2 + 5x)$
(f) $\ln x\sqrt{x + 3}$
(g) $(\ln x)^2$
(h) $\ln (x - 3)^2(x + 2)$
(i) $\ln ((a + t)/(a - t))$

4. Differentiate logarithmically (see Example 4):

(a) $y = (x - a)(x - b)^2(x - c)^3$
(b) $y = (x^2 - 3)^5 \sqrt{x + 1}$

5. Integrate as indicated:

(a) $\displaystyle\int \frac{dx}{x + a}$ (b) $\displaystyle\int \frac{x\,dx}{x^2 - 3}$ (c) $\displaystyle\int \frac{x\,dx}{x - 1}$

6. Evaluate:

(a) $\displaystyle\int_{-1}^{2} \frac{x\,dx}{x + 2}$ (b) $\displaystyle\int_{-4}^{-2} \frac{dx}{x}$

7. Evaluate: (a) $\displaystyle\lim_{x \to \infty} \frac{\ln x}{x}$ (b) $\displaystyle\lim_{x \to \frac{\pi}{4}} \frac{\ln \tan x}{x - \pi/4}$

Exercises 10.1 [B]

1. Sketch: (a) $y = \ln (1/x)$; (b) $y = \ln |x|$.
2. Solve for x:

(a) $\ln x = 2 \ln 10 - \ln 2$
(b) $\ln x = \frac{2}{5} \ln 32 - \frac{1}{3} \ln 27$
(c) $\ln x = 2 + \ln 3$
(d) $\ln x + \ln (5 - x) = \ln 2 + \ln 3$
(e) $\ln (x - 2) - \ln 2 = \ln 3 - \ln (x - 1)$

3. Differentiate:

(a) $\log_{10} x$ (see (13a))
(b) $(\ln z)^3$
(c) $\ln \sec \theta$
(d) $\ln (y^2 - y^4)$
(e) $\ln \sqrt{(x - 1)(x + 5)}$
(f) $\ln (\sec x + \tan x)$
(g) $\sqrt{\ln t}$
(h) $\ln (\ln x)$
(i) $\ln \tan 3x$
(j) $x \ln x - x$

4. Differentiate logarithmically:

(a) $y = \dfrac{2\,ax}{x^2 - a^2}$ (b) $y = \sqrt{\dfrac{x+1}{x-1}}$ (c) $y = \sqrt[5]{(x+1)(x-3)^2}$

5. Integrate as indicated:

(a) $\displaystyle\int \frac{dx}{2x+7}$ (b) $\displaystyle\int \frac{x\,dx}{x+3}$ (c) $\displaystyle\int \frac{(x-1)dx}{x^2-2x-3}$

6. Evaluate:

(a) $\displaystyle\int_0^3 \frac{x^2\,dx}{x+2}$ (b) $\displaystyle\int_{-1}^0 \frac{dx}{x-1}$ (c) $\displaystyle\int_e^{e^2} \frac{\ln x\,dx}{x}$

7. Evaluate: (a) $\displaystyle\lim_{x\to\infty} \frac{\ln(\ln x)}{\ln x}$ (b) $\displaystyle\lim_{x\to 1} \frac{\ln x}{x - x^2}$

Miscellaneous Exercises 10.1

1. Sketch: (a) $y = \ln(x+2)$; (b) $y = \ln(5-x)$.

2. On the same set of axes, sketch $y = x^2 - 9$ and $y = \ln(x^2 - 9)$.

3. Sketch $y = \ln\cos x$, $-2\pi \le x \le 2\pi$.

4. (a) In elementary algebra, the equation $y = \log_b x$ is meaningful only if y is rational. Write this equation in exponential form.

(b) Take the natural logarithm of each side and eliminate y to obtain

$$\log_b x = \frac{\ln x}{\ln b}.$$

Since this result is our definition (14), you have proved that (14) is consistent with the elementary definition whenever both apply, that is, whenever y is rational.

5. Differentiate:

(a) $\log_a x$ (see Ex. 4)

(b) $\ln\sin\theta$

(c) $\log_3(2x+1)$

(d) $\ln\csc^2 z$

(e) $\ln(\csc x - \cot x)$

(f) $\ln\tan\left(\dfrac{\pi}{4} + \dfrac{\theta}{2}\right)$

(g) $\ln\left(x + \sqrt{x^2+1}\right)$

(h) $\ln\left(x/\sqrt{x^2+1}\right)$

(i) $\log_{10}(x^2+3)$

(j) $\ln\left(\left(1 + \sqrt{1-y^2}\right)/y\right)$

(k) $\tan x \ln\tan x$

(l) $\ln\sqrt{\dfrac{1-\sin\phi}{1+\sin\phi}}$

6. Integrate:

(a) $\int \frac{x^2 - 3x}{x + 1}\,dx$ (c) $\int \cot\theta\,d\theta$

(b) $\int \frac{dx}{ax + b}$ (d) $\int \frac{(2ax + b)dx}{ax^2 + bx + c}$

7. Find dy/dx if (a) $y = \sqrt[3]{x^2 + 1}/\sqrt{x^3 - 4}$.

(b) $y = \frac{(x - 1)(x + 3)^{2/3}}{\sqrt{(x + 2)(x^2 - 6)}}$.

(c) $y = \sqrt[4]{\frac{x^2(x^2 - 3)^3}{x^2 - 3x + 5}}$.

8. Evaluate:

(a) $\int_{-1}^{0} \frac{x^2\,dx}{x - 1}$ (b) $\int_{a}^{3a} \frac{(x + a)dx}{x^2 + 2ax}$

9. Find the area of the region bounded by $(x^2 + 1)y = 2x$, the x-axis, $x = 1$, and $x = 3$.

10. Find the volume of the solid generated by rotating about the x-axis the region bounded by $xy^2 = 4$, the x-axis, $x = 1$, and $x = 2$.

11. Let A be the area of the region bounded by $y = x^n$, the x-axis, $x = 1$, and $x = b$ $(b > 1)$. Find A and, if it exists, $\lim_{b\to\infty} A$, considering three cases: (a) $n > -1$, (b) $n = -1$, (c) $n < -1$.

12. Let V be the volume of the solid generated by rotating the region of Ex. 11 about the x-axis. Find V, and, if it exists, $\lim_{b\to\infty} V$, considering three cases: (a) $n > -\frac{1}{2}$, (b) $n = -\frac{1}{2}$, (c) $n < -\frac{1}{2}$.

13. Find the area of the region bounded by $xy = 18$ and $x + y = 9$.

14. Find the volume of the solid generated by rotating about the line $y = -1$ the region bounded by $xy = 8$, $x = 2$, $x = 4$, and $y = -1$.

15. (a) Show that the function $L = \{(x, \ln x)\}$ is increasing throughout its domain.

(b) Use (a) to show that if $\ln x_1 = \ln x_2$, then $x_1 = x_2$.

(c) Conclude that L has an inverse L^{-1} (§ 2.5).

10.2 The exponential function

We turn now to the inverse of the natural-logarithm function ln defined and studied in § 10.1. A proof that such an inverse exists is developed in

Ex. 15, page 376, and we shall therefore not prove it here; if, however, you have not done that exercise, you are urged at least to review the definition of inverse function in § 2.5. The inverse of the natural-logarithm function is called the exponential function, for reasons that we shall investigate, and is therefore often denoted "exp." It follows, from the definition of inverse, that exp (ln x) $= x$ for each x in the domain of ln (that is, for each *positive* number x), and that ln (exp y) $= y$ for each y in the range of ln (that is, for each *real* number y). In other words,

$$y = \ln x \quad \text{if and only if} \quad x = \exp y.$$

It follows that the domain of exp is the set of all real numbers and the range is the set of positive numbers.

(Note that we often omit parentheses after the symbols "ln" and "exp," as we do, for example, after "sin" and "tan," when we wish to indicate the value of the function at a particular point; thus, we write "ln 3," "exp 10," "sin π," and "tan x," where, to be consistent with the notation "$f(x)$," we really should write "ln (3)," "exp (10)," "sin (π)," and "tan (x).")

Let us now see what we can discover about this function exp. First we note that exp 0 is, by definition of exp, the number whose natural logarithm is 0, namely 1:

$$\exp 0 = 1. \tag{1}$$

Again by definition, exp 1 is the number whose natural logarithm is 1, which is, by definition, e:

$$\exp 1 = e. \tag{2}$$

We shall now show that all values of exp can be expressed as powers of e, that, in fact,

▶ $$\mathbf{exp}\ \mathbf{x} = \mathbf{e}^{x} \tag{3}$$

for each real number x; it is this relationship that motivates the use of the word "exponential" to describe this particular function.

Equation (3) not only explains the word "exponential," it also explains the word "base" used to describe the number e in its relation to the function ln. The logarithm of elementary algebra is defined as the inverse of an exponent: $\log_b y = x$ if and only if $b^x = y$. In this context, the number b is called the *base* of the logarithms. If we set exp $x = y$ in (3), we then have, by definition of exp, ln $y = x$ if and only if $y = e^x$.

We turn now to the task of establishing (3). We first note, from (1) and (2), that it is valid when $x = 0$ and when $x = 1$. Now, from equation (8) of § 10.1, we have (for $x > 0$, $y > 0$)

$$\ln x + \ln y = \ln xy.$$

Hence
$$\begin{aligned}\exp(\ln x + \ln y) &= \exp(\ln xy)\\ &= xy\\ &= \exp(\ln x)\cdot\exp(\ln y),\end{aligned}$$

where the last two lines follow from the definition of exp. But, since the range of ln is the set of all real numbers, it follows that for any real numbers p and q there are (positive) numbers x and y such that $\ln x = p$ and $\ln y = q$. Hence, for any real numbers p and q,

$$\exp(p + q) = (\exp p)(\exp q). \tag{4}$$

In particular,

$$\exp 2x = \exp(x + x) = (\exp x)(\exp x) = (\exp x)^2,$$
$$\exp 3x = \exp(2x + x) = (\exp 2x)(\exp x) = (\exp x)^3,$$

and the argument proceeds by induction to give us

$$\exp nx = (\exp x)^n \tag{5}$$

for any real number x and positive integer n. We note from (1) and the definition of the zero exponent that (5) is valid when $n = 0$, and we shall show that it is valid when n is a negative integer. First, however, we need the following:

$$\begin{aligned}(\exp x)(\exp(-x)) &= \exp(x - x) &&\text{from (4)}\\ &= \exp 0 = 1 &&\text{from (1)},\end{aligned}$$

and therefore

$$\exp(-x) = 1/\exp x = (\exp x)^{-1} \tag{6}$$

by definition of the negative exponent. Now, if n is a negative integer, $-n$ is then a positive integer, and

$$\begin{aligned}\exp nx &= \exp(-n)(-x)\\ &= (\exp(-x))^{-n} &&\text{by (5)}\\ &= ((\exp x)^{-1})^{-n} &&\text{by (6)}\\ &= (\exp x)^n\end{aligned}$$

by the laws of exponents. Hence (5) is valid for all integers n. In particular, setting $x = 1$ in (5) and using (2), we get

$$\exp n = e^n, \tag{7}$$

thus establishing (3) for all integers. Our next step is to extend this result to all rational numbers.

From (2) and (5) we get, for any integer $n \neq 0$,

$$e = \exp 1 = \exp\left(n \cdot \frac{1}{n}\right) = \left(\exp \frac{1}{n}\right)^n,$$

and therefore

$$\exp \frac{1}{n} = \sqrt[n]{e} = e^{1/n} \tag{8}$$

by definition of the fractional exponent. Hence, for any integers s and t, $t \neq 0$,

$$\begin{aligned} \exp \frac{s}{t} &= \exp\left(s \cdot \frac{1}{t}\right) = \left(\exp \frac{1}{t}\right)^s \qquad &\text{by (5)} \\ &= (e^{1/t})^s &\text{by (8)} \\ &= e^{s/t} \end{aligned}$$

by the laws of exponents. But any rational number r can, by definition, be represented as a quotient of integers, $r = s/t$, and we have therefore established (3) for all rational numbers.

This essentially completes the argument, since the elementary definition of exponent extends only to the rational numbers. The function exp, however, has for domain the set of all real numbers, irrational as well as rational, and we therefore take (3) as *defining* e^x in case x is irrational. It should be noted that all the familiar rules for operating with exponents, such as (4), carry over to this extended concept, though we shall not take the time to verify them all here. Because the form is more familiar, we shall generally use e^x rather than $\exp x$ in what follows.

This concept of an arbitrary real exponent can be extended to bases other than e. If x is rational and a is positive, then a^x is defined by elementary definitions and is positive; hence $\ln a^x$ is defined. But $\ln a^x = x \ln a$, $\exp(\ln a^x) = a^x$, and therefore

▶ $$\mathbf{a^x = \exp(x \ln a) = e^{x \ln a}.} \tag{9}$$

This result, which is then a theorem if x is rational, we take as the definition of a^x if x is irrational; once again, these irrational exponents conform just as do rational ones to the laws of exponents. We note, for example, using (9) and (6), that $a^{-x} = e^{-x \ln a} = \dfrac{1}{e^{x \ln a}} = \dfrac{1}{a^x}$; using (9) and (4), that $a^x a^y = e^{x \ln a} e^{y \ln a} = e^{(x+y) \ln a} = a^{x+y}$; using (9) and equation (11) of § 10.1, that $(a^x)^y = e^{y \ln a^x} = e^{xy \ln a} = a^{xy}$.

Because of the inverse relationship between ln and exp, the graph of

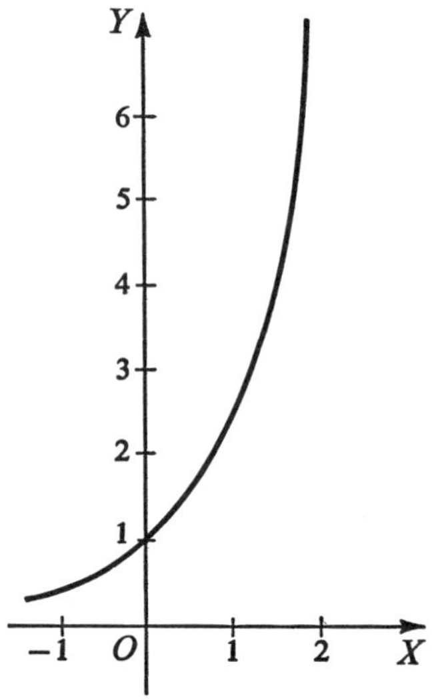

Fig. 4

$y = e^x$ can be obtained from the graph of $y = \ln x$ by a simple reflection in the line $y = x$ (see § 8.1).

To find the derivative of exp, we use the technique of logarithmic differentiation:

$$y = e^x, \quad \ln y = x, \quad \frac{1}{y}\frac{dy}{dx} = 1,$$

and

$$\frac{dy}{dx} = y = e^x.$$

Thus we come to the extraordinary result that the exponential function is its own derivative: $\exp' = \exp$; in geometrical terms, the slope of the graph at any point is equal to the ordinate of that point. This important result gives us another formula for differentiation:

▶ **Rule 16.** $\frac{d}{dx}(e^x) = e^x.$ **Rule 16a.** $d(e^x) = e^x\,dx.$

If u is a differentiable function of x, then Rule 16 can be extended, with the help of the chain rule, to

$$\frac{d}{dx}(e^u) = e^u \frac{du}{dx}.$$

There is, of course, a formula for integration corresponding to Rule 16:

▶ $$\int e^x\,dx = e^x + C.$$

Example 1. Differentiate $e^{\sin x}$.

Solution: Applying Rule 16a and the chain rule, we have

$$d(e^{\sin x}) = e^{\sin x}\, d(\sin x) = e^{\sin x} \cos x\, dx.$$

Example 2. Integrate: $\int xe^{-x^2}\, dx$.

Solution: We find, by an inverse application of Rule 16a,

$$\int xe^{-x^2}\, dx = -\tfrac{1}{2}\int e^{-x^2}\, d(-x^2) = -\tfrac{1}{2}e^{-x^2} + C.$$

The same result can of course be obtained with the explicit substitution $u = e^{-x^2}$.

Example 3. Differentiate a^x.

Solution: METHOD 1. From (2), using Rule 16a and the chain rule, we have

$$d(a^x) = d(e^{x \ln a}) = e^{x \ln a}\, d(x \ln a) = e^{x \ln a} \ln a\, dx = a^x \ln a\, dx.$$

METHOD 2. If $y = a^x$, then $\ln y = x \ln a$, and

$$\frac{1}{y}\frac{dy}{dx} = \ln a, \quad \frac{dy}{dx} = y \ln a = a^x \ln a.$$

Example 4. Differentiate $y = x^x$.

Solution: We must use the technique of logarithmic differentiation as in Method 2 of the preceding example. We have $\ln y = x \ln x$, and

$$\frac{1}{y}\frac{dy}{dx} = x \cdot \frac{1}{x} + \ln x = 1 + \ln x, \quad \frac{dy}{dx} = y(1 + \ln x) = x^x(1 + \ln x).$$

The student must take care to distinguish between x^n, a variable base to a constant power, and e^x or a^x or x^x, in which the exponent is variable. The first case is the elementary one and can be differentiated by Rule 1; the others require logarithmic differentiation or Rule 16.

Exercises 10.2 [A]

1. Solve for x:

(a) $e^x = 1$

(b) $e^x = 2$

(c) $e^{2x} = 4$

(d) $e^{\ln x} = 3$

(e) $\ln x = 1$

(f) $\ln x = 2$

(g) $\ln x = -5$

(h) $e^{2x} - 3e^x + 2 = 0$

(i) $\dfrac{e^x + 1}{e^x - 1} = 3$

(j) $e^x - 2e^{-x} = 2$

2. In the following equations, (i) solve for x in terms of y, (ii) state the values of y for which your solution is valid, and (iii) state the corresponding values of x:

(a) $y = \ln x$ (b) $y = 2 \ln x$ (c) $y = \ln x^2$

3. Differentiate:

(a) e^{2x} (b) e^{-x} (c) $\frac{1}{e^{3x}}$ (d) $\frac{e^{2x}}{e^{4x}}$

(e) $\sin e^x$ (f) $e^{\cos x}$ (g) $e^{\ln x}$ (h) $\frac{e^x}{x}$

(i) 10^{3x} (j) $x^{1/x}$ (k) $e^{-x} \sin x$ (l) $\ln \frac{e^x}{e^x + 1}$

4. Integrate:

(a) $\int e^{3x}\, dx$ (b) $\int e^{-kx}\, dx$

(c) $\int xe^{x^2}\, dx$ (d) $\int \cos x\, e^{\sin x}\, dx$

(e) $\int \frac{e^x\, dx}{e^x + 1}$ (f) $\int a^x\, dx \quad (a > 0)$

5. Evaluate: (a) $\lim\limits_{x \to \infty} \frac{x}{e^x}$ (b) $\lim\limits_{x \to \infty} \frac{x^3}{e^x}$

Exercises 10.2 [B]

1. Find all real x such that

(a) $\ln x = 2$

(b) $\ln \sqrt{3x} = 2$

(c) $e^x = -1$

(d) $e^{2x} - 3e^x - 4 = 0$

(e) $e^x - e^{-x} = 2$

(f) $e^{3x} - 10e^x - 12 = 0$

(g) $\ln \sin x = 0$

(h) $\sin \ln x = 0$

2. In the following equations, (i) solve for x in terms of y, (ii) state the values of y for which your solution is valid, and (iii) state the corresponding values of x:

(a) $y = e^x$ (b) $y = e^{\ln x}$ (c) $y = \sin x$

3. Differentiate:

(a) e^{-4x} (b) e^{x^2} (c) $x^2 e^x$

(d) $\tan e^x$ (e) $e^x \ln x$ (f) $e^{\sqrt{x}}$

(g) x^{x^2} (h) $(1 + t)^{1/t}$ (i) $\ln \frac{e^x + 1}{e^x - 1}$

4. Integrate:

(a) $\int x^2 e^{x^3}\, dx$ (c) $\int \sec^2 x\, e^{\tan x}\, dx$ (e) $\int e^x \sqrt{e^x + 1}\, dx$

(b) $\int e^{1/x} \frac{dx}{x^2}$ (d) $\int 10^x\, dx$ (f) $\int \frac{e^x\, dx}{e^{2x} + 1}$

5. Find the volume of the solid generated by rotating about the x-axis the region bounded by $y = e^x$, $x = \ln 3$, and the coordinate axes.

6. Evaluate: (a) $\lim_{x \to 0} \frac{e^x - 1}{x}$ (b) $\lim_{x \to 0} \frac{a^x - b^x}{x}$

Miscellaneous Exercises 10.2

1. Analyze and sketch:

(a) $y = e^{-x^2}$ (c) $y = x^2 e^{-x}$ (e) $y = 1/(1 + e^{-x})$

(b) $y = xe^{-x^2}$ (d) $y = e^{1/x}$ (f) $y = e^{-x} \cos x,\ x \in [-2, 3]$

2. (a) Sketch $y = e^{-x}$.
   (b) Find the area of the region in the first quadrant bounded by the coordinate axes, $y = e^{-x}$, and $x = b$.
   (c) Find the limit of this area as $b \to \infty$.
   (d) Find the volume of the solid obtained by rotating the region described in (b) about the x-axis, and find the limit of this volume as $b \to \infty$.

3. (a) Find the area of the region enclosed by $y = e^x$, the y-axis, and $y = 2$.
   (b) From your answer to (a), infer the area of the region enclosed by $y = \ln x$, the x-axis, and $x = 2$.

4. If $y = xe^x$, find dy/dx, d^2y/dx^2, and d^3y/dx^3, and guess an expression for d^ny/dx^n.

5. A rectangle has two vertices on the x-axis and two on the curve $y = e^{-x^2}$. Find the maximum area of the rectangle.

6. The line tangent to $y = e^x$ at $P(x_1, y_1)$ meets the x-axis at Q. Show that the length of the projection of PQ on the x-axis (the subtangent) is independent of the choice of P.

7. (a) Find the area of the region bounded by $y = \frac{1}{2}(e^x + e^{-x})$, the x-axis, and the lines $x = \pm 1$.
   (b) Find the volume of the solid generated by rotating this region about the x-axis.

8. Show that the curves $y = e^{-x}$ and $y = e^{-x} \cos x$ are tangent at each point common to both curves. Sketch.

10.3 Indeterminate forms

We saw in Section 5.5 how L'Hôpital's Rule is used to evaluate a limit that can be symbolized by one of the quotient indeterminate forms 0/0 or ∞/∞. There are limits that correspond to other indeterminate forms; many of these can be transformed into quotient forms, thus permitting the use of L'Hôpital's Rule.

1. The product form $0\cdot\infty$ Because multiplication by $f(x)$ is equivalent to division by $1/f(x)$, this form is usually easy to transform into the form ∞/∞.

Example 1. $$\lim_{x\to 0+} x \ln x = \lim_{x\to 0+} \frac{\ln x}{1/x} \quad \left(\frac{\infty}{\infty};\ \text{use L'Hôpital's Rule}\right)$$
$$= \lim_{x\to 0+} \frac{1/x}{-1/x^2} = \lim_{x\to 0+} (-x) = 0.$$

2. The difference form $\infty - \infty$. As in the $0\cdot\infty$ case, this can usually be transformed into a quotient by the use of a little algebra.

Example 2. $$\lim_{x\to 0+} (x^{-1} - \csc x) = \lim_{x\to 0+} \left[\frac{1}{x} - \frac{1}{\sin x}\right]$$
$$= \lim_{x\to 0+} \frac{\sin x - x}{x \sin x} \quad \left(\frac{0}{0}\right)$$
$$= \lim_{x\to 0+} \frac{\cos x - 1}{\sin x + x\cos x} \quad \left(\frac{0}{0}\right)$$
$$= \lim_{x\to 0+} \frac{-\sin x}{2\cos x - x \sin x}$$
$$= 0.$$

3. The exponential forms 0^0, ∞^0, 1^∞. These occur as limits of expressions in the form

$$F(x) = f(x)^{g(x)}.$$

In such cases we take the logarithm,

$$\ln F(x) = g(x) \ln f(x),$$

and find the limit L of $\ln F(x)$ if it exists. The limit of the original expression is then e^L, for the following reason. Since for all $F(x) > 0$, it is true that

$$F(x) = \exp(\ln F(x)),$$

we have
$$\lim F(x) = \lim \exp(\ln F(x))$$
$$= \exp(\lim \ln F(x))$$

by the Chain Rule for Limits, Theorem 5.3-6. The limit in question may be any of the types we have considered.

Example 3 (0^0). Find $\lim_{x\to 0+} x^x$.

Solution: $\ln x^x = x \ln x$, and we saw in Example 1 that $\lim_{x\to 0+} x \ln x = 0$. Thus

$$\lim_{x\to 0+} x^x = \exp(\lim_{x\to 0+} \ln x^x) = \exp 0 = e^0 = 1.$$

Example 4 (∞^0). Find $\lim_{x\to\infty} (e^x + e^{3x})^{1/x}$.

Solution:
$$\begin{aligned}\lim_{x\to\infty} \ln (e^x + e^{3x})^{1/x} &= \lim_{x\to\infty} \frac{1}{x} \ln (e^x + e^{3x}) \\ &= \lim_{x\to\infty} \frac{\ln (e^x + e^{3x})}{x} \qquad \left(\frac{\infty}{\infty}\right) \\ &= \lim_{x\to\infty} \frac{\dfrac{e^x + 3e^{3x}}{e^x + e^{3x}}}{1} \\ &= \lim_{x\to\infty} \frac{e^x + 3e^{3x}}{e^x + e^{3x}} \\ &= \lim_{x\to\infty} \frac{1 + 3e^{2x}}{1 + e^{2x}} \qquad \left(\frac{\infty}{\infty}\right) \\ &= \lim_{x\to\infty} \frac{6e^{2x}}{2e^{2x}} \\ &= 3.\end{aligned}$$

Thus $\lim_{x\to\infty} (e^x + e^{3x})^{1/x} = e^3$.

Example 5 (1^∞). Find $\lim_{x\to 0} (1 + x)^{1/x}$.

Solution:
$$\begin{aligned}\lim_{x\to 0} \ln (1 + x)^{1/x} &= \lim_{x\to 0} \frac{\ln (1 + x)}{x} \qquad \left(\frac{0}{0}\right) \\ &= \lim_{x\to 0} \frac{1/(1 + x)}{1} \\ &= \lim_{x\to 0} \frac{1}{1 + x} \\ &= 1\end{aligned}$$

Thus $\lim_{x\to 0} (1 + x)^{1/x} = e^1 = e$.

Exercises 10.3[A]

Find the limit if it exists.

1. $\lim_{x \to 0+} x^{2x}$
2. $\lim_{x \to 0+} x^2 \ln x$
3. $\lim_{x \to \frac{\pi}{2}^-} (\sec x - \tan x)$
4. $\lim_{x \to 0+} x \ln \sin x$
5. $\lim_{x \to \infty} x \ln \dfrac{x}{x + 2}$
6. $\lim_{x \to \infty} x^{1/x}$
7. $\lim_{x \to 0} (1 + x)^{r/x}$
8. $\lim_{x \to \infty} \left(1 + \dfrac{r}{x}\right)^x$
9. $\lim_{x \to \infty} (e^x + e^{-x})^{2/x}$
10. $\lim_{x \to 0} \left(\dfrac{1}{x^2} - \dfrac{1}{x \sin x}\right)$

Exercises 10.3[B]

Find the limit if it exists.

1. $\lim_{x \to 0+} (\sin x)^x$
2. $\lim_{x \to 0+} \sqrt{x} \ln x$
3. $\lim_{x \to 0+} (\csc x - \cot x)$
4. $\lim_{x \to \infty} x \sin \dfrac{1}{x}$
5. $\lim_{x \to 0+} \left(\dfrac{1}{x} - \dfrac{1}{e^x - 1}\right)$
6. $\lim_{x \to 0} (1 + kx)^{1/x}$
7. $\lim_{x \to 0+} \left(\ln \dfrac{1}{x}\right)^x$
8. $\lim_{x \to \infty} \dfrac{(\ln x)^n}{x}$, n a positive integer.
9. $\lim_{x \to 0+} (1 - e^{-x})^{-x}$
10. $\lim_{x \to 0+} \left(\dfrac{\sin x}{x}\right)^{1/x}$

10.4 Exponential growth

One dollar put at interest at an annual rate r, compounded annually, will amount to $A = (1 + r)^t$ dollars after t years. If the interest is compounded semiannually, the amount in t years will be $A = (1 + r/2)^{2t}$, and if x times a year, $A = (1 + r/x)^{xt}$. Many banks now follow the practice of continuous compounding, which they sometimes refer to as "instant interest." This implies that the amount after t years will be $\lim_{x \to \infty} A$ dollars. Does this practice offer great wealth overnight, or is some more moderate outcome to be expected? To find out, we must evaluate the limit.

$$\lim_{x \to \infty} A = \lim_{x \to \infty} (1 + r/x)^{xt} = \lim_{x \to \infty} [(1 + r/x)^x]^t = [\lim_{x \to \infty} (1 + r/x)^x]^t,$$

where the last step uses Theorem 5.3-6. The last limit is a straightforward application of L'Hôpital's Rule, as you may have discovered in doing Exercise 8 of Exercises 10.3 [A]. Using the result, we have

$$\lim_{x\to\infty} A = [e^r]^t = e^{rt}.$$

To see how "instant interest" works out in practice, consider the case of one dollar at 6% for 10 years. Compounded annually, it amounts to $(1.06)^{10} = 1.79$ dollars; compounded continuously, to $e^{0.6} = 1.82$ dollars. There is no sudden wealth to be found here.

It is instructive to approach the same problem from a different direction. Under the assumption of continuous compounding, the rate at which the amount A is increasing is at all times proportional to the amount A already present, or, if we measure the time t in years,

$$\frac{dA}{dt} = rA.$$

Separating the variables, we have

$$\frac{dA}{A} = r\,dt, \quad \ln|A| = rt + C.$$

If $A = 1$ when $t = 0$, this gives us $\ln 1 = 0 + C$, $C = 0$; hence $\ln|A| = rt$ and $A = e^{rt}$, where we have been able to drop the absolute-value notation because A is necessarily positive. The result is, of course, just the same as our previous one.

The exponential law of growth that we have found for money continuously compounded also applies, approximately, to the growth of many living organisms. Yeast cells, for example, with plenty of food available, will multiply at a uniform rate, and the rate of increase is therefore proportional to the number present. Human populations sometimes show similar characteristics, but epidemics, wars, famines, and changing economic conditions complicate the problem.

Example 1. A culture of yeast doubles itself every hour. What will be its percentage increase in 20 min.?

Solution: If the amount of yeast at time t is x, then $\frac{dx}{dt} = kx$, $\frac{dx}{x} = k\,dt$, and $\ln|x| = kt + C$. If $x = x_0$ when $t = 0$, $\ln x_0 = 0 + C$, $\ln x = kt + \ln x_0$, $\ln x - \ln x_0 = \ln\frac{x}{x_0} = kt$, $\frac{x}{x_0} = e^{kt}$, and $x = x_0e^{kt}$. (Since x is necessarily positive, $|x| = x$.) To evaluate the constant k, we use the fact that $x = 2x_0$ when $t = 1$: $2x_0 = x_0e^k$, $2 = e^k$, $k = \ln 2$. Hence $x = x_0e^{t\ln 2} = x_0\,2^t$ (see equation (9), § 10.2). When $t = \frac{1}{3}$, $x = x_0\,2^{1/3} = 1.26x_0$, and the increase is about 26%.

Example 2. The rate of decay of a radioactive substance is proportional to the amount present. If half of a gram of radium decays in 1600 years, what part of a gram will be left after 2000 years?

Solution: If x grams of radium are present after t years, then $dx/dt = -kx$. This differential equation leads, as in Example 1, to the solution $x = x_0e^{-kt}$. We have $x_0 = 1$, and since $x = \frac{1}{2}$ when $t = 1600$, we have $\frac{1}{2} = e^{-1600k}$ and $-1600k = \ln \frac{1}{2} = -\ln 2$, $k = \dfrac{\ln 2}{1600}$, and $x = e^{-\frac{\ln 2}{1600}t} = 2^{-t/1600}$. When $t = 2000$, $x = 2^{-5/4} = 0.42$, approximately.

Exercises 10.4 [C]

1. If the population of a city increases at a rate which is proportional to the current population, and if the population of a certain city was 1,000,000 in 1970 and 1,200,000 in 1980, express in terms of t its population t years after 1970.
2. A hot object loses heat to its surroundings at a rate which is proportional to the temperature difference (Newton's Law of Cooling). If a bucket of water in a 70° room cools from 150° to 110° in 10 min., express in terms of t the temperature of the water t min. after being placed in the room.
3. The rotation of a flywheel is opposed by a frictional force which produces a negative acceleration proportional to the velocity. If a wheel slows from 1200 r.p.m. to 900 r.p.m. in 3 min., how fast will it be going after another 3 min.?
4. Work Ex. 3 on the assumption that the retardation is proportional to the *square* of the velocity.
5. A condenser discharges through a resistance at a rate which is proportional to the charge remaining. Express the charge Q at time t in terms of t, if $Q = Q_0$ when $t = 0$.
6. A current of i amperes flows through a coil of inductance L henries and resistance R ohms subject to the law $L\,(di/dt) + Ri = E$, where E is the electromotive force in volts. Assuming that E is constant and that $i = 0$ when $t = 0$, show that $i = \dfrac{E}{R}(1 - e^{-Rt/L})$. What is $\lim_{t\to\infty} i$?
7. A cylindrical water tank is 10 ft. high and has a 4-foot radius. It develops a leak at the bottom, and the water escapes at a rate proportional to the pressure at the bottom. If 19% of the water escapes in the first hour, how deep is the water t hours after the leak develops?
8. Work Ex. 7 on the assumption that the rate of escape is proportional to the square root of the pressure.

9. The size of a population may have an upper bound b imposed on it, perhaps by the available space or supply of food. In such a case, the rate of growth may be jointly proportional to the number already present, x, and to the "room to grow in," $b - x$; hence $dx/dt = kx(b - x)$. Show that this leads to a growth law $x = b/(1 + ae^{-bkt})$, and sketch its graph for $a = b = k = 1$.

10. A tank contains 50 gallons of a salt solution which has 40 lb. of dissolved salt. Water runs in at the rate of 1 gal./min., the concentration is kept uniform by stirring, and the mixture runs out at the same rate at which the water runs in. How much salt is left in the tank after t minutes?

Chapter Review

1. Differentiate:

(a) $\ln \dfrac{e^x - 1}{e^x + 1}$

(b) $\sqrt{\dfrac{x^2 - 5}{x(x + 3)}}$

(c) $\dfrac{1}{2} \ln \dfrac{1 + x}{1 - x}$

(d) $\frac{1}{5}e^{2x}(2 \sin x - \cos x)$

(e) x^{2x-3}

(f) $\ln \sin^{-1} x$. (*Hint:* Set $y = \ln \sin^{-1} x$, solve for x in terms of y, and then differentiate.)

2. Integrate:

(a) $\displaystyle\int \frac{x\,dx}{ax^2 + b}$

(b) $\displaystyle\int 7^x\,dx$

(c) $\displaystyle\int_0^4 \frac{x^3\,dx}{x^2 + 4}$

(d) $\displaystyle\int \frac{x\,dx}{(x^2 + a^2)^3}$

3. Find the area of the region in the first quadrant enclosed by $y = \dfrac{x}{x^2 + 1}$ and $x - 5y = 0$.

4. Find, in simplified form, the equation of the family of curves satisfying the differential equation

$$\frac{dy}{dx} = -\frac{y}{x},$$

and find the particular solution which goes through $(-2, 1)$.

5. Find the volume of the solid generated by rotating about the line $y = 1$ the region bounded by $xy = 2$ and $x + y - 3 = 0$.

6. Find the length of $y = \frac{1}{2}(e^x + e^{-x})$ from $x = 0$ to $x = \ln 2$.

7. Evaluate: (a) $\displaystyle\lim_{x \to 0^+} (\tan x)^x$ (b) $\displaystyle\lim_{x \to 0} \left(\frac{1}{\ln(1 + x)} - \frac{1}{x}\right)$

11

Some Techniques of Integration

11.1 Substitutions

Integration is an art. Differentiation proceeds by a set of rules which can be applied mechanically, almost without thought, but there are no universally applicable rules for integration. The problem is simply stated: Given a function f, find a function F which has the property (at least on some interval) that $F' = f$. If f is continuous, there exists such a function F, though we may have to invent it, as, in effect, we invent the function ln, in § 10.1, to be able to integrate $\int(1/x)dx$. The chance that the required function F is one we already know about is, however, small indeed; there is no familiar function F, for example, corresponding to $f(x) = \sqrt{1 + x^4}$ or $f(x) = e^{-x^2}$, so that, in this sense, we must regard these functions as being not integrable. There are, however, many functions which can be integrated in terms of known functions, and it is with this important minority that we are concerned here.

The technique of integration consists in manipulating integrands into forms that are recognizable as known derivatives. By far the most important of these manipulations are the changes of form produced by substitutions. The more common substitutions, most of them already familiar from § 7.4, are summarized here.

1. In algebraic integrands, substitutions like $u = \sqrt{ax + b}$ or $u = (ax^n + b)^{2/3}$ or $u = ax^2 + bx + c$ are often useful. Many problems involving these can be done by an implicit, mental substitution, without explicitly introducing another letter.

Example 1.

$$\int (x + 1)\sqrt{x^2 + 2x - 5}\, dx = \tfrac{1}{2}\int \sqrt{x^2 + 2x - 5}\, d(x^2 + 2x - 5)$$

$$= \tfrac{1}{3}(x^2 + 2x - 5)^{3/2} + C.$$

The substitution $u = x^2 + 2x - 5$ is implied here.

2. In integrands that contain a sum or a difference of squares, certain trigonometric substitutions may be useful. (We assume $a > 0$.)

If $\theta = \sin^{-1}(x/a)$, then $x = a \sin \theta$ and $a^2 - x^2 = a^2 \cos^2 \theta \quad (|x| \leq a)$.
If $\theta = \tan^{-1}(x/a)$, then $x = a \tan \theta$ and $x^2 + a^2 = a^2 \sec^2 \theta$.
If $\theta = \cos^{-1}(a/x)$, then $x = a \sec \theta$ and $x^2 - a^2 = a^2 \tan^2 \theta \quad (|x| \geq a)$.

It may be necessary to complete a square before introducing one of these substitutions. The student may find it desirable, at this point, to review the ranges of the inverse circular functions (§ 2.5).

Example 2. Evaluate: $I = \int \frac{dx}{\sqrt{5 + 4x - x^2}}$.

Solution: Since $5 + 4x - x^2 = 9 - (x - 2)^2$, we set $\theta = \sin^{-1}(x - 2)/3$; then $x - 2 = 3 \sin \theta$, $5 + 4x - x^2 = 9 - 9 \sin^2 \theta = 9 \cos^2 \theta$, and $dx = 3 \cos \theta \, d\theta$.

Hence $$I = \int \frac{3 \cos \theta \, d\theta}{3 \cos \theta} = \int d\theta = \theta + C = \sin^{-1} \frac{x - 2}{3} + C.$$

3. Substitutions such as $u = \sin x$, $u = \tan x$, $u = e^x + 1$, are often convenient and usually fairly obvious.

Example 3. $\int \tan^2 x \sec^2 x \, dx = \int \tan^2 x \, d(\tan x) = \frac{1}{3} \tan^3 x + C$. The substitution $u = \tan x$ is implied here.

Example 4. Evaluate: $I = \int \frac{e^{2x} \, dx}{\sqrt{e^x + 1}}$.

Solution: Set $u = \sqrt{e^x + 1}$; then $u^2 - 1 = e^x$, $2u \, du = e^x \, dx$, and

$$I = \int \frac{e^x \cdot e^x \, dx}{\sqrt{e^x + 1}} = \int \frac{(u^2 - 1) \cdot 2u \, du}{u} = 2 \int (u^2 - 1) du$$
$$= 2(\tfrac{1}{3}u^3 - u) + C = \tfrac{2}{3}u(u^2 - 3) + C = \tfrac{2}{3}(e^x - 2)\sqrt{e^x + 1} + C.$$

4. The basic identities which relate the circular functions are often useful (§§ 2.2, 2.3). In particular, the identities $\sin^2 \theta = \frac{1}{2}(1 - \cos 2\theta)$, $\cos^2 \theta = \frac{1}{2}(1 + \cos 2\theta)$ are standard devices for changing the square of a sine or a cosine to a first-degree expression.

Example 5. $$\int \sin^5 x \, dx = \int \sin^4 x \, (\sin x \, dx)$$
$$= -\int (1 - \cos^2 x)^2 \, d(\cos x)$$
$$= -\int (1 - 2 \cos^2 x + \cos^4 x) d(\cos x)$$
$$= -\cos x + \tfrac{2}{3} \cos^3 x - \tfrac{1}{5} \cos^5 x + C.$$

Example 6. $\int \tan^2 \theta \, d\theta = \int (\sec^2 \theta - 1) d\theta = \tan \theta - \theta + C.$

5. If $t = \tan \frac{1}{2}\theta$, then $\sin \theta = \dfrac{2t}{1 + t^2}$, $\cos \theta = \dfrac{1 - t^2}{1 + t^2}$, and $d\theta = \dfrac{2\,dt}{1 + t^2}$.
The demonstration of these relations is left to the student as an exercise. Their application to integration permits the transformation of any trigonometric integrand into an algebraic integrand, and this is sometimes advantageous.

Example 7.
$$\int \frac{dx}{1 + \cos x} = \int \frac{2\,dt/(1 + t^2)}{1 + (1 - t^2)/(1 + t^2)} = \int dt = t + C$$
$$= \tan \tfrac{1}{2}x + C.$$

This integral can be attacked in other ways:

$$\int \frac{dx}{1 + \cos x} = \int \frac{dx}{2 \cos^2 \frac{1}{2}x} = \int \sec^2 \tfrac{1}{2}x \, d(\tfrac{1}{2}x) = \tan \tfrac{1}{2}x + C,$$

or
$$\int \frac{dx}{1 + \cos x} = \int \frac{(1 - \cos x)dx}{1 - \cos^2 x} = \int \frac{(1 - \cos x)dx}{\sin^2 x}$$
$$= \int (\csc^2 x - \csc x \cot x)dx$$
$$= -\cot x + \csc x + C.$$

It is left to the student as an exercise to show that these two results are equivalent.

You will undoubtedly think of substitutions other than those in this brief list, and you should feel free to try them. Skill, ingenuity, and luck all play a part in integration. Often several substitutions must be tried to find one that works, and sometimes one that works must be followed by another and yet another before the job is done. You must be alert to simplify expressions whenever possible and to recognize a known derivative when you see it.

Exercises 11.1 [A]

1. If $\tan \frac{1}{2}\theta = t$ show that

(a) $\sin \theta = \dfrac{2t}{1 + t^2}$; (b) $\cos \theta = \dfrac{1 - t^2}{1 + t^2}$; (c) $d\theta = \dfrac{2\,dt}{1 + t^2}$.

Perform the following integrations:

2. $\int \frac{\sec \theta \tan \theta \, d\theta}{\sec \theta - 1}$

3. $\int \frac{(x + 3)dx}{x^2 + 6x + 5}$

4. $\int \frac{dx}{x^2 + 4x + 13}$

5. $\int \frac{dx}{1 + \sin x}$

6. $\int \frac{dx}{1 + \sqrt{x}}$

7. $\int \frac{dx}{x\sqrt{x^2 - a^2}} \quad (x > a)$

8. $\int \frac{x \, dx}{\sqrt{x^2 - a^2}} \quad (x > a)$

9. $\int \sqrt{4x - x^2} \, dx$

10. $\int \frac{\sin z \cos z \, dz}{\sqrt{\cos^2 z + 3}}$

11. $\int \frac{dy}{5 + 4 \cos y}$

12. $\int \frac{x \, dx}{x^2 + 2x + 5}$

13. $\int \frac{x^2 \, dx}{\sqrt{9 - x^2}}$

14. $\int \frac{dx}{\sqrt{21 - 4x - x^2}}$

15. $\int \frac{dx}{(x^2 + a^2)^2}$

Exercises 11.1 [B]

1. Show that the two answers to Example 7 are equivalent.

Perform the following integrations:

2. $\int \sin^2 x \cos^2 x \, dx$

3. $\int \frac{(2y - 3)dy}{4y^2 - 12y + 7}$

4. $\int \frac{dx}{x^2 + 2x + 10}$

5. $\int \frac{dx}{1 - \cos x}$

6. $\int \sqrt{e^{2x} - 1} \, dx$

7. $\int \frac{dx}{\sqrt{4 - 9x^2}}$

8. $\int \frac{dx}{x\sqrt{x^2 - 16}} \quad (x > 4)$

9. $\int \frac{dx}{\sqrt{-x^2 - x}}$

10. $\int \frac{\sqrt{x^2 - a^2}}{x} \, dx \quad (x > a)$

11. $\int \frac{dx}{3 - \cos x}$

12. $\int \frac{x \, dx}{x^2 - 6x + 18}$

13. $\int \frac{dx}{\sqrt{2 - 4x - 4x^2}}$

14. $\int \frac{\cos x \, dx}{\sin^2 x + 1}$

15. $\int \frac{dx}{x^2\sqrt{a^2 - x^2}}$

11.2 Partial fractions

The process by which a sum of two or more fractions is combined into a single fraction is one of the chief concerns of elementary algebra. The reverse process, in which a fraction with a factorable denominator is split into the sum of several fractions (called *partial fractions*) whose denominators are the factors of the original denominator, is a useful tool in integration.

Example 1. To integrate $\int \frac{(x-4)dx}{x(x-1)(x+2)}$, we first split the integrand into partial fractions. To do this we assume that there exist constants A, B, and C such that

$$\frac{x-4}{x(x-1)(x+2)} = \frac{A}{x} + \frac{B}{x-1} + \frac{C}{x+2}.$$

If we clear this equation of fractions we obtain

$$x - 4 = A(x-1)(x+2) + Bx(x+2) + Cx(x-1).$$

We wish to find A, B, C such that this equation will be true for *all* x. In particular, then, it must be true for $x = 0$, and substituting 0 for x we obtain $-4 = -2A$, from which we obtain $A = 2$. Similarly, substituting $x = 1$ yields $B = -1$, and $x = -2$ yields $C = -1$. Thus

$$\begin{aligned} \int \frac{(x-4)dx}{x(x-1)(x+2)} &= \int \left(\frac{2}{x} - \frac{1}{x-1} - \frac{1}{x+2}\right)dx \\ &= 2 \ln |x| - \ln |x-1| - \ln |x+2| + C \\ &= \ln \frac{x^2}{|(x-1)(x+2)|} + C. \end{aligned}$$

The method illustrated in Example 1 can be generalized to any rational fraction $p(x)/q(x)$, where $p(x)$ and $q(x)$ are polynomials with real coefficients and the degree of $p(x)$ is less than the degree of $q(x)$. It works as follows.

We begin by expressing $q(x)$ as a product of factors (with real coefficients) of degree not more than 2, that is, factors of the form $ax + b$ or $ax^2 + bx + c$. (It is a theorem of algebra that such a factorization always exists.) Each factor may appear to a power higher than the first. If $ax + b$ appears to the power n but to no higher power, then this factor contributes the sum

$$\frac{A_1}{ax+b} + \frac{A_2}{(ax+b)^2} + \cdots + \frac{A_n}{(ax+b)^n}$$

to the partial-fraction decomposition of $p(x)/q(x)$. (Of course, some of the A_i may turn out to be 0 in particular cases.) If the factor $ax^2 + bx + c$ appears to the power m, but to no higher power, then its contribution is

$$\frac{A_1x + B_1}{ax^2 + bx + c} + \frac{A_2x + B_2}{(ax^2 + bx + c)^2} + \cdots + \frac{A_mx + B_m}{(ax^2 + bx + c)^m}.$$

Having found the form of this decomposition, we then determine the unknown coefficients A_i and B_i by the method illustrated in Example 1 or by the method introduced in our next example.

Example 2. When we integrate $\int \frac{(2x^3 - 4x)dx}{(x^2 + 1)(x - 1)^2}$, our partial-fraction decomposition takes the form

$$\frac{2x^3 - 4x}{(x^2 + 1)(x - 1)^2} = \frac{Ax + B}{x^2 + 1} + \frac{C}{x - 1} + \frac{D}{(x - 1)^2}.$$

Clearing this equation of fractions, we have

$$\begin{aligned} 2x^3 - 4x &= (Ax + B)(x - 1)^2 + C(x^2 + 1)(x - 1) + D(x^2 + 1) \\ &= (A + C)x^3 + (-2A + B - C + D)x^2 + \\ &\quad (A - 2B + C)x + (B - C + D). \end{aligned}$$

Since this equation must be satisfied by every real x, it follows that the coefficients of like powers of x on the two sides of the equation must be the same:

$$\begin{aligned} 2 &= A + C, \\ 0 &= -2A + B - C + D, \\ -4 &= A - 2B + C, \\ 0 &= B - C + D. \end{aligned}$$

Solving these equations, we find $A = 0$, $B = 3$, $C = 2$, and $D = -1$. Hence

$$\begin{aligned} \int \frac{(2x^3 - 4x)dx}{(x^2 + 1)(x - 1)^2} &= \int \left(\frac{3}{x^2 + 1} + \frac{2}{x - 1} - \frac{1}{(x - 1)^2}\right)dx \\ &= 3 \tan^{-1} x + 2 \ln |x - 1| + \frac{1}{x - 1} + C. \end{aligned}$$

Example 3. Integrate $I = \int \frac{2\,dx}{x(x^2 + 2x + 2)}$.

In this case the partial-fraction decomposition of the integrand is

$$\frac{2}{x(x^2 + 2x + 2)} = \frac{A}{x} + \frac{Bx + C}{x^2 + 2x + 2};$$

$$\begin{aligned} 2 &= A(x^2 + 2x + 2) + (Bx + C)x \\ &= (A + B)x^2 + (2A + C)x + 2A. \end{aligned}$$

Thus $\qquad A + B = 0,\ 2A + C = 0,\ 2A = 2,$

and it follows that $A = 1$, $B = -1$, $C = -2$. Hence

$$I = \int\left(\frac{1}{x} - \frac{x+2}{x^2+2x+2}\right)dx = \ln|x| - \int\frac{(x+2)dx}{x^2+2x+2}.$$

To evaluate this last integral, we observe that

$$\frac{d}{dx}(x^2+2x+2) = 2x+2 = 2(x+1).$$

Thus
$$\begin{aligned}\int\frac{(x+2)dx}{x^2+2x+2} &= \int\frac{(x+1)dx}{x^2+2x+2} + \int\frac{dx}{x^2+2x+2}\\ &= \frac{1}{2}\int\frac{d(x^2+2x+2)}{x^2+2x+2} + \int\frac{dx}{(x+1)^2+1}\\ &= \tfrac{1}{2}\ln|x^2+2x+2| + \tan^{-1}(x+1) + C.\end{aligned}$$

Combining these results, we have

$$I = \ln|x| - \tfrac{1}{2}\ln(x^2+2x+2) - \tan^{-1}(x+1) + C.$$

Query: Why is it unnecessary to write $\ln|x^2+2x+2|$?

Example 4. $\displaystyle\int\frac{x^3+6x^2-4}{x^3+2x^2}\,dx.$

We first observe that the integrand is an improper fraction because the degree of the numerator is not less than the degree of the denominator. We therefore transform it into a mixed expression, that is, a polynomial plus a proper fraction. This can be done by long division or, more easily, as follows:

$$\frac{x^3+6x^2-4}{x^3+2x^2} = \frac{(x^3+2x^2)+(4x^2-4)}{x^3+2x^2} = 1 + \frac{4x^2-4}{x^3+2x^2}.$$

We now decompose the remaining fraction into partial fractions:

$$\frac{4x^2-4}{x^2(x+2)} = \frac{A}{x} + \frac{B}{x^2} + \frac{C}{x+2},$$
$$4x^2-4 = Ax(x+2) + B(x+2) + Cx^2.$$

Setting $x = 0$, we obtain $-4 = 2B$, $B = -2$; setting $x = -2$, we obtain $12 = 4C$, $C = 3$. Finally, since the coefficient of x^2 is 4 on the left and $A + C$ on the right, we have $A + C = 4$, $A = 4 - C = 1$. It now follows that

$$\begin{aligned}\int\frac{x^3+6x^2-4}{x^3+2x^2}\,dx &= \int\left(1+\frac{1}{x}-\frac{2}{x^2}+\frac{3}{x+2}\right)dx\\ &= x + \ln|x| + \frac{2}{x} + 3\ln|x+2| + C.\end{aligned}$$

In theory, the method of partial fractions completely solves the problem of integrating

$$\int \frac{p(x)}{q(x)}\,dx$$

for arbitrary polynomials $p(x)$ and $q(x)$. The first step, in case the integrand is an improper fraction, is to reduce it to a mixed expression, as in Example 4. The next step is to split the remaining fraction into partial fractions, a process that will lead to a set of integrals that can be handled by familiar techniques. We emphasize, however, that our solution of this problem is complete in theory only, because in practice it is not always possible to discover the factors of the denominator $q(x)$.

Example 5. Evaluate $I = \int \sec\theta\,d\theta$.

Solution: We first set $t = \tan\dfrac{\theta}{2}$ (see suggestion 5, § 11.1). Then $\sec\theta = \dfrac{1}{\cos\theta} = \dfrac{1+t^2}{1-t^2}$, and $d\theta = \dfrac{2\,dt}{1+t^2}$. Therefore $I = \displaystyle\int \frac{1+t^2}{1-t^2}\cdot\frac{2\,dt}{1+t^2} = \int \frac{2\,dt}{1-t^2}$.

Now we set $\dfrac{2}{1-t^2} = \dfrac{A}{1+t} + \dfrac{B}{1-t}$; then $2 = A(1-t) + B(1+t)$. If $t = -1$, $2 = 2A$ and $A = 1$; if $t = 1$, $2 = 2B$ and $B = 1$. Hence

$$\begin{aligned} I &= \int\left(\frac{1}{1+t} + \frac{1}{1-t}\right)dt = \ln|1+t| - \ln|1-t| + C = \ln\left|\frac{1+t}{1-t}\right| + C \\ &= \ln\left|\frac{1+\tan\theta/2}{1-\tan\theta/2}\right| + C = \ln\left|\frac{\tan\pi/4 + \tan\theta/2}{1-\tan\pi/4\tan\theta/2}\right| + C \\ &= \ln\left|\tan\left(\frac{\pi}{4}+\frac{\theta}{2}\right)\right| + C. \end{aligned}$$

This important integral, of common occurrence in applications of calculus, can also be evaluated as follows:

$$I = \int \sec\theta\,d\theta = \int \frac{d\theta}{\cos\theta} = \int \frac{\cos\theta\,d\theta}{\cos^2\theta} = \int \frac{d(\sin\theta)}{1-\sin^2\theta} = \int \frac{du}{1-u^2},$$

where $u = \sin\theta$. But we have just shown that $\dfrac{2}{1-u^2} = \dfrac{1}{1+u} + \dfrac{1}{1-u}$; hence

$$\begin{aligned} I &= \frac{1}{2}\int\left(\frac{1}{1+u} + \frac{1}{1-u}\right)du = \frac{1}{2}\ln\left|\frac{1+u}{1-u}\right| + C = \frac{1}{2}\ln\left|\frac{1+\sin\theta}{1-\sin\theta}\right| + C \\ &= \ln|\sec\theta + \tan\theta| + C. \end{aligned}$$

The last transformation can be verified by the student.

Miscellaneous Exercises 11.2

1. Show that $\frac{1}{2} \ln \left|\frac{1 + \sin\theta}{1 - \sin\theta}\right| = \ln|\sec\theta + \tan\theta| = \ln\left|\tan\left(\frac{\pi}{4} + \frac{\theta}{2}\right)\right|.$

2. $\int \frac{dx}{x^2 + x}$

3. $\int \frac{dx}{x^3 + x}$

4. $\int \frac{dx}{x^2 - 2x - 3}$

5. $\int \frac{x\,dx}{x^2 - 2x - 3}$

6. $\int \frac{x^2\,dx}{x^2 - 2x - 3}$

7. $\int \frac{(x - 16)dx}{x^2 - 2x - 8}$

8. $\int \frac{dx}{x^2 - a^2}$

9. $\int \frac{dx}{x^2 + a^2}$

10. $\int \frac{(x + 1)dx}{x^2 + 2x - 8}$

11. $\int \frac{dx}{x^3 + 2x}$

12. $\int \frac{x^3 + x^2 - 8x - 3}{x^2 - 2x - 3}\,dx$

13. $\int \frac{x^3\,dx}{(x - 1)(x + 1)(x + 2)}$

14. $\int \frac{dx}{(x + 1)(x^2 + 2)}$

15. $\int \frac{dx}{x^4 - 13x^2 + 36}$

16. $\int \frac{dx}{x^4 - 1}$

17. $\int \frac{dx}{x^3 + 2x^2 + 5x}$

18. $\int \frac{(x^2 - 2x - 1)dx}{(x + 1)^2(x^2 + 1)}$

19. $\int \frac{(4x + 5)dx}{x^4 + 4x^3 + 5x^2}$

20. $\int \csc x\,dx$

21. $\int \frac{dx}{3 - 5\sin x}$

22. $\int \frac{\tan x\,dx}{1 + \cos x}$

23. $\int \frac{dx}{\sqrt{x^2 - a^2}}\ (x > a)$

24. $\int \frac{dx}{x\sqrt{x^2 + a^2}}$

25. $\int \frac{dx}{x\sqrt{x^4 - 1}}$

11.3 Integration by parts

If we integrate, term by term, the formula for the differential of the product of two functions,

$$d(uv) = u\,dv + v\,du,$$

we get a result which is sometimes very useful:

$$\int d(uv) = \int u\,dv + \int v\,du, \quad uv = \int u\,dv + \int v\,du,$$

or

$$\int u\,dv = uv - \int v\,du. \tag{1}$$

(We have omitted a constant of integration C; this can be supplied at a later stage without loss of generality.) If an integrand can be split into two parts u and dv such that (i) dv is the differential of a known function v, and (ii) $\int v\,du$ can be evaluated, then, clearly, (1) gives us a method of evaluating $\int u\,dv$.

Example 1. Evaluate $\int \ln x\,dx$.

Solution: We set $u = \ln x$, $dv = dx$; then $du = \frac{1}{x}\,dx$, $v = x$. (Again, we postpone introducing the constant of integration.) If we now substitute these expressions in (1), we have

$$\begin{aligned}\int \ln x\,dx &= x \ln x - \int x\left(\frac{1}{x}\right) dx = x \ln x - \int dx = x \ln x - x + C \\ &= x(\ln x - 1) + C.\end{aligned}$$

Example 2. Evaluate $\int x \sin x\,dx$.

Solution: We set $u = x$, $dv = \sin x\,dx$; then $du = dx$, $v = -\cos x$, and

$$\int x \sin x\,dx = -x \cos x - \int (-\cos x)dx = -x \cos x + \sin x + C.$$

Example 3. Evaluate $I = \int e^x \sin x\,dx$.

Solution: We set $u = e^x$, $dv = \sin x\,dx$; then $du = e^x\,dx$, $v = -\cos x$, and $I = -e^x \cos x + \int e^x \cos x\,dx$. The problem is not yet solved, and it may indeed seem as though no progress has been made. But see what happens when we repeat the process: if we apply this method to the remaining integral, $\int e^x \cos x\,dx$, we have $U = e^x$, $dU = e^x\,dx$; $dV = \cos x\,dx$, $V = \sin x$;

$$\int e^x \cos x\,dx = e^x \sin x - \int e^x \sin x\,dx,$$

and therefore

$$I = -e^x \cos x + e^x \sin x - \int e^x \sin x\,dx = -e^x \cos x + e^x \sin x - I,$$

$$2I = e^x(\sin x - \cos x),$$
$$I = \tfrac{1}{2}e^x(\sin x - \cos x).$$

Thus $\int e^x \sin x \, dx = \frac{1}{2}e^x(\sin x - \cos x) + C.$

Example 4. Evaluate $I = \int \sec^3 \theta \, d\theta.$

Solution: We set $u = \sec \theta$, $dv = \sec^2 \theta \, d\theta$; then $du = \sec \theta \tan \theta \, d\theta$, $v = \tan \theta$, and

$$I = \sec \theta \tan \theta - \int \sec \theta \tan^2 \theta \, d\theta = \sec \theta \tan \theta - \int \sec \theta \, (\sec^2 \theta - 1) d\theta$$

$$= \sec \theta \tan \theta - \int \sec^3 \theta \, d\theta + \int \sec \theta \, d\theta.$$

Therefore $$2I = \sec \theta \tan \theta + \int \sec \theta \, d\theta$$

and $$I = \tfrac{1}{2} \sec \theta \tan \theta + \tfrac{1}{2} \int \sec \theta \, d\theta,$$

where the value of $\int \sec \theta \, d\theta$ can be supplied from § 11.2, Example 4. This integral occurs in many problems, and the result is therefore an important one.

The technique exhibited in these examples is called *integration by parts.* It is likely to seem, on first acquaintance, like a way out of all difficulties, but it isn't, and the student should not be too quick to try it. It is useful in problems like Example 1, in which the function $\ln x$ being integrated has the property of being simplified by differentiation. It is useful in problems like Example 2 and Example 3, which involve the product of two different varieties of function: algebraic and circular in Example 2, exponential and circular in Example 3. Clearly, the success of this method depends in large part on the selection of the elements to be called u and dv. Thus, if in Example 2 we take $u = \sin x$, $dv = x \, dx$, we have $du = \cos x \, dx$, $v = \frac{1}{2}x^2$, $\int x \sin x \, dx = \frac{1}{2}x^2 \sin x - \frac{1}{2}\int x^2 \cos x \, dx$, and we are worse off than we were at the beginning.

Miscellaneous Exercises 11.3

Integrate:

1. $\int xe^x \, dx$
2. $\int x \ln x \, dx$
3. $\int x \cos x \, dx$
4. $\int e^\theta \cos \theta \, d\theta$

5. $\int z \tan^2 z\, dz$

6. $\int \ln (a^2 + x^2)dx$

7. $\int x \sec^2 x\, dx$

8. $\int x\sqrt{x+1}\, dx$

9. $\int e^{-x} \sin x\, dx$

10. $\int x^2 \sin x\, dx$

11. $\int y^2 \cos y\, dy$

12. $\int x^3 \sin x\, dx$

13. $\int \sin (\ln x)dx$

14. $\int \csc^3 x\, dx$

15. (a) Use integration by parts to obtain the *reduction formula*

$$\int \sin^n x\, dx = -\frac{1}{n} \sin^{n-1} x \cos x + \frac{n-1}{n} \int \sin^{n-2} x\, dx,$$

where n is an integer greater than 1. (*Hint*: Take $u = \sin^{n-1} x$, $dv = \sin x\, dx$, and use $\cos^2 x = 1 - \sin^2 x$.)

(b) Use the reduction formula found in (a) to integrate $\int \sin^5 x\, dx$.

16. Obtain the following reduction formulas:

(a) $\int \cos^n x\, dx = \frac{1}{n} \cos^{n-1} x \sin x + \frac{n-1}{n} \int \cos^{n-2} x\, dx.$

(b) $\int x^n e^x\, dx = x^n e^x - n \int x^{n-1} e^x\, dx.$

(c) $\int x^n \sin x\, dx = -x^n \cos x + n \int x^{n-1} \cos x\, dx.$

(d) $\int \tan^n x\, dx = \frac{1}{n-1} \tan^{n-1} x - \int \tan^{n-2} x\, dx.$

(*Hint*: Take $u = \tan^{n-2} x$, $dv = \tan^2 x\, dx$.)

(e) $\int x^n\sqrt{x+1}\, dx = \frac{2}{2n+3} x^n (x+1)^{3/2} - \frac{2n}{2n+3} \int x^{n-1}\sqrt{x+1}\, dx.$

(*Hint*: $(x+1)^{3/2} = (x+1)\sqrt{x+1}$.)

11.4 Improper integrals*

The definite integral, $\int_a^b f(x)dx$, was defined in § 4.1 for a function f which is continuous on the bounded interval $[a, b]$. In this section, we shall generalize the concept of definite integral by showing one way in which the

* None of the material in this section will be required in subsequent sections.

requirement of continuity can be relaxed, and by admitting unbounded intervals of integration. The mathematical objects which we thus obtain are known as improper integrals.

The mark

$$\int_0^1 \frac{dx}{\sqrt{1-x^2}} \tag{1}$$

looks very much like a definite integral, but in fact it does not represent anything we have met so far, for if $f(x) = 1/\sqrt{1-x^2}$, then f is not defined at 1 and therefore fails to meet the condition of continuity on the interval of integration, $[0, 1]$. Since, however, the substitution $\theta = \sin^{-1} x$ gives us

$$\int \frac{dx}{\sqrt{1-x^2}} = \sin^{-1} x + C,$$

it is plausible to argue that the value of (1) *ought* to be

$$\int_0^1 \frac{dx}{\sqrt{1-x^2}} = \sin^{-1} x \Big|_0^1 = \sin^{-1} 1 - \sin^{-1} 0 = \frac{\pi}{2}. \tag{2}$$

What we must do here is define (1) in some way that will be consistent with our previous definition of definite integral, and therefore consistent with our conjecture (2).

The process is rather simple. For any $t \in [0, 1)$, we have

$$\int_0^t \frac{dx}{\sqrt{1-x^2}} = \sin^{-1} t,$$

and since

$$\lim_{t\to1^-} \int_0^t \frac{dx}{\sqrt{1-x^2}} = \lim_{t\to1^-} \sin^{-1} t = \sin^{-1} 1 = \frac{\pi}{2},$$

we will get just what we want if we define (1) to be a symbol designating this limit.

▶ **Definition.** If f is a function continuous on the interval $[a, b)$ but discontinuous at b, and if

$$\lim_{t\to b^-} \int_a^t f(x)dx$$

exists, then this limit is called the improper integral of f from a to b and is designated

$$\int_a^b f(x)dx.$$

In cases where this improper integral exists (the word "improper" is often omitted), we say that the integral converges, or is convergent; the fact that it does not exist is sometimes expressed by saying that the integral diverges, or is divergent. It is left to the student to formulate a definition of $\int_a^b f(x)dx$ in case f is discontinuous at a.

Example 1. $\int_1^2 \frac{dx}{(x-2)^2}$ diverges, because

$$\int_1^t \frac{dx}{(x-2)^2} = \frac{-1}{x-2}\Big|_1^t = \frac{-1}{t-2} + \frac{1}{1-2} = \frac{1-t}{t-2},$$

and this has no limit as t approaches 2 from the left.

Example 2. $\int_0^1 x^{-1/2}\,dx = 2$ because $\int_t^1 x^{-1/2}\,dx = 2x^{1/2}\Big|_t^1 = 2 - 2\sqrt{t}$, and this has the limit 2 as t approaches zero from the right.

A slight extension of the preceding argument enables us to deal with a discontinuity at an interior point of the interval of integration:

▶ **Definition.** If f is a function continuous on $[a, c)$ and on $(c, b]$ but discontinuous at c, then the improper integral of f from a to b is

$$\int_a^b f(x)dx = \int_a^c f(x)dx + \int_c^b f(x)dx,$$

provided both the latter improper integrals exist.

Example 3. $\int_{-1}^1 x^{-2}\,dx$ diverges because

$$\int_{-1}^0 x^{-2}\,dx = \lim_{t\to 0-} \int_{-1}^t x^{-2}\,dx = \lim_{t\to 0-} -x^{-1}\Big|_{-1}^t = \lim_{t\to 0-}\left(1 - \frac{1}{t}\right)$$

diverges, as does $\int_0^1 x^{-2}\,dx$.

Example 4. $\int_{-1}^1 x^{-2/3}\,dx = 6$ because

$$\int_{-1}^0 x^{-2/3}\,dx = \lim_{t\to 0-} \int_{-1}^t x^{-2/3}\,dx = \lim_{t\to 0-} 3x^{1/3}\Big|_{-1}^t = \lim_{t\to 0-} (3t^{1/3} + 3) = 3 \quad \text{and also}$$

$\int_0^1 x^{2/3}\,dx = 3.$

We have used the concept of limit to take care of discontinuities of the integrand; we can use the same concept to introduce unbounded intervals of integration:

Definition. If f is a function continuous on $[a, \infty)$, and if

$$\lim_{t \to \infty} \int_a^t f(x)dx$$

exists, then this limit is called the improper integral of f from a to ∞, and is designated

$$\int_a^{\infty} f(x)dx.$$

The student should have no difficulty in formulating definitions for

$$\int_{-\infty}^{a} f(x)dx \quad \text{and} \quad \int_{-\infty}^{\infty} f(x)dx.$$

Example 5.

$$\int_1^{\infty} x^{-2}\,dx = \lim_{t \to \infty} \int_1^t x^{-2}\,dx = \lim_{t \to \infty} -x^{-1}\Big|_1^t = \lim_{t \to \infty} \left(1 - \frac{1}{t}\right) = 1.$$

Example 6. $\displaystyle\int_1^{\infty} x^{-1/2}\,dx = \lim_{t \to \infty} 2x^{1/2}\Big|_1^t = \lim_{t \to \infty} (2\sqrt{t} - 2)$ diverges.

The student should sketch $y = x^{-2}$ and $y = x^{-1/2}$ and arrive at an intuitive explanation of the reason for convergence of the integral in Example 5 and divergence in Example 6.

Miscellaneous Exercises 11.4

Either evaluate each of the following integrals or show that it diverges:

1. $\displaystyle\int_0^1 x^{-1/3}\,dx$
2. $\displaystyle\int_0^1 x^{-3}\,dx$
3. $\displaystyle\int_1^{\infty} x^{-1/3}\,dx$
4. $\displaystyle\int_1^{\infty} x^{-3}\,dx$
5. $\displaystyle\int_0^1 \frac{dx}{\sqrt{1-x}}$
6. $\displaystyle\int_{-1}^1 x^{-1/5}\,dx$
7. $\displaystyle\int_{-1}^1 x^{-2/5}\,dx$
8. $\displaystyle\int_1^2 \frac{dx}{x-1}$
9. $\displaystyle\int_0^2 (x-1)^{-2/3}\,dx$
10. $\displaystyle\int_0^1 \frac{x\,dx}{\sqrt{1-x^2}}$
11. $\displaystyle\int_0^{\infty} e^{-x}\,dx$
12. $\displaystyle\int_0^{\pi/2} \tan x\,dx$
13. $\displaystyle\int_0^{\pi/2} \sec x\,dx$
14. $\displaystyle\int_{-\infty}^{\infty} \frac{dx}{x^2+1}$

15. $\int_0^\infty \sin x\, dx$

16. $\int_0^\infty \cos x\, dx$

17. $\int_0^a \frac{dx}{a^2 - x^2}$

18. $\int_0^a \frac{dx}{\sqrt{a^2 - x^2}}$

19. $\int_0^a \frac{x\, dx}{\sqrt{a^2 - x^2}}$

20. $\int_0^a \frac{x^2\, dx}{\sqrt{a^2 - x^2}}$

21. Determine the values of k for which the following converge:

(a) $\int_0^1 x^k\, dx$ (b) $\int_1^\infty x^k\, dx$

22. A function f is defined on $[0, 2]$ by the rule

$$f(x) = \begin{cases} 2, & x = 0 \\ x, & 0 < x < 1 \\ 0, & x = 1 \\ 1/x, & 1 < x < 2 \\ -1, & x = 2 \end{cases}$$

Sketch the graph of f and evaluate $\int_0^2 f(x)dx$.

23. Show how $\int_a^b f(x)dx$ can be defined if f is continuous at all but a finite set of points of $[a, b]$.

24. Find (**a**) the area of the region bounded by $x = 1$, $xy = 1$, and the x-axis; (**b**) the volume of the solid generated by rotating this region about the x-axis.

25. Find (**a**) the area of the region between the witch $y = \frac{8a^2}{x^2 + 4a^2}$ and the x-axis; (**b**) the volume of the solid generated by rotating this region about the x-axis.

Evaluate:

26. $\int_0^\infty \frac{dx}{1 + e^x}$

27. $\int_0^\infty e^{-x} \sin x\, dx$

28. $\int_{-\infty}^\infty \frac{dx}{\sqrt{x^2 + a^2}}$

29. $\int_a^\infty \frac{dx}{x(x^2 + a^2)}$ $(a > 0)$

30. $\int_a^\infty \frac{dx}{x\sqrt{x^2 - a^2}}$ $(a > 0)$

31. $\int_a^\infty \frac{dx}{x\sqrt{x^2 + a^2}}$ $(a > 0)$

Chapter Review

Perform the following integrations:

1. $\int \sin(x+2)dx$
2. $\int \sin^2 x \cos^3 x \, dx$
3. $\int \frac{x^3\, dx}{\sqrt{1+x^2}}$
4. $\int \cos^5 \theta \, d\theta$
5. $\int \cos 2x \cos x \, dx$
6. $\int \frac{dx}{x\sqrt{x^2-a^2}}$
7. $\int \frac{dx}{x(3+\ln x)}$
8. $\int x^3(3x^2-1)^{2/3}\, dx$
9. $\int \frac{x^3\, dx}{x^2-4}$
10. $\int (1-\sqrt{x})^{1/2}\, dx$
11. $\int \sqrt{e^x-1}\, dx$
12. $\int \frac{(2x-3)dx}{4x^3+8x^2-x-2}$
13. $\int \frac{\cos x\, dx}{\sin^2 x - \sin x - 6}$
14. $\int xe^{-x}\, dx$
15. $\int \frac{dx}{\sqrt{x^2+a^2}}$
16. $\int \cos\left(\frac{\pi}{2}+x\right)\cos\left(\frac{\pi}{2}-x\right)dx$
17. $\int \sin\left(\frac{3\pi}{2}+x\right)\cos(\pi-x)dx$
18. $\int \frac{dx}{3+\cos x}$
19. $\int \frac{(7x^2+4x)dx}{x^3+x^2-4x-4}$
20. $\int \sin(A+x)\cos(A-x)dx$

    (*Hint*: Change the product to a sum.)
21. $\int \frac{\sec^2\theta\, d\theta}{\sec^2\theta - \tan\theta - 3}$
22. $\int \sin 5z \cos 3z \, dz$
23. $\int \frac{x^2\, dx}{(x^3+a^2)^2}$
24. $\int \frac{\sqrt{4-x^2}\, dx}{x}$
25. $\int \sin(5x+3B)\sin(x-B)dx$
26. $\int \frac{dx}{4x^4-17x^2+4}$
27. $\int \frac{dx}{x\sqrt{a^2-x^2}}$
28. $\int \frac{dx}{e^x-1}$
29. $\int \sin 3\theta \sin\theta \, d\theta$
30. $\int \frac{dx}{2x^2+6x+5}$
31. $\int \frac{\sqrt{x^2-a^2}}{x}\, dx \quad (x \geq a)$

32. $\int x^2(x+1)^{-1/3}\,dx$

33. $\int \left(\sqrt{\frac{a-x}{a+x}} - \sqrt{\frac{a+x}{a-x}}\right) dx$

34. $\int \sqrt{x^2+1}\,dx$

35. $\int \sin \sqrt{x}\,dx$

36. $\int x \sin^2 x\,dx$

37. $\int x^2\sqrt{9-x^2}\,dx$

38. $\int \ln(1+x^2)dx$

39. $\int \frac{dx}{1+2\cos x}$

40. $\int \cos(\ln x)dx$

41. $\int x^n \ln x\,dx \quad (n \neq -1)$

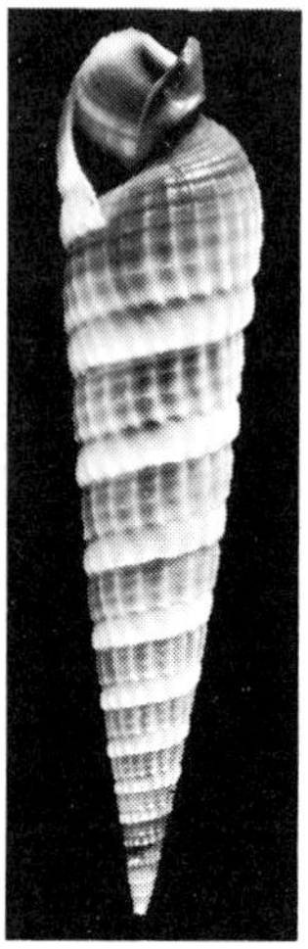

People, like nature, find the spiral a pleasing and useful form.

spiral staircase in Supreme Court Building

12

Polar Coordinates

12.1 Definition; graphs

The position in a plane of a point P can be specified by its Cartesian coordinates (x, y), which fix the point by giving its distance and direction from the coordinate axes OX and OY. There are, however, many other ways of specifying the position of P: we may, for example, give the distance $OP = r$ and the angle $XOP = \theta$, and these then fix the point by giving its distance from the origin and its direction from a certain fixed line through the origin. If we do this, the numbers (r, θ) are called polar coordinates of P; r is called the radius vector, and θ, the vectorial angle. If a locus bears a special relationship to a particular point, as the orbit of a planet bears a special relationship to the sun, for example, then an equation of the locus in polar coordinates may take a particularly simple and useful form.

As can be seen from Fig. 1, polar and Cartesian coordinates are connected by the equations

$$x = r\cos\theta, \quad y = r\sin\theta \tag{1}$$

and

$$r^2 = x^2 + y^2, \quad \tan\theta = \frac{y}{x}. \tag{2}$$

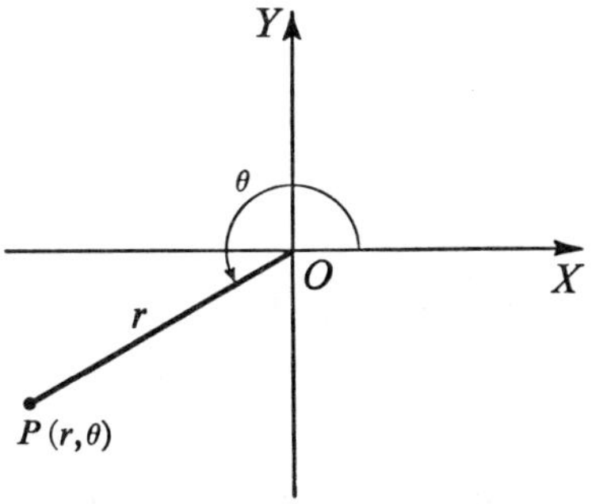

Fig. 1

The Cartesian coordinates of a point are unique, but this is not true of the polar coordinates; we see from Fig. 2 that the point (r_1, θ_1) is the same as the point $(r_1, \theta_1 \pm 2\pi)$, and, since we interpret negative values of r as being measured on the backward extension, through the origin, of the terminal side of the vectorial angle, $(-r_1, \theta_1 \pm \pi)$ are also coordinates of the same point (Fig. 3).

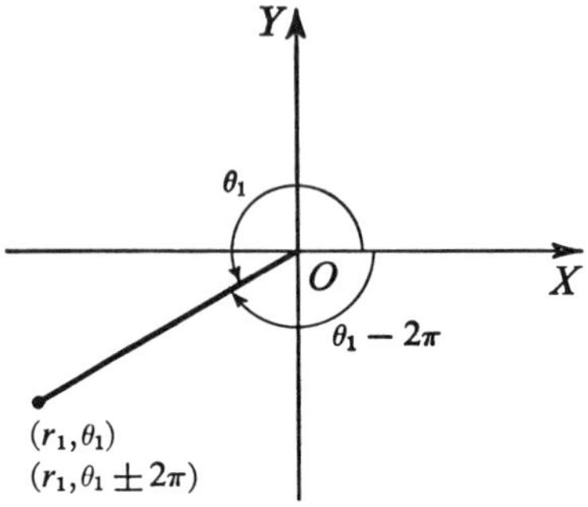

Fig. 2

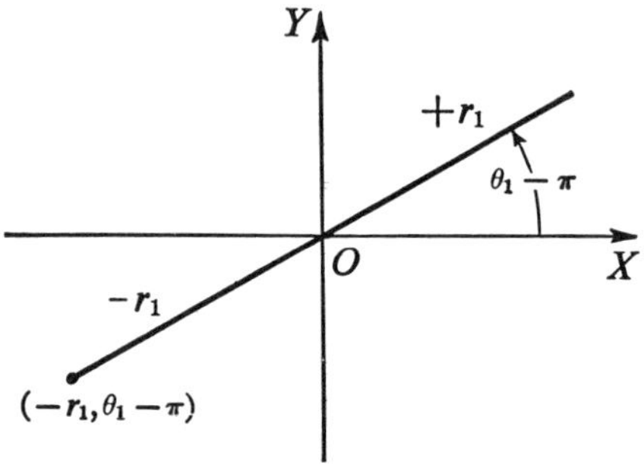

Fig. 3

The polar equations $r = r_1$ and $\theta = \theta_1$ describe, respectively, a circle with center at the origin and radius r_1, and a line through the origin making an angle θ_1 with the x-axis. Polar coordinate graph paper has drawn on it such circles and lines, spaced at regular intervals, just as Cartesian coordinate graph paper is ruled with the lines $x = x_1$ and $y = y_1$, at regular intervals.

Many polar graphs are symmetrical. It is easy to verify that a curve is symmetric in the x-axis if the substitution of $-\theta$ for θ leaves its equation essentially unchanged, and is symmetric in the y-axis if the substitution of $\pi - \theta$ for θ makes no difference. It is worth noting, in this connection, that $\cos(-\theta) = \cos\theta$ and $\sin(\pi - \theta) = \sin\theta$.

Example 1. **(a)** Draw the graph of $r = 10\cos\theta$; **(b)** obtain a Cartesian equation.

Solution:

(a) Making a table of values, we find

$\theta =$ 0	$\pi/6$	$\pi/3$	$\pi/2$	$2\pi/3$	$5\pi/6$	π
$r =$ 10	8.7	5	0	-5	-8.7	-10

The curve is symmetric in the x-axis, so there is no need to compute values of r for $\pi < \theta < 2\pi$. The graph is the circle indicated by the heavy line in Fig. 4.

(b) From (1) and (2), we have

$$r = \sqrt{x^2 + y^2}, \quad \cos\theta = \frac{x}{r};$$

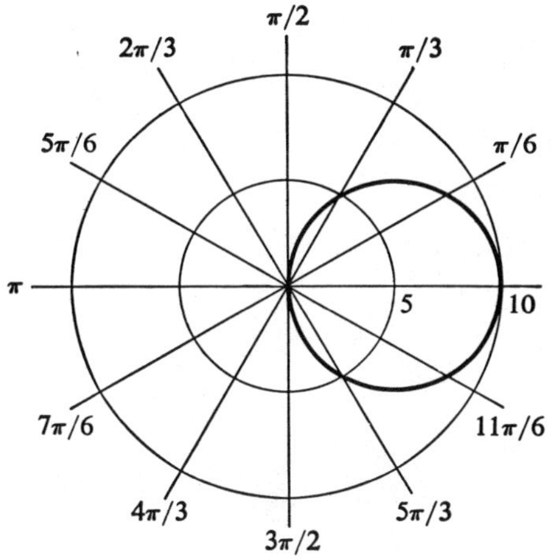

Fig. 4

hence the equation of the curve becomes

$$\sqrt{x^2 + y^2} = \frac{10x}{\sqrt{x^2 + y^2}},$$

$$x^2 + y^2 = 10x,$$

or
$$(x - 5)^2 + y^2 = 25,$$

a circle with center at (5, 0) and radius 5.

Example 2. Graph $r = 1 + \cos\theta$.

Solution: This curve also is symmetric in the x-axis, so we need compute values of r only for $0 \le \theta \le \pi$.

$\theta = 0$	$\pi/6$	$\pi/3$	$\pi/2$	$2\pi/3$	$5\pi/6$	π
$r = 2$	1.9	1.5	1	0.5	0.1	0

This curve, shown in Fig. 5, is called a cardioid.

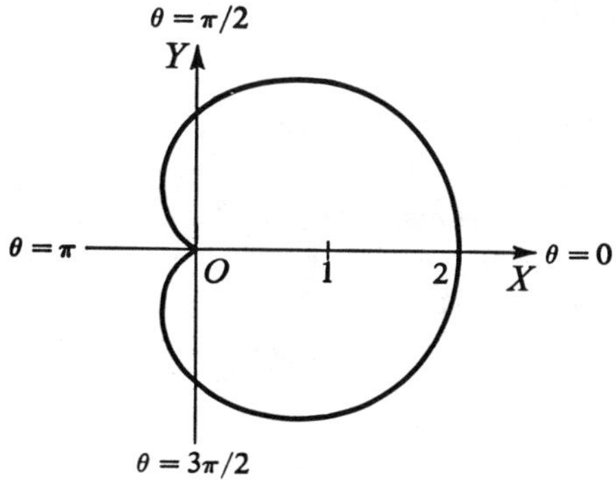

Fig. 5

Exercises 12.1 [A]

1. Give the Cartesian coordinates and two other sets of polar coordinates for

   (a) $(2, \pi/6)$. (b) $(-1, 3\pi/4)$. (c) $(10, -40°)$. (d) $(0, 1.2)$.

Plot the curves in Exs. 2–7.

2. $r = a \sec \theta$
3. $r = 2a \sin \theta$
4. $r = 1 + 2 \cos \theta$
5. $r = k\theta$
6. $r = a \sin 2\theta$
7. $r = \dfrac{10}{1 - \cos \theta}$
8. Write and simplify Cartesian equations for Exs. 2, 4, and 7.
9. Write and simplify polar equations for

   (a) $x^2 + y^2 - 6y = 0$. (b) $2xy = k^2$. (c) $(x^2 + y^2)^2 = a^2(x^2 - y^2)$.

10. Find all intersections of

    (a) $r^2 = a^2 \sin^2 \theta$ and $r = a$. (b) $r = 1 + \sin \theta$ and $r = 2 \cos \theta$.

Exercises 12.1 [B]

1. Give two sets of polar coordinates for the point with Cartesian coordinates

   (a) $(1, \sqrt{3})$. (b) $(-2, -2)$. (c) $(3, -4)$. (d) $(0, 5)$.

Plot the curves in Exs. 2–7.

2. $r = a \csc \theta$
3. $r\theta = k$
4. $r = \sec \theta - 2$
5. $r = a \cos 2\theta$
6. $r = a \sin \frac{1}{2}\theta$
7. $r = 2 - \cos \theta$
8. Write and simplify Cartesian equations for Exs. 2 and 5.
9. Write and simplify polar equations for

   (a) $y^2 = 2px + p^2$. (b) $(x^2 + y^2)^2 = 2a^2xy$. (c) $x^3 + y^3 = 3axy$.

10. Find all intersections of

    (a) $r^2 = 2a^2 \cos 2\theta$ and $r = a$.

    (b) $r = \sin 2\theta$ and $r = 1 - \cos 2\theta$.

Miscellaneous Exercises 12.1

Sketch the curves in Exs. 1–12.

1. $r^2 = a^2 \cos 2\theta$
2. $r^2 = a^2 \cos \theta$
3. $r = a \sin 3\theta$
4. $r = a \cos 3\theta$
5. $r = a \sec^2 (\frac{1}{2}\theta)$
6. $r = a \csc^2 (\frac{1}{2}\theta)$
7. $r(1 + 2 \cos \theta) = 4$
8. $r(2 - \cos \theta) = a$
9. $r = a(\sin \theta + \cos \theta)$
10. $r \cos (\theta + \alpha) = a$
11. $r = a(\sec \theta - \cos \theta)$
12. $r = a(\sec \theta + \tan \theta)$

13. Change to a Cartesian equation:

(a) Ex. 5 (b) Ex. 10 (c) $r^2 = 2a^2 \sec^2 \theta$

14. Show that, if $\theta + \pi$ is in the domain of f if and only if θ is in the domain of f, then $r = f(\theta)$ and $-r = f(\theta + \pi)$ represent the same curve, and use this to obtain a different equation for the graph of $r = a + b \cos \theta$.

15. (a) Find all intersections of $r = \sin \theta$ and $r = \cos \theta$.

(b) Sketch both curves.

(c) Explain why the usual techniques of solving simultaneous equations fail to discover the complete answer to (a).

16. Find all intersections of

(a) $r = a(1 - \cos \theta)$, $r = a(1 + \sin \theta)$.

(b) $r = a \tan \theta$, $r = -a \tan \theta$.

12.2 Loci

A locus can be described by a polar equation, just as by a Cartesian equation, and if the locus bears a special relation to a particular point or is defined in terms of a particular angle, the polar equation may be simpler than the Cartesian equation. Compare the equations $x^2 + y^2 = a^2$ and $r = a$ for the same locus.

Example 1. Find a polar equation of the straight line.

Solution: We have seen in § 12.1 that a line through the origin has the equation $\theta = \theta_1$. In any other case, let the normal from the origin meet the line at (p, α). Then we see from Fig. 6 that the point (r, θ) will be on the line if and only if

$$\frac{p}{r} = \cos(\theta - \alpha), \quad \text{or} \quad r = p \sec(\theta - \alpha).$$

This, then, is the required equation.

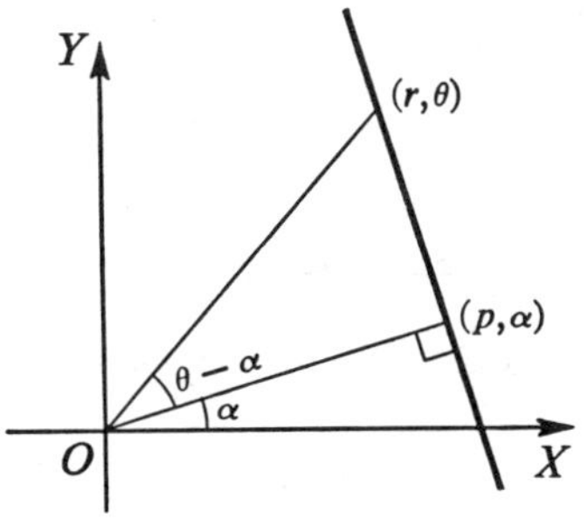

Fig. 6

Example 2. Find a polar equation of the circle with center at $A(p, \alpha)$ and radius a.

Solution: The point P will be on the circle if and only if $AP = a$ (see Fig. 7). But, by the law of cosines,

$$(AP)^2 = r^2 + p^2 - 2pr \cos(\theta - \alpha).$$

Hence the required equation is $r^2 - 2pr \cos(\theta - \alpha) + p^2 = a^2$.

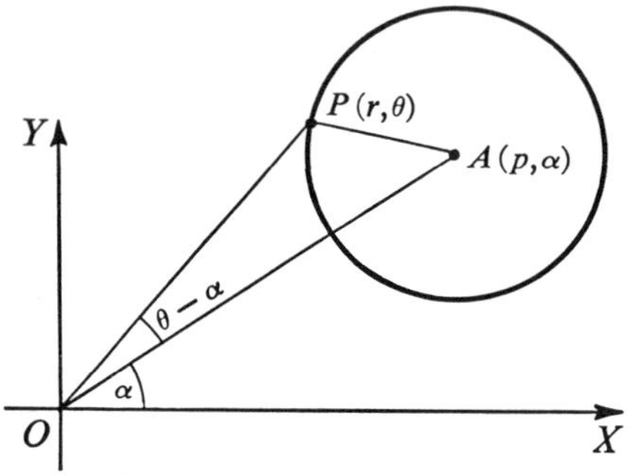

Fig. 7

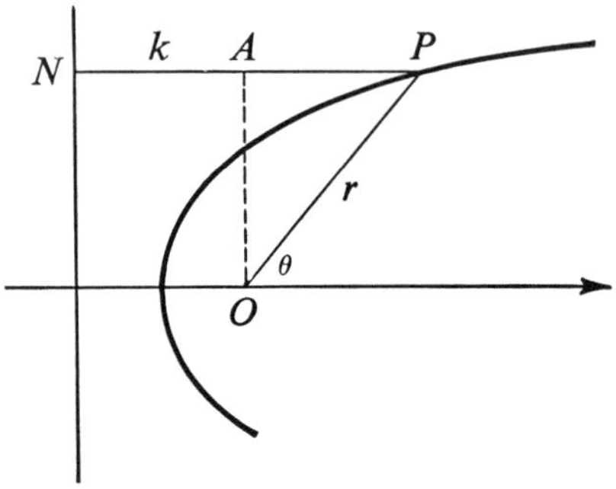

Fig. 8

Example 3. Find a polar equation of a conic with focus at the origin and eccentricity e. (See Example 2, § 8.8.)

Solution: Let us take the focus at the origin and the directrix k units to the left. Then, in Fig. 8, $AN = k$, $PA = r \cos \theta$, and the focus-directrix definition of the conic, $PO = e(PN)$, then becomes

$$r = e(k + r \cos \theta),$$

$$r(1 - e \cos \theta) = ek,$$

or

$$r = \frac{ek}{1 - e \cos \theta}.$$

Thus we have one simple equation for all the conics except the circle, and the circle $r = a$ can be obtained as a limit by letting $e \to 0$ while increasing k so as to hold the product $ek = a$ constant. The advantage of the polar equation in this case is apparent.

Exercises 12.2 [A]

1. Write a formula for the distance between $P_1(r_1, \theta_1)$ and $P_2(r_2, \theta_2)$, assuming $r_1 > 0$, $r_2 > 0$.
2. Find a polar equation of a line a units from the origin if it is perpendicular to (**a**) the positive x-axis; (**b**) the positive y-axis.

3. Find a polar equation of a circle of radius a if it passes through the origin and has its center on (**a**) the positive x-axis; (**b**) the positive y-axis; (**c**) the ray $\theta = \pi/4$, $r \geq 0$.
4. Find a polar equation of a parabola with focus at the origin and vertex at (**a**) $(3, \pi)$; (**b**) $(5, 0)$; (**c**) $(6, \pi/2)$.
5. Find an angle between the lines $r = a \sec(\theta - \alpha)$ and $r = b \sec(\theta - \beta)$.
6. Find the center of the ellipse $r = \dfrac{8}{2 - \cos\theta}$.
7. Find the vertices of the hyperbola $r = \dfrac{6}{1 - 2\cos\theta}$.
8. Show that if a focal chord of a parabola makes an angle $\pi/6$ with the axis of the curve, its length is 4 times that of the latus rectum.
9. A focal chord of the conic $r = \dfrac{ek}{1 - e\cos\theta}$ is divided by the origin into segments r_1 and r_2. Show that $\dfrac{1}{r_1} + \dfrac{1}{r_2} = \dfrac{2}{ek}$.
10. A line through the origin O intersects at Q the line $r = a \sec\theta$. A point P is taken on OQ such that
$$(OP)(OQ) = a^2.$$
Find a polar equation of the locus of P.

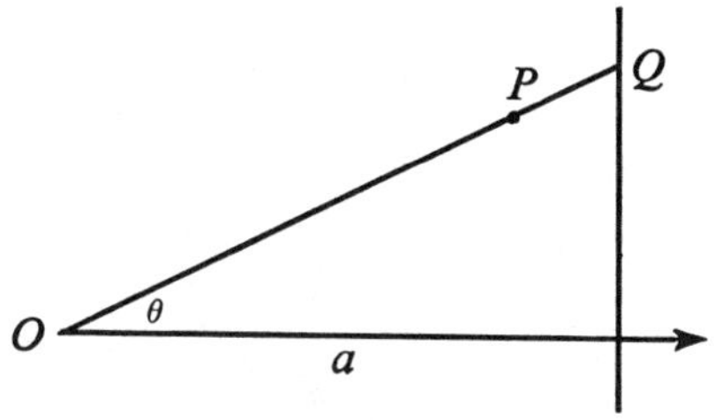

Fig. 9

Exercises 12.2 [C]

In the following problems, find a polar equation of the locus of P.

1. A line of length $2a$ slides with its ends on the x- and y-axes, and OP is perpendicular at P to this line.
2. OP is perpendicular at P to a line that is tangent to the circle $r = 2a\cos\theta$.

3. The cissoid of Diocles (Ex. 5, page 302): MN is tangent at $(2a, 0)$ to the circle $r = 2a \cos \theta$, a variable secant through O meets the circle at B and MN at C, and $OP = BC$.

4. The limaçon of Pascal: A line through O meets the circle $r = a \cos \theta$ at the point Q, and points P_1 and P_2 are taken on OQ such that $QP_1 = QP_2 = b$. Find the locus of P_1 and P_2. If $a = b$, this is the cardioid.

5. The lemniscate of Bernoulli: Given the points $F_1(a, 0)$ and $F_2(a, \pi)$, then $PF_1 \cdot PF_2 = a^2$.

6. The conchoid of Nicomedes: A line through O meets the line $r = a \sec \theta$ at Q, and points P_1 and P_2 are taken on OQ such that $QP_1 = QP_2 = b$. Find the locus of P_1 and P_2.

7. The strophoid: A is the point $(a, 0)$. A line through O meets the line $r = a \sec \theta$ at Q, and points P_1 and P_2 are taken on OQ such that $QP_1 = QP_2 = QA$. Find the locus of P_1 and P_2.

8. A line through the origin meets the circle $r^2 - 2pr \cos \theta + p^2 = a^2$ in Q_1 and Q_2, and P is the midpoint of Q_1Q_2.

9. The ovals of Cassini: Given F_1 and F_2 as in Ex. 5, then $PF_1 \cdot PF_2 = b^2$. Sketch for $b < a$; for $b > a$.

10. P lies on the line through (r_1, θ_1) and (r_2, θ_2).

12.3 Tangent line and length of arc

The direction of a curve at a given point is commonly described by giving the slope of the tangent line at that point. Some other measure of direction—the inclination, for example—would be fully as informative, but the slope is preferred because it is so easily determined from the Cartesian equation of the curve. In polar coordinates, the slope is not particularly easy to compute (it should be noted that $dr/d\theta$, the rate at which r changes with θ, is *not* the slope), and it is therefore customary to use another measure of direction, comparatively easy to determine, an angle between the tangent line and the radius vector to the point of tangency. This angle is commonly denoted ψ (the Greek letter psi), and, as we shall see, the quantity $\tan \psi$ is easily determined from the equation $r = f(\theta)$ of a curve.

Consider Fig. 10, which shows the vectorial angle θ of a point P on the curve $r = f(\theta)$, the angle ψ at P, and the angle $\alpha = \theta + \psi$; we have $\psi = \alpha - \theta$ and therefore

$$\tan \psi = \tan (\alpha - \theta) = \frac{\tan \alpha - \tan \theta}{1 + \tan \alpha \tan \theta}. \tag{1}$$

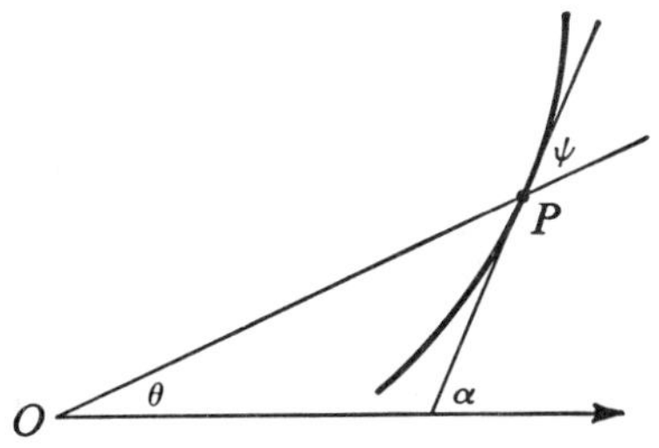

Fig. 10

Now, the Cartesian coordinates of any point on the curve can be determined by introducing its polar coordinates into equations (1) of § 12.1. In fact, since r is given as $f(\theta)$, the equations

$$x = r\cos\theta = f(\theta)\cos\theta, \quad y = r\sin\theta = f(\theta)\sin\theta \tag{2}$$

may be regarded as parametric equations, in Cartesian coordinates, of the given curve, in terms of a parameter θ. If we differentiate (2) with respect to θ, we find

$$dx = dr\cos\theta - r\sin\theta\,d\theta, \quad dy = dr\sin\theta + r\cos\theta\,d\theta, \tag{3}$$

and therefore

$$\tan\alpha = \frac{dy}{dx} = \frac{dr\sin\theta + r\cos\theta\,d\theta}{dr\cos\theta - r\sin\theta\,d\theta} = \frac{dr\tan\theta + r\,d\theta}{dr - r\tan\theta\,d\theta}. \tag{4}$$

Substituting this value of $\tan\alpha$ in (1), and simplifying, we find

$$\tan\psi = \frac{r\,d\theta}{dr}. \tag{5}$$

Example 1. Find the angle ψ for an arbitrary point (r, θ) of the cardioid $r = a(1 - \cos\theta)$.

Solution: We have $dr = a\sin\theta\,d\theta$, and therefore

$$\tan\psi = \frac{a(1 - \cos\theta)\cdot d\theta}{a\sin\theta\,d\theta} = \frac{1 - \cos\theta}{\sin\theta} = \tan\tfrac{1}{2}\theta.$$

Example 2. Find the angle ψ for an arbitrary point (r, θ) of the spiral $r = e^{k\theta}$.

Solution: Since $dr = ke^{k\theta}\,d\theta$, $\tan\psi = \dfrac{e^{k\theta}\,d\theta}{ke^{k\theta}\,d\theta} = \dfrac{1}{k}$. The spiral therefore crosses each radius vector at the same angle $\psi = \cot^{-1} k$. For this reason, it is called the equiangular spiral.

We have found (§ 9.5) that $(ds)^2 = (dx)^2 + (dy)^2$. If we substitute in this equation the values of dx and dy from (3), we find

$$(ds)^2 = (dr)^2 + r^2(d\theta)^2. \tag{6}$$

The details are left to the student as an exercise.

Example 3. Find the length of the cardioid $r = 1 + \cos\theta$.

Solution: Since $dr = -\sin\theta\, d\theta$, we have

$$\begin{aligned}(ds)^2 &= \sin^2\theta\,(d\theta)^2 + (1 + \cos\theta)^2(d\theta)^2 \\ &= (\sin^2\theta + 1 + 2\cos\theta + \cos^2\theta)(d\theta)^2 \\ &= 2(1 + \cos\theta)(d\theta)^2 = 4\cos^2\tfrac{1}{2}\theta(d\theta)^2.\end{aligned}$$

Hence, exploiting the symmetry of the curve, we find

$$s = 2\int_0^\pi 2\cos\tfrac{1}{2}\theta\, d\theta = 8[\sin\tfrac{1}{2}\theta]_0^\pi = 8.$$

The following device may help the student remember the principal results of this section, formulas (5) and (6). Figure 11 shows the points $P_1(r, \theta)$ and

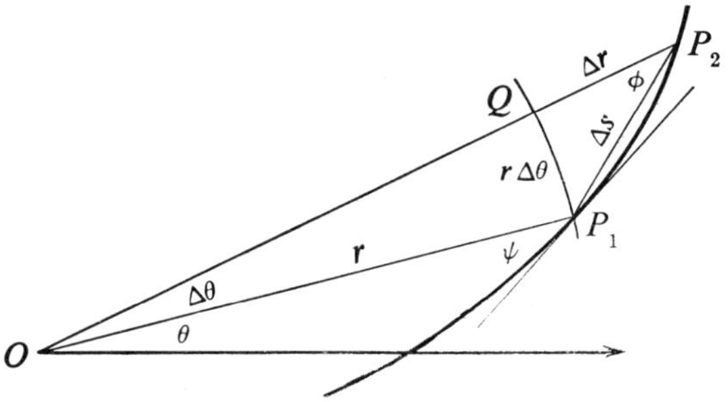

Fig. 11

$P_2(r + \Delta r, \theta + \Delta\theta)$ on the curve $r = f(\theta)$. With the origin O as center, we swing a circular arc of radius OP_1, cutting OP_2 at Q. This arc is orthogonal to the radius vector OP_2, and the figure P_1P_2Q is therefore virtually a right triangle. Arc $P_1Q = r\,\Delta\theta$, and $QP_2 = \Delta r$. Moreover, the angle $P_1P_2Q = \phi$ is almost equal to ψ and will approach ψ as $\Delta\theta \to 0$, and the chord P_1P_2 is approximately equal to the arc $P_1P_2 = \Delta s$. In the "right triangle" P_1P_2Q, we have, approximately, $\tan\phi = \dfrac{r\,\Delta\theta}{\Delta r}$ (compare this with (5)), and $(\Delta s)^2 = (\Delta r)^2 + r^2(\Delta\theta)^2$ (compare this with (6)).

Exercises 12.3 [A]

1. Carry through the derivation of (5) and (6).
2. Find the angle ψ for
   (a) the spiral $r = a\theta$, at $\theta = \pi/2$ and at $\theta = \pi$.
   (b) the rose $r = a \sin 3\theta$, at $\theta = \pi/4$.
   (c) the conic $r = \dfrac{ek}{1 - e \cos \theta}$, at $\theta = \theta_1$.
3. Show that an angle A between two curves is given by
$$\tan A = \frac{\tan \psi_1 - \tan \psi_2}{1 + \tan \psi_1 \tan \psi_2},$$
   and that the curves are orthogonal if $\tan \psi_1 \tan \psi_2 = -1$.
4. Show that the parabolas $r = a \sec^2 \frac{1}{2}\theta$ and $r = b \csc^2 \frac{1}{2}\theta$ are orthogonal.
5. At what angle does the limaçon $r = 2 + \cos \theta$ meet the circle $r = 2$? Sketch.
6. At what points is the tangent to the cardioid $r = a(1 - \cos \theta)$ parallel to (a) the x-axis? (b) the y-axis? (*Hint:* Consider $dx/d\theta$ and $dy/d\theta$.)
7. Find the length of $r = ae^{k\theta}$ from $\theta = 0$ to $\theta = 2\pi$.

Exercises 12.3 [B]

1. Find the angle ψ for
   (a) the lemniscate $r^2 = 2a^2 \cos 2\theta$, at $\theta = \pi/6$.
   (b) the circle $r = a$, at $\theta = 2\pi/3$.
   (c) the circle $r = a \cos \theta$, at $\theta = \pi/4$.
2. Find the angle of intersection of
$$r = 2a \sec\left(\theta - \frac{\pi}{2}\right) \quad \text{and} \quad r = a \sec\left(\theta - \frac{\pi}{6}\right).$$
3. Show that the cardioids $r = a(1 - \cos \theta)$ and $r = b(1 + \cos \theta)$ are orthogonal.
4. Find the angle between $r = 1 + 2 \cos \theta$ and $r = 1$.
5. Find the length of $r = a \sec \theta$ from $\theta = -\pi/4$ to $\theta = \pi/3$.
6. A point is tracing the cardioid $r = a(1 - \cos \theta)$ with constant angular velocity $d\theta/dt = k$. Find (a) dr/dt; (b) ds/dt.

Miscellaneous Exercises 12.3

1. Find the length of the spiral $r = a\theta^2$ between $\theta = 0$ and $\theta = 2\pi$.

2. Sketch the curve $r = a \sin^3 \dfrac{\theta}{3}$ and find its entire length.

3. A particle moves toward the origin along the spiral $r = e^\theta$.
   (a) At what angle does it cross each radius vector?
   (b) After how many turns about the origin will it reach the origin?
   (c) How far will it then have traveled, if it starts at (1, 0)?

4. Find the length of $r = a(1 - \cos\theta)$.

5. (a) Find an angle of intersection of the circle $r = a$ and the rose $r = 2a \cos n\theta$.
   (b) What is the limit of this angle as n increases without bound?

12.4 Area

To find the area A of the region enclosed by the curve $r = f(\theta)$ and the radius vectors at vectorial angles α and β, $\beta > \alpha$, we divide the angular region thus bounded into n sectors of angle $\Delta\theta = \dfrac{\beta - \alpha}{n}$, as shown in Fig. 12.

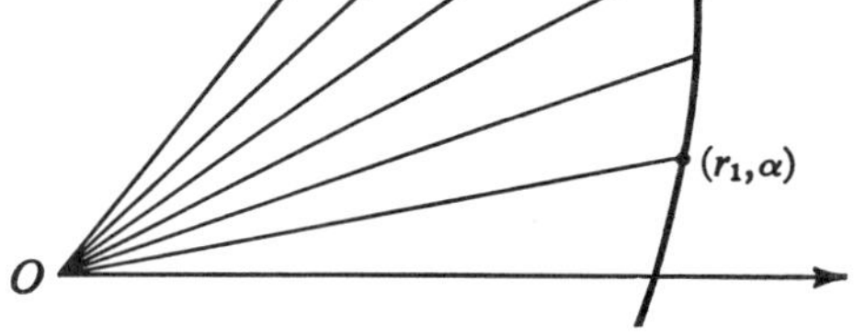

Fig. 12

If we assume that f is continuous, then in each sector there will be a greatest value of r, say r_g, and a least value, say r_l (this is Lemma 5.4-1), and the area ΔA of the sector will then be intermediate between the area of an inner circular sector, of radius r_l, and that of an outer one of radius r_g (Fig. 13).

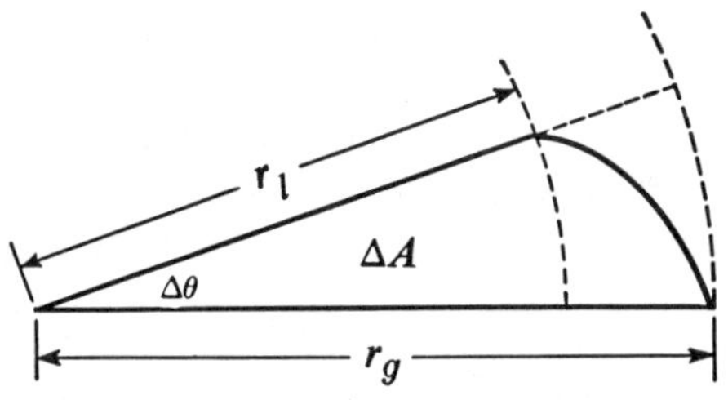

Fig. 13

Since the area of a circular sector of radius r and angle θ radians is $\frac{1}{2}r^2\theta$, this gives us

$$\tfrac{1}{2}r_l^{\,2}\,\Delta\theta \le \Delta A \le \tfrac{1}{2}r_g^{\,2}\,\Delta\theta.$$

The argument now assumes a familiar form (see § 4.1). The sum of the inner areas yields a lower sum S_L, the sum of the ΔA's is the required area A, the sum of the outer areas is an upper sum S_U, and a term-by-term comparison of these three sums yields

$$S_L \le A \le S_U.$$

As in § 4.1, we can show that $\lim\limits_{\Delta\theta\to 0} S_L = \lim\limits_{\Delta\theta\to 0} S_U$ (the details are left to the student), and the pinching theorem (Theorem 5.3-5) tells us that this common limit must be the required area A. If, moreover, we form a sum S by picking *any* r_i^* in the ith sector and setting $S = \frac{1}{2}\sum\limits_{i=1}^{n} r_i^{*2}\,\Delta\theta$, we have $S_L \le S \le S_U$, and the intermediate sum S will therefore have the same limit as S_L and S_U, namely A. But since $r = f(\theta)$, then also r^2 is a function of θ, and the limit of the sum S is therefore, by definition, a definite integral. Hence

$$A = \lim_{\Delta\theta\to 0} \frac{1}{2}\sum_{i=1}^{n} r_i^{*2}\,\Delta\theta = \frac{1}{2}\int_{\alpha}^{\beta} r^2\,d\theta. \tag{1}$$

Example: Find the area enclosed by one loop of $r^2 = a^2 \sin 2\theta$.

Solution: In order to determine the limits of integration, we must determine the extent of one loop. Considering only the non-negative square root, $r = a\sqrt{\sin 2\theta}$, we see that r is zero when $\theta = 0$, increases to a when $\theta = \pi/4$, and is again zero when $\theta = \pi/2$. This, then, completes a loop, and therefore

$$A = \frac{1}{2}\int_0^{\pi/2} r^2\,d\theta = \frac{1}{2}\int_0^{\pi/2} a^2 \sin 2\theta\,d\theta = -\frac{1}{4}a^2\cos 2\theta\Big|_0^{\pi/2} = \frac{1}{2}a^2.$$

Exercises 12.4 [A]

Find, by (1), the area of the region enclosed by each of the following curves:

1. $r = a$
2. $r = 2a \cos \theta$
3. $r = a(1 - \cos \theta)$
4. $r = 2 + \cos \theta$
5. $r^2 = a^2 \cos \theta$
6. $r = a \sin 2\theta$ (one loop)
7. Find the area of the region bounded by $\theta = -\pi/4$, $\theta = \pi/3$, and $r = a \sec \theta$.

Exercises 12.4 [B]

Find, by (1), the area of the region enclosed by each of the following:

1. $r = a(1 + \sin \theta)$
2. $r^2 = a^2 \cos 2\theta$
3. $r = 2 \cos \theta + 3$
4. $r = 2a \cos^2 \theta$
5. One loop of $r = a \sin n\theta$.
6. Find the ratio of the area of $r = a \cos 3\theta$ to the area of its circumscribed circle.

Miscellaneous Exercises 12.4

1. Find the area inside $r = a(1 + \cos \theta)$ and outside $r = 2a \cos \theta$.
2. Find the area of the region enclosed by the parabola $r = a \sec^2 \frac{1}{2}\theta$ and its latus rectum. (*Hint:* The focus is at the origin.)
3. Find the ratio of the area of $r = a \sin n\theta$ to the area of its circumscribed circle if **(a)** n is odd; **(b)** n is even.
4. Find the area enclosed between the loops of $r = 1 + 2 \cos \theta$.
5. Find the area of the region common to $r = \dfrac{1}{1 + \cos \theta}$ and $r = \dfrac{1}{1 - \cos \theta}$.

Chapter Review

1. Write and simplify a Cartesian equation of $r = a(1 + \cos \theta)$.
2. Find the angles of intersection of $r = a(1 + \cos \theta)$ and $r = a(1 - \cos \theta)$.
3. Sketch the curve $r = a \sin^2 (\theta/2)$ and find its length.
4. Find the area common to the circles $r = a$ and $r = 2a \cos \theta$.
5. Show that the tangent line to the parabola $r = \dfrac{k}{1 - \cos \theta}$ makes equal angles with the axis of the curve and the radius vector to the point of tangency.
6. Find the mean value, with respect to θ, of the radius vectors of the curve $r = a(1 - \cos \theta)$.

Cumulative Review Exercises

GROUP 1 [through Chapter 2]

1. Show on a number line $\{x: 0 < |x - 1| < 3\} \cap \{x: |x - 2| \leq 2\}$.
2. Find the angles of the triangle with vertices (1, 1), (4, 5), and (5, −2).
3. If $f(x) = \sqrt{x + a}$, use the definition of derivative to find $f'(a)$.
4. If $f = \{(1, 3), (2, 4), (3, 5)\}$, describe, by listing their elements, the function ff^{-1} and the function $f^{-1}f$.
5. Simplify: $\dfrac{\cos(\pi/6 + x) - \cos(\pi/6 - x)}{\cos(2\pi/3 + x) + \cos(4\pi/3 + x)}$.
6. If $y = f(x)$, find $f'(x)$:

   (a) $y = \dfrac{x + 3}{\sqrt{3x}}$ (c) $2xy - 3x^2 = 7$

   (b) $y = (2x^2 - 3x)^2$ (d) $y = \sqrt{x}\,(x - 2)(x + 2)$
7. Prove that, in any quadrilateral, the midpoints of a pair of opposite sides and the midpoints of the diagonals are the vertices of a parallelogram.

GROUP 2 [through Chapter 3]

1. If $f(x) = \dfrac{1 - x}{1 + x}$, use the definition of derivative to find $f'(x)$.
2. If $f(x) = \sqrt{1 + x^2}$, $g(x) = \tan x$, and $h(x) = \sec x$, compare the ranges of fg and h. Are fg and h the same function?
3. $A(6, 5)$, $B(2, 1)$, and $C(3, 8)$ are vertices of parallelogram $ABCD$. Find D.
4. Show that the graph of $y = ax^4 + bx^2 + c$ $(a > 0)$ has an inflection point only if $b < 0$.
5. A rectangular bin, with square ends and no top, is to have a capacity of 36 cu. ft. What dimensions will use the least lumber?
6. Find the length of the chord of $y = x^2 - 5$ that is normal to the curve at the point (1, −4).

GROUP 3 [through Chapter 4]

1. (a) Find any extreme or inflection points on the graph of $f(x) = x^3 + 3x + 7$.

   (b) On what intervals is f increasing?
2. The abscissa of a particle moving on the x-axis is $x = 18t - t^3$, at time $t \geq 0$. Find the acceleration of the particle when x is a maximum.

3. Find the area of the region that is enclosed by the curve $y = x^3 - 3x + 6$ and the line tangent to it at $(-1, 8)$.

4. Find the function f that has $f''(x) = 8x^2 - 12$ and has a relative minimum value of 108 at 3.

5. Find the volume of the solid formed by rotating about the x-axis the region bounded by the x-axis, the curve $x^2 - x^2y + 1 = 0$, and the lines $x = 1$ and $x = 2$.

6. Prove: In the parabola $y = ax^2 + bx + c$, the midpoints of all chords of constant slope m lie on a straight line.

7. Find an angle, at each intersection, between $y = 3 + (2/x)$ and $x + 2y - 11 = 0$.

GROUP 4 [through Chapter 4]

1. Find an equation of the line through $(2, -3)$ (**a**) parallel, and (**b**) perpendicular, to $2x + y - 7 = 0$.

2. What is the range of $f = \{(x, 2x - 3): -1 \leq x \leq 4\}$?

3. Solve the differential equation $f'(x) = x^3 - x$, if $f(2) = 1$.

4. Find $F'(t)$ if $F(t) = \sqrt{2t + 3}$.

5. What line parallel to the y-axis bisects the area bounded by $y = x^2$, the x-axis, and the line $x = 6$?

6. If $y = x^4 - 4x^3 - 18x^2 + 7$, find the abscissas of all maximum, minimum, and inflection points, and draw a rough graph.

7. A cylindrical metal can is to have a slip-on lid a inches deep. If the total area of metal available for the can is fixed, show that the volume will be a maximum when $h = a + 2r$.

GROUP 5 [through Chapter 5]

1. Write an equation of the line that is met at (h, k) by the perpendicular from the origin.

2. Find, if it exists, $\lim_{x \to 0} \csc x \tan \dfrac{x}{2}$.

3. (**a**) Sketch $f = \{(x, y): y = |x|\}$.

   (**b**) For what x does $f'(x)$ not exist? Why?

   (**c**) Evaluate $\int_{-1}^{2} f(x)dx$.

4. Find the area of the region enclosed by $y = 3 + 2x - x^2$ and $3x - 2y + 3 = 0$.

5. A function f is defined for all real numbers x by

$$f(x) = \begin{cases} x^2, \ x < 1, \\ x, \ x \geq 1. \end{cases}$$

At the point $x = 1$, (a) is f continuous? (b) is f' defined?

6. What part of a sphere is cut off by a plane that bisects a radius at right angles?

7. (a) Write an equation of the line tangent at (x_1, y_1) to $y = 3x^3 - 13x^2 - x - 4$.

   (b) Show that there are three tangents to this curve that pass through the origin, and find the point of tangency of each.

GROUP 6 [through Chapter 5]

1. Find the area of the region enclosed by $y^2 = 8x$ and $y = 2x$.
2. Given $A(-1, -2)$, $B(3, 0)$, and $C(2, 4)$, find $\angle BAC$.
3. Prove that $f(x) = x^3 + x^2 + 2x - 3$ has neither a maximum nor a minimum, and determine whether $f'(x)$ has a maximum or has a minimum.
4. Find the number c of the mean value theorem (Theorem 5.4-3), in terms of the numbers a and b $(a < 0, b < 0)$, for the function $f(x) = 1/x$, and illustrate by a sketch.
5. A particle that is moving in a straight line AB passes a fixed point P at 40 cm./sec. As it passes P, it begins to slow down with a deceleration $a = -15t^2$ cm./sec.^{2}, where t is the time in seconds. How far from P will the particle come to a stop?
6. A function f is defined by $f(x) = \begin{cases} ax^2 + bx, \ x < 1, \\ 2ax + 1, \ x \geq 1. \end{cases}$ If f' is defined at 1, find a and b and sketch the graph of f.

GROUP 7 [through Chapter 6]

1. If $y = 4 - x^2$, find dy and Δy when $x = 1$ and $\Delta x = 0.5$, and show these quantities on a figure.
2. Find dy if (a) $y = \sqrt{\dfrac{a - x}{a + x}}$; (b) $y = \sec x \sec 2x$.
3. Find $\dfrac{d^2y}{dx^2}$ if (a) $3xy + y^2 = 6$; (b) $y = \sin (a + x) \sin (a - x)$.
4. Find and classify the extreme values of $\dfrac{x^2 - 3}{x - 2}$.
5. Prove that $\lim_{x \to 1} (3x - 5) = -2$.

6. An isosceles triangle has its vertex at the origin, and its base, parallel to and above the x-axis, is a chord of the parabola $y = 12 - x^2$. Find the maximum area of the triangle.

GROUP 8 [through Chapter 6]

1. Find dy if (a) $x^3 + y^3 = 3axy$; (b) $y = x \sin \sqrt{x}$.
2. Find and classify the extreme and inflection points of

$$y = \sin x + \tfrac{1}{2} \sin 2x, \quad 0 \leq x < 2\pi.$$

3. Find the volume of the solid generated by rotating about the x-axis the region enclosed by $y = x^2 - 2$ and the x-axis.
4. Find the highest and the lowest point on the ellipse $x^2 + 2xy + 3y^2 = 8$.
5. At what angle do the circles $x^2 + y^2 = a^2$ and $x^2 + y^2 - 2ax = 0$ intersect?

GROUP 9 [through Chapter 7]

1. If $\frac{1}{16}$ in. is planed off each face of an 8-inch cube, use differentials to find the approximate decrease in volume.
2. Show that $xy = k^2$ and $x^2 - y^2 = a^2$ are orthogonal.
3. Find the area of the region enclosed by the graph of $|x| + |y| = 1$.
4. Integrate:

(a) $\int \tan \theta \sec^3 \theta \, d\theta$ (b) $\int (x + a)\sqrt{x + b} \, dx$

(c) $\int \sqrt{x^4 + x^7} \, dx$

5. Find the area of the region bounded by the x-axis, the lines $x = \pm 1$, and the curve $y = 1/(x^2 + 1)$.
6. A light is on the ground, and a 6-foot man walks from it toward a building 50 ft. away. If he walks at 4 ft./sec., how fast is the height of his shadow on the building changing when he is 30 ft. from the light?
7. If $f(x) = x^2\sqrt{5 - x}$, $0 \leq x \leq 5$, (a) find the maximum value of f; (b) find the area of the region enclosed by $y = f(x)$ and $y = 0$.

GROUP 10 [through Chapter 7]

1. Use differentials to approximate $\sqrt{145}$.
2. Find an equation of the curve that goes through (6, 2) and has the slope $x/4y$ at the point (x, y).
3. For what x does $f(x) = x^4 - 4x^3 + 4x^2 + 1$ define an increasing function?
4. Find the maximum and minimum points on the hyperbola $2x^2 - 5xy + 3y^2 = -2$.

5. The hour hand of a clock is 2 in. long and the minute hand is 3 in. How fast are their tips separating at 9 o'clock? (*Hint:* Use the law of cosines.)

6. Evaluate: (a) $\int_3^6 \frac{x\,dx}{\sqrt{x-2}}$ (b) $\int_0^1 x^2\sqrt{1-x^2}\,dx$

7. A 6-inch square has one side on the base of an isosceles triangle and the other two vertices on the equal sides of the triangle. Find the minimum area of the triangle.

GROUP 11 [through Chapter 8]

1. Find an equation of the parabola with focus $(2, -1)$ and directrix $y + 2 = 0$.

2. Analyze and sketch: $y = \dfrac{x^3}{x^2 - 1}$.

3. Prove that there is no line through the origin tangent to the hyperbola $xy = k^2$.

4. Find the derivatives $\dfrac{dy}{dx}$ and $\dfrac{d^2y}{dx^2}$ in terms of the parameter t, and derive a Cartesian equation: $x = t + \dfrac{1}{t}$, $y = t - \dfrac{1}{t}$.

5. A particle moves on a line with acceleration $4 - 18t$. If its distance from a fixed point of the line is 3 when $t = 1$ and is 12 when $t = 2$, what is its velocity when $t = 0$?

6. If A and B are $(1, 1)$ and $(-1, -1)$, respectively, and P is a point such that $PA + PB = 4$, find an equation of the locus of P.

7. A tangent to the ellipse $4x^2 + 9y^2 = 36$ cuts a triangle out of the first quadrant. Find the point of tangency if the area of this triangle is a minimum. (A parametric equation is convenient.)

GROUP 12 [through Chapter 8]

1. Find the center, vertices, foci, and eccentricity, and sketch: $x^2 - y^2 - 4x = 0$.

2. Sketch $x = t^2 - 4$, $y = t^3 - 4t$, and find the area of the loop.

3. If $\dfrac{1}{x} + \dfrac{1}{y} = \dfrac{1}{10}$, find to two significant figures the change in y when x changes from 30.00 to 30.12.

4. The normal to a parabola at P meets the axis of the curve at N. Prove that P and N are equidistant from the focus.

5. Water is pouring at 2 cu. ft./min. into a conical container the height of which is twice the radius of the top. How fast is the water level rising when the water is 3 ft. deep?

6. Show that the length of the segment cut by the coordinate axes on the tangent to $x^{2/3} + y^{2/3} = a^{2/3}$ is constant.

GROUP 13 [through Chapter 9]

1. Find the centroid of the region bounded by $y = \sin 2x$, $0 \leq x \leq \pi/2$, and the x-axis.

2. Find and classify the extreme points of $y = x \sin x + \cos x$, $0 \leq x < 2\pi$.

3. Below the surface of the earth, the attraction of gravity is directly proportional to the distance from the center of the earth. Find the work done in lifting to the surface of the earth from a depth of h miles an object which weighs w pounds at the surface. State your answer in terms of the radius R of the earth.

4. (a) For the ellipse $x = a \cos \theta$, $y = b \sin \theta$, show that

$$ds = a\sqrt{1 - e^2 \cos^2 \theta}\, d\theta.$$

(b) A point P is tracing the ellipse $x = 5 \cos \theta$, $y = 3 \sin \theta$ in such a way that $d\theta/dt = 3$ rad./min. Find how fast P is moving when it goes through $(\frac{5}{2}\sqrt{3}, \frac{3}{2})$.

5. Find the volume of the solid generated by rotating about the line $x = 2p$ the region bounded by the parabola $y^2 = 4px$ and its latus rectum.

6. For $0 < x \leq \pi/2$, show that:

(a) $x > \sin x$ (consider $f(x) = x - \sin x$).

(b) $\sin x > x - \frac{1}{6}x^3$ (consider $F(x) = \sin x - x + \frac{1}{6}x^3$).

GROUP 14 [through Chapter 9]

1. Show that the normal to the cycloid $x = a(\theta - \sin \theta)$, $y = a(1 - \cos \theta)$ at the point $\theta = \theta_1$ meets the x-axis at $(a\theta_1, 0)$.

2. Find $\dfrac{dy}{dx}$ in terms of x if $y = \tan^{-1} \dfrac{x}{\sqrt{a^2 - x^2}}$.

3. The base of a solid is the ellipse $x^2 + 4y^2 = 4$, and every section by a plane perpendicular to the x-axis is an equilateral triangle. Find the volume of the solid.

4. Evaluate: (a) $\displaystyle\int_0^{\pi/8} \tan^2 2x\, dx$ (b) $\displaystyle\int_0^4 \frac{x\, dx}{\sqrt{x^2 + 9}}$

5. A sector is to be cut out of a circular piece of paper and then formed into a cone. What must be the angle of the sector if the volume of the cone is to be a maximum?

6. A water tank is a right circular cylinder, 6 ft. in diameter, with its axis horizontal. Find the force on one end when the water is 3 ft. deep.

GROUP 15 [through Chapter 10]

1. Find the minimum value of $ae^x + be^{-x}$ $(a > 0,\ b > 0)$.

2. Integrate: $\int \frac{x(x-1)^2}{x-2}\,dx$. For what x is the result valid?

3. Find the volume of the solid generated by rotating about the y-axis the region bounded by $x^2 - y^2 = a^2$ and $x = 2a$.

4. A particle starts from rest at time $t = 0$ with acceleration $a = 16 \sin \pi t$. How far has it gone when $t = \frac{1}{2}$?

5. Analyze and sketch: $y = e^{-1/x^2}$.

6. The curvature of the curve $y = f(x)$ is defined as

$$\frac{d^2y/dx^2}{[1 + (dy/dx)^2]^{3/2}}.$$

Find an expression for the curvature of: **(a)** $x^2 + y^2 = a^2$; **(b)** $y = \ln \cos x$; **(c)** $x = a(\cos t + t \sin t)$, $y = a(\sin t - t \cos t)$.

GROUP 16 [through Chapter 10]

1. Show that the extreme values of $e^{-x} \sin x$ are in geometric progression, and find the value of the common ratio.

2. Find dy/dx :

(a) $e^y = \sec x$ **(b)** $y = x \cdot 10^x$ **(c)** $y = \frac{(x^2 - 2)(x + 1)^2}{x\sqrt{x + 3}}$

3. Find the length of $y = \frac{1}{4}x^2 - \frac{1}{2} \ln x$ from $x = \frac{1}{2}$ to $x = 2$.

4. Show that each line of the family $y = mx + 2$ cuts orthogonally each circle of the family $x^2 + (y - 2)^2 = r^2$.

5. Prove that the midpoint of a focal chord of a parabola is the same distance from the directrix as from the ends of the chord.

6. Find the volume of the solid generated by rotating about the x-axis the hypocycloid $x = a \cos^3 \theta$, $y = a \sin^3 \theta$.

GROUP 17 [through Chapter 11]

1. Find the length of $y = \ln \sec x$ from $x = 0$ to $x = \pi/4$.
2. Integrate:

(a) $\int (1 + \sqrt{x})^{-3/2}\, dx$ (b) $\int \frac{dx}{x^2 + 4x}$ (c) $\int \frac{dx}{\sqrt{2 + x - x^2}}$

3. (a) Show that $d(\sin^{-1} x) = \frac{dx}{\sqrt{1 - x^2}}$. (See Ex. 5, page 217. Why need we consider only the positive square root?)

(b) Integrate: $\int \sin^{-1} x\, dx$.

4. Prove that the product of the distances from the foci of an ellipse to any tangent is constant.
5. The illumination from a source of light varies inversely as the square of the distance from the source. Two lights are 10 ft. apart, and one of them is twice as strong as the other. At what point in the line segment joining them is the total illumination a minimum?

GROUP 18 [through Chapter 11]

1. A point is tracing the parabola $y^2 = 4x$ in such a way that its ordinate is increasing at 2 units/min. When the point is at $(9, -6)$, find the rate of change of (a) its abscissa; (b) its distance from the origin.
2. Integrate:

(a) $\int \frac{dx}{x^4 + 2x^2 + 1}$ (b) $\int \frac{dx}{1 + \sin x + \cos x}$

(c) $\int \sqrt{x} \ln x\, dx$

3. Find the length of the parabola $x = pt^2$, $y = 2pt$ from $t = 0$ to $t = \sqrt{3}$.
4. Find the volume of the solid generated by rotating about the y-axis the region bounded by the x-axis and $y = \sin x$, $0 \leq x \leq \pi$.
5. Prove that the area of the triangle bounded by the asymptotes and a tangent to a hyperbola is independent of the point of tangency.

GROUP 19 [through Chapter 12]

1. Find the area of the region enclosed by the hyperbola $r^2 \cos 2\theta = a^2$ and the rays $\theta = 0$, $\theta = \pi/6$ $(r \geq 0)$.
2. If $|f(x + h) - f(x)| \leq 3h^2$ for all real x and h, prove that $f'(x) = 0$ and describe the function f.

3. Find the angle ψ at an arbitrary point of the lituus $r^2\theta = a^2$.

4. Write an equation of the line tangent at $(1, -2)$ to $x^2 + 3xy + y^2 - 2x + y + 5 = 0$.

5. Integrate: (a) $\int \frac{dx}{x\sqrt{1 + 2x}}$ (b) $\int \frac{dx}{\sqrt{x + 1} - \sqrt{x - 1}}$

GROUP 20 [through Chapter 12]

1. Find the length of $x = e^{-t} \cos t$, $y = e^{-t} \sin t$ from $t = 0$ to $t = 2\pi$.

2. Find the area of the region that is inside both $r = a(1 - \cos\theta)$ and $r = a(1 + \cos\theta)$.

3. Show that an angle between the normal and the radius vector at the point (r_1, θ_1) on the lemniscate $r^2 = 2a^2 \cos 2\theta$ is $2\theta_1$.

4. If $y = e^x \sin x$,

   (a) show that $\frac{dy}{dx} = \sqrt{2}\, e^x \sin\left(x + \frac{\pi}{4}\right)$;

   (b) obtain a similar expression for $\frac{d^2y}{dx^2}$;

   (c) guess a formula for $\frac{d^ny}{dx^n}$.

5. Prove that, if a line is tangent to a hyperbola, the point of tangency is the midpoint of the segment included between the asymptotes.

6. The position vector of a moving particle $P(x, y)$ is $\mathbf{r}(t) = (e^t, e^{-t})$. Find a Cartesian equation of the path of P. Sketch, showing the velocity and acceleration vectors at the time $t = 0$.

Appendix: Vectors in the Plane

The velocity of a moving object has a magnitude, or size, and a direction. For example, the velocity of the wind might have a magnitude of 15 mi./hr. and a direction of northeast. Any quantity characterized by a magnitude and a direction is called a vector quantity, or simply a vector. Besides velocity, some other vectors are force, acceleration, and displacement.

Vectors are represented in diagrams by arrows. The direction of each arrow shows the direction of its vector, and the length of the arrow its magnitude. These arrows are themselves commonly called vectors, though strictly speaking they are only symbols, or pictures, representing vectors. This usage is much like the use of the words "number" and "numeral" mentioned in the footnote on page 4. The arrow from a point X to a point Y is written $\mathbf{XY}$; if it represents a vector $\mathbf{v}$, we write $\mathbf{v} = \mathbf{XY}$. The order of the letters is significant: $\mathbf{XY}$ and $\mathbf{YX}$ represent different vectors because they have opposite directions. Every vector can be represented by infinitely many arrows, all having the same length and direction and therefore said to be equivalent. Figure 1 shows equivalent arrows $\mathbf{AB}$ and $\mathbf{CD}$; we write $\mathbf{AB} = \mathbf{CD}$. Because $ABDC$ is a parallelogram it is also true that $\mathbf{AC} = \mathbf{BD}$.

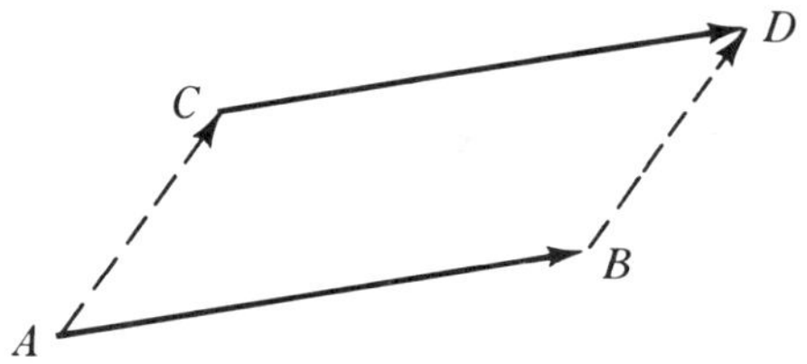

Fig. 1

The arrow $\mathbf{OP}$ from the origin O of a coordinate system to any point $P(x,y)$ represents the position vector $\mathbf{r}$ of P: $\mathbf{r} = \mathbf{OP}$. Note that $\mathbf{r}$ is completely determined by the coordinate pair of P. These coordinates are called the components of $\mathbf{r}$, which is therefore determined by its component-pair. Thus, in Figure 2, the component form of $\mathbf{r}$ is

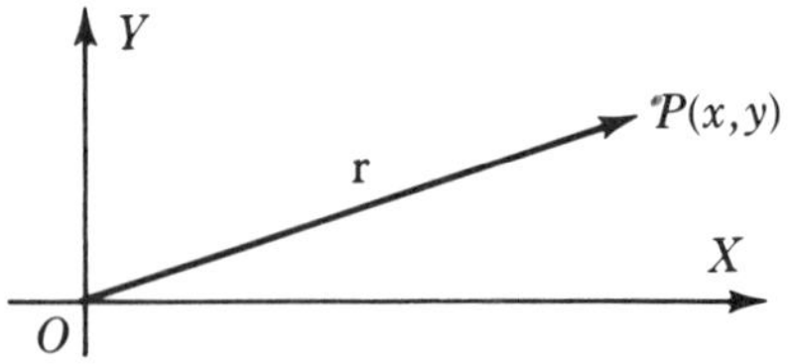

Fig. 2

$$\mathbf{r} = (x,y).$$

Since the point (x, y) has the position vector (x, y) we take the view that a point and a vector are abstractly the same object, an ordered pair of real numbers. The words you use to describe such pairs, "points" or "vectors," depend on how you want to use the pairs and on what kinds of pictures you want to draw, dots or arrows.

Length. The length (also called the magnitude or norm) of a vector $\mathbf{r} = (x, y)$, written $|\mathbf{r}|$ or $|(x, y)|$, can be found by the distance formula (see Figure 2):

$$|\mathbf{r}| = |(x, y)| = \sqrt{x^2 + y^2}. \tag{1}$$

Addition. The sum $\mathbf{a} + \mathbf{b}$ of vectors $\mathbf{a} = (a_1, a_2)$ and $\mathbf{b} = (b_1, b_2)$ is defined to be

$$\mathbf{a} + \mathbf{b} = (a_1, a_2) + (b_1, b_2) = (a_1 + b_1, a_2 + b_2); \tag{2}$$

to add vectors, just add their corresponding components. The arrow diagram in Figure 3 shows vector addition: If **OA** and **OB** represent **a** and **b**, respectively, then the diagonal **OP** of parallelogram $OAPB$ represents $\mathbf{a} + \mathbf{b}$, as the student may show (Exercise 1). For this reason, Formula (2) for the addition of vectors is commonly called the Parallelogram Rule. We write

$$\mathbf{OA} + \mathbf{OB} = \mathbf{OP}. \tag{3}$$

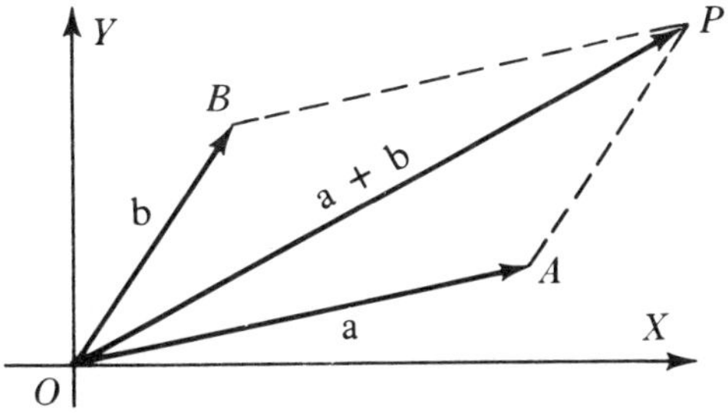

Fig. 3

Since $\mathbf{OB} = \mathbf{AP}$, it is also true that

$$\mathbf{OA} + \mathbf{AP} = \mathbf{OP}. \tag{4}$$

Formula (4) is sometimes called the Shortcut Rule for vector addition.

In the addition of numbers there is a neutral element, 0, such that $x + 0 = x$ for every number x. Vector addition also has a neutral element, the zero vector,

$$\mathbf{0} = (0, 0).$$

The zero vector has length 0, by (1), and has no direction.

Just as every number x has an additive inverse $-x$, so every vector $\mathbf{v} = (v_1, v_2)$ has the additive inverse (or negative, or opposite) $-\mathbf{v}$:

$$-(v_1, v_2) = (-v_1, -v_2).$$

By (1), $|-\mathbf{v}| = |\mathbf{v}|$, and a figure will show that $\mathbf{v}$ and $-\mathbf{v}$ have opposite directions. Thus if an arrow **AB** represents **v**, then **BA** represents $-\mathbf{v}$, and we therefore write

$$\mathbf{BA} = -\mathbf{AB}.$$

We use additive inverses to define vector subtraction. Just as with numbers, "subtract" means "add the inverse":

$$\mathbf{a} - \mathbf{b} = \mathbf{a} + (-\mathbf{b}).$$

If neither A nor B is the origin, what vector does $\mathbf{AB}$ represent? The question is easily answered using vector subtraction. Let $A = (a_1, a_2)$, $B = (b_1, b_2)$. Then (see Figure 4)

$$\mathbf{OA} + \mathbf{AB} = \mathbf{OB},$$

and,
$$\begin{aligned}\mathbf{AB} &= \mathbf{OB} - \mathbf{OA}\\ &= (b_1, b_2) - (a_1, a_2)\\ &= (b_1 - a_1, b_2 - a_2).\end{aligned}$$

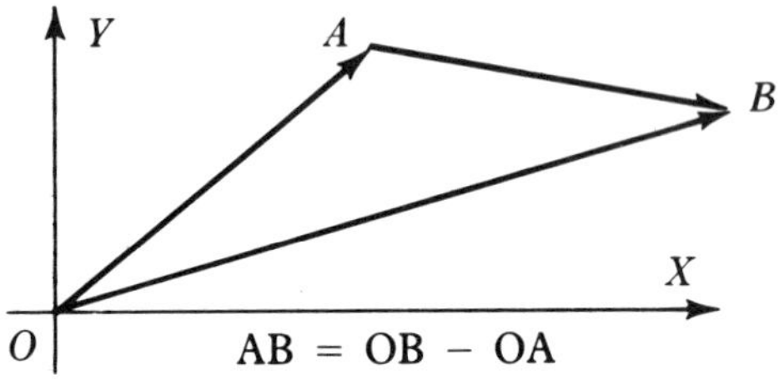

AB = OB − OA

Fig. 4

Scalar multiplication. If we write $\mathbf{v} + \mathbf{v} = 2\mathbf{v}$, then by definition of vector addition we have

$$2\mathbf{v} = (v_1, v_2) + (v_1, v_2) = (v_1 + v_1, v_2 + v_2) = (2v_1, 2v_2)$$

or
$$2(v_1, v_2) = (2v_1, 2v_2).$$

This suggests the generalization

$$k(v_1, v_2) = (kv_1, kv_2) \tag{5}$$

for any number k and any vector (v_1, v_2). In this context a number is usually called a scalar, and the operation of multiplying a vector by a scalar defined in (5) is called scalar multiplication.

Lines. For any value of the scalar t let the vector $\mathbf{r}$ be given by

$$\mathbf{r}(t) = t(2, 1) + (0, 3). \tag{6}$$

We write $\mathbf{r}(t)$ because if t is variable then $\mathbf{r}$ is a function, a vector-valued function of the scalar t. The domain of $\mathbf{r}$ is the set of all numbers t, and the range is the set of all vectors given by (6). For example,

$$\mathbf{r}(0) = 0(2, 1) + (0, 3) = (0, 3), \qquad \mathbf{r}(1) = 1(2, 1) + (0, 3) = (2, 4),$$

$$\mathbf{r}(2) = 2(2, 1) + (0, 3) = (4, 5),$$

and so on. Figure 5 shows some of the values of $\mathbf{r}$, identified by the corresponding values of t. As the figure suggests, the values of $\mathbf{r}$ (plotted as points) lie on a line L. In this context the variable t is called a parameter, and (6) is called a vector parametric equation of L. The vectors (2, 1) and (0, 3) that you see in (6) are called, respectively, a direction vector and a position vector for L.

If $\mathbf{r} = (x, y)$, we have

$$(x, y) = t(2, 1) + (0, 3)$$
$$= (2t, t) + (0, 3)$$
$$= (2t, t + 3),$$

and therefore

$$x = 2t, \ y = t + 3. \qquad (7)$$

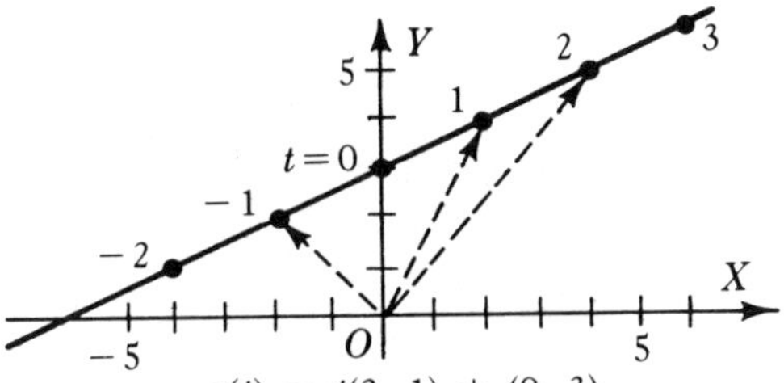

$\mathbf{r}(t) = t(2, 1) + (0, 3)$

Fig. 5

These are called scalar parametric equations for L. (The word "scalar" is frequently omitted; compare § 8.2.) If we eliminate t from the system (7), we obtain the Cartesian equation $x - 2y + 6 = 0$.

The preceding example can clearly be generalized. Given vectors $\mathbf{v} = (v_1, v_2)$ and $\mathbf{p} = (p_1, p_2)$ with $\mathbf{v} \neq \mathbf{0}$, then

$$\mathbf{r}(t) = t\mathbf{v} + \mathbf{p}$$

is a vector parametric equation of the line with direction vector $\mathbf{v}$ and position vector $\mathbf{p}$. This line has parametric equations

$$x = v_1 t + p_1, \ y = v_2 t + p_2,$$

and a Cartesian equation

$$v_2 x - v_1 y = v_2 p_1 - v_1 p_2.$$

A further generalization to $\mathbf{r}(t) = (f_1(t), f_2(t))$, in which the scalar parametric functions f_1 and f_2 need not be linear and the graph is not necessarily a line, is considered in § 8.2, though without the convenience of vector notation.

Exercises

1. Show that $O(0,0)$, $A\ (a_1, a_2)$, $P(a_1 + b_1, a_2 + b_2)$ and $B(b_1, b_2)$ are the vertices of a parallelogram.

2. Let $\mathbf{a} = (2, 3)$, $\mathbf{b} = (5, -1)$. Find

(a) $\mathbf{a} + \mathbf{b}$. (b) $2\mathbf{a} - 3\mathbf{b}$. (c) $|\mathbf{a} - \mathbf{b}|$.

3. Give the component form of **AB**, given

(a) $A = (0, 0)$, $B = (3, 1)$. (b) $A = (1, 5)$, $B = (3, 1)$.

4. Write a vector parametric equation, scalar parametric equations, and a Cartesian equation of the line

(a) with direction vector $(2, 1)$ and position vector $(-1, 2)$.

(b) with direction vector $(4, 2)$ and position vector $(3, 4)$.

(c) with direction vector $(-1, 3)$, through the point $(2, 0)$.

(d) through the points $(3, -3)$ and $(2, 0)$.

Table 1. Powers, Roots, Reciprocals

n	n^2	$\sqrt{n}$	$\sqrt{10n}$	n^3	$\sqrt[3]{n}$	$\sqrt[3]{10n}$	$\sqrt[3]{100n}$	$1/n$
1.0	1.00	1.00 000	3.16 228	1.000	1.00 000	2.15 443	4.64 159	1.00 000
1.1	1.21	1.04 881	3.31 662	1.331	1.03 228	2.22 398	4.79 142	.90 909
1.2	1.44	1.09 545	3.46 410	1.728	1.06 266	2.28 943	4.93 242	.83 333
1.3	1.69	1.14 018	3.60 555	2.197	1.09 139	2.35 133	5.06 580	.76 923
1.4	1.96	1.18 322	3.74 166	2.744	1.11 869	2.41 014	5.19 249	.71 429
1.5	2.25	1.22 474	3.87 298	3.375	1.14 471	2.46 621	5.31 329	.66 667
1.6	2.56	1.26 491	4.00 000	4.096	1.16 961	2.51 984	5.42 884	.62 500
1.7	2.89	1.30 384	4.12 311	4.913	1.19 348	2.57 128	5.53 966	.58 824
1.8	3.24	1.34 164	4.24 264	5.832	1.21 644	2.62 074	5.64 622	.55 556
1.9	3.61	1.37 840	4.35 890	6.859	1.23 856	2.66 840	5.74 890	.52 632
2.0	4.00	1.41 421	4.47 214	8.000	1.25 992	2.71 442	5.84 804	.50 000
2.1	4.41	1.44 914	4.58 258	9.261	1.28 058	2.75 892	5.94 392	.47 619
2.2	4.84	1.48 324	4.69 042	10.648	1.30 059	2.80 204	6.03 681	.45 455
2.3	5.29	1.51 658	4.79 583	12.167	1.32 001	2.84 387	6.12 693	.43 478
2.4	5.76	1.54 919	4.89 898	13.824	1.33 887	2.88 450	6.21 447	.41 667
2.5	6.25	1.58 114	5.00 000	15.625	1.35 721	2.92 402	6.29 961	.40 000
2.6	6.76	1.61 245	5.09 902	17.576	1.37 507	2.96 250	6.38 250	.38 462
2.7	7.29	1.64 317	5.19 615	19.683	1.39 248	3.00 000	6.46 330	.37 037
2.8	7.84	1.67 332	5.29 150	21.952	1.40 946	3.03 659	6.54 213	.35 714
2.9	8.41	1.70 294	5.38 516	24.389	1.42 604	3.07 232	6.61 911	.34 483
3.0	9.00	1.73 205	5.47 723	27.000	1.44 225	3.10 723	6.69 433	.33 333
3.1	9.61	1.76 068	5.56 776	29.791	1.45 810	3.14 138	6.76 790	.32 258
3.2	10.24	1.78 885	5.65 685	32.768	1.47 361	3.17 480	6.83 990	.31 250
3.3	10.89	1.81 659	5.74 456	35.937	1.48 881	3.20 753	6.91 042	.30 303
3.4	11.56	1.84 391	5.83 095	39.304	1.50 369	3.23 961	6.97 953	.29 412
3.5	12.25	1.87 083	5.91 608	42.875	1.51 829	3.27 107	7.04 730	.28 571
3.6	12.96	1.89 737	6.00 000	46.656	1.53 262	3.30 193	7.11 379	.27 778
3.7	13.69	1.92 354	6.08 276	50.653	1.54 668	3.33 222	7.17 905	.27 027
3.8	14.44	1.94 936	6.16 441	54.872	1.56 049	3.36 198	7.24 316	.26 316
3.9	15.21	1.97 484	6.24 500	59.319	1.57 406	3.39 121	7.30 614	.25 641
4.0	16.00	2.00 000	6.32 456	64.000	1.58 740	3.41 995	7.36 806	.25 000
4.1	16.81	2.02 485	6.40 312	68.921	1.60 052	3.44 822	7.42 896	.24 390
4.2	17.64	2.04 939	6.48 074	74.088	1.61 343	3.47 603	7.48 887	.23 810
4.3	18.49	2.07 364	6.55 744	79.507	1.62 613	3.50 340	7.54 784	.23 256
4.4	19.36	2.09 762	6.63 325	85.184	1.63 864	3.53 035	7.60 590	.22 727
4.5	20.25	2.12 132	6.70 820	91.125	1.65 096	3.55 689	7.66 309	.22 222
4.6	21.16	2.14 476	6.78 233	97.336	1.66 310	3.58 305	7.71 944	.21 739
4.7	22.09	2.16 795	6.85 565	103.823	1.67 507	3.60 883	7.77 498	.21 277
4.8	23.04	2.19 089	6.92 820	110.592	1.68 687	3.63 424	7.82 974	.20 833
4.9	24.01	2.21 359	7.00 000	117.649	1.69 850	3.65 931	7.88 374	.20 408
5.0	25.00	2.23 607	7.07 107	125.000	1.70 998	3.68 403	7.93 701	.20 000
5.1	26.01	2.25 832	7.14 143	132.651	1.72 130	3.70 843	7.98 957	.19 608
5.2	27.04	2.28 035	7.21 110	140.608	1.73 248	3.73 251	8.04 145	.19 231
5.3	28.09	2.30 217	7.28 011	148.877	1.74 351	3.75 629	8.09 267	.18 868
5.4	29.16	2.32 379	7.34 847	157.464	1.75 441	3.77 976	8.14 325	.18 519
5.5	30.25	2.34 521	7.41 620	166.375	1.76 517	3.80 295	8.19 321	.18 182

n | n^2 | $\sqrt{n}$ | $\sqrt{10n}$ | n^3 | $\sqrt[3]{n}$ | $\sqrt[3]{10n}$ | $\sqrt[3]{100n}$ | $1/n$

Table 1. Powers, Roots, Reciprocals

n	n^2	$\sqrt{n}$	$\sqrt{10n}$	n^3	$\sqrt[3]{n}$	$\sqrt[3]{10n}$	$\sqrt[3]{100n}$	$1/n$
5.5	30.25	2.34 521	7.41 620	166.375	1.76 517	3.80 295	8.19 321	.18 182
5.6	31.36	2.36 643	7.48 331	175.616	1.77 581	3.82 586	8.24 257	.17 857
5.7	32.49	2.38 747	7.54 983	185.193	1.78 632	3.84 850	8.29 134	.17 544
5.8	33.64	2.40 832	7.61 577	195.112	1.79 670	3.87 088	8.33 955	.17 241
5.9	34.81	2.42 899	7.68 115	205.379	1.80 697	3.89 300	8.38 721	.16 949
6.0	36.00	2.44 949	7.74 597	216.000	1.81 712	3.91 487	8.43 433	.16 667
6.1	37.21	2.46 982	7.81 025	226.981	1.82 716	3.93 650	8.48 093	.16 393
6.2	38.44	2.48 998	7.87 401	238.328	1.83 709	3.95 789	8.52 702	.16 129
6.3	39.69	2.50 998	7.93 725	250.047	1.84 691	3.97 906	8.57 262	.15 873
6.4	40.96	2.52 982	8.00 000	262.144	1.85 664	4.00 000	8.61 774	.15 625
6.5	42.25	2.54 951	8.06 226	274.625	1.86 626	4.02 073	8.66 239	.15 385
6.6	43.56	2.56 905	8.12 404	287.496	1.87 578	4.04 124	8.70 659	.15 152
6.7	44.89	2.58 844	8.18 535	300.763	1.88 520	4.06 155	8.75 034	.14 925
6.8	46.24	2.60 768	8.24 621	314.432	1.89 454	4.08 166	8.79 366	.14 706
6.9	47.61	2.62 679	8.30 662	328.509	1.90 378	4.10 157	8.83 656	.14 493
7.0	49.00	2.64 575	8.36 660	343.000	1.91 293	4.12 129	8.87 904	.14 286
7.1	50.41	2.66 458	8.42 615	357.911	1.92 200	4.14 082	8.92 112	.14 085
7.2	51.84	2.68 328	8.48 528	373.248	1.93 098	4.16 017	8.96 281	.13 889
7.3	53.29	2.70 185	8.54 400	389.017	1.93 988	4.17 934	9.00 411	.13 699
7.4	54.76	2.72 029	8.60 233	405.224	1.94 870	4.19 834	9.04 504	.13 514
7.5	56.25	2.73 861	8.66 025	421.875	1.95 743	4.21 716	9.08 560	.13 333
7.6	57.76	2.75 681	8.71 780	438.976	1.96 610	4.23 582	9.12 581	.13 158
7.7	59.29	2.77 489	8.77 496	456.533	1.97 468	4.25 432	9.16 566	.12 987
7.8	60.84	2.79 285	8.83 176	474.552	1.98 319	4.27 266	9.20 516	.12 821
7.9	62.41	2.81 069	8.88 819	493.039	1.99 163	4.29 084	9.24 434	.12 658
8.0	64.00	2.82 843	8.94 427	512.000	2.00 000	4.30 887	9.28 318	.12 500
8.1	65.61	2.84 605	9.00 000	531.441	2.00 830	4.32 675	9.32 170	.12 346
8.2	67.24	2.86 356	9.05 539	551.368	2.01 653	4.34 448	9.35 990	.12 195
8.3	68.89	2.88 097	9.11 043	571.787	2.02 469	4.36 207	9.39 780	.12 048
8.4	70.56	2.89 828	9.16 515	592.704	2.03 279	4.37 952	9.43 539	.11 905
8.5	72.25	2.91 548	9.21 954	614.125	2.04 083	4.39 683	9.47 268	.11 765
8.6	73.96	2.93 258	9.27 362	636.056	2.04 880	4.41 400	9.50 969	.11 628
8.7	75.69	2.94 958	9.32 738	658.503	2.05 671	4.43 105	9.54 640	.11 494
8.8	77.44	2.96 648	9.38 083	681.472	2.06 456	4.44 796	9.58 284	.11 364
8.9	79.21	2.98 329	9.43 398	704.969	2.07 235	4.46 475	9.61 900	.11 236
9.0	81.00	3.00 000	9.48 683	729.000	2.08 008	4.48 140	9.65 489	.11 111
9.1	82.81	3.01 662	9.53 939	753.571	2.08 776	4.49 794	9.69 052	.10 989
9.2	84.64	3.03 315	9.59 166	778.688	2.09 538	4.51 436	9.72 589	.10 870
9.3	86.49	3.04 959	9.64 365	804.357	2.10 294	4.53 065	9.76 100	.10 753
9.4	88.36	3.06 594	9.69 536	830.584	2.11 045	4.54 684	9.79 586	.10 638
9.5	90.25	3.08 221	9.74 679	857.375	2.11 791	4.56 290	9.83 048	.10 526
9.6	92.16	3.09 839	9.79 796	884.736	2.12 532	4.57 886	9.86 485	.10 417
9.7	94.09	3.11 448	9.84 886	912.673	2.13 267	4.59 470	9.89 898	.10 309
9.8	96.04	3.13 050	9.89 949	941.192	2.13 997	4.61 044	9.93 288	.10 204
9.9	98.01	3.14 643	9.94 987	970.299	2.14 723	4.62 607	9.96 655	.10 101
10.0	100.00	3.16 228	10.00 000	1000.000	2.15 443	4.64 159	10.00 000	.10 000
n	n^2	$\sqrt{n}$	$\sqrt{10n}$	n^3	$\sqrt[3]{n}$	$\sqrt[3]{10n}$	$\sqrt[3]{100n}$	$1/n$

Table 2. Circular Functions

x	sin x	tan x	cot x	cos x
.00	.0000	.0000	.....	1.0000
.01	.0100	.0100	99.997	1.0000
.02	.0200	.0200	49.993	.9998
.03	.0300	.0300	33.323	.9996
.04	.0400	.0400	24.987	.9992
.05	.0500	.0500	19.983	.9988
.06	.0600	.0601	16.647	.9982
.07	.0699	.0701	14.262	.9976
.08	.0799	.0802	12.473	.9968
.09	.0899	.0902	11.081	.9960
.10	.0998	.1003	9.967	.9950
.11	.1098	.1104	9.054	.9940
.12	.1197	.1206	8.293	.9928
.13	.1296	.1307	7.649	.9916
.14	.1395	.1409	7.096	.9902
.15	.1494	.1511	6.617	.9888
.16	.1593	.1614	6.197	.9872
.17	.1692	.1717	5.826	.9856
.18	.1790	.1820	5.495	.9838
.19	.1889	.1923	5.200	.9820
.20	.1987	.2027	4.933	.9801
.21	.2085	.2131	4.692	.9780
.22	.2182	.2236	4.472	.9759
.23	.2280	.2341	4.271	.9737
.24	.2377	.2447	4.086	.9713
.25	.2474	.2553	3.916	.9689
.26	.2571	.2660	3.759	.9664
.27	.2667	.2768	3.613	.9638
.28	.2764	.2876	3.478	.9611
.29	.2860	.2984	3.351	.9582
.30	.2955	.3093	3.233	.9553
.31	.3051	.3203	3.122	.9523
.32	.3146	.3314	3.018	.9492
.33	.3240	.3425	2.920	.9460
.34	.3335	.3537	2.827	.9428
.35	.3429	.3650	2.740	.9394
.36	.3523	.3764	2.657	.9359
.37	.3616	.3879	2.578	.9323
.38	.3709	.3994	2.504	.9287
.39	.3802	.4111	2.433	.9249
.40	.3894	.4228	2.365	.9211
.41	.3986	.4346	2.301	.9171
.42	.4078	.4466	2.239	.9131
.43	.4169	.4586	2.180	.9090
.44	.4259	.4708	2.124	.9048
.45	.4350	.4831	2.070	.9004
.46	.4439	.4954	2.018	.8961
.47	.4529	.5080	1.969	.8916
.48	.4618	.5206	1.921	.8870
.49	.4706	.5334	1.875	.8823
.50	.4794	.5463	1.830	.8776
.51	.4882	.5594	1.788	.8727
.52	.4969	.5726	1.747	.8678
.53	.5055	.5859	1.707	.8628
.54	.5141	.5994	1.668	.8577
.55	.5227	.6131	1.631	.8525
.56	.5312	.6269	1.595	.8473
.57	.5396	.6410	1.560	.8419
.58	.5480	.6552	1.526	.8365
.59	.5564	.6696	1.494	.8309
.60	.5646	.6841	1.462	.8253

x	sin x	tan x	cot x	cos x
.60	.5646	.6841	1.462	.8253
.61	.5729	.6989	1.431	.8196
.62	.5810	.7139	1.401	.8139
.63	.5891	.7291	1.372	.8080
.64	.5972	.7445	1.343	.8021
.65	.6052	.7602	1.315	.7961
.66	.6131	.7761	1.288	.7900
.67	.6210	.7923	1.262	.7838
.68	.6288	.8087	1.237	.7776
.69	.6365	.8253	1.212	.7712
.70	.6442	.8423	1.187	.7648
.71	.6518	.8595	1.163	.7584
.72	.6594	.8771	1.140	.7518
.73	.6669	.8949	1.117	.7452
.74	.6743	.9131	1.095	.7385
.75	.6816	.9316	1.073	.7317
.76	.6889	.9505	1.052	.7248
.77	.6961	.9697	1.031	.7179
.78	.7033	.9893	1.011	.7109
.79	.7104	1.009	.9908	.7038
.80	.7174	1.030	.9712	.6967
.81	.7243	1.050	.9520	.6895
.82	.7311	1.072	.9331	.6822
.83	.7379	1.093	.9146	.6749
.84	.7446	1.116	.8964	.6675
.85	.7513	1.138	.8785	.6600
.86	.7578	1.162	.8609	.6524
.87	.7643	1.185	.8437	.6448
.88	.7707	1.210	.8267	.6372
.89	.7771	1.235	.8100	.6294
.90	.7833	1.260	.7936	.6216
.91	.7895	1.286	.7774	.6137
.92	.7956	1.313	.7615	.6058
.93	.8016	1.341	.7458	.5978
.94	.8076	1.369	.7303	.5898
.95	.8134	1.398	.7151	.5817
.96	.8192	1.428	.7001	.5735
.97	.8249	1.459	.6853	.5653
.98	.8305	1.491	.6707	.5570
.99	.8360	1.524	.6563	.5487
1.00	.8415	1.557	.6421	.5403
1.01	.8468	1.592	.6281	.5319
1.02	.8521	1.628	.6142	.5234
1.03	.8573	1.665	.6005	.5148
1.04	.8624	1.704	.5870	.5062
1.05	.8674	1.743	.5736	.4976
1.06	.8724	1.784	.5604	.4889
1.07	.8772	1.827	.5473	.4801
1.08	.8820	1.871	.5344	.4713
1.09	.8866	1.917	.5216	.4625
1.10	.8912	1.965	.5090	.4536
1.11	.8957	2.014	.4964	.4447
1.12	.9001	2.066	.4840	.4357
1.13	.9044	2.120	.4718	.4267
1.14	.9086	2.176	.4596	.4176
1.15	.9128	2.234	.4475	.4085
1.16	.9168	2.296	.4356	.3993
1.17	.9208	2.360	.4237	.3902
1.18	.9246	2.427	.4120	.3809
1.19	.9284	2.498	.4003	.3717
1.20	.9320	2.572	.3888	.3624

Table 2. Circular Functions

x	sin x	tan x	cot x	cos x
1.20	.9320	2.572	.3888	.3624
1.21	.9356	2.650	.3773	.3530
1.22	.9391	2.733	.3659	.3436
1.23	.9425	2.820	.3546	.3342
1.24	.9458	2.912	.3434	.3248
1.25	.9490	3.010	.3323	.3153
1.26	.9521	3.113	.3212	.3058
1.27	.9551	3.224	.3102	.2963
1.28	.9580	3.341	.2993	.2867
1.29	.9608	3.467	.2884	.2771
1.30	.9636	3.602	.2776	.2675
1.31	.9662	3.747	.2669	.2579
1.32	.9687	3.903	.2562	.2482
1.33	.9711	4.072	.2456	.2385
1.34	.9735	4.256	.2350	.2288
1.35	.9757	4.455	.2245	.2190
1.36	.9779	4.673	.2140	.2092
1.37	.9799	4.913	.2035	.1994
1.38	.9819	5.177	.1931	.1896
1.39	.9837	5.471	.1828	.1798
1.40	.9854	5.798	.1725	.1700

x	sin x	tan x	cot x	cos x
1.40	.9854	5.798	.1725	.1700
1.41	.9871	6.165	.1622	.1601
1.42	.9887	6.581	.1519	.1502
1.43	.9901	7.055	.1417	.1403
1.44	.9915	7.602	.1315	.1304
1.45	.9927	8.238	.1214	.1205
1.46	.9939	8.989	.1113	.1106
1.47	.9949	9.887	.1011	.1006
1.48	.9959	10.983	.0910	.0907
1.49	.9967	12.350	.0810	.0807
1.50	.9975	14.101	.0709	.0707
1.51	.9982	16.428	.0609	.0608
1.52	.9987	19.670	.0508	.0508
1.53	.9992	24.498	.0408	.0408
1.54	.9995	32.461	.0308	.0308
1.55	.9998	48.078	.0208	.0208
1.56	.9999	92.620	.0108	.0108
1.57	1.0000	1255.8	.0008	.0008
1.58	1.0000	−108.65	−.0092	−.0092
1.59	.9998	−52.067	−.0192	−.0192
1.60	.9996	−34.233	−.0292	−.0292

Table 3. Degrees and Radians

1. To Change Degrees, Minutes, and Seconds into Radians

°	Radians	′	Radians	″	Radians
1	0.01745 33	1	0.00029 09	1	0.00000 48
2	0.03490 66	2	0.00058 18	2	0.00000 97
3	0.05235 99	3	0.00087 27	3	0.00001 45
4	0.06981 32	4	0.00116 36	4	0.00001 94
5	0.08726 65	5	0.00145 44	5	0.00002 42
6	0.10471 98	6	0.00174 53	6	0.00002 91
7	0.12217 30	7	0.00203 62	7	0.00003 39
8	0.13962 63	8	0.00232 71	8	0.00003 88
9	0.15707 96	9	0.00261 80	9	0.00004 36
10	0.17453 29	10	0.00290 89	10	0.00004 85
20	0.34906 59	20	0.00581 78	20	0.00009 70
30	0.52359 88	30	0.00872 66	30	0.00014 54
40	0.69813 17	40	0.01163 55	40	0.00019 39
50	0.87266 46	50	0.01454 44	50	0.00024 24
60	1.04719 76	60	0.01745 33	60	0.00029 09

2. To Change Radians to Degrees, Minutes, and Seconds

Rad.	Radians			Tenths			Hundredths			Thousandths			Ten-Thousandths	
	°	′	″	°	′	″	°	′	″	°	′	″	′	″
1	57	17	45	5	43	46	0	34	23	0	3	26	0	21
2	114	35	30	11	27	33	1	8	45	0	6	53	0	41
3	171	53	14	17	11	19	1	43	8	0	10	19	1	2
4	229	10	59	22	55	6	2	17	31	0	13	45	1	22
5	286	28	44	28	38	52	2	51	53	0	17	11	1	43
6	343	46	29	34	22	39	3	26	16	0	20	38	2	4

Table 4. Natural Logarithms

x	$\ln x$	x	$\ln x$	x	$\ln x$	x	$\ln x$	x	$\ln x$
0	$-\infty$	**2.0**	0.69 315	**4.0**	1.38 629	**6.0**	1.79 176	**8.0**	2.07 944
0.1	7.697 − 10	2.1	0.74 194	4.1	1.41 099	6.1	1.80 829	8.1	2.09 186
0.2	8.391 − 10	2.2	0.78 846	4.2	1.43 508	6.2	1.82 455	8.2	2.10 413
0.3	8.796 − 10	2.3	0.83 291	4.3	1.45 862	6.3	1.84 055	8.3	2.11 626
0.4	9.084 − 10	2.4	0.87 547	4.4	1.48 160	6.4	1.85 630	8.4	2.12 823
0.5	9.307 − 10	2.5	0.91 629	4.5	1.50 408	6.5	1.87 180	8.5	2.14 007
0.6	9.489 − 10	2.6	0.95 551	4.6	1.52 606	6.6	1.88 707	8.6	2.15 176
0.7	9.643 − 10	2.7	0.99 325	4.7	1.54 756	6.7	1.90 211	8.7	2.16 332
0.8	9.777 − 10	2.8	1.02 962	4.8	1.56 862	6.8	1.91 692	8.8	2.17 475
0.9	9.895 − 10	2.9	1.06 471	4.9	1.58 924	6.9	1.93 152	8.9	2.18 605
1.0	0.00 000	**3.0**	1.09 861	**5.0**	1.60 944	**7.0**	1.94 591	**9.0**	2.19 722
1.1	0.09 531	3.1	1.13 140	5.1	1.62 924	7.1	1.96 009	9.1	2.20 827
1.2	0.18 232	3.2	1.16 315	5.2	1.64 866	7.2	1.97 408	9.2	2.21 920
1.3	0.26 236	3.3	1.19 392	5.3	1.66 771	7.3	1.98 787	9.3	2.23 001
1.4	0.33 647	3.4	1.22 378	5.4	1.68 640	7.4	2.00 148	9.4	2.24 071
1.5	0.40 547	3.5	1.25 276	5.5	1.70 475	7.5	2.01 490	9.5	2.25 129
1.6	0.47 000	3.6	1.28 093	5.6	1.72 277	7.6	2.02 815	9.6	2.26 176
1.7	0.53 063	3.7	1.30 833	5.7	1.74 047	7.7	2.04 122	9.7	2.27 213
1.8	0.58 779	3.8	1.33 500	5.8	1.75 786	7.8	2.05 412	9.8	2.28 238
1.9	0.64 185	3.9	1.36 098	5.9	1.77 495	7.9	2.06 686	9.9	2.29 253

Table 5. Exponential Function

x	e^x	e^{-x}
0	1.0000	1.0 000
.1	1.1052	.90 484
.2	1.2214	.81 873
.3	1.3499	.74 082
.4	1.4918	.67 032
.5	1.6487	.60 653
.6	1.8221	.54 881
.7	2.0138	.49 659
.8	2.2255	.44 933
.9	2.4596	.40 657
1.0	2.7183	.36 788
1.1	3.0042	.33 287
1.2	3.3201	.30 119
1.3	3.6693	.27 253
1.4	4.0552	.24 660
1.5	4.4817	.22 313
1.6	4.9530	.20 190
1.7	5.4739	.18 268
1.8	6.0496	.16 530
1.9	6.6859	.14 957
2.0	7.3891	.13 534
2.1	8.1662	.12 246
2.2	9.0250	.11 080
2.3	9.9742	.10 026
2.4	11.023	.09 072
2.5	12.182	.08 208
2.6	13.464	.07 427
2.7	14.880	.06 721
2.8	16.445	.06 081
2.9	18.174	.05 502
3.0	20.086	.04 979
3.1	22.198	.04 505
3.2	24.533	.04 076
3.3	27.113	.03 688
3.4	29.964	.03 337

x	e^x	e^{-x}
3.5	33.115	.03 020
3.6	36.598	.02 732
3.7	40.447	.02 472
3.8	44.701	.02 237
3.9	49.402	.02 024
4.0	54.598	.01 832
4.1	60.340	.01 657
4.2	66.686	.01 500
4.3	73.700	.01 357
4.4	81.451	.01 228
4.5	90.017	.01 111
4.6	99.484	.01 005
4.7	109.95	.00 910
4.8	121.51	.00 823
4.9	134.29	.00 745
5.0	148.41	.00 674
5.1	164.02	.00 610
5.2	181.27	.00 552
5.3	200.34	.00 499
5.4	221.41	.00 452
5.5	244.69	.00 409
5.6	270.43	.00 370
5.7	298.87	.00 335
5.8	330.30	.00 303
5.9	365.04	.00 274
6.0	403.43	.00 248
6.1	445.86	.00 224
6.2	492.75	.00 203
6.3	544.57	.00 184
6.4	601.85	.00 166
6.5	665.14	.00 150

Summary of Formulas

Geometry

Circumference of circle: $C = 2\pi r$.
Length of circular arc: $s = r\theta$, where θ = number of radians in central angle.
Hypotenuse of right triangle (Pythagorean theorem): $c = \sqrt{a^2 + b^2}$.

Areas:

Parallelogram: $K = bh$.
Square: $K = e^2$.
Trapezoid: $K = \frac{1}{2}h(b_1 + b_2)$.
Triangle: $K = \frac{1}{2}bh$.

$$K = \sqrt{s(s-a)(s-b)(s-c)}, \text{ where } s = \tfrac{1}{2}(a+b+c).$$
$$K = \tfrac{1}{2}bc \sin A = \tfrac{1}{2}ab \sin C = \tfrac{1}{2}ca \sin B.$$

Circle: $K = \pi r^2$.
Sector of circle: $K = \frac{1}{2}r^2\theta$, where θ = number of radians in central angle.
Segment of circle: $K = \frac{1}{2}r^2(\theta - \sin\theta)$.
Surface of sphere: $S = 4\pi r^2$.
Curved surface of circular cylinder: $S = 2\pi rh$.
Total surface of circular cylinder: $T = 2\pi rh + 2\pi r^2$.
Curved surface of circular cone: $S = \pi re$, where slant height $e = \sqrt{h^2 + r^2}$.
Total surface of circular cone: $T = \pi re + \pi r^2$.
Regular polygon: $K = \frac{1}{2}ap$, where a and p are lengths of apothem and perimeter, respectively.
Ellipse: $K = \pi ab$ (see Ex. 7 of Exercises 8.4[A], page 281).

Volumes:

Prism: $V = Bh$, where B = number of square units in area of base.
Cube: $V = e^3$.
Pyramid: $V = \frac{1}{3}Bh$.
Circular cylinder: $V = \pi r^2 h$.
Circular cone: $V = \frac{1}{3}\pi r^2 h$.
Sphere: $V = \frac{4}{3}\pi r^3$.

Analytic Geometry

Distance between two points: $P_1P_2 = \sqrt{(x_2 - x_1)^2 + (y_2 - y_1)^2}$.

Midpoint: $P\left(\dfrac{x_1 + x_2}{2}, \dfrac{y_1 + y_2}{2}\right)$.

Slope of a line: $m = \dfrac{y_2 - y_1}{x_2 - x_1} = \dfrac{y_1 - y_2}{x_1 - x_2}$, or $m = \dfrac{\Delta y}{\Delta x}$.

Condition for parallel lines: $m_1 = m_2$.

Condition for perpendicular lines: $m_1 m_2 = -1$.

Angle between two lines: $\tan\theta = \dfrac{m_2 - m_1}{1 + m_2 m_1}$.

Distance between a point and a line: $PQ = \dfrac{|ax_1 + by_1 + c|}{\sqrt{a^2 + b^2}}$.

Equation of straight line:

Two-point form: $y - y_1 = \dfrac{y_2 - y_1}{x_2 - x_1}(x - x_1)$.
Point-slope form: $y - y_1 = m(x - x_1)$.
Slope-intercept form: $y = mx + b$.
Intercept form: $\dfrac{x}{a} + \dfrac{y}{b} = 1$.
Normal form: $x\cos\theta + y\sin\theta - p = 0$.

Circle:

General equation: $x^2 + y^2 + ax + by + c = 0$.
Center at origin, radius r: $x^2 + y^2 = r^2$, or $x = r\cos\theta$, $y = r\sin\theta$.
  Equation of line tangent at (x_1, y_1): $x_1 x + y_1 y = r^2$.
Center at (h, k), radius r: $(x - h)^2 + (y - k)^2 = r^2$.

Ellipse:

Center at origin: $x^2/a^2 + y^2/b^2 = 1$, or $x = a\cos\theta$, $y = b\sin\theta$.
  Equation of line tangent at (x_1, y_1): $x_1 x/a^2 + y_1 y/b^2 = 1$.
  If $a > b$: $c^2 = a^2 - b^2$; eccentricity, $e = c/a$; foci, $(-c, 0)$, $(c, 0)$.
  If $b > a$: $c^2 = b^2 - a^2$; eccentricity, $e = c/b$; foci, $(0, -c)$, $(0, c)$.
Center at (h, k): $(x - h)^2/a^2 + (y - k)^2/b^2 = 1$.

Hyperbola:

Center at origin: $x^2/a^2 - y^2/b^2 = 1$, or $x = a\sec\theta$, $y = b\tan\theta$.
  Equation of line tangent at (x_1, y_1): $x_1 x/a^2 - y_1 y/b^2 = 1$.
  Asymptotes: $y = \dfrac{b}{a}x$, $y = -\dfrac{b}{a}x$. Foci: $(-c, 0)$, $(c, 0)$.
  Eccentricity: $e = c/a$, where $c^2 = a^2 + b^2$.
Conjugate, center at origin: $y^2/b^2 - x^2/a^2 = 1$, or $y = b\sec\theta$, $x = a\tan\theta$.
  Equation of line tangent at (x_1, y_1): $y_1 y/b^2 - x_1 x/a^2 = 1$.
  Asymptotes: $y = \dfrac{b}{a}x$, $y = -\dfrac{b}{a}x$. Foci: $(0, -c)$, $(0, c)$.
  Eccentricity: $e = c/b$, where $c^2 = a^2 + b^2$.
Center at (h, k): $(x - h)^2/a^2 - (y - k)^2/b^2 = \pm 1$.
With asymptotes $x = a$, $y = b$: $(x - a)(y - b) = k$.

Parabola:

Vertex at origin, opening in direction of positive x: $y^2 = 4px$ $(p > 0)$, or $x = pt^2$, $y = 2pt$.

Equation of line tangent at (x_1, y_1): $y_1 y = 2p(x + x_1)$.

Focus: $(p, 0)$. Directrix: $x + p = 0$.

Vertex at (h, k), opening in direction of positive x: $(y - k)^2 = 4p(x - h)$.

Trigonometry

In triangle ABC, with $\angle C$ a right angle:

$$\sin A = \frac{a}{c}. \qquad \csc A = \frac{1}{\sin A} = \frac{c}{a}.$$

$$\cos A = \frac{b}{c}. \qquad \sec A = \frac{1}{\cos A} = \frac{c}{b}.$$

$$\tan A = \frac{a}{b} = \frac{\sin A}{\cos A}. \qquad \cot A = \frac{1}{\tan A} = \frac{\cos A}{\sin A} = \frac{b}{a}.$$

$\cos(-x) = \cos x.$ $\qquad \sin(-x) = -\sin x.$

$\cos(\pi - x) = -\cos x.$ $\qquad \sin(\pi - x) = \sin x.$

$\cos(\pi + x) = -\cos x.$ $\qquad \sin(\pi + x) = -\sin x.$

$\cos^2 x + \sin^2 x = 1.$

$\sec^2 x = \tan^2 x + 1.$

$\csc^2 x = \cot^2 x + 1.$

$\cos(x + y) = \cos x \cos y - \sin x \sin y.$

$\cos(x - y) = \cos x \cos y + \sin x \sin y.$

$\sin(x + y) = \sin x \cos y + \cos x \sin y.$

$\sin(x - y) = \sin x \cos y - \cos x \sin y.$

$$\tan(x + y) = \frac{\tan x + \tan y}{1 - \tan x \tan y} \quad (\cos x \cos y \neq 0).$$

$$\tan(x - y) = \frac{\tan x - \tan y}{1 + \tan x \tan y} \quad (\cos x \cos y \neq 0).$$

$\sin 2x = 2 \sin x \cos x.$

$\cos 2x = \cos^2 x - \sin^2 x = 2\cos^2 x - 1 = 1 - 2\sin^2 x.$

$$\tan 2x = \frac{2 \tan x}{1 - \tan^2 x}.$$

$$\left|\sin \frac{x}{2}\right| = \sqrt{\frac{1 - \cos x}{2}}. \qquad \left|\cos \frac{x}{2}\right| = \sqrt{\frac{1 + \cos x}{2}}.$$

$$\tan \frac{x}{2} = \frac{1 - \cos x}{\sin x} = \frac{\sin x}{1 + \cos x}.$$

$$\sin x + \sin y = 2 \sin \frac{x + y}{2} \cos \frac{x - y}{2}.$$

$$\sin x - \sin y = 2 \cos \frac{x + y}{2} \sin \frac{x - y}{2}.$$

$$\cos x + \cos y = 2 \cos \frac{x + y}{2} \cos \frac{x - y}{2}.$$

$$\cos x - \cos y = -2 \sin \frac{x + y}{2} \sin \frac{x - y}{2}.$$

$$\sin x \cos y = \tfrac{1}{2}[\sin (x + y) + \sin (x - y)].$$
$$\cos x \cos y = \tfrac{1}{2}[\cos (x + y) + \cos (x - y)].$$
$$\sin x \sin y = -\tfrac{1}{2}[\cos (x + y) - \cos (x - y)].$$

Law of sines: $\dfrac{a}{\sin A} = \dfrac{b}{\sin B} = \dfrac{c}{\sin C}.$

Law of cosines: $c^2 = a^2 + b^2 - 2ab \cos C;$
$a^2 = b^2 + c^2 - 2bc \cos A;$
$b^2 = c^2 + a^2 - 2ca \cos B.$

Differentiation

If $f(x) =$	then $f'(x) =$
1. c	0
2. x^n	nx^{n-1}
3. $cg(x)$	$cg'(x)$
4. $f_1(x) + f_2(x)$	$f_1'(x) + f_2'(x)$
5. $g(x)h(x)$	$g(x)h'(x) + g'(x)h(x)$
6. $\dfrac{g(x)}{h(x)}$	$\dfrac{h(x)g'(x) - g(x)h'(x)}{[h(x)]^2}$
7. $g[h(x)]$	$g'[h(x)] \cdot h'(x)$
8. $[g(x)]^n$	$n[g(x)]^{n-1}g'(x)$
9. $\sin x$	$\cos x$
10. $\cos x$	$-\sin x$
11. $\tan x$	$\sec^2 x$
12. $\cot x$	$-\csc^2 x$
13. $\sec x$	$\sec x \tan x$
14. $\csc x$	$-\csc x \cot x$
15. $\ln x$	$1/x$
16. e^x	e^x

Integration

1. $\int x^n\,dx = \frac{1}{n+1}x^{n+1} + C \ (n \neq -1)$.

2. $\int \frac{1}{x}\,dx = \ln|x| + C$.

3. $\int \sin x\,dx = -\cos x + C$.

4. $\int \cos x\,dx = \sin x + C$.

5. $\int \tan x\,dx = \ln|\sec x| + C$.

6. $\int \sec x\,dx = \ln\left|\tan\left(\frac{\pi}{4} + \frac{x}{2}\right)\right| + C$

$$= \ln|\sec x + \tan x| + C = \frac{1}{2}\ln\frac{1 + \sin x}{1 - \sin x} + C.$$

7. $\int e^x\,dx = e^x + C$.

8. $\int \frac{dx}{\sqrt{a^2 - x^2}} = \sin^{-1}\frac{x}{a} + C$.

9. $\int \frac{dx}{x^2 + a^2} = \frac{1}{a}\tan^{-1}\frac{x}{a} + C$.

10. $\int \sqrt{a^2 - x^2}\,dx = \frac{1}{2}x\sqrt{a^2 - x^2} + \frac{1}{2}a^2\sin^{-1}\frac{x}{a} + C$.

11. $\int \frac{dx}{a^2 - x^2} = \frac{1}{2a}\ln\left|\frac{a + x}{a - x}\right| + C$.

Answers to Selected Exercises

Exercises 1.2 [A], pages 9–10

1. **(a)** $\{22, 24, 26, 28\}$; **(b)** $\{20, 22, 24, 26, 28, 30\}$; **(c)** $\{-1, 0, 1\}$; **(d)** $\{-1, 0, 1\}$; **(e)** {Robert Louis Stevenson}.

2. Many possibilities, for example: **(a)** $\{x: x^2 = 1\}$; **(b)** $\{x: x$ is prime and $10 < x < 20\}$; **(c)** $\{x: 17x = 0\}$; **(d)** $\{x: x$ is a New England state$\}$.

3. **(a)** $x = \pm 4$; **(b)** no solutions; **(c)** $x = 1, -5$; **(d)** $x = \frac{7}{2}, -\frac{3}{2}$.

4. **(a)** $A \cup B = \{1, 2, 3, 4\}$; $B \cup C = \{2, 3, 4, 5\}$; $A \cup C = \{1, 3, 5\}$. **(b)** $(0, 1) \cup (2, 3)$ is not an interval. **(c)** A is a part (a subset) of B.

5. **(a)** $[0, 2]$; **(b)** $(-\infty, 1) \cup [3, \infty)$; **(c)** $(-1, 5)$; **(d)** $(-1, 0) \cup (2, 3)$; **(e)** $(0, 1) \cup (1, 2)$; **(f)** $(0, 2)$.

6. **(a)** $x = \pm 4$; **(b)** no solutions; **(c)** $x = 0, \pm 2$.

9. $x = 2, -1$. **10.** $(-\infty, 0) \cup (\frac{1}{2}, \infty)$.

Exercises 1.2 [B], pages 10–11

1. **(a)** $\{-3, -2, -1, 0, 1, 2, 3\}$; **(b)** $\{-2, -1, 0, 1, 2\}$; **(c)** $\{-5, -4, -3, -2, -1, 0, 1\}$; **(d)** $\{3 + \sqrt{3}, 3 - \sqrt{3}\}$; **(e)** $\{1, \frac{1}{2}\}$.

2. **(a)** $(-4, -2)$; **(b)** $(-3, -1) \cup (-1, 1)$; **(c)** $[0, \infty)$; **(d)** $(1, 4)$; **(e)** $(-\infty, 0) \cup (4, \infty)$; **(f)** $(1, 5) \cup (5, 9)$.

3. **(a)** $y = \pm\frac{3}{5}$; **(b)** $x = \pm(-1 + \sqrt{2})$; **(c)** $s = \pm 1, \pm 3$.

5. $4, -2$.

6. **(a)** $[1, 2)$; **(b)** $[-2, 2]$; **(c)** A is a part (a subset) of B; **(d)** No. There are *no* points in $(0, 1) \cap (1, 2)$, for example.

7. $2 < x < \frac{7}{3}$. **8.** $-\frac{7}{2} < x < \frac{3}{2}$.

9. $x < (x + b)/2 < b$, $a < (x + a)/2 < x$.

Exercises 1.3 [A], pages 20–21

1. **(a)** $3.2\sqrt{5}$ units; **(b)** $\sqrt{h^2 + k^2}$ units; **(c)** $\sqrt{12610}/35$ units.

2. **(a)** $-\frac{48}{37}$; **(b)** $-\frac{8}{5}$; **(c)** no slope; **(d)** $b(b - a)/a(b + a)$.

4. **(a)** Yes; **(b)** yes; **(c)** no. **5.** ± 4.

6. $\tan^{-1}\frac{29}{26} = 48° 7'$; $\tan^{-1}\frac{29}{15} = 62° 39'$; $\tan^{-1}\frac{29}{11} = 69° 14'$.

Exercises 1.3 [B], page 21

1. $(\frac{7}{4}, 0)$. **2.** $(0, -3)$.

4. **(a)** $\tan^{-1}(-\frac{5}{3}) = 120° 58'$; $\tan^{-1}\frac{5}{11} = 24° 26'$; $\tan^{-1}\frac{20}{29} = 34° 35'$. **(b)** $45°, 45°, 90°$.

5. $(\frac{7}{3}, \frac{2}{3})$. **6.** $(-1, 5), (3, 7), (5, 1)$.

Miscellaneous Exercises 1.3, pages 21–22

2. 45°, 45°, 135°, 135°. **3. (a)** (2, −4), (14, 4), (−2, 4). **4.** 3, $\frac{1}{3}$.
5. $(2 \pm 5\sqrt{5})/11$. **7.** $(\frac{5}{3}, \frac{14}{3})$.

Exercises 1.4 [A], pages 25–26

1.

	slope	x-int.	y-int.
(a)	$\frac{1}{5}$	-15	3
(b)	$-\frac{3}{2}$	4	6
(c)	none	$\frac{3}{5}$	none
(d)	-1	$-k$	$-k$

2. (a) $x + 10y - 27 = 0$; **(b)** $y + 5 = 0$; **(d)** $2x - 3y + 12 = 0$;
(e) $x - 2y - 6 = 0$; **(f)** $x - y\sqrt{3} + 3 = 0$; **(g)** $2x - y - 5 = 0$;
(h) $2x - 3y = 0$.

3. $x - 8y + 53 = 0$; $7x - 3y - 53 = 0$; $8x - 11y = 0$.

4. $\left.\begin{matrix} x - y - 2 = 0 \\ x + y - 2 = 0 \\ x - y + 2 = 0 \\ x + y + 2 = 0 \end{matrix}\right\} |x| \le 2,\ |y| \le 2;$ *or* $|x| + |y| = 2$.

5. $d = |y_1 - k|$, or $d = |k - y_1|$. **7.** $\tan^{-1}(\pm 7)$.

Exercises 1.4 [B], pages 26–27

1. (a) $21x + 26y - 72 = 0$; **(b)** $x = 0$; the y-axis; **(c)** $2x + y = 0$;
(d) $y - 5 = 0$; **(e)** $x + y - 4 = 0$; **(f)** $2x + y - 7 = 0$;
(g) $ax + by - (ax_1 + by_1) = 0$; **(h)** $bx - ay - (bx_1 - ay_1) = 0$.

2. $(0, 4\sqrt{3} - 1)$.

3. (a) $4x - 3y + 2 = 0$; $x + 3y - 7 = 0$; $x - 1 = 0$.

5. $\pm 2\sqrt{2}$.

6. (a) (−4, 2), (4, −5), (3, 4). **(b)** $(0, -\frac{3}{2})$, $(\frac{7}{2}, -\frac{1}{2})$, $(-\frac{1}{2}, 3)$.

7. (a) $\tan^{-1} \dfrac{a_1b_2 - a_2b_1}{a_1a_2 + b_1b_2}$. **(b) (i)** $a_1b_2 - a_2b_1 = 0$ and $c_1 \neq c_2$;
(ii) $a_1a_2 + b_1b_2 = 0$.

8. $-\cot\theta$, $|p|$.

Miscellaneous Exercises 1.4, pages 27–29

2. (a) 5; **(b)** 0; **(c)** −2; **(d)** −7.

3. (b) $2x + 3y - 13 = 0$; **(c)** $172x - 313y = 0$; **(d)** $3x + 4y = 0$;
(e) $ax + by - (ax_1 + by_1) = 0$.

4. (a) −3; **(b)** −11; **(c)** 0; **(d)** 0.

5. (a) $3x - 2y - 4 = 0$; **(b)** $x - y + 7 = 0$; **(c)** $192x + 43y = 0$.

6. **(a)** $4x + 3y - 26 = 0$; **(b)** $(5, 2)$; **(c)** 5 units.

7. **(a)** $1/\sqrt{5}$ units; **(b)** $1/\sqrt{29}$ units; **(c)** $|c|/\sqrt{a^2 + b^2}$ units.

8. **(a)** **(i)** $m = 1$, $x - y + 5 = 0$; **(ii)** $m = 1$, $x - y + 5 = 0$; **(iii)** $m = 0$, $y - 2 = 0$; **(iv)** $m = \frac{1}{2}$, $x - 2y + 7 = 0$. **(b)** $x + 3 = 0$.

9. **(a)** **(i)** $k = -4$, $2x + 19y + 19 = 0$; **(ii)** $k = \frac{3}{4}$, $11x + 19y = 0$; **(iii)** $k = -\frac{4}{11}$, $18x - 9y - 49 = 0$; **(iv)** $k = -\frac{1}{5}$, $9x - 19 = 0$. **(b)** $x + 5y + 4 = 0$.

10. $(2k + 1)x + (3k + 1)y - (k + 1) = 0$. **(a)** $x + 2y = 0$; **(b)** $x - y - 3 = 0$; **(c)** $2x - y - 5 = 0$.

Exercises 1.4 [C], pages 29–31

5. **(a)** $y - y_1 = \frac{b}{a}(x - x_1)$, or $bx - ay = bx_1 - ay_1$.

7. **(a)** 6 units; **(b)** $2\sqrt{10}$ units; **(c)** $|c|/\sqrt{a^2 + b^2}$ units; **(d)** 0.

8. **(a)** $10/\sqrt{13}$ units; **(b)** $5/2\sqrt{26}$ units; **(c)** $|c_1 - c_2|/\sqrt{a^2 + b^2}$ units.

9. **(c)** $4x - 3y - 3 = 0$, $2x - 7y + 3 = 0$; **(d)** $(3 - 2\sqrt{2})x + (3 + \sqrt{2})y - (15 + 3\sqrt{2}) = 0$, $(3 + 2\sqrt{2})x + (3 - \sqrt{2})y - (15 - 3\sqrt{2}) = 0$.

10. **(a)** $\dfrac{|3X + 4Y - 7|}{5} = \dfrac{|5X - 12Y + 4|}{13}$; **(b)** the pair of lines bisecting the vertical angles formed by the given lines; **(c)** $14x + 112y - 111 = 0$, $64x - 8y - 71 = 0$.

11. **(a)** $y - 2 = 0$, $x - 1 = 0$; **(b)** $2y + 11 = 0$, $4x + 3 = 0$; **(c)** $(\cos\alpha \pm \cos\beta)x + (\sin\alpha \pm \sin\beta)y + (p_1 \pm p_2) = 0$.

12. **(a)** $\frac{5}{2}$ sq. units; **(b)** $\frac{1}{2}$ sq. unit; **(c)** 35 sq. units; **(d)** 0; **(e)** $\frac{1}{2}|x_1y_2 - x_2y_1|$ sq. units.

13. **(a)** $\cos A = \dfrac{x_1x_2 + y_1y_2}{\sqrt{(x_1^2 + y_1^2)(x_2^2 + y_2^2)}}$.

Exercises 2.1 [A], pages 38–39

1. **(a)** D = $\{3, 7\}$, R = $\{3, 7, 14\}$; not a fn. **(b)** D = R = $(-\infty, \infty)$; fn. **(c)** D = R = $(-\infty, \infty)$; not a fn. **(d)** D = $(-\infty, \infty)$, R = $\{1\}$; fn. **(e)** D = $\{1\}$, R = $(-\infty, \infty)$; not a fn. **(f)** D = R = $(-\infty, \infty)$; fn.

2. **(a)** D = $(-\infty, \infty)$, R = $[0, \infty)$. **(b)** D = $[1, \infty)$, R = $[0, \infty)$. **(c)** D = $[-2, 2]$, R = $[0, 2]$. **(d)** D = $(0, \infty)$, R = $(-\infty, \infty)$. **(e)** D = $\{x: x \neq 2\} = (-\infty, 2) \cup (2, \infty)$, R = $\{x: x \neq 4\} = (-\infty, 4) \cup (4, \infty)$.

3. Various possibilities, for example: **(a)** $\{(x, y): y = 1/x\}$; **(b)** $\{(x, y): y = x^2 - 5\}$; **(c)** $\{(x, y): y = \cos x\}$.

4. Various possibilities, for example: **(a)** $f(x) = 2x$; **(b)** $f(x) = x - 1$; **(c)** $f(x) = x^2$.

5. $5, -5, 2 - \sqrt{3}, 2a - 1.$ **6.** $\dfrac{2(x-1)^2}{x^2(x-2)}.$

7. $f(g(t)) = 2t^2 + 1,\ g(f(t)) = (2t+1)^2;\ \ t = 0, -2.$

8. $(1, -1) \in g,\ (1, -1) \notin f.$ **9.** $-9, 1, 1, 9;\ \ D = R = (-\infty, \infty).$

Exercises 2.1 [B], pages 39–40

1. **(a)** $D = \{-6, \pi, 4\}$, $R = \{2, 4\}$; fn. **(b)** $D = R = (-\infty, \infty)$; fn. **(c)** $D = R = (-\infty, \infty)$; not a fn. **(d)** $D = [0, \infty)$, $R = (-\infty, \infty)$; not a fn. **(e)** $D = (-\infty, \infty)$, $R = [1, \infty)$; fn. **(f)** $D = (0, \infty)$, $R = (-8, \infty)$; fn.

2. **(a)** $D = \{x: x \neq 0\} = (-\infty, 0) \cup (0, \infty)$, $R = \{-1, 1\}$. **(b)** $D = (-\infty, \infty)$, $R = [-1, 1]$. **(c)** $D = \{x: 0 < |x| \leq 1\} = [-1, 0) \cup (0, 1]$, $R = (-\infty, \infty)$. **(d)** $D = \{x: x \neq n\pi/2$, n an odd integer$\}$, $R = (-\infty, \infty)$. **(e)** $D = \{x: x \neq 2\} = (-\infty, 2) \cup (2, \infty)$, $R = \{x: x \neq 6\} = (-\infty, 6) \cup (6, \infty)$.

3. Various possibilities, including **(a)** $\{(x, y): 2x - 3y - 5 = 0\}$; **(b)** $\{(x, y): y = x^3 - 5x\}$; **(c)** $\{(x, y): y = \log x\}$.

4. **(a)** $f(x) = 1 - x$; **(b)** $f(x) = 1/(1 + x)$; **(c)** $f(x) = 2x - 1$.

5. $1, z^2 + 4z + 2$; $\{-1 \pm \sqrt{2}\}$. **6.** $4\sqrt{2}$. **7.** x.

8. $(0, 4) \in g,\ (0, 4) \notin f.$

Miscellaneous Exercises 2.2 (1), pages 44–46

1.

x	$W(x)$	$\sin x$	$\cos x$	$\tan x$	$\cot x$	$\sec x$	$\csc x$
0	$(1, 0)$	0	1	0	—	1	—
$\pi/6$	$(\tfrac12\sqrt3, \tfrac12)$	$\tfrac12$	$\tfrac12\sqrt3$	$\tfrac13\sqrt3$	$\sqrt3$	$\tfrac23\sqrt3$	2
$\pi/4$	$(\tfrac12\sqrt2, \tfrac12\sqrt2)$	$\tfrac12\sqrt2$	$\tfrac12\sqrt2$	1	1	$\sqrt2$	$\sqrt2$
$\pi/3$	$(\tfrac12, \tfrac12\sqrt3)$	$\tfrac12\sqrt3$	$\tfrac12$	$\sqrt3$	$\tfrac13\sqrt3$	2	$\tfrac23\sqrt3$
$\pi/2$	$(0, 1)$	1	0	—	0	—	1
$2\pi/3$	$(-\tfrac12, \tfrac12\sqrt3)$	$\tfrac12\sqrt3$	$-\tfrac12$	$-\sqrt3$	$-\tfrac13\sqrt3$	-2	$\tfrac23\sqrt3$
$3\pi/4$	$(-\tfrac12\sqrt2, \tfrac12\sqrt2)$	$\tfrac12\sqrt2$	$-\tfrac12\sqrt2$	-1	-1	$-\sqrt2$	$\sqrt2$
$5\pi/6$	$(-\tfrac12\sqrt3, \tfrac12)$	$\tfrac12$	$-\tfrac12\sqrt3$	$-\tfrac13\sqrt3$	$-\sqrt3$	$-\tfrac23\sqrt3$	2
π	$(-1, 0)$	0	-1	0	—	-1	—
$7\pi/6$	$(-\tfrac12\sqrt3, -\tfrac12)$	$-\tfrac12$	$-\tfrac12\sqrt3$	$\tfrac13\sqrt3$	$\sqrt3$	$-\tfrac23\sqrt3$	-2
$5\pi/4$	$(-\tfrac12\sqrt2, -\tfrac12\sqrt2)$	$-\tfrac12\sqrt2$	$-\tfrac12\sqrt2$	1	1	$-\sqrt2$	$-\sqrt2$
$4\pi/3$	$(-\tfrac12, -\tfrac12\sqrt3)$	$-\tfrac12\sqrt3$	$-\tfrac12$	$\sqrt3$	$\tfrac13\sqrt3$	-2	$-\tfrac23\sqrt3$
$3\pi/2$	$(0, -1)$	-1	0	—	0	—	-1
$5\pi/3$	$(\tfrac12, -\tfrac12\sqrt3)$	$-\tfrac12\sqrt3$	$\tfrac12$	$-\sqrt3$	$-\tfrac13\sqrt3$	2	$-\tfrac23\sqrt3$
$7\pi/4$	$(\tfrac12\sqrt2, -\tfrac12\sqrt2)$	$-\tfrac12\sqrt2$	$\tfrac12\sqrt2$	-1	-1	$\sqrt2$	$-\sqrt2$
$11\pi/6$	$(\tfrac12\sqrt3, -\tfrac12)$	$-\tfrac12$	$\tfrac12\sqrt3$	$-\tfrac13\sqrt3$	$-\sqrt3$	$\tfrac23\sqrt3$	-2
2π	$(1, 0)$	0	1	0	—	1	—

3.	Domain	Range
(a)	$\{x: x \neq (2n-1)\pi/2, n \text{ an integer}\}$	$(-\infty, \infty)$
(b)	$\{x: x \neq n\pi, n \text{ an integer}\}$	$(-\infty, \infty)$
(c)	same as tan	$\{y: \|y\| \geq 1\}$
(d)	same as cot	same as sec

5. -5. **6.** **(b)** $(\frac{1}{2}, 1)$; $(\pi - 3, 0)$; **(c)** f. period is 4; **(d)** $(1, 0)$.
7. F. period is 3; $(0, 0)$, $(\frac{1}{4}, \sqrt{3}/4)$, $(0, 0)$, $(0, 0)$.
15. **(a)** π; **(b)** $\pi/3$; **(c)** $2\pi/k$.

Miscellaneous Exercises 2.2 (2), pages 46–47

1. $\sin x$. **2.** 1. **3.** 1. **4.** $\sec x$. **5.** $\sin x$. **6.** $\csc x$. **7.** 1.
8. 1. **9.** $\cot x$. **10.** $\sin x$. **11.** $\sec x$. **12.** $\sec x$. **13.** 2.
14. $\sin x$. **15.** $\csc x$. **16.** 2. **17.** 0. **18.** $\cot x$. **19.** $\cos x$.
20. $\sin x$. **21.** $2(\sec x + \tan x)$. **22.** $2 \csc x$. **23.** $\tan x$. **24.** $\sec^2 x$.
25. $2 \sec x$. **26.** 1. **27.** 2. **28.** 1. **29.** 1. **30.** $\cos^2 x$.
31. $2 \tan x$. **32.** $\sec x$. **33.** 1. **34.** 1. **35.** $\cos^2 x$.

Miscellaneous Exercises 2.2 (3), pages 47–48

1. **(a)** $\pi/4, 5\pi/4$; **(b)** $\pi/4, 3\pi/4, 5\pi/4, 7\pi/4$; **(c)** $0, \pi; \pi/2$;
(d) $\pi/2, 3\pi/2; \pi$; **(e)** $\pi/2$; **(f)** $\pi/2, 3\pi/2; \pi/3, 5\pi/3$; **(g)** no roots;
(h) $\pi/3, 2\pi/3, 4\pi/3, 5\pi/3$; **(i)** $\pi/4, 3\pi/4, 5\pi/4, 7\pi/4$.

2.	sin	cos	tan	cot	sec	csc
(a)	$3/5$	$4/5$	$3/4$	$4/3$	$5/4$	$5/3$
(b)	$2\sqrt{5}/5$	$-\sqrt{5}/5$	-2	$-1/2$	$-\sqrt{5}$	$\sqrt{5}/2$
(c)	$-12/13$	$5/13$	$-12/5$	$-5/12$	$13/5$	$-13/12$
(d)	0	1	0	—	1	—
(e)	$-1/2$	$-\sqrt{3}/2$	$\sqrt{3}/3$	$\sqrt{3}$	$-2\sqrt{3}/3$	-2
(f)	$-7/25$	$-24/25$	$7/24$	$24/7$	$-25/24$	$-25/7$
(g)	$\frac{m^2 - n^2}{m^2 + n^2}$	$\frac{-2mn}{m^2 + n^2}$	$\frac{m^2 - n^2}{-2mn}$	$\frac{-2mn}{m^2 - n^2}$	$\frac{m^2 + n^2}{-2mn}$	$\frac{m^2 + n^2}{m^2 - n^2}$

19. $(x, y) = (\pi/4, \pi/6), (3\pi/4, 5\pi/6), (5\pi/4, 7\pi/6), (7\pi/4, 11\pi/6)$.
20. **(a)** $x \in (\pi/4, 5\pi/4)$; **(b)** $x \in (\pi/4, 3\pi/4) \cup (5\pi/4, 7\pi/4)$.

Miscellaneous Exercises 2.3 (1), pages 51–53

3. **(a)** $\frac{1}{4}(\sqrt{6} - \sqrt{2})$; **(b)** $\frac{1}{4}(\sqrt{6} + \sqrt{2})$.
4. **(a)** $\frac{1}{4}(\sqrt{6} + \sqrt{2})$; **(b)** $\frac{1}{4}(\sqrt{6} - \sqrt{2})$; **(c)** $\frac{1}{4}(\sqrt{6} + \sqrt{2})$;
(d) $\frac{1}{4}(\sqrt{2} - \sqrt{6})$; **(e)** $\frac{1}{4}(\sqrt{6} - \sqrt{2})$; **(f)** $-\frac{1}{4}(\sqrt{6} + \sqrt{2})$;
(g) $2 + \sqrt{3}$.
5. **(a)** $\frac{4}{5}, -\frac{5}{13}$; **(b)** $\frac{16}{65}$; **(c)** $\frac{63}{65}$.

6. (a) $\frac{3}{5}$; (b) $-\frac{4}{5}$.
7. (b) $\sin y$; (c) $\sin 2x$.
8. (a) $2 \sin x \cos y$; (b) $2 \sin x \sin y$.
13. (a) $2 + \sqrt{3}$; (b) $-2 - \sqrt{3}$; (c) $-2 + \sqrt{3}$.
15. $3\pi/4$, $7\pi/4$. **16.** $-44/117$.
18. (b) $\sqrt{2} - 1$; (c) $\sqrt{2} + 1$.
21. (a) $4/5$; (b) $24/25$; (c) $7/25$; (d) $336/625$.
22. (a) $5/12$, $120/119$.
23. (a) $\sin \frac{\pi}{6} = \frac{1}{2}$; (b) $\frac{1}{2} \sin \frac{\pi}{4} = \sqrt{2}/4$; (c) $\cos \frac{\pi}{6} = \sqrt{3}/2$; (d) $\tan \frac{\pi}{4} = 1$; (e) $\cos \frac{\pi}{6} = \sqrt{3}/2$; (f) $\cos \frac{7\pi}{4} = \sqrt{2}/2$.
24. (a) $\sin x$; (b) $\frac{1}{2} \sin 34x$; (c) $\tan z/3$; (d) $\cos 10t$; (e) $\cos y$; (f) $\cos 6x$.
26. (a) $\frac{3}{5}, \frac{4}{5}$; (b) $\frac{3}{5}, -\frac{4}{5}$.
27. $\sqrt{6}/3$, $-\sqrt{3}/3$.
28. (a) $\frac{1}{2}\sqrt{2 - \sqrt{2}}$; (b) $\frac{1}{2}\sqrt{2 + \sqrt{2}}$.

Miscellaneous Exercises 2.3 (2), pages 54–55

1. (a) $\pi/8$, $5\pi/8$, $9\pi/8$, $13\pi/8$. (b) $\pi/6$, $5\pi/6$, $3\pi/2$. (c) $\pi/12$, $11\pi/12$, $13\pi/12$, $23\pi/12$. (d) 0, π; $\pi/3$, $5\pi/3$. (e) 0; $2\pi/3$, $4\pi/3$. (f) $\pi/12$, $5\pi/12$, $13\pi/12$, $17\pi/12$; $3\pi/4$, $7\pi/4$. (g) $\pi/6$, $5\pi/6$, $7\pi/6$, $11\pi/6$. (h) $\pi/8$, $3\pi/8$, $5\pi/8$, $7\pi/8$, $9\pi/8$, $11\pi/8$, $13\pi/8$, $15\pi/8$. (i) 0, π; $\pi/3$, $2\pi/3$, $4\pi/3$, $5\pi/3$. (j) $\pi/6$, $5\pi/6$, $7\pi/6$, $11\pi/6$.

Miscellaneous Exercises 2.3 (3), pages 55–57

5. $\sqrt{2} \cos x$. **7.** $-\tan 2x$. **8.** $-\cot (x - y)$. **13.** 1.
14. (a) 0, $\pi/2$, π, $3\pi/2$; $\pi/3$, $5\pi/3$. (b) $\pi/4$, $3\pi/4$, $5\pi/4$, $7\pi/4$; $2\pi/3$, $4\pi/3$. (c) 0, $\pi/3$, $2\pi/3$, π, $4\pi/3$, $5\pi/3$; $\pi/6$, $5\pi/6$, $7\pi/6$, $11\pi/6$.
23. (a) 0, $\pi/3$, π, $4\pi/3$. (b) 0, $\pi/2$, π, $3\pi/2$.

Miscellaneous Exercises 2.4 (1), pages 60–62

1. $36°$, $75°$, $210°$, $600°$, $-45°$, $3600°$, $360°/\pi$, or $115°$ approx., $-900°/\pi$, or $-286°$ approx.
2. $5\pi/18$, $5\pi/6$, $-5\pi/3$, $5\pi/2$, $163\pi/1080$, or 0.47 approx.
4. (a) 1; (b) 1.28; (c) 0.14; (d) 0.72; (e) 1.15; (f) 0.42; (g) 0.29; (h) 0.17.
5. (a) -0.87; (b) 3.08; (c) -0.87; (d) -1.41; (e) undefined; (f) 0.91; (g) -3.38; (h) -1.01; (i) 2.43; (j) -0.65; (k) 1.18; (l) -2.23; (m) -0.90; (n) 0.99; (o) -0.98.

6. **(a)** 0.64, 2.50; **(b)** 2.03, 5.18; **(c)** 1.23, 5.05; **(d)** 0.94, 5.34; **(e)** 1.98, 4.30; **(f)** 0.34, 2.80; **(g)** 2.50, 5.64; **(h)** 3.42, 6.00; **(i)** 0.52, 2.62; **(j)** 2.71, 3.57.

7. 2.68, 7.64.

8. **(a)** 1.11, 4.25. **(b)** 0, 3.14; 1.25, 4.39. **(c)** 1.89, 5.03. **(d)** 0.62, 2.53, 3.76, 5.67. **(e)** 1.57, 4.71; 0.40, 3.54. **(f)** 0.52, 2.62; 0.73, 2.41. **(g)** 1.05, 5.24; 1.23, 5.05. **(h)** 1.57; 3.67, 5.76. **(i)** 0.46, 3.61; 1.11, 4.25. **(j)** 0.79, 3.93; 1.11, 4.25. **(k)** 1.57, 4.71; 2.42, 3.86. **(l)** 0.43, 2.71. **(m)** 0.63, 3.77; 1.92, 5.06. **(n)** 0.64, 4.71.

9. **(a)** 0, 3.14; 1.05, 5.24. **(b)** 0; 2.09, 4.19. **(c)** 0.26, 1.31, 2.36, 3.40, 4.45, 5.50. **(d)** 4.71; 0.52, 2.62. **(e)** 0, 3.14; 0.52, 2.62. **(f)** 0.79, 3.93; 1.83, 2.88, 4.97, 6.02. **(g)** 3.14; 0.59, 5.70. **(h)** 0, 3.14. **(i)** 0.55, 2.12, 3.70, 5.27; 1.18, 2.75, 4.32, 5.89. **(j)** 0, 1.57, 3.14, 4.71; 1.05, 5.24. **(k)** 0.79, 2.36, 3.93, 5.50; 1.05, 5.24. **(l)** 0; 1.57, 3.67, 5.76. **(m)** 0, 1.05, 2.09, 3.14, 4.20, 5.24.

11. **(a)** 2.2 rad.; **(b)** 15 in.

12. 1800π, or approximately 5650 (ft./min.).

15. **(a)** 45 sq. cm.; **(b)** 30 sq. cm.

Miscellaneous Exercises 2.4 (2), pages 62–65

2. $q = r \sin Q = p \tan Q$.

4. **(a)** $A = 34°$, $B = 56°$, $c = 5.8$ cm. **(b)** $X = 11°$, $Y = 79°$, $z = 0.66$ in. **(c)** $p = 59$ yd., $r = 43$ yd., $R = 47°$. **(d)** $b = 47$ ft., $c = 30$ ft., $C = 33°$. **(e)** $P = 40°$, $Q = 50°$, $r = 7.8$ m. **(f)** $B = 70°$, $C = 20°$, $c = 5.6$ in.

5. 40 ft. **6.** 23 ft., 11 ft. **7.** 71°, 71°, 39°.

8. **(a)** **(i)** $2nR \sin (180/n)°$; **(ii)** $2nR \tan (180/n)°$. **(b)** 3.14, 3.14.

10. 540 ft. **11.** 185 yd.

13. **(a)** 48 sq. ft.; **(b)** 410 sq. in.; **(c)** 0.052 sq. m.; **(d)** 18 sq. in.

14. 32 sq. in. **15.** 29 sq. in. **16.** 10 in., 30 in.; 154 sq. in.

17. 188 sq. in. **18.** **(a)** **(i)** $\frac{1}{2}nR^2 \sin (360/n)°$; **(ii)** $nR^2 \tan (180/n)°$. **(b)** **(i)** 3.14 sq. in., 3.14 sq. in.; **(ii)** 3.14 sq. in. **19.** **(a)** $\frac{5}{8}d^2 \csc 72°$; **(b)** 66 sq. in.

Miscellaneous Exercises 2.4 (3), pages 65–67

2. **(a)** $A = 21°$, $C = 134°$, $c = 20$. **(b)** $b = 3.3$, $c = 2.7$, $C = 47°$. **(c)** $B = 26°$, $C = 42°$, $c = 12$. **(d)** $b = 0.33$, $B = 91°$, $C = 36°$. **(e)** $a = 3.3$, $b = 1.0$, $C = 26°$.

3. $Q = 53°, 127°$; $r = 8.7$ in., 3.8 in. **4.** $Q = 61°, 119°$; $r = 4.2$ in., 0.72 in. **5.** 30°, 150°.

7. **(a)** $a = 3.6$ in.; **(b)** $a = 6.1$ in.; **(c)** $c = 6.9$ cm.; **(d)** $b = 11$ in.

8. 83°, 56°, 41°. **9.** 117°. **11.** $-\frac{5}{16}, \frac{37}{40}$. **12.** $\sqrt{80}$ in.

Exercises 2.5 [A], pages 73–74

1. **(a)** $(fg)(x) = 2x - 7$; $D = R = (-\infty, \infty)$. $(gf)(x) = 2x - 2$; $D = R = (-\infty, \infty)$. **(b)** $(fg)(x) = (gf)(x) = abx$; $D = R = (-\infty, \infty)$ if $ab \neq 0$; $R = \{0\}$ if $ab = 0$. **(c)** $(fg)(x) = \log(x + 2)$; $D = (-2, \infty)$, $R = (-\infty, \infty)$. $(gf)(x) = 2 + \log x$; $D = (0, \infty)$, $R = (-\infty, \infty)$. **(d)** $(fg)(x) = \sin 2x$; $D = (-\infty, \infty)$, $R = [-1, 1]$. $(gf)(x) = 2 \sin x$; $D = (-\infty, \infty)$, $R = [-2, 2]$. **(e)** $(fg)(x) = 3\sqrt{1 - x^2}$; $D = [-1, 1]$, $R = [0, 3]$. $(gf)(x) = 3\sqrt{9 - x^2}$; $D = [-3, 3]$, $R = [0, 9]$.

2. **(a)** $f^{-1}(x) = x - 2$; **(b)** $f^{-1}(x) = (x + 7)/2$; **(c)** $f^{-1}(x) = x$; **(d)** $f^{-1}(x) = -x$; **(e)** $f^{-1} = \{(3, 1), (4, 2), (5, 3)\}$; **(f)** $f^{-1} = f$.

3. Various possibilities, including **(a)** $f(x) = \sqrt{x}$, $g(x) = x - 1$; **(b)** $f(x) = \sin x$, $g(x) = 2x + 3$; **(c)** $f(x) = 1/x$, $g(x) = x^2 + 3$; **(d)** $f(x) = |x|$, $g(x) = 5x - 7$; **(e)** $f(x) = x + 1$, $g(x) = x - 1$; **(f)** $f(x) = 2x$, $g(x) = \log x$.

5. **(a)** 0; **(b)** π; **(c)** 0; **(d)** $\frac{2}{3}\sqrt{2}$; **(e)** $\frac{4}{9}\sqrt{2}$; **(f)** $-\frac{7}{9}$; **(g)** $\frac{1}{9}(\sqrt{5} + 4\sqrt{2})$; **(h)** 1.

Exercises 2.5 [B], pages 74–75

1. **(a)** $(fg)(x) = 3x - 2$; $D = R = (-\infty, \infty)$. $(gf)(x) = 3x - 1$; $D = R = (-\infty, \infty)$. **(b)** $(fg)(x) = (x - 1)^2$; $D = (-\infty, \infty)$, $R = [0, \infty)$. $(gf)(x) = x^2 - 1$; $D = (-\infty, \infty)$, $R = [-1, \infty)$. **(c)** $(fg)(x) = 3 \cos 3x$; $D = (-\infty, \infty)$, $R = [-3, 3]$. $(gf)(x) = 9 \cos x$; $D = (-\infty, \infty)$, $R = [-9, 9]$. **(d)** $(fg)(x) = \sqrt{25 - x^4}$; $D = [-\sqrt{5}, \sqrt{5}]$, $R = [0, 5]$. $(gf)(x) = 25 - x^2$; $D = [-5, 5]$, $R = [0, 25]$. **(e)** $(fg)(x) = -x$; $D = (-\infty, 0]$, $R = [0, \infty)$. $(gf)(x) = \sqrt{-x^2}$; $D = R = \{0\}$. **(f)** $(fg)(x) = -(\log x)^2$; $D = (0, \infty)$, $R = (-\infty, 0]$. gf does not exist.

2. **(a)** $f^{-1}(x) = (x + 5)/3$; **(b)** $f^{-1}(x) = 10^x$; **(c)** $f^{-1}(x) = 2 + \sqrt{x + 4}$; **(d)** $f^{-1}(x) = 2 - \sqrt{x + 4}$; **(e)** $f^{-1} = \{(7, 2), (3, 3), (-5, 1)\}$; **(f)** $f^{-1} = f$.

3. Various possibilities, including **(a)** $f(x) = \log x$, $g(x) = \sin x$, $h(x) = 5x$; **(b)** $f(x) = \sqrt{x}$, $g(x) = x + 1$, $h(x) = x^2$; **(c)** $f(x) = 1/x$, $g(x) = 2 + x$, $h(x) = \sin x$; **(d)** $f(x) = x^2$, $g(x) = \tan x$, $h(x) = 5x$.

5. **(a)** $-\pi/2$; **(b)** $\pi/2$; **(c)** not defined; **(d)** 4/3; **(e)** 4/3; **(f)** $-24/7$; **(g)** 1/7; **(h)** 120/119; **(i)** $2\sqrt{2}/3$; **(j)** $(\sqrt{3}/4) + (\sqrt{5}/10)$. **6.** $-\pi/2$.

Miscellaneous Exercises 2.5, page 75

1. $k \neq 0$.

6. **(a)** $ff^{-1} = \{(7, 7), (-4, -4), (2, 2)\}$; $f^{-1}f = \{(2, 2), (3, 3), (6, 6)\}$. **(b)** $ff^{-1} = \{(1, 1), (2, 2), (7, 7)\} = f^{-1}f$. **(c)** Domain = range.

Exercises 2.6 [A], page 82

1. (a)

$\Delta x =$	-1	-0.1	-0.01	0.01	0.1	1
$\Delta y =$	4	0.31	0.0301	-0.0299	-0.29	-2
$\Delta y/\Delta x =$	-4	-3.1	-3.01	-2.99	-2.9	-2

(b) -3. (c) $\Delta y/\Delta x = -3 + \Delta x$.

2. (a)

$\Delta x =$	-1	-0.1	-0.01	0.01	0.1	1
$\Delta y =$	-3	-0.39	-0.0399	0.0401	0.41	5
$\Delta y/\Delta x =$	3	3.9	3.99	4.01	4.1	5

(b) 4. (c) $\Delta y/\Delta x = 4 + \Delta x$.

3. (a)

$\Delta x =$	-1	-0.1	-0.01	0.01	0.1	1
$\Delta y =$	1	0.01	0.0001	0.0001	0.01	1
$\Delta y/\Delta x =$	-1	-0.1	-0.01	0.01	0.1	1

(b) 0. (c) $\Delta y/\Delta x = \Delta x$.

4. (a)

$\Delta x =$	-1	-0.1	-0.01	0.01	0.1	1
$\Delta y =$	3	0.399	0.039999	-0.039999	-0.399	-3
$\Delta y/\Delta x =$	-3	-3.99	-3.9999	-3.9999	-3.99	-3

(b) -4. (c) $\Delta y/\Delta x = -4 + (\Delta x)^2$.

5. (a)

$\Delta x =$	-1	-0.1	-0.01	0.01	0.1	1
$\Delta y =$	-1	-0.05127	-0.00502	0.0049	0.0488	0.4142
$\Delta y/\Delta x =$	1	0.5127	0.502	0.49	0.488	0.4142

(b) $\frac{1}{2}$. (c) $\dfrac{\Delta y}{\Delta x} = \dfrac{\sqrt{1 + \Delta x} - 1}{\Delta x} = \dfrac{1}{\sqrt{1 + \Delta x} + 1}$.

Exercises 2.7 [A], pages 84–85

1. 32 ft./sec., -32 ft./sec., $96 - 32t$ ft./sec.; $t = 3$; 144 ft.
2. (a) 2.5 sec.; (b) -104 ft./sec. 3. -3 yd./min.

Exercises 2.8 [A], page 88

1. $6x - 7$; 5; -25. 2. $3x^2$; 12, 3, 0, 3, 12. 3. $3x^2 - 12x$; 0, 4.
4. (a) $4x^3$; (b) $5x^4$. 5. $48 - 32t$; 16 ft./sec., -16 ft./sec.
6. $-1/(x - 1)^2$; -1, -1.
7. (a) $a/(x + a)^2$; (b) $-2x/(x^2 - 4)^2$; (c) $2/(x + 1)^2$; (d) $4x/(x^2 + 1)^2$.
8. $1/2\sqrt{x}$; $[0, \infty)$, $(0, \infty)$; $\frac{1}{2}$, 1, 2, 5; $\pi/2$. 9. $-x/\sqrt{25 - x^2}$; $-\frac{3}{4}$.

Exercises 2.9 [A], pages 92–93

1. (a) $3x^2$; (b) $12x^3 - 6$; (c) $42z^5 + 6z^{-3}$; (d) $27x^8 - 8x^3 - \frac{1}{5}x^{-2}$; (e) $4x^{13} - 3x^2$; (f) $\frac{4}{9}x^{-1/3} + 49x^{-8}$; (g) $y^{-1/2} - \frac{4}{3}y^{-2/3}$; (h) $-\frac{3}{8}x^{-4} + 8x^{-2}$; (i) $-6x^{-4} - \frac{4}{3}x^{1/3}$; (j) $\frac{1}{2}\sqrt{5}(t^{-1/2} + t^{-3/2})$.

2. **(a)** $x^4 + 4x^3y + 6x^2y^2$; **(b)** $x^7 + 7x^6\,\Delta x + 21x^5(\Delta x)^2$; **(c)** $t^{10} - 10t^7 + 40t^4$; **(d)** $1 + 10 + 45$.

3. **(a)** $1 + 3x^{-2}$; **(b)** $24y^2 - 24y + 6$; **(c)** $\frac{7}{6}x^{5/2} - \frac{25}{6}x^{3/2} + \frac{3}{2}x^{1/2} + \frac{7}{6}x^{-3/2}$; **(d)** $\sqrt{2}(\frac{5}{2}z^{3/2} + \frac{3}{10}z^{-3/2})$.

4. $6x + 2$; $20, -10, 0$. **5.** $24.5 - 9.8t$; 14.7 m./sec., 4.9 m./sec.

6. Various possibilities, including **(a)** x^3; **(b)** x^5; **(c)** $\frac{1}{4}x^4$.

Exercises 2.9 [B], pages 93–94

1. **(a)** $12x^3 - 7$; **(b)** $8t + 10t^{-3}$; **(c)** $3z^2 - 12z + 12$; **(d)** $6 - 15\sqrt{2}y^{-1/2}$; **(e)** $4x^3 - 4x^{-5}$; **(f)** $\frac{5}{3}t^{2/3} + \frac{4}{3}t^{-1/3} + \frac{5}{3}t^{-4/3}$; **(g)** $6 - 3\sqrt{6}x^{-1/2}$; **(h)** $(7\sqrt{2}/4)y^{-1/2} + (\sqrt{2}/4)y^{-3/2}$; **(i)** $3x^2 - 4x - 15$; **(j)** $3z^{1/2} + \frac{5}{6}z^{-1/6}$.

2. **(a)** $y = f(x) = -\frac{2}{3}x + \frac{7}{3}$; $f'(x) = -\frac{2}{3}$. **(b)** $y = f(x) = 4x^{-1/2}$; $f'(x) = -2x^{-3/2}$. **(c)** $y = f(x) = -x^{-1} + x^{-2}$; $f'(x) = x^{-2} - 2x^{-3}$. **(d)** $y = f(x) = a - 2a^{1/2}x^{1/2} + x$; $f'(x) = 1 - a^{1/2}x^{-1/2}$.

3. 96, 36, 0, −12, 0. **4.** **(a)**, **(b)**, and **(c)**: $2x + 7$. **5.** 3, 0, −9; at the origin. **6.** Various possibilities, including **(a)** x^8; **(b)** $1/x$.

Miscellaneous Exercises 2.9, page 94

1. **(a)** $16x^7 - 6x + 7$; **(b)** $16x^3 - 40x$; **(c)** $2x + 4x^{-2} - 39x^{-4}$; **(d)** $\frac{5}{2}x^{3/2} + \frac{7}{2}x^{-1/2}$; **(e)** $1 - 4t^{-1/3} + 4t^{-2/3}$; **(f)** $\sqrt{5}(\frac{5}{2}x^{3/2} - \frac{5}{2}x^{-1/2})$; **(g)** $-8x^{-3} - 28x^{-2}$; **(h)** $2(3x)^{-1/3} - 10(3x)^{-2/3}$; **(i)** $\frac{4}{3}x^{1/3} - \frac{4}{3}x^{-7/3}$; **(j)** $(2x)^{-3/2}(2x - 1)$.

2. **(a)** $y = f(x) = x - 2x^{-1}$; $f'(x) = 1 + 2x^{-2}$. **(b)** $y = f(x) = a - 3a^{2/3}x^{1/3} + 3a^{1/3}x^{2/3} - x$; $f'(x) = -a^{2/3}x^{-2/3} + 2a^{1/3}x^{-1/3} - 1$. **(c)** $y = f(x) = 4x - 12 + 9x^{-1}$; $f'(x) = 4 - 9x^{-2}$.

3. $x^2 - 4x$; 0, 4. **4.** Position, 3, 0, −1; velocity, −4, −2, 0.

5. **(a)** −13, −7, −1, 5; **(c)** $(\frac{7}{6}, -\frac{49}{12})$; **(d)** $-\frac{49}{12}$. **6.** $x > -\frac{5}{4}$, $x < -\frac{5}{4}$; $(-\frac{5}{4}, -\frac{121}{8})$.

Exercises 3.1 [A], page 102

1. $f'(x) = 4x + 3$.

$x =$	−4	−3	−2	−1	0	1	2	3
$f(x) =$	15	4	−3	−6	−5	0	9	22
$f'(x) =$	−13	−9	−5	−1	3	7	11	15

Stationary point: $x = -\frac{3}{4}$. Increasing on $[-\frac{3}{4}, \infty)$; decreasing on $(-\infty, -\frac{3}{4}]$.

2. $f'(x) = 3x^2 - 6x$.

$x =$	−3	−2	−1	0	1	2	3	4
$f(x) =$	−52	−18	−2	2	0	−2	2	18
$f'(x) =$	45	24	9	0	−3	0	9	24

Stationary points: $x = 0, 2$. Increasing on $(-\infty, 0]$, $[2, \infty)$; decreasing on $[0, 2]$.

3. $f'(x) = 4x^3 - 16x$.

$x =$	-3	-2	-1	0	1	2	3
$f(x) =$	19	-6	3	10	3	-6	19
$f'(x) =$	-60	0	12	0	-12	0	60

Stationary points: $x = -2, 0, 2$. Increasing on $[-2, 0]$, $[2, \infty)$; decreasing on $(-\infty, -2]$, $[0, 2]$.

4. $f'(x) = 2x + 3 - x^{-2}$.

$x =$	-5	-4	-3	-2	-1	0	1	2	3
$f(x) =$	$9\frac{4}{5}$	$3\frac{3}{4}$	$-\frac{1}{3}$	$-2\frac{1}{2}$	-3	—	5	$10\frac{1}{2}$	$18\frac{1}{3}$
$f'(x) =$	$-7\frac{1}{25}$	$-5\frac{1}{16}$	$-3\frac{1}{9}$	$-1\frac{1}{4}$	0	—	4	$6\frac{3}{4}$	$8\frac{8}{9}$

Stationary points: $x = -1, \frac{1}{2}$. Increasing on $[\frac{1}{2}, \infty)$; decreasing on $(-\infty, 0)$, $(0, \frac{1}{2}]$.

Exercises 3.1 [B], page 102

1. $f'(x) = -2x + 1$.

$x =$	-3	-2	-1	0	1	2	3
$f(x) =$	-13	-7	-3	-1	-1	-3	-7
$f'(x) =$	7	5	3	1	-1	-3	-5

Stationary point: $x = \frac{1}{2}$. Increasing on $(-\infty, \frac{1}{2}]$; decreasing on $[\frac{1}{2}, \infty)$.

2. $f'(x) = 9x^2 + 1$.

$x =$	-3	-2	-1	0	1	2	3
$f(x) =$	-83	-25	-3	1	5	27	85
$f'(x) =$	82	37	10	1	10	37	82

No stationary point. Increasing on $(-\infty, \infty)$.

3. $f'(x) = 12x^3 - 12x^2 - 24x$.

$x =$	-2	-1	0	1	2	3
$f(x) =$	37	0	5	-8	-27	32
$f'(x) =$	-96	0	0	-24	0	144

Stationary points: $x = -1, 0, 2$. Increasing on $[-1, 0]$, $[2, \infty)$; decreasing on $(-\infty, -1]$, $[0, 2]$.

4. $f'(x) = 1 - 20x^{-3}$.

$x =$	-4	-3	-2	-1	0	1	2	3	4
$f(x) =$	$-3\frac{3}{8}$	$-1\frac{8}{9}$	$\frac{1}{2}$	9	—	11	$4\frac{1}{2}$	$4\frac{1}{9}$	$4\frac{5}{8}$
$f'(x) =$	$1\frac{5}{16}$	$1\frac{20}{27}$	$3\frac{1}{2}$	21	—	-19	$-1\frac{1}{2}$	$\frac{7}{27}$	$\frac{11}{16}$

Stationary point: $x = \sqrt[3]{20} = 2.7$, approximately. Increasing on $(-\infty, 0)$, $[\sqrt[3]{20}, \infty)$; decreasing on $(0, \sqrt[3]{20}]$.

Exercises 3.2 [A], pages 106–107

1. **(a)** $f(-3/10) = 191/20$, min. **(b)** $f(-1) = 6$, max.; $f(1) = 2$, min. **(c)** $g(-\sqrt{2}) = -2\sqrt{2}$, max.; $g(\sqrt{2}) = 2\sqrt{2}$, min.
2. **(a)** $\phi(0) = 0$, min. **(b)** $F(-2) = 28/3$, max.; $F(1) = 29/6$, min. **(c)** $f(-2) = -11$, min.; $f(0) = 5$, max.; $f(2) = -11$, min.
3. $2'' \times 8'' \times 8''$; 128 cu. in. **4.** $5 + 5$.
6. **(a)** $V = 72x^2 - 4x^3, 0 \leq x \leq 18$; 3456 cu. in.; **(b)** $V = 36x^2 - \frac{3}{2}x^3$, $0 \leq x \leq 24$; 3072 cu. in.

Exercises 3.2 [B], page 107

1. **(a)** $F(0) = -7$, max. **(b)** $g(-2) = 15$, max.; $g(\frac{2}{3}) = 149/27$, min. **(c)** $f(-2.5) = 0.9$, max.; $f(-0.1) = -6.1$, min.
2. **(a)** $f(-4) = 40$, max.; $f(\frac{2}{3}) = -292/27$, min. **(b)** $F(-1) = 0$, max.; $F(1) = -4$, min. **(c)** None.
3. $6 + 6$. **4.** $6 + 3$. **6.** $5'' \times 10''$. **7.** $25' \times 50'$.

Miscellaneous Exercises 3.2, page 108

1. **(a)** $f(-3) = 23$, max.; $f(1) = -9$, min. **(b)** None. **(c)** $f(27) = -27$, min. **(d)** $g(1) = -1$, min. **(e)** $f(-1) = 2$, min.; $f(1) = 2$, min. **(f)** $h(-5/6) = 2573/108$, max.; $h(4) = -202$, min. **(g)** $r(1) = 2$, min. **(h)** $F(-1) = 31/12$, min.; $F(0) = 3$, max.; $F(2) = 1/3$, min. **(i)** $g(2) = -61$, min. **(j)** $f(3/2) = -27/16$, min.
2. $\frac{1}{2}$.
3. **(a)** 250,000/27 cu. in., or approximately 9260 cu. in.; **(b)** 2,000,000/243 cu. in., or approximately 8230 cu. in.
4. **(a)** 2048 in./sec.; **(b)** 16 sec.

Exercises 3.3 [A], page 113

1. **(a)** 107/12, min.; **(b)** $-39/8$, max.; **(c)** $-37/4$, min.; **(d)** 0, min.
2. **(a)** $(-1, 12)$, max.; $(1/2, -3/2)$, infl.; $(2, -15)$, min. **(b)** $(\pm 3, -25)$, mins.; $(\pm\sqrt{3}, -13)$, infls.; $(0, 2)$, max. **(c)** None.
4. $f = \{(x, y): y = 2x^3 - 6x^2 - 48x + 39\}$.
5. $f = \{(x, y): y = x^2 + x + 3\}$.

Exercises 3.3 [B], page 114

1. **(a)** $(-1, 1)$, max.; $(1, -15)$, infl.; $(3, -31)$, min. **(b)** $(-3, -2)$, min.; $(-2.2, 5.1)$, infl.; $(-1, 14)$, max.; $(0.2, 5.1)$, infl.; $(1, -2)$, min. **(c)** $(-2, 109/15)$, max.; $(-1.4, 5.6)$, infl.; $(0, 3)$, infl.; $(1.4, 0.4)$, infl.; $(2, -19/15)$, min. **(d)** $(-4, -\frac{1}{8})$, min.; $(-6, -\frac{1}{9})$, infl.

3. $f = \{(x, y): y = -\frac{3}{2}x^3 - \frac{9}{2}x^2 + 12\}$. **4.** $f = \{(x, y): y = -\frac{1}{2}x^3 + 5x\}$.
5. (a) $3b^2 - 8ac > 0$.

Exercises 3.5 [A], page 118

1. (a) $x + y + 1 = 0, x - y - 3 = 0$; **(b)** $4x - y + 2 = 0, x + 4y + 9 = 0$.
2. $y + 4 = 0, 8x - y - 4 = 0$. **3.** $\tan^{-1} \frac{8}{15}$. **4.** $Q(-\frac{4}{3}, -\frac{8}{9})$; $PQ = 10\sqrt{10}/9$.

Exercises 3.5 [B], page 119

1. (a) $y - 4 = 0, x - 1 = 0$; **(b)** $3x - 2y + 4 = 0, 2x + 3y + 7 = 0$.
2. $6x - y - 5 = 0, 2x + y - 3 = 0$. **3.** $\tan^{-1} (8/11), \tan^{-1} (12/109), \tan^{-1} (24/107)$.

Miscellaneous Exercises 3.5, page 119

1. (a) $y + 19 = 0, x - 3 = 0$; **(b)** $3x - 16y + 28 = 0, 32x + 6y - 143 = 0$.
2. (a) $x - y = 0, 13x - y - 24 = 0$; **(b)** $x + y - 6 = 0, 25x + y - 30 = 0$.
3. (a) $(-2, -24), (\frac{1}{2}, \frac{13}{8}), (2, -4)$; **(b)** $(1, 0)$.
4. (a) $\frac{\pi}{2}, \tan^{-1} \frac{3}{4}$; **(b)** $\tan^{-1} 3$. **5.** $(-2, -8)$.

Exercises 3.6 [A], page 124

1. 4 yd. × 6 yd. **2.** Base, 8″; sides, 8″ each. **3.** 1.58′ × 1.16′ × 0.42′.
4. 4′ × 4′ × 2′. **5.** $32\sqrt{3}/9$ sq. units. **6.** $32\pi a^3/81$ cu. in.

Exercises 3.6 [B], page 125

1. $12\frac{1}{2}' \times 8\frac{1}{3}'$. **2.** $15'' \times 15'' \times 7\frac{1}{2}''$. **3.** $(\frac{5}{2}, 0)$. **4.** $\frac{9}{2}$ sq. units.
5. 8″ × 10″. **6.** 8 : 24 : 9.

Miscellaneous Exercises 3.6, pages 125–126

1. 24 sq. in. **2.** $128\pi/3$ cu. in. **3.** $\frac{1}{8}BH$ sq. units.
4. $\pi B^2H/81$ cu. units. **5.** $h = 4'', r = 4''$. **6. (a)** $\frac{3}{2}'' \times 2''$;
(b) $\frac{5}{2}'' \times \frac{6}{5}''$. Areas are equal. **7.** $\frac{63}{16}$ sq. units.
8. $\frac{20}{3}' \times \frac{20}{3}' \times \frac{45}{2}'$. **9.** $(1, 1)$. **10.** $250\pi/3$ cu. in.
11. $r = \sqrt{12h - h^2}$; $h = 8$. **12. (a)** 83 articles; **(b)** 26 articles.
13. 35. **14.** $300v^{-1} + v^2$; $\sqrt[3]{150}$ knots, or approximately 5.3 knots.
15. 12 knots.

Exercises 3.6 [C], page 127

1. $h = 2r$. **2.** $h = 8r/\pi$. **3.** $h = r$. **4.** Total height = width.
5. $2\pi b^3\sqrt{3}/27$ cu. in. **6.** $2\sqrt{30}x + 3y - 40 = 0$. **7.** $W : D = 1 : \sqrt{2}$.

Exercises 4.2 [A], pages 138–139

1. $x^3 - \frac{5}{2}x^2 + x + C$. **2.** $x^2 - 5x - 3x^{-1} + C$.
3. $\frac{1}{4}x^4 + 4x + 5x^{-1} + C$. **4.** $\frac{1}{2}x^2 - \frac{1}{6}x^{-2} + C$. **5.** $\frac{2}{3}x^{3/2} + 2x^{1/2} + C$.
6. $\frac{2}{3}s^{3/2} + 4s^{1/2} + 2s^{-1/2} + C$. **7.** $\dfrac{3\sqrt[3]{5}}{5}\,x^{5/3} - 3x + C$.
8. $-t^{-1} + \frac{3}{2}t^{-2} + \frac{10}{3}t^{-3} + C$. **9.** $\frac{1}{3}ax^3 + \frac{1}{2}bx^2 + cx + C$.
10. $-1\frac{1}{3}$. **11.** 9. **12.** $5\frac{1}{3}$.

Exercises 4.2 [B], page 139

1. $\frac{1}{3}x^3 - ax^2 + a^2x + C$. **2.** $\frac{2}{3}x\sqrt{ax} + C$. **3.** $\frac{3}{2}x + \frac{5}{2}x^{-1} + C$.
4. $\frac{2}{5}x^{5/2} - 2x^{3/2} - 2x^{1/2} + C$. **5.** $\frac{3}{4}y^{4/3} - 6y^{1/3} + C$.
6. 6. **7.** 8. **8.** $5\frac{1}{3}$. **9.** $\frac{8}{15}h^5$. **10.** $\dfrac{-a^3\sqrt{2}}{3}$.

Exercises 4.3 [A], pages 143–144

1. (a) 30 sq. units; **(b)** $\frac{3}{8}$ sq. unit; **(c)** 12 sq. units; **(d)** $8\frac{2}{3}$ sq. units.
2. (a) $10\frac{2}{3}$ sq. units; **(b)** $52\frac{1}{12}$ sq. units; **(c)** 8 sq. units.
3. $\frac{1}{6}$ sq. unit. **4.** $20\frac{5}{6}$ sq. units. **5.** hb.

Exercises 4.3 [B], page 144

1. (a) $13\frac{1}{3}$ sq. units; **(b)** $42\frac{2}{3}$ sq. units; **(c)** 36 sq. units; **(d)** $46\frac{4}{5}$ sq. units.
2. (a) 36 sq. units; **(b)** $40\frac{1}{2}$ sq. units; **(c)** $34\frac{2}{15}$ sq. units.
3. $20\frac{5}{6}$ sq. units. **4.** $a^2/6$ sq. units. **5.** $\dfrac{ab}{2}$.

Miscellaneous Exercises 4.3, pages 144–145

1. $\frac{16}{3}$ sq. units. **2.** $6\frac{3}{4}$ sq. units. **3.** $\frac{8}{3}a\sqrt{ap}$ sq. units.
4. $10\frac{2}{3}$ sq. units. **5.** $4\frac{1}{2}$ sq. units. **6.** $\frac{4}{3}$ sq. units.
7. (a) $\left(1 - \dfrac{1}{k}\right)$ sq. units; **(b)** $(2\sqrt{k} - 2)$ sq. units.

Exercises 4.4 [A], page 148

1. (a) $f(x) = x^2 + 3x - 5$; **(b)** $F(y) = \frac{2}{3}y\sqrt{5y} - \frac{25}{3}$;

(c) $g(x) = x^3 - 2x^2 + 7x + 23$; **(d)** $f(t) = 3t^3 - 15t^2 + 25t$;
(e) $F(x) = \frac{1}{6}x^2 + \frac{1}{3}x^{-1} - \frac{3}{4}$.

2. $4x^2y - 17x^2 + 4 = 0$. **3.** $y = x^3 - 3x^2 + 5x$. **4.** $y = \frac{1}{2}x^2 + 2x + 2$.
5. $(-5, 0)$, $(2, 0)$, $(3, 0)$. **6.** $F(x) = 2x^3 - 7x$.

Exercises 4.4 [B], page 149

1. (a) $F(x) = \frac{1}{3}x^3 - x^2 + 5x - 14$; **(b)** $h(y) = \frac{1}{5}y^5 - y^4 + 2y^3 - 2y^2 + y + 13$; **(c)** $f(x) = x^{5/3} + x^{2/3} + 33$.

2. $13\frac{1}{2}$. **3.** $-\frac{1}{6}a^2$. **4.** ± 1, $(\pm i)$. **5.** $f(x) = 5 + 4x - x^2$.

6. Straight lines not parallel to the y-axis. **7.** -1, $1 \pm \sqrt{3}$.

8. $g(x) = \frac{4}{3}x^{3/2} + x - 5$. **9.** $y = \frac{1}{2}x^3 + 2x$.

Exercises 4.5 [A], pages 150–151

1. (a) 8π cu. units; **(b)** $166\pi/15$ cu. units; **(c)** $4\pi/5$ cu. units.

2. (a) $1296\pi/5$ cu. units; **(b)** $2^{14}\pi/105$ cu. units.

3. 36π cu. units. **4.** 27π cu. units. **5.** $\frac{1}{3}\pi r^2 h$ cu. units.

6. $81\pi/2$ cu. units. **7.** $640\pi/3$ cu. units. **8.** $V = \frac{1}{3}\pi h(r_1^2 + r_1 r_2 + r_2^2)$.

Exercises 4.5 [B], page 151

1. (a) $81\pi/2$ cu. units; **(b)** $258\pi/5$ cu. units.

2. $512\pi/15$ cu. units. **3.** $\frac{4}{3}\pi r^3$ cu. units. **4.** 144π cu. units.

5. 54π cu. units. **6.** $\frac{4\pi a^3}{3}(2\sqrt{2} - 1)$ cu. units. **7.** 63π cu. units.

8. $\frac{1}{3}\pi(b - a)(3r^2 - a^2 - ab - b^2)$ cu. units.

Exercises 4.6 [A], page 155

1. (a) $\sqrt{5} = 2.2$ sec; **(b)** $32\sqrt{5} = 72$ ft./sec. **2.** $8\sqrt{105} = 82$ ft./sec.

3. $16\sqrt{15} = 62$ ft./sec. **4. (a)** 50 ft.; **(b)** 25 ft.; **(c)** 5 sec. **5.** 20 units.

6. (a) 12 units/sec.^{2}; **(b)** $10\frac{1}{3}$ units. **7. (b)** 32 units. **8. (a)** $t = \frac{1}{4}$;
(b) none.

Exercises 4.6 [B], pages 155–156

1. (a) 22,500 ft.; **(b)** 75 sec. **2.** 113 ft./sec. **3.** $16\sqrt[4]{75} = 47$ ft./sec.

4. At $t = 0$, 10 units/sec.; at $t = 2$, -6 units/sec.; at $t = 5$, 15 units/sec.

5. Returns at $t = 3$; 15 units/sec. **6.** 8 units/sec. **7.** $145\frac{4}{5}$ units.

Miscellaneous Exercises 5.1, page 162

1. **(a)** 0 **(b)** none **(c)** 0 **(d)** 0 **(e)** none **(f)** each integer **(g)** each integer **(h)** none **(i)** each integer **(j)** $\pm\sqrt{n}$, for each positive integer n **(k)** $n/2$, for each integer n.

2. **(a)** 0 **(b)** nowhere **(c)** everywhere **(d)** everywhere except 0.

3. **(a)** $1 \leq x < 4$ **(b)** $2 \leq x < 3$ **(c)** $2 \leq x < 3$

Miscellaneous Exercises 5.2, pages 165–166

7. **(a)** $x \neq 0$ **(b)** $x \neq 2$ **(c)** $x \neq c$ **8.** **(c)** $\frac{1}{2} < x < 1,\ 1 < x < 1\frac{1}{2}$ **(d)** $\leq \frac{1}{2}$

9. **(b)** $\frac{2}{3} < x < 1,\ 1 < x < 1\frac{1}{3}$ **(c)** $\leq \frac{1}{6}$ **(e)** $\leq \frac{1}{3}\epsilon$ **10.** $h \neq 0$

Miscellaneous Exercises 5.3, pages 173–174

3. **(c)** $L \geq 0,\ L = 0,\ L \leq 0.$

5. **(a)** $\frac{1}{2}$ **(b)** 1 **(c)** no limit **(d)** $\frac{1}{2}$ **6.** **(a)** 0 **(b)** 1 **(c)** $\frac{1}{2}$

Miscellaneous Exercises 5.4, pages 179–180

1. **(a)** $c = 2$; **(b)** $c = \pm\sqrt{3}$; **(c)** $c = 2$.

2. **(a)** $c = \sqrt{ab}$; **(b)** $c = -\sqrt{ab}$; **(c)** $c = \dfrac{b + a}{2}$.

Miscellaneous Exercises 5.5, page 185

1. **(a)** **(i)** none; **(ii)** none; **(iii)** none. **(c)** **(i)** -1; **(ii)** 1; **(iii)** none. **(e)** **(i)** 0; **(ii)** 0; **(iii)** 0. **(g)** **(i)** -1; **(ii)** 1; **(iii)** none. **(i)** **(i)** 1; **(ii)** 0; **(iii)** none. **(k)** **(i)** -1; **(ii)** 0; **(iii)** none.

5. No limit (oscillates).

6. **(a)** 4 **(b)** 9/13 **(c)** 5/8 **(d)** ∞ **(e)** 1/5 **(f)** 0 **(g)** 2 **(h)** 1/3 **(i)** 0 **(j)** $-1/2$

7. **(b)** **(i)** $1,\ -1$; **(ii)** 0, none; **(iii)** none, none.

Exercises 6.1 [A], pages 190–191

1. **(a)** $dy = (2x - 5)dx$; **(b)** $dy = \left(-6x^{-3} + \dfrac{\sqrt{7}}{2}x^{-\frac{1}{2}}\right)dx$;

(c) $dy = (\frac{3}{2}x^{1/2} + \frac{3}{2}x^{-3/2})dx$; **(d)** $dy = 6x(x^2 - 2)^2\, dx$.

2. (a) -220; **(b)** -0.02. **4.** $-z^{-2}, 2z^{-3}, -6z^{-4}$; $d^n u/dz^n = (-1)^n n!\ z^{-(n+1)}$. **5.** $x < 2$. **6. (a)** $dy = 0$; **(b)** those which make $f'(x) = 0$. **7.** $y = 2x^4 + C$. **8.** $n!$, 0.

Exercises 6.1 [B], page 191

1. (a) $dy = (3x^2 + 6x)dx$; **(b)** $dy = (\frac{1}{2}\sqrt{2}x^{-1/2} + x^{-3/2})dx$; **(c)** $dy = (1 - x^{-2})dx$; **(d)** $dy = (1 - x^{-2/3})dx$. **2.** $y = x^3 - 2x + C$.

3.

$dx =$	10	1	0.1	0.01
$dy =$	60	6	0.6	0.06
$\Delta y =$	360	9	0.63	0.0603
$\Delta y/dy =$	6	1.5	1.05	1.005

4. $-2t^{-1} + 6t^{-3}$. **5.** $x^2 + C$. **6.** $x^{n-k}/(n-k)!$.

Exercises 6.2 [A], page 196

3. (a) $2x(x+3) + (x^2 - 2) = 3x^2 + 6x - 2$;
(b) $(2x - 3)(x^2 - 3x^{-1}) + (2x + 3x^{-2})(x^2 - 3x) = 4x^3 - 9x^2 - 3$;
(c) $(2x + 1)(3x^5 - 7) + 15x^4(x^2 + x) = 21x^6 + 18x^5 - 14x - 7$;
(d) $(x - 1)(x + 3) + x(x + 3) + x(x - 1) = 3x^2 + 4x - 3$.

Exercises 6.2 [B], pages 196–197

2. (a) $(2x + 3)(x^3 - 17) + 3x^2(x^2 + 3x - 5)$;
(b) $(1/2\sqrt{x})(x^2 + 3x^{-2}) + (2x - 6x^{-3})(\sqrt{x} - 5)$; **(c)** $3a(ax + b)^2$;
(d) $2(2ax + b)(ax^2 + bx + c)$.

3. (d) $-a/(ax + b)^2$.

4. (a) $-2a(ax + b)^{-3}$; **(b)** $-(2ax + b)/(ax^2 + bx + c)^2$.

6. $f(-2) = -1$, min.; $f(5) = \frac{1}{6}$, max.

7. (a) $f(0) = 2$, max.; $(\pm 2\sqrt{3}/3, 3/2)$, infl. **(b)** None; none; all. **(c)** 0; 0.

Exercises 6.3 [A], page 203

1. (a) $dy/du = 9u^8$; $du/dx = 2x - 7$; $dy/dx = 9u^8(2x - 7)$.
(b) $dy/du = 6u^2 - 1$; $du/dx = 2x - 2x^{-3}$; $dy/dx = (6u^2 - 1)(2x - 2x^{-3})$.

2. (a) $1/2\sqrt{y + 2}$; **(b)** $10(2x - 7)(x^2 - 7x)^9$;
(c) $-2t(t + 2)(t^3 + 3t^2 - 6)^{-5/3}$; **(d)** $5\left(x - \frac{1}{x}\right)^4\left(1 + \frac{1}{x^2}\right)$.

3. (a) $\dfrac{4+3x}{2\sqrt{2+x}}$; (b) $\dfrac{3x^2+6x+4}{2(x+1)^{3/2}}$; (c) $\dfrac{-2(2x-1)}{(x-1)^3}$;
(d) $-1/(x+1)^{1/2}(x-1)^{3/2}$.

4. (a) $10(9x^2+18x+10)(x^2+2x+2)^3$; (b) $-\dfrac{t+8}{4(t+2)^{5/2}}$.

5. $6(x+\sqrt{a_2-x^2})^5\left(1-\dfrac{x}{\sqrt{a^2-x^2}}\right)$.

6. $4x-5y+9=0$; $5x+4y-40=0$.

Exercises 6.3 [B], page 204

1. (a) $dy/dx=-x^{-2}$; $dx/dt=2+2t^{-2}$; $dy/dt=-x^{-2}(2+2t^{-2})$.
(b) $dy/dx=10x^9+26x-3x^{-4}$; $dx/dt=3t^2+4$;
$dy/dt=(10x^9+26x-3x^{-4})(3t^2+4)$.

2. (a) $-6x^2/\sqrt{2-4x^3}$; (b) $8(2x+3)(x^2+3x-5)^7$;
(c) $n(2ax+b)(ax^2+bx+c)^{n-1}$; (d) $\frac{1}{2}(x^{1/2}+x^{-1/2})^{-1/2}(\frac{1}{2}x^{-1/2}-\frac{1}{2}x^{-3/2})$.

3. (a) $(3x+1)/2\sqrt{x-1}$; (b) $-12t(1-t^2)^2(1+t^2)^{-4}$;
(c) $a^2(a^2-x^2)^{-3/2}$; (d) $\dfrac{3(1+4x\sqrt{x+2})}{\sqrt{x+2}}(\sqrt{x+2}+x^2)^5$.

4. $\tan^{-1}\frac{24}{7}$.

Miscellaneous Exercises 6.3, pages 204–205

1. $an(ax+b)^{n-1}$ 3. $5(6x^2-7)(2x^3-7x+3)^4$ 5. $x/\sqrt{x^2+a^2}$
7. $(2ax+b)/2\sqrt{ax^2+bx+c}$ 9. $2(x+2)(2x+1)(7x+5)$
11. $(3x+2)/\sqrt{2x-1}$ 13. $\frac{2}{3}(x-1)(4x-5)(x-3)^{-1/3}$
15. $2(5x+14)(x+3)(2x+5)^2$ 17. $6(x-2)(4x-x^2)^{-4}$ 19. $23/(3x+4)^2$
21. $(ps-qr)/(rx+s)^2$ 23. $x(x-2)/(x-1)^2$
25. $-2(7x^3-6x^2-14)/(x^3+4)^2$ 27. $x(a^2-x^2)^{-3/2}$
29. $(-2x^2+13x-8)/2(x+3)^2(x-1)^{3/2}$ 31. $-31/2(2x+3)^{1/2}(7x-5)^{3/2}$
33. $1/4\sqrt{x}\sqrt{1+\sqrt{x}}$ 35. $(x-1)^{-1/2}(x+1)^{-3/2}$ 37. $-a^2/x^2\sqrt{a^2+x^2}$
39. $-a^2/x^2\sqrt{a^2-x^2}$

Exercises 6.4 [A], page 211

1. $x/(4-y)$. 2. $-x/2y$. 3. $-x^2/y^2$. 4. $-2y/3x$.
5. $-y(2x^{1/2}+y^{1/2})/x(2y^{1/2}+x^{1/2})$. 6. $-4/y^3$. 7. $-21/25y^3$.
8. $x+3y-5=0$. 9. (b) $0, \pi/2$; (c) $\pi/2$.

Exercises 6.4 [B], page 212

1. $-y/x$. **2.** $-\left(\frac{y}{x}\right)^{1/2}$. **3.** $(ay - x^2) / (y^2 - ax)$.

4. $-(3x^2 + 2xy + y^2) / (x^2 + 2xy + 3y^2)$. **5.** $(x^2 - 8x) / [2y(x - 4)^2]$.

6. $3x/4ay$. **7.** $-8/9y^3$. **8.** $-3, \frac{7}{3}$. **9.** $(-2, 1)$, max.; $(2, -1)$, min.

Exercises 6.5 [A], pages 217–218

2. **(a)** $-3 \sin 3x$
(b) $\sec (x - s) \tan (x - s)$
(c) $k \sec^2 kx$
(d) $-\frac{a}{x^2} \cos \frac{a}{x}$
(e) $-3 \csc^2 (3x + 10)$
(f) $4 \sin (7 - 4x)$
(g) $6 \sin x \cos x$
(h) $\frac{1}{2}(\tan x)^{-1/2} \sec^2 x$
(i) $\frac{1}{2\sqrt{x}} \sec^2 \sqrt{x}$
(j) $\frac{1}{\sqrt{x}} \tan \sqrt{x} \sec^2 \sqrt{x}$
(k) $\sin x + x \cos x$
(l) $-\sin x \cos 2x - 2 \cos x \sin 2x$
(m) $\frac{\cos x + x \sin x}{\cos^2 x}$
(n) $2x \cos (4x - 3) - 4x^2 \sin (4x - 3)$
(o) $\frac{-5 \cos 5x}{2\sqrt{1 - \sin 5x}}$
(p) -1
(q) $2 \cos^2 xy - \frac{y}{x}$
(r) $\frac{y \cos x - \sin y}{x \cos y - \sin x}$

3. **(a)** 1; **(b)** 2; **(c)** 0.

4. **(a)** $\pi/4$, infl.; $3\pi/4$, max.; $5\pi/4$, infl.; $7\pi/4$, min. **(b)** 0, min.; $\pi/3$, infl.; π, max.; $5\pi/3$, infl. **6.** 2 sq. units. **8.** $\tan^{-1} \frac{4}{3}$.

9. **(a)** 1 **(b)** a/b **(c)** 1/6

Exercises 6.5 [B], pages 218–219

1. **(a)** $6 \sec^2 2x$
(b) $\sin (\pi - x) = \sin x$
(c) $-5 \csc (5x - 3) \cot (5x - 3)$
(d) $-2 \csc^2 \frac{x}{2}$
(e) $\frac{1}{2} \sec^2 \frac{x - 3}{4} \tan \frac{x - 3}{4}$
(f) $(1 - x^{-2}) \cos (x + x^{-1})$
(g) $\frac{\sin \sqrt{a - x}}{2\sqrt{a - x}}$
(h) $\frac{-\sin 2x}{\sqrt{1 + \cos 2x}}$
(i) $\frac{x \sec^2 x - \tan x}{x^2}$
(j) $\cos 2x$
(k) $\cos 2x \sec y$
(l) -1

2. **(a)** $\sec \theta (\sec^2 \theta + \tan^2 \theta)$; **(b)** $-9 \sin 3x$; **(c)** $-2 \sin t - t \cos t$; **(d)** $8 \sec^2 2\phi \tan 2\phi$.

3. (a) 3; **(b)** 0; **(c)** 1; **(d)** $\frac{1}{2}$.

4. (a) 0.98, max.; 2.55, infl.; 4.12, min.; 5.70, infl. **(b)** $\pi/6$, min.; $5\pi/6$, max.

5. $2\sqrt{2}$ sq. units.

8. $V = \frac{1}{3}\pi e^3 \sin^2 \theta \cos \theta$; $\tan^{-1} \sqrt{2}$. **10. (a)** 1 **(b)** 1/2 **(c)** 1/3

Exercises 7.1 [A], pages 226–227

1. (a) $24x$ cu. ft.; **(b)** dV/dt, dx/dt; **(c)** $dV/dt = 24\,dx/dt$; **(d)** $\frac{1}{8}$ ft./sec.

2. (a) (i) dA/dt; **(ii)** dr/dt; **(iii)** $dA/dt = 8\pi r(dr/dt)$. **(b) (i)** $1/20\pi$ ft./min.; **(ii)** $1/40\pi$ ft./min.

3. (a) $\dfrac{dy}{dt} = f'(x)\dfrac{dx}{dt}$; **(b)** 9.

4. (a) $\dfrac{dx}{dt} = 2t, \dfrac{dy}{dt} = 1 - 3t^2, \dfrac{dy}{dx} = \dfrac{1-3t^2}{2t}$. **(b)** $\dfrac{dx}{dt} = -a \sin t$, $\dfrac{dy}{dt} = b \cos t, \dfrac{dy}{dx} = -\dfrac{b}{a} \cot t$.

5. (a) $\frac{1}{4}\sqrt{25t^2 - 320t + 1600}$ mi.; **(b)** when $t = 6.4$; **(c)** 6 mi.

6. 5 rad./hr.

Exercises 7.1 [B], pages 227–228

1. (a) $\dfrac{b}{a} \csc (kt - n)$; **(b)** $(t^2 - 1)/2t$.

2. (a) $V = \pi r^2 h$; **(b)** $dV/dr = 2\pi rh$.

3. (b) $V = \frac{4}{3}\pi r^3, A = 4\pi r^2$; **(c)** $\dfrac{dV}{dt} = 4\pi r^2 \cdot \dfrac{dr}{dt}, \dfrac{dA}{dt} = 8\pi r \cdot \dfrac{dr}{dt}$; **(d)** $1/18\pi$ in./sec.; **(e)** $\frac{4}{3}$ sq. in./sec.

4. (a) $V = \frac{1}{3}\pi r^2 h$; **(c)** $r/h = 4/5$; **(d)** $\dfrac{16}{75}\pi \cdot 3h^2 \cdot \dfrac{dh}{dt} = \dfrac{16}{25}\pi h^2 \cdot \dfrac{dh}{dt}$; **(e)** $75/256\pi$ cm./min.

5. (b) $z^2 = x^2 + 64, 2z\dfrac{dz}{dt} = 2x\dfrac{dx}{dt}$; **(c)** 27.2 ft./min.; **(d)** 64/85 rad./min.

6. $169/\sqrt{170}$, or 13^-, ft./sec.

Miscellaneous Exercises 7.1, pages 228–230

1. πr^2. **2.** 64π mi./min. **3. (a)** $1/200\pi$ ft./sec.; **(b)** $\frac{1}{5}$ sq. ft./sec.

4. (a) $16\sqrt{2}/3$, or approximately 7.54, ft./sec.; **(b)** $-2/225$ rad./sec.

5. (a) 2.4 sq. ft./sec.; **(b)** $6\sqrt{3}/5\sqrt{7}$, or approximately 0.79, ft./sec.

6. **(a)** 4 ft./sec.; **(b)** 10 ft./sec. **7.** -30 rad./hr. **8.** $29/\sqrt{73}$, or approximately 3.4, mi./hr. **9.** 125 ft./sec. **11.** **(a)** 4; **(b)** $\frac{1}{4}$ rad./unit time. **12.** -16. **13.** $2\pi a^3/3$. **14.** $a^2/2$ sq. units/min. **15.** $\frac{25}{32}$ unit/sec. **16.** $6\sqrt{3}$, or approximately 10.4, knots. **17.** Shortening at 20/9 ft./sec.

Exercises 7.2 [A], page 233

1. **(a)** 0.2; **(b)** 2. **2.** 5.2. **3.** **(a)** $5\frac{5}{6}$; **(b)** $4\frac{1}{48}$; **(c)** $9\frac{2}{9}$. **4.** **(a)** 0.01; **(b)** $0.01\sqrt{2}/2 = 0.007$; **(c)** 0. **5.** $\pi/2, 3\pi/2$. **6.** 7640 c.c. **7.** 1.55; none.

Exercises 7.2 [B], page 234

1. **(a)** 1771.2; **(b)** $2\frac{53}{54}$; **(c)** $4\frac{73}{75}$. **2.** $2\pi r\,\Delta r$. **3.** **(a)** $\pi/180$; **(b)** $\pi/90$; **(c)** $\pi/90$. **4.** $-\Delta x/x^2$. **5.** 0.005π cu. in. **6.** $0.01\pi\sqrt{3}/6$, or 0.91%. **7.** Roots are 0.15, 1.42, -4.57.

Miscellaneous Exercises 7.2, pages 234–235

1. **(a)** $7\frac{1}{14}$; **(b)** $2\frac{25}{27}$; **(c)** $1\frac{31}{32}$; **(d)** $1\frac{637}{640}$; **(e)** 0.01001; **(f)** 7.7; **(g)** 0.099; **(h)** 0.0106; **(i)** 1; **(j)** $1 + nh$; **(k)** 1.06.

2. $2\pi rh\,\Delta r$ cu. in. **3.** Within 1/135, or approximately 0.0074, in.

6. **(a)** 90°; **(b)** independent of θ. **7.** **(a)** $x \geq 2500$; **(b)** $x \geq 50$. **8.** **(a)** $0 \pm n\pi$; **(b)** no maximum exists. **9.** 0.031 yd.

10. $\{x\colon |\cos x| \leq 0.1\}$.

11. **(a)** 0.60; **(b)** 1.35; **(c)** -0.14; **(d)** -0.29; **(e)** 1.25; **(f)** 1.35, 1.69, -3.05.

12. **(b)** 2.00; **(c)** $x_{n+1} = \dfrac{1}{r}[(r-1)x_n + a/x_n^{\,r-1}]$.

13. **(a)** 0.51; **(b)** 1.18; **(c)** 0.62; **(d)** 1.06.

Exercises 7.3 [A], page 240

1. 15; $-16\sqrt{3}/9 = -3.1$. **2.** a^2 sq. units. **3.** 70°, 70°. **4.** **(a)** $h = 0$; **(b)** $h = 2r$. **5.** 1; $3\sqrt{6}/2$.

Exercises 7.3 [B], page 240

1. $f(0) = 1$, min.; $f(\pi/6) = \frac{3}{2}$, max.; $f(\pi/2) = 1$, min. **2.** 12 sq. units. **3.** Height, $h/3$; radius, $2r/3$. **4.** 3 ft. **5.** 6 mi. up the beach from the point nearest the ship.

Miscellaneous Exercises 7.3, page 241

1. 30 sq. units. **2. (a)** 0; **(b)** $a/(\pi + 4)$ in.

3. $y = \sqrt{x(x^2 - 9)}$; $[-3, 0] \cup [3, \infty)$; $f(-3) = f(0) = f(3) = 0$, all minima; $f(-\sqrt{3}) = \sqrt{6\sqrt{3}}$, max. $y = -\sqrt{x(x^2 - 9)}$; same domain; $f(-3) = f(0) = f(3) = 0$, all maxima; $f(-\sqrt{3}) = -\sqrt{6\sqrt{3}}$, min.

4. (a) $\sqrt{6}/3$, 3; **(b)** 1, 5. **5.** 128/3 sq. units. **6.** 24 sq. in.
7. $512\pi/9$ cu. in.
8. (a) At the camp; **(b)** $20/\sqrt{21}$, or approximately 4.4, mi. up the beach.
9. 4 sq. units. **10.** $3a^2\sqrt{3}/4$ sq. units. **11.** 120°.
12. $8a^3\sqrt{3}/9$ cu. units.

Exercises 7.3 [C], page 242

1. $8\pi a^3/3$ cu. units. **2.** $\pi a^2(2\sqrt{2} + 3)$ sq. units. **3.** $a^2/2$, $3a^2/2$.
4. $(a^{2/3} + b^{2/3})^{3/2}$. **5.** $(-1, 0)$, infl.; $(0, 2^{2/3})$, max.; $(2, 0)$, min.
6. $(\pm 2\sqrt{\sqrt[3]{2} - 1}, \sqrt[3]{4})$. **7.** $13^{3/2}$, or approximately 47, ft. **8.** 12 ft.
9. $5^{3/2}$, or approximately 11.2, ft. **10.** $10/\pi$, or approximately 3.2, in.

Exercises 7.4 (1) [A], page 245

1. $\frac{2}{3}(x - 2)^{3/2} + C$. **2.** $\frac{1}{26}(2x + 3)^{13} + C$. **3.** $\frac{2}{9}(3t + 2)^{3/2} + C$.
4. $\frac{1}{88}(2x - 5)^{44} + C$. **5.** $\frac{9}{20}(4x - 5)^{5/3} + C$. **6.** $-\frac{1}{4}(3 - 5x)^{8/5} + C$.
7. $\frac{3}{16}(4x + 7)^{4/3} + C$. **8.** $-\frac{1}{6}(2x - 3)^{-3} + C$. **9.** $\frac{3}{10}(y^2 + 4)^{5/3} + C$.
10. $-\frac{5}{3}(1 - x^2)^{3/2} + C$. **11.** $\frac{1}{6}(2x^2 + 7)^{3/2} + C$. **12.** $\frac{1}{9}(2x^3 - 7)^{3/2} + C$.
13. $-\frac{5}{8}(3 - 2x^3)^{4/3} + C$. **14.** $\frac{1}{6}(x^4 - 16)^{3/2} + C$. **15.** $\frac{1}{3}\sqrt{3t^2 + 1} + C$.
16. $\frac{3}{28}(x^{4/3} + 81)^7 + C$. **17.** $\frac{4}{3}(1 + \sqrt{x})^{3/2} + C$. **18.** $\frac{5}{3}(x^{1/2} + 2)^{6/5} + C$.
19. $\frac{2}{3}(x^2 + x - 3)^{3/2} + C$. **20.** $\frac{1}{3}(x^2 + 10x + 24)^{3/2} + C$.
21. $\frac{2}{3}(x - 8)\sqrt{x + 1} + C$. **22.** $\frac{1}{135}(9x^2 - 2)(3x^2 + 1)^{3/2} + C$.
23. $\frac{2}{15}(3x + 1)(x + 2)^{3/2} + C$. **24.** $\frac{2}{5}(x^2 + 4x + 24)\sqrt{x - 3} + C$.

Exercises 7.4 (2) [A], pages 247–248

1. $-\cos u + C$. **2.** $\sin u + C$. **3.** $-\frac{1}{5}\cos 5x + C$. **4.** $\tan z + C$.
5. $-\cot\theta + C$. **6.** $\frac{1}{2}\sec 2x + C$. **7.** $-\frac{1}{3}\csc 3t + C$.
8. $\frac{1}{2}\tan^2 x + C_1 = \frac{1}{2}\sec^2 x + C_2$. **9.** $-\frac{1}{4}\cos^4 x + C$.
10. $-\frac{1}{20}\cos 10\theta + C_1 = \frac{1}{10}\sin^2 5\theta + C_2 = -\frac{1}{10}\cos^2 5\theta + C_3$.

11. $\frac{1}{6}\sin^3 2x + C.$ **12.** $\frac{1}{a}\sin(ax+b) + C.$ **13.** $\frac{2}{3}\sin^{3/2} x + C.$
14. $-\frac{3}{5}(2+\cos x)^{5/3} + C.$ **15.** $-1/(1-\cos x) + C.$
16. $-\frac{1}{15}(3 - 2\sin 5\theta)^{3/2} + C.$ **17.** $\frac{1}{3}\tan^3 x + C.$ **18.** $\tan x - x + C.$
19. $\frac{1}{3}\tan^3 x - \tan x + x + C.$ **20.** $-\cot x - x + C.$
21. $-\frac{1}{3}\cot^3 t + \cot t + t + C.$ **22.** $\frac{1}{3}\tan^3 x + \tan x + C.$
23. $-\frac{1}{3}\cot^3 x - \cot x + C.$ **24.** $\frac{1}{2}x - \frac{1}{4}\sin 2x + C.$
25. $\frac{1}{2}\theta + \frac{1}{16}\sin 8\theta + C.$ **26.** $\frac{1}{n+1}\sin^{n+1} x + C.$
27. $\frac{-1}{n+1}\cos^{n+1} x + C.$ **28.** $\frac{1}{5}\cos^5 x - \frac{1}{3}\cos^3 x + C.$
29. $-\frac{1}{5}\sin^5 x + \frac{1}{3}\sin^3 x + C.$ **30.** $\frac{1}{n+1}\tan^{n+1} x + C.$
31. $\frac{1}{n}\sec^n x + C.$ **32.** $\frac{1}{3}\cos^3 x - \cos x + C.$
33. $-\frac{1}{3}\sin^3 x + \sin x + C.$ **34.** $\frac{3}{8}x - \frac{1}{4}\sin 2x + \frac{1}{32}\sin 4x + C.$
35. $\frac{1}{5}\tan^5 x + \frac{2}{3}\tan^3 x + \tan x + C.$ **36.** $-\frac{1}{10}\cos 5x + \frac{1}{2}\cos x + C.$
37. $-\frac{1}{12}\cos 6x - \frac{1}{8}\cos 4x + C.$
38. $\frac{-1}{2(a+b)}\cos(a+b)x - \frac{1}{2(a-b)}\cos(a-b)x + C.$
39. $\frac{-1}{2(a+b)}\sin(a+b)x + \frac{1}{2(a-b)}\sin(a-b)x + C.$
40. $\frac{1}{2(a+b)}\sin(a+b)x + \frac{1}{2(a-b)}\sin(a-b)x + C.$

Exercises 7.4 (3) [A], page 249

1. $\sin^{-1} x + C.$ **2.** $\frac{1}{2}\sin^{-1} x + \frac{x}{2}\sqrt{1-x^2} + C.$ **3.** $\sin^{-1}\frac{x}{2} + C.$
4. $x - \frac{1}{3}x^3 + C.$ **5.** $\frac{1}{2}\sin^{-1}\frac{2x}{3} + C.$ **6.** $-\sqrt{4-x^2} + C.$
7. $-\frac{x}{2}\sqrt{1-x^2} + \frac{1}{2}\sin^{-1} x + C.$ **8.** $\tan^{-1} x + C.$

Exercises 7.4 (4) [A], page 250

1. $\frac{1}{22}(2x-5)^{11} + C.$ **2.** $-\frac{1}{14}(3-4x)^{7/2} + C.$ **3.** $-\frac{1}{4}\cos 4\theta + C.$
4. $\frac{1}{3}(x^2+3)^{3/2} + C.$ **5.** $\frac{1}{5}\tan 5\theta + C.$ **6.** $\frac{1}{3}(3x^{2/3} - 5)^{3/2} + C.$
7. $-\cot x + C.$ **8.** $\frac{3}{2}x - 2\sin x + \frac{1}{4}\sin 2x + C.$
9. $\frac{1}{30}(3x^2-1)(2x^2+1)^{3/2} + C.$ **10.** $\frac{1}{2}\sin x^2 + C.$
11. $\frac{1}{2}y - \frac{1}{12}\sin 6y + C.$ **12.** $\sin^{-1} x + C.$ **13.** $\frac{2}{3}(x+19)\sqrt{x-5} + C.$
14. $\frac{1}{2}\tan^{-1}(x/2) + C.$

Exercises 7.4 (4) [B], page 251

1. $\frac{2}{3}\sqrt{3x-7}+C.$ 2. $\frac{1}{5}(x^3-5)^{5/3}+C.$ 3. $\frac{1}{3}\sqrt{3x^2-7}+C.$
4. $2\sin\frac{1}{2}v+C.$ 5. $-\frac{1}{a}\cos(ax+b)+C.$ 6. $3\sec\frac{x+1}{3}+C.$
7. $\frac{-1}{x+1}+C.$ 8. $-\frac{1}{2}(x^2+3x-5)^{-2}+C.$ 9. $-\frac{1}{4}\cos^4 z+C.$
10. $-\frac{1}{3}\cot^3\theta+C.$ 11. $\frac{6}{5}(\sqrt{x}+\sqrt{a})^{5/3}+C.$ 12. $\frac{2}{3}(1+\tan x)^{3/2}+C.$
13. $-2\cos\sqrt{x}+C.$ 14. $-\sqrt{a^2-x^2}+C.$ 15. $\sin^{-1}\frac{x}{a}+C.$
16. $\frac{2}{5}(x+4)(x-1)^{3/2}+C.$

Miscellaneous Exercises 7.4, pages 251–252

1. $\frac{1}{7}\sin 7\theta+C.$ 2. $-\frac{2}{33}(14-11x)^{3/2}+C.$ 3. $\sin\left(x-\frac{\pi}{3}\right)+C.$
4. $\frac{1}{5}(2x-1)^{5/2}+C.$ 5. $2\sqrt{x+2}+C.$ 6. $\frac{1}{a(n+1)}(ax+b)^{n+1}+C.$
7. $\frac{1}{2}\sin^2\theta+C_1=-\frac{1}{2}\cos^2\theta+C_2=-\frac{1}{4}\cos 2\theta+C_3.$
8. $\frac{1}{6}(4x^3-1)^{3/2}+C.$ 9. $-\frac{1}{12}(x^2-1)^{-6}+C.$
10. $-a^4x^{-1}-2a^2x+\frac{1}{3}x^3+C.$
11. $-\frac{1}{8}(x^4-5)^{-2}+C.$ 12. $\frac{1}{2}\tan^2 x+C_1=\frac{1}{2}\sec^2 x+C_2.$
13. $-\csc x+C.$ 14. $-\sqrt{\cos 2z}+C.$
15. $-\frac{2}{15}(3x^2+31x+233)\sqrt{4-x}+C.$ 16. $1/(2+\csc y)+C.$
17. $\sin^{-1}\frac{x}{\sqrt{3}}+C.$ 18. $3\tan\frac{x}{3}+C.$ 19. $\frac{1}{6}\tan^{-1}\frac{2x}{3}+C.$
20. $\frac{2}{5}(2x+13)(x-1)^{3/2}+C.$
21. $-\frac{1}{4}\cot^4\theta+C.$ 22. $\frac{1}{3}\sin^{-1}3x+C.$ 23. $\frac{1}{3}\tan 3z-z+C.$
24. $\frac{1}{5}(x^2+2x+6)\sqrt{2x-3}+C.$ 25. $1/(1-x)+C.$
26. $\frac{-1}{a^2x}\sqrt{a^2-x^2}+C.$ 27. $\frac{1}{2}\theta-\frac{1}{16}\sin 8\theta+C.$ 28. $-2\sqrt{2}\cos\frac{x}{2}+C.$
29. $\frac{1}{2}\sin^2 x+C_1=-\frac{1}{2}\cos^2 x+C_2=-\frac{1}{4}\cos 2x+C_3.$
30. $\sin^{-1}\frac{x}{\sqrt{2}}+C.$
31. $\sin x+C.$ 32. $\frac{1}{3}(x+14)\sqrt{2x+1}+C.$ 33. $\frac{1}{4}\sin 2x-\frac{x}{2}+C.$
34. $-\frac{x}{2}\sqrt{9-x^2}+\frac{9}{2}\sin^{-1}\frac{x}{3}+C.$ 35. $\frac{19\theta}{2}+3\cos 2\theta-\frac{1}{8}\sin 4\theta+C.$
36. $\frac{3z}{8}+\frac{1}{4}\sin 2z+\frac{1}{32}\sin 4z+C.$ 37. $-\cos t+\frac{2}{3}\cos^3 t-\frac{1}{5}\cos^5 t+C.$
38. $-\frac{2}{3}\cos^3\theta+\cos\theta+C.$

Exercises 7.5 [A], page 254

1. $-10\frac{2}{3}$. **2.** $4\frac{2}{3}$. **3.** $\frac{21}{128}$. **4.** $\frac{4}{15}(6\sqrt{3}+1)$. **5.** $\pi/2$.
6. $43\frac{5}{28}$. **7.** $\pi/6$. **8.** $\frac{9}{2}\sin^{-1}\frac{1}{3}-\sqrt{2}$.
9. $5\frac{1}{3}$ sq. units. **10.** $\frac{1}{2}\pi^2$ cu. units. **11.** πa^2 sq. units.
12. 8π sq. units.

Exercises 7.5 [B], pages 254–255

1. $\frac{1}{2}(9-\sqrt{3})$. **2.** $4\frac{2}{3}$. **3.** $13\frac{5}{9}$. **4.** 2. **5.** $\pi/8$. **6.** $a^2/6$.
7. $\pi/4$. **8.** $\dfrac{4\sqrt{2}-\sqrt{5}}{18}$.
9. 4 sq. units. **10.** $\pi/3$ cu. units. **11.** 12π sq. units.
12. $\dfrac{a^2}{12}(4\pi-3\sqrt{3})$ sq. units.

Exercises 7.6 [A], pages 256–257

1. $y=-\frac{2}{3}$. **2.** $4\sqrt{y}=x^2+8$. **3.** $-\dfrac{2}{x}=(y+3)^2$.
4. (a) $\sin^2 x+\sin^2 y=C$; **(b)** $y^{3/2}=3x^{1/2}+C$;
(c) $\sin^{-1}x+\sin^{-1}y=C$, or $x^2+2cxy+y^2=1-c^2$ $(|x|\le 1, |y|\le 1)$;
(d) $\tan^{-1}x-\tan^{-1}y=C$, or $y-x=c(1+xy)$;
(e) $(x+1)^{3/2}=3(y-1)^{1/2}+C$.
6. $y^2=\tan x+1$. **7.** $(x+C)y=-1$. **8.** $y=\sin\left(x+\dfrac{\pi}{6}\right)$.
9. $v=\frac{1}{10}$. **10.** 5 hr.

Exercises 8.2 [A], pages 269–270

1. (a) $bx-ay=0$, $|x|\le|a|$; **(b)** $bx-ay=0$, $|x|\ge|a|$;
(c) $bx-ay=0$.
2. (a) $-\cot\theta$, $-\dfrac{\csc^3\theta}{a}$; $x^2+y^2=a^2$. **(b)** $1/t$, $-\dfrac{1}{2pt^3}$; $y^2=4px$.
(c) $-\cot 2\phi$, $-\csc^3 2\phi$; $(x-1)^2+y^2=1$.
3. $x=1-t^2$, $y=t-t^3$. **4.** 36 sq. units; $1296\pi/5$ cu. units.
5. $x=3t+t^2$, $y=5t$.

Exercises 8.2 [B], page 270

1. (a) $2x-y+1=0$, $x\ge 0$; **(b)** $2x-y+1=0$, $|x|\le 1$;
(c) $2x-y+1=0$, $x\ne 0$.
2. (a) $\csc t$, $-\dfrac{1}{a}\cot^3 t$; $x^2-y^2=a^2$. **(b)** $t/2$, $-t^3/16p$; $y^2=4px$,
$x\ne 0$. **(c)** $2t/(1-2t)$, $2/(1-2t)^3$; $x^2+2xy+y^2+2x+y=0$.

4. πa^2 sq. units; $\frac{4}{3}\pi a^3$ cu. units. **5.** $x = t^2 - t + 2$, $y = \frac{1}{3}t^3 - \frac{1}{2}t^2 + 2t + 5$.

Exercises 8.3 [A], page 273

1. (a) $(1, 5), 4$; **(b)** $(-\frac{1}{2}, 3), \frac{1}{2}$; **(c)** no locus; **(d)** $(\frac{3}{2}, -\frac{1}{3}), \frac{1}{3}\sqrt{37}$.
2. $x = h + r\cos\theta,\ y = k + r\sin\theta$.
5. $x - y = 0,\ 23x - 7y = 0$. **7.** $r^2 \sin 2\theta;\ r^2$.
8. $(x - h)^2 + (y - k)^2 = C;\ (x - h)^2 + (y - k)^2 = h^2 + k^2$.
9. $x^2 + (y + 1)^2 = 18$.

Exercises 8.3 [B], pages 273–274

1. $\left(-\frac{a}{2}, -\frac{b}{2}\right), \frac{1}{2}\sqrt{a^2 + b^2 - 4c};\ a^2 + b^2 - 4c < 0$.
3. $3x^2 + 3y^2 - 13x - 15y + 12 = 0$. **4.** $\pm\sqrt{3}$.
5. $(x - 4)^2 + (y - 4)^2 = 16,\ (x + 12)^2 + (y - 12)^2 = 144$. **8.** $\tan^{-1} 7$.

Exercises 8.4 [A], pages 280–281

1. (a) Foci $(\pm 1, 0)$; vertices $(\pm\sqrt{5}, 0)$; $e = \frac{1}{5}\sqrt{5}$. **(b)** Foci $(0, \pm 4)$; vertices $(0, \pm 5)$; $e = \frac{4}{5}$. **(c)** Foci $(\pm\sqrt{3}, 0)$; vertices $(\pm 2, 0)$; $e = \frac{1}{2}\sqrt{3}$. **(d)** Foci $(3, 2), (-5, 2)$; vertices $(4, 2), (-6, 2)$; $e = \frac{4}{5}$.
2. (a) $\frac{x^2}{13} + \frac{y^2}{4} = 1$; **(b)** $\frac{x^2}{5} + \frac{y^2}{9} = 1$; **(c)** $\frac{(x - 2)^2}{16} + \frac{(y - 3)^2}{12} = 1$.
3. $\frac{1}{2}$. **4.** $\pm e$. **6.** $\left(\pm\frac{a^2}{\sqrt{a^2 + b^2}}, \pm\frac{b^2}{\sqrt{a^2 + b^2}}\right)$.
7. πab sq. units. **8.** $2ab$ sq. units. **9.** $-\sqrt{3}$ units/min.; -2 rad./min.

Exercises 8.4 [B], page 281

1. (a) Foci $(\pm\frac{3}{2}\sqrt{2}, 0)$; vertices $(\pm 3, 0)$; $e = \frac{1}{2}\sqrt{2}$. **(b)** Foci $(3, \pm\sqrt{5})$; vertices $(3, \pm 3)$; $e = \frac{1}{3}\sqrt{5}$. **(c)** The locus is the point $(2, -4)$. **(d)** Foci $(0, \pm 2\sqrt{2})$; vertices $(0, \pm 3)$; $e = \frac{2}{3}\sqrt{2}$.
2. (a) $\frac{(x - 1)^2}{16} + \frac{(y - 2)^2}{25} = 1$; **(b)** $\frac{(x + 2)^2}{16} + \frac{(y + 2)^2}{4} = 1$; **(c)** $\frac{x^2}{15} + \frac{y^2}{10} = 1$.
3. $\left(\frac{a}{2}\sqrt{2}, \frac{b}{2}\sqrt{2}\right)$; $\left(-\frac{a}{2}\sqrt{2}, -\frac{b}{2}\sqrt{2}\right)$. **5. (a)** $\frac{4}{3}\pi ab^2$ cu. units; **(b)** $\frac{4}{3}\pi a^2 b$ cu. units. **8.** $\left(\pm\frac{a}{2}\sqrt{2}, \pm\frac{b}{2}\sqrt{2}\right)$.

Exercises 8.5 [A], page 284

1. (a) Foci $(\pm 5, 0)$; vertices $(\pm\sqrt{5}, 0)$; $e = \sqrt{5}$; $2x \pm y = 0$.
(b) Foci $(0, \pm\sqrt{2})$; vertices $(0, \pm 1)$; $e = \sqrt{2}$; $x \pm y = 0$.
(c) Foci $(\pm 13, 0)$; vertices $(\pm 5, 0)$; $e = 13/5$; $12x \pm 5y = 0$.
(d) Foci $(2 \pm \sqrt{5}, -3)$; vertices $(1, -3)$, $(3, -3)$; $e = \sqrt{5}$; $2x + y - 1 = 0$, $2x - y - 7 = 0$.
2. (a) $\dfrac{x^2}{4} - \dfrac{y^2}{5} = 1$; **(b)** $\dfrac{(x-3)^2}{9} - \dfrac{y^2}{16} = -1$; **(c)** $x^2 - y^2 = 9$.
3. 2; $\pm\sqrt{3}$. **4.** $2b^2/a$; $\pm e$.

Exercises 8.5 [B], page 285

1. (a) Foci $(\pm\frac{4}{3}\sqrt{3}, 0)$; vertices $(\pm 2, 0)$; $e = \frac{2}{3}\sqrt{3}$; $x \pm y\sqrt{3} = 0$.
(b) Foci $\left(\pm\dfrac{k}{6}\sqrt{13}, 0\right)$; vertices $\left(\pm\dfrac{k}{3}, 0\right)$; $e = \frac{1}{2}\sqrt{13}$; $3x \pm 2y = 0$.
(c) Foci $(1 \pm \frac{1}{2}\sqrt{13}, -3)$; vertices $(0, -3)$, $(2, -3)$; $e = \frac{1}{2}\sqrt{13}$; $3x + 2y + 3 = 0$, $3x - 2y - 9 = 0$. **(d)** Foci $(2, \pm 3\sqrt{5})$; vertices $(2, \pm 3)$; $e = \sqrt{5}$; $x \pm 2y - 2 = 0$.
2. $\pm\sqrt{e^2 - 1}$. **4.** $\pm\frac{5}{3}$. **5.** $18\pi(\sqrt{2} - 1)$ cu. units.

Exercises 8.6 [A], pages 288–289

1. (a) Focus $(0, 2)$; vertex $(0, 0)$; $y = -2$. **(b)** Focus $(-3, 0)$; vertex $(0, 0)$; $x = 3$. **(c)** Focus $(\frac{5}{6}, 0)$; vertex $(0, 0)$; $x = -\frac{5}{6}$.
(d) Focus $(2, 0)$; vertex $(0, 0)$; $x = -2$. **(e)** Focus $(-1, 4)$; vertex $(-3, 4)$; $x = -5$. **(f)** Focus $(-2, \frac{1}{2})$; vertex $(-2, 1)$; $y = \frac{3}{2}$.
2. (a) $x^2 = -8y$; **(b)** $y^2 = 12x$; **(c)** $(y - 5)^2 = -4(x - 1)$.
6. $\pi/2$; $\tan^{-1}\frac{3}{4}$. **7.** $16p^2/3$ sq. units. **8.** $2\pi pa^2$ cu. units; cylinder : paraboloid : cone = 6 : 3 : 2.

Exercises 8.6 [B], page 289

1. Vertex $\left(-\dfrac{b}{2a}, \dfrac{4ac - b^2}{4a}\right)$; focus $\left(-\dfrac{b}{2a}, \dfrac{4ac - b^2 + 1}{4a}\right)$.
2. $x - y = 0$; $\left(\dfrac{a}{4}, \dfrac{a}{4}\right)$. **6. (a)** 2; **(b)** 1 rad./unit time. **7.** $\frac{64}{27}p^2$ sq. units.

Exercises 8.7 [C], pages 295–296

1. (a) $(x'^2/4) + (y'^2/36) = 1$; ellipse. (b) $(x'^2/4) + (y'^2/36) = 0$; the point $(0, 0)$. (c) $(x'^2/4) + (y'^2/36) = -1$; no locus. (d) $(x'^2/4) - (y'^2/36) = 1$; hyperbola. (e) $(x'^2/4) - (y'^2/36) = 0$; two lines. (f) $(x'^2/4) - (y'^2/36) = -1$; hyperbola. (g) $x'^2 = 9$; two lines. (h) $x'^2 = 0$; one line.

2. $(2a + b)x'^2 + (2a - b)y'^2 = 2k$.

3. (a) $(x'^2/12) + (y'^2/28) = 1$; ellipse. (b) $(x'^2/4) - (y'^2/44) = 1$; hyperbola. (c) $(x''^2/4) - (y''^2/16) = -1$; hyperbola. (d) $(x''^2/9) + (y''^2/4) = 1$; ellipse. (e) $x' = \frac{2}{5}, -\frac{3}{5}$; two lines.

Exercises 8.8 [A], page 300

1. Circle; $x^2 + y^2 = 25$. 2. Circle; $(x + 5)^2 + y^2 = 16$.

3. Ellipse; $\dfrac{(x-4)^2}{4} + \dfrac{y^2}{3} = 1$. 4. Ellipse; $\dfrac{(x-2)^2}{16} + \dfrac{(y-3)^2}{25} = 1$.

5. Parabola; $x^2 = 4(y + 4)$, $|x| \neq 4$.

Exercises 8.8 [B], page 300

1. Parabola; $x^2 = -2a\left(y - \dfrac{a}{2}\right)$, $|x| \neq a$.

2. Hyperbola; $2x^2 - y^2 = 2a^2$, $|x| \neq a$.

3. Branch of hyperbola; $x = \frac{1}{4}\sqrt{2y^2 + 16}$.

4. Ellipse; $3x^2 + 2xy + 3y^2 - 8x - 8y = 0$.

Exercises 8.8 [C], pages 301–302

1. $x = a\theta - b \sin\theta$, $y = a - b\cos\theta$. 2. $x = 2a\tan\theta$, $y = 2a\cos^2\theta$; $y = 8a^3/(x^2 + 4a^2)$. 3. $x = a(3\cos\theta + \cos 3\theta) = 4a\cos^3\theta$, $y = a(3\sin\theta - \sin 3\theta) = 4a\sin^3\theta$. 4. $x = a\sec\theta$, $y = b\tan\theta$; $(x^2/a^2) - (y^2/b^2) = 1$. 5. $x = 2a\sin^2\theta$, $y = 2a\sin^2\theta\tan\theta$; $y^2 = x^3/(2a - x)$.

Miscellaneous Exercises 9.1 pages 308–309

1. (a) $2\frac{1}{12}$; (b) $6\frac{2}{3}$; (c) 5; (d) 5; (e) 36; (f) 216; (g) 36; (h) 36.

2. (a) (i) 6; (ii) 34; (iii) $\frac{7}{10}$; (iv) 2.

4. (a) $\sum_{i=2}^{17} \sqrt{i}$; (b) $\sum_{i=1}^{n} \sqrt{i}$; (c) $\sum_{i=1}^{2n} \sqrt{i}$

(d) $\sum_{i=3}^{n+1} \sqrt{i}$; (e) $\sum_{i=n}^{2n} \sqrt{i}$; (f) $\sum_{i=1}^{2n} \sqrt{i} - \sum_{i=1}^{n} \sqrt{i} = \sum_{i=n+1}^{2n} \sqrt{i}$.

5. (a) $\sum_{k=1}^{n} i$; (b) $\sum_{k=7}^{43} k^3$; (c) $\sum_{k=1}^{n} 2k$; (d) $\sum_{k=1}^{n} (2k-1)$

(e) $\sum_{k=1}^{n} 2^k$; (f) $\sum_{k=1}^{n} (-1)^k \cdot k$; (g) $\sum_{k=1}^{n} (-1)^k \cdot \frac{1}{k}$.

6. (a) $\sum_{n=10}^{100} 0.01n^2$; (b) $\sum_{k=10}^{100} \frac{10}{k}$.

7. $\frac{2}{3}$. **8.** (a) $2\frac{1}{2}$; (b) $\frac{2}{\pi}$; (c) $\frac{7}{3}$; (d) $\frac{4}{\pi}$; (e) $\frac{1}{4}$; (f) $6\frac{1}{3}$.

Exercises 9.1 [C], pages 309–310

1. (a) $\frac{1}{2}$; (b) $2(2\sqrt{2}-1)/3$; (c) 14/9.
4. (a) 3,330; (b) 204; (c) 646; (d) 274.
5. 103. **6.** 573.

Miscellaneous Exercises 9.2, pages 319–322

7. (b) (i) $\frac{52}{3}$; (ii) 306; (iii) -12 **9.** 44π.
10. (a) 1.896; (b) 2.052; (c) 2.005. **11.** (a) 3.151; (b) 3.147; (c) 3.142.
12. 0.6932. **13.** (a) 84 miles; (b) 93 miles.
14. (a) 0.6932; (b) 0.6932; (c) 1.151; (d) 10.07; (e) 3.820; (f) 1.402.

Exercises [A], page 324

1. (i) $256\pi/15$ cu. units; (ii) 8π cu. units. **2.** (i) $96\pi/5$ cu. units; (ii) $576\pi/7$ cu. units. **3.** $512\pi/15$ cu. units.

Exercises [A], page 326

1. (i) $56\pi/15$ cu. units; (ii) $32\pi/5$ cu. units. **2.** $704\pi/5$ cu. units. **3.** $4\pi h^5/5$ cu. units.

Exercises [A], page 328

1. (i) $128\pi/3$ cu. units; (ii) 64π cu. units. **2.** $512\pi/21$ cu. units. **3.** $128\pi/3$ cu. units. **4.** $512\pi\sqrt{2}/105$ cu. units.

Exercises [A], pages 329–330

1. **(i)** 128/3 cu. units; **(ii)** $32\sqrt{3}/3$ cu. units; **(iii)** 32/3 cu. units.
2. 128/3 cu. units. **3.** $128\pi/15$ cu. units.

Exercises 9.3 [A], page 330

1. **(a)** 8π cu. units; **(b)** $256\pi/15$ cu. units; **(c)** $128\pi/5$ cu. units; **(d)** $40\pi/3$ cu. units.
2. **(a)** and **(b)** $32\pi/3$ cu. units. **3.** $128\pi/15$ cu. units. **4.** π cu. units.
5. $\frac{1}{2}\pi r^2 b$ cu. units.

Exercises 9.3 [B], page 330

1. **(a)** $2\pi/5$ cu. units; **(b)** 2π cu. units; **(c)** $2\pi/3$ cu. units; **(d)** $26\pi/15$ cu. units; **(e)** $10\pi/3$ cu. units.
2. **(a)** and **(b)** $128\pi p^3/15$ cu. units. **3.** 4π cu. units.
4. $\frac{1}{3}\pi h^2(3r - h)$ cu. units. **5.** $8a^3/3$ cu. units.

Miscellaneous Exercises 9.3, page 331

1. $V = \frac{1}{3}\pi r^2 h$. **2.** $\pi a^3/15$ cu. units. **3.** 36π cu. units.
4. $2a^3/3$ cu. units. **5.** $\frac{2}{3}\pi a^3(1 - \cos\theta)$ cu. units. **6.** $32\pi a^3/105$ cu. units.
7. π cu. units. **8.** **(a)** 2048/15 cu. units; **(b)** $512\sqrt{3}/15$ cu. units.
9. $45\pi/2$ cu. units.

Exercises 9.3 [C], pages 331–332

1. $5\pi^2 a^3$ cu. units. **2.** $3\pi\sqrt{3}/16$ cu. units. **3.** 36π cu. in.
4. $500\pi^2$ cu. units. **5.** $8\pi a^3\left(\tan^{-1}\frac{b}{2a} + \frac{2ab}{b^2 + 4a^2}\right)$ cu. units; $4\pi^2 a^3$ cu. units.
6. $V = 2\pi^2 a^2 b$. **7.** $16r^3/3$ cu. units. **8.** $8r^3(2 - \sqrt{2})$ cu. units.

Exercises 9.4 [A], page 334

1. **(a)** $3200\pi\rho$ ft.-lb., or 310 ft.-T. approx.; **(b)** $11{,}200\pi\rho$ ft.-lb., or 1100 ft.-T. approx.
2. $3204\pi\rho$ ft.-lb., or 310 ft.-T. approx. **3.** 120 in.-lb.
4. **(a)** 196 ergs; **(b)** $200(1 - 10^{-6})$ ergs; **(c)** 200 ergs.
5. 50,000 ft.-lb.

Exercises 9.4 [B], pages 334–335

1. $79{,}200\pi$ ft.-lb. = 124 ft.-T.; $(39{,}600\pi - 14{,}400)$ ft.-lb. = 55 ft.-T.
2. 1750 ft.-T. **3.** $\dfrac{4000h}{h + 4000}$ lb.-mi.; 4000 lb.-mi. **4.** 864 in.-lb.
5. $2200\rho/3$ ft.-lb., or 23 ft.-T. **6.** $102{,}400\pi\rho/3$ ft.-lb., or 3300 ft.-T.

Exercises 9.5 [A], page 342

1. $\frac{2}{3}(10^{3/2} - 1)$. **2.** 53/6. **3.** 22/3. **4. (a)** $(-3, 2)$, $\sqrt{13}$, $(-2, 0)$; **(b)** $(1, 2)$, $\sqrt{5}$, $(-2, 0)$. **5. (a)** $(a\sqrt{2}, 2a)$, $a\sqrt{6}$; **(b)** $x^2 - y^2 = a^2$.
6. (a) and **(b)** $6a$.

Exercises 9.5 [B], pages 342–343

1. $\frac{2}{3}(2\sqrt{2} - 1)$. **2.** 14/3. **3.** $\frac{4}{3}\sqrt{3}$. **4. (a)** $(0, 2)$, 2; $(-1, 0)$, 1. **(b)** $\pi/2$. **5.** $4\,pt^{-3}\sqrt{t^2 + 4}$. **6.** 16.

Miscellaneous Exercises 9.5, page 343

1. (a) $2\sqrt{2}$; **(b)** $x - y + 2 = 0$, $-1 \leq x \leq 1$.
2. $2\pi^2$. **3.** $(-9a/4, 3a\sqrt{3}/4)$; $3a\sqrt{3}/2$; $(-3a\sqrt{3}/2, 15a/2)$.
4. (a) $x^2 + y^2 = r^2$; **(b)** $(-r \sin t, r \cos t)$; $(-r \cos t, -r \sin t)$; **(c)** $ds/dt = r$.
5. (a) $\sqrt{2}/2$; **(b)** k. **6.** $\mathbf{r}(t) = (1 - \cos t, \sin t)$; $ds/dt = 1$

Miscellaneous Exercises 9.6, pages 348–349

1. 8/3; 4(8/3) = 32/3 (sq. units). **2. (a)** $-27/4$; **(b)** $-36/5$. **3.** $(y_1 + y_2)/2$. **4. (a)** $4p/3$; **(b)** p. **5. (a)** 48 ft./sec.; **(b)** $v = 8\sqrt{s}$; **(c)** 64 ft./sec. **6. (a)** $2/\pi$; **(b)** $\frac{1}{2}$. **7. (a)** 901; **(b)** $2/\pi$; **(c)** $1802/\pi$. **8. (a)** $4ab/3$; **(b)** $4ab/3$. **9. (a)** a; **(b)** $3a/2$; **(c)** $4a/3$. **10. (a)** $\pi a/2$; **(b)** $4a/\pi$.

Miscellaneous Exercises 9.7, pages 350–351

1. $\dfrac{\pi}{27}(10^{3/2} - 1)$ sq. units. **2. (a)** and **(b)** $\dfrac{8\pi p^2}{3}(2^{3/2} - 1)$ sq. units.
3. $16\pi/3$ sq. units. **4.** $2\pi rh$ sq. units. **5.** $12\pi a^2/5$ sq. units.
6. $4\pi^2 ab$ sq. units. **7.** 3π sq. units. **8.** $\dfrac{2\pi}{1215}(247\sqrt{13} + 64)$ sq. units.
9. $2\pi[b^2 + (ab/e)\sin^{-1} e]$ sq. units.

Exercises 9.8 [A], page 354

1. 56ρ, or 3500, lb. **2.** $27\rho/8$, or 211, lb. **3.** $(9\sqrt{3} + \frac{27}{8})\rho$, or 1185, lb. **4.** $64\rho/15$, or 267, lb. **5.** $62{,}000\rho/9$ lb., or 215 T.

6. $\left(\frac{8\pi}{3} + 3\sqrt{3}\right)\rho$, or 679, lb. **7.** $4\rho/3$, or 83, lb.

Exercises 9.8 [B], pages 354–355

1. 72ρ, or 4500, lb. **2.** $40\rho/3$, or 833, lb. **3.** 7 ft. 4 in. **4.** $16\rho/3$, or 333, lb. **5.** 174ρ lb., or 5.4 T. **6. (a)** $(4\pi + \frac{9}{2}\sqrt{3})\rho$ lb.; **(b)** $12\pi\rho$ lb.

Miscellaneous Exercises 9.9, page 360

1. $(2a/\pi, 2a/\pi)$. **2.** $(2a/5, 2a/5)$. **3.** $(4a/3\pi, 4a/3\pi)$. **4.** $(4a/3\pi, 0)$. **5.** $(\pi/2, \pi/8)$. **6.** $(0, \frac{2}{5}a^2)$. **7.** $(53/50, 7/20)$. **8.** $(16/15, 128/105)$. **9.** $(\frac{1}{2}, \frac{8}{5})$. **10.** $(\frac{5}{4}, \frac{16}{3})$.

Exercises 9.9 [C], pages 360–361

1. $2a/3$. **2. (a)** $h/2$; **(b)** $3h/5$; **(c)** $3h/7$. **3.** $3\pi a/16$. **5.** $h = a/2$. **6.** $h/3$. **7.** A solid generated by rotating about the x-axis a rectangular region in the xy-plane. **8.** $3a/8$. **9.** $h/4$. **10.** $h/4$.

Exercises 10.1 [A], pages 373–374

1. (a) Reading down, missing values are: -2.30, -1.61, -1.20, -0.92, -0.69, -0.23; 0, 1.38, 2.07; 2.30, 2.99, 3.40, 3.68, 3.91, 4.37, 4.60.

2. (a) 6; **(b)** $\frac{2}{5}$; **(c)** $\frac{9}{8}$; **(d)** 1; **(e)** $6\sqrt{2}$; **(f)** e; **(g)** 5; **(h)** $3\sqrt{e}$.

3. (a) $1/x$; **(b)** $7/x$; **(c)** $1/(x + 7)$; **(d)** $2x/(x^2 + 5)$; **(e)** $(2x + 5)/(x^2 + 5x)$; **(f)** $3(x + 2)/2x(x + 3)$; **(g)** $(2 \ln x)/x$; **(h)** $(3x + 1)/(x - 3)(x + 2)$; **(i)** $2a/(a - t)(a + t)$.

4. (a) $y\left(\frac{1}{x - a} + \frac{2}{x - b} + \frac{3}{x - c}\right)$; **(b)** $y\left(\frac{10x}{x^2 - 3} + \frac{1}{2(x + 1)}\right)$.

5. (a) $\ln |x + a| + C$; **(b)** $\frac{1}{2} \ln |x^2 - 3| + C$; **(c)** $x + \ln |x - 1| + C$.

6. (a) $3 - \ln 16$; (b) $-\ln 2$. 7. (a) 0 (b) 2

Exercises 10.1 [B], pages 374–375

2. (a) 50; **(b)** $\frac{4}{3}$; **(c)** $3e^2$; **(d)** 2, 3; **(e)** 4.

3. (a) $1/(x \ln 10)$; **(b)** $3(\ln z)^2/z$; **(c)** $\tan \theta$; **(d)** $2(1 - 2y^2)/y(1 - y^2)$; **(e)** $(x + 2)/(x - 1)(x + 5)$; **(f)** $\sec x$; **(g)** $1/2t\sqrt{\ln t}$; **(h)** $1/(x \ln x)$; **(i)** $6 \csc 6x$; **(j)** $\ln x$.

4. **(a)** $y\left(\frac{1}{x} - \frac{1}{x+a} - \frac{1}{x-a}\right)$;
(b) $-y/(x-1)(x+1) = -1/(x-1)^{3/2}(x+1)^{1/2}$;
(c) $y(3x-1)/5(x+1)(x-3)$.

5. **(a)** $\frac{1}{2}\ln|2x+7| + C$; **(b)** $x - 3\ln|x+3| + C$; **(c)** $\frac{1}{2}\ln|x^2 - 2x - 3| + C$.

6. **(a)** $-1.5 + 4\ln 2.5$; **(b)** $-\ln 2 = \ln\frac{1}{2}$; **(c)** $\frac{3}{2}$.

7. **(a)** 0 **(b)** -1

Miscellaneous Exercises 10.1, pages 375–376

5. **(a)** $1/(x\ln a)$; **(b)** $\cot\theta$; **(c)** $2/(2x+1)\ln 3$; **(d)** $-2\cot z$; **(e)** $\csc x$; **(f)** $\sec\theta$; **(g)** $1/\sqrt{x^2+1}$; **(h)** $1/x(x^2+1)$; **(i)** $2x/(x^2+3)\ln 10$; **(j)** $-1/y\sqrt{1-y^2}$; **(k)** $(1 + \ln\tan x)\sec^2 x$; **(l)** $-\sec\phi$.

6. **(a)** $\frac{x^2}{2} - 4x + 4\ln|x+1| + C$; **(b)** $\frac{1}{a}\ln|ax+b| + C$; **(c)** $\ln|\sin\theta| + C$; **(d)** $\ln|ax^2+bx+c| + C$.

7. **(a)** $y\left[\frac{2x}{3(x^2+1)} - \frac{3x^2}{2(x^3-4)}\right]$;
(b) $y\left[\frac{1}{x-1} + \frac{2}{3(x+3)} - \frac{1}{2(x+2)} - \frac{x}{x^2-6}\right]$;
(c) $\frac{y}{4}\left[\frac{2}{x} + \frac{6x}{x^2-3} - \frac{2x-3}{x^2-3x+5}\right]$.

8. **(a)** $\frac{1}{2} - \ln 2$; **(b)** $\frac{1}{2}\ln 5$. **9.** $\ln 5$ sq. units. **10.** $4\pi\ln 2$ cu. units.

11. **(a)** $A = (b^{n+1} - 1)/(n+1)$ sq. units; no limit. **(b)** $A = \ln b$ sq. units; no limit. **(c)** $A = (b^{n+1} - 1)/(n+1)$ sq. units; limit $= -1/(n+1)$.

12. **(a)** $V = \pi(b^{2n+1} - 1)/(2n+1)$ cu. units; no limit.
(b) $V = \pi\ln b$ cu. units; no limit.
(c) $V = \pi(b^{2n+1} - 1)/(2n+1)$ cu. units; limit $= -\pi/(2n+1)$.

13. $\frac{27}{2} - 18\ln 2$ sq. units. **14.** $\pi(18 + 16\ln 2)$ cu. units.

Exercises 10.2 [A], pages 381–382

1. **(a)** 0; **(b)** $\ln 2$; **(c)** $\ln 2$; **(d)** 3; **(e)** e; **(f)** e^2; **(g)** e^{-5}; **(h)** 0, $\ln 2$; **(i)** $\ln 2$; **(j)** $\ln(1+\sqrt{3})$.

2. **(a)** $x = e^y$, all y, $x > 0$. **(b)** $x = e^{y/2}$, all y, $x > 0$. **(c)** $x = e^{y/2}$, all y, $x > 0$; or $x = -e^{y/2}$, all y, $x < 0$.

3. **(a)** $2e^{2x}$; **(b)** $-e^{-x}$; **(c)** $-3e^{-3x}$; **(d)** $-2e^{-2x}$; **(e)** $e^x\cos e^x$;
(f) $-\sin x\cdot e^{\cos x}$; **(g)** 1; **(h)** $x^{-2}e^x(x-1)$; **(i)** $3\cdot 10^{3x}\ln 10$;
(j) $x^{-2}x^{1/x}(1-\ln x)$; **(k)** $e^{-x}(\cos x - \sin x)$; **(l)** $1/(e^x+1)$.

4. **(a)** $\frac{1}{3}e^{3x} + C$; **(b)** $-\frac{1}{k}e^{-kx} + C$; **(c)** $\frac{1}{2}e^{x^2} + C$; **(d)** $e^{\sin x} + C$; **(e)** $\ln(e^x + 1) + C$; **(f)** $a^x/\ln a + C$.

5. **(a)** 0 **(b)** 0

Exercises 10.2 [B], pages 382–383

1. **(a)** e^2; **(b)** $e^4/3$; **(c)** no root; **(d)** $\ln 4$; **(e)** $\ln(1 + \sqrt{2})$; **(f)** $\ln(1 + \sqrt{7})$; **(g)** $\pi/2 + 2n\pi$, n any integer; **(h)** $e^{n\pi}$, n any integer.

2. **(a)** $x = \ln y$, $y > 0$, all x. **(b)** $x = y$, $y > 0$, $x > 0$. **(c)** $x = \sin^{-1}y$, $-1 \leq y \leq 1$, $-\pi/2 \leq x \leq \pi/2$, and many other solutions. **(d)** $x = \ln(\sin^{-1} y)$, $0 < y \leq 1$, $x \leq \ln(\pi/2)$, and other solutions.

3. **(a)** $-4e^{-4x}$; **(b)** $2xe^{x^2}$; **(c)** $xe^x(x + 2)$; **(d)** $e^x \sec^2 e^x$; **(e)** $e^x(\ln x + 1/x)$; **(f)** $e^{\sqrt{x}}/2\sqrt{x}$; **(g)** $x^{x^2+1}(2\ln x + 1)$; **(h)** $\left(\frac{1}{t(t+1)} - \frac{\ln(t+1)}{t^2}\right)(1 + t)^{1/t}$; **(i)** $-2e^x/(e^{2x} - 1)$.

4. **(a)** $e^{x^3}/3 + C$; **(b)** $-e^{1/x} + C$; **(c)** $e^{\tan x} + C$; **(d)** $10^x/\ln 10 + C$; **(e)** $\frac{2}{3}(e^x + 1)^{3/2} + C$; **(f)** $\tan^{-1} e^x + C$.

5. 4π cu. units.

6. **(a)** 1 **(b)** $\ln(a/b)$

Miscellaneous Exercises 10.2, page 383

2. **(b)** $1 - e^{-b}$ sq. units; **(c)** 1 sq. unit; **(d)** $\frac{\pi}{2}(1 - e^{-2b})$ cu. units, $\frac{\pi}{2}$ cu. units.

3. **(a)** $(2\ln 2 - 1)$ sq. units; **(b)** $(2\ln 2 - 1)$ sq. units.

4. $(x + 1)e^x$; $(x + 2)e^x$; $(x + 3)e^x$; $(x + n)e^x$. **5.** $\sqrt{2}e^{-1/2}$ sq. units.

7. **(a)** $(e - e^{-1})$ sq. units; **(b)** $\frac{\pi}{4}(e^2 + 4 - e^{-2})$ cu. units.

Exercises 10.3 [A], page 386

1. 1 2. 0 3. 0 4. 0 5. -2
6. 1 7. e^r 8. e^r 9. e^2 10. $-1/6$

Exercises 10.3 [B], page 386

1. 1 2. 0 3. 0 4. 1 5. 1/2
6. e^k 7. 1 8. 0 9. 1 10. 1

Exercises 10.4 [C], pages 388–389

1. $1{,}000{,}000e^{0.1t\ln 1.2} = 1{,}000{,}000(1.2)^{t/10}$.
2. $70 + 80e^{-0.1t\ln 2} = 70 + 80\cdot 2^{-t/10}$ (degrees). **3.** 675 r.p.m.
4. 720 r.p.m. **5.** $Q = Q_0e^{-kt}$. **6.** $\lim_{t\to\infty} i = E/R$.
7. $10e^{t\ln 0.81} = 10(0.81)^t$ (ft.). **8.** $0.1(10 - t)^2$ (ft.). **10.** $40e^{-0.02t}$ (lb.).

Exercises 11.1 [A], pages 392–393

2. $\ln|\sec\theta - 1| + C$. **3.** $\frac{1}{2}\ln|x^2 + 6x + 5| + C$.
4. $\frac{1}{3}\tan^{-1}\frac{x+2}{3} + C$. **5.** $\tan x - \sec x + C$.
6. $2\sqrt{x} - 2\ln(1 + \sqrt{x}) + C$. **7.** $\frac{1}{a}\cos^{-1}\frac{a}{x} + C$. **8.** $\sqrt{x^2 - a^2} + C$.
9. $2\sin^{-1}\frac{x-2}{2} + \frac{x-2}{2}\sqrt{4x - x^2} + C$. **10.** $-\sqrt{\cos^2 z + 3} + C$.
11. $\frac{2}{3}\tan^{-1}(\frac{1}{3}\tan\frac{1}{2}y) + C$. **12.** $\ln\sqrt{x^2 + 2x + 5} - \frac{1}{2}\tan^{-1}\frac{x+1}{2} + C$.
13. $\frac{9}{2}\sin^{-1}\frac{x}{3} - \frac{1}{2}x\sqrt{9 - x^2} + C$. **14.** $\sin^{-1}\frac{x+2}{5} + C$.
15. $\frac{1}{2a^3}\left(\tan^{-1}\frac{x}{a} + \frac{ax}{x^2 + a^2}\right) + C$.

Exercises 11.1 [B], page 393

2. $\frac{x}{8} - \frac{1}{32}\sin 4x + C$. **3.** $\frac{1}{4}\ln|4y^2 - 12y + 7| + C$.
4. $\frac{1}{3}\tan^{-1}\frac{x+1}{3} + C$. **5.** $-\cot\frac{1}{2}x + C$.
6. $\sqrt{e^{2x} - 1} - \cos^{-1} e^{-x} + C = \sqrt{e^{2x} - 1} - \tan^{-1}\sqrt{e^{2x} - 1} + C$.
7. $\frac{1}{3}\sin^{-1}\frac{3x}{2} + C$. **8.** $\frac{1}{4}\cos^{-1}\frac{4}{x} + C$. **9.** $\sin^{-1}(2x + 1) + C$.
10. $\sqrt{x^2 - a^2} - a\cos^{-1}\frac{a}{x} + C$. **11.** $\frac{\sqrt{2}}{2}\tan^{-1}(\sqrt{2}\tan\frac{1}{2}x) + C$.
12. $\ln\sqrt{x^2 - 6x + 18} + \tan^{-1}\frac{x-3}{3} + C$. **13.** $\frac{1}{2}\sin^{-1}\frac{2x+1}{\sqrt{3}} + C$.
14. $\tan^{-1}(\sin x) + C$. **15.** $-\frac{\sqrt{a^2 - x^2}}{a^2x} + C$.

Miscellaneous Exercises 11.2, page 398

2. $\ln\left|\dfrac{x}{x+1}\right|+C.$ **3.** $\ln|x|-\frac{1}{2}\ln(x^2+1)+C.$

4. $\frac{1}{4}\ln\left|\dfrac{x-3}{x+1}\right|+C.$ **5.** $\frac{3}{4}\ln|x-3|+\frac{1}{4}\ln|x+1|+C.$

6. $x+\frac{9}{4}\ln|x-3|-\frac{1}{4}\ln|x+1|+C.$ **7.** $3\ln|x+2|-2\ln|x-4|+C.$

8. $\dfrac{1}{2a}\ln\left|\dfrac{x-a}{x+a}\right|+C.$ **9.** $\dfrac{1}{a}\tan^{-1}\dfrac{x}{a}+C.$

10. $\frac{1}{2}\ln|x^2+2x-8|+C.$ **11.** $\frac{1}{2}\ln|x|-\frac{1}{4}\ln(x^2+2)+C.$

12. $\frac{1}{2}x^2+3x+\frac{9}{4}\ln|x-3|-\frac{5}{4}\ln|x+1|+C.$

13. $x-\frac{8}{3}\ln|x+2|+\frac{1}{2}\ln|x+1|+\frac{1}{6}\ln|x-1|+C.$

14. $\frac{1}{3}\ln|x+1|-\frac{1}{3}\left\{\frac{1}{2}\ln(x^2+2)+\dfrac{\sqrt{2}}{2}\tan^{-1}\dfrac{x\sqrt{2}}{2}\right\}+C.$

15. $\frac{1}{30}\ln\left|\dfrac{x-3}{x+3}\right|+\frac{1}{20}\ln\left|\dfrac{x+2}{x-2}\right|+C.$ **16.** $\frac{1}{4}\ln\left|\dfrac{x-1}{x+1}\right|-\frac{1}{2}\tan^{-1}x+C.$

17. $\frac{1}{5}\ln|x|-\frac{1}{10}\,ln\,(x^2+2x+5)-\frac{1}{10}\tan^{-1}\dfrac{x+1}{2}+C.$

18. $\frac{1}{2}\ln(x^2+1)-\ln|x+1|-\tan^{-1}x-\dfrac{1}{x+1}+C.$

19. $-\dfrac{1}{x}-\tan^{-1}(x+2)+C.$ **20.** $\ln\left|\tan\dfrac{x}{2}\right|+C.$ **21.** $\frac{1}{4}\ln\left|\dfrac{\tan\frac{x}{2}-3}{3\tan\frac{x}{2}-1}\right|+C.$

22. $\ln|\sec x+1|+C.$ **23.** $\ln|x+\sqrt{x^2-a^2}|+C.$

24. $-\dfrac{1}{a}(\ln|a+\sqrt{x^2+a^2}|-\ln|x|)+C.$ **25.** $\frac{1}{2}\cos^{-1}(1/x^2)+C.$

Miscellaneous Exercises 11.3, pages 400–401

1. $e^x(x-1)+C.$ **2.** $\frac{1}{4}x^2(2\ln x-1)+C.$ **3.** $x\sin x+\cos x+C.$
4. $\frac{1}{2}e^{\theta}(\sin\theta+\cos\theta)+C.$ **5.** $z\tan z+\ln|\cos z|-\frac{1}{2}z^2+C.$

6. $x\ln(a^2+x^2)-2x+2a\tan^{-1}\dfrac{x}{a}+C.$ **7.** $x\tan x+\ln|\cos x|+C.$

8. $\frac{2}{15}(x+1)^{3/2}(3x-2)+C.$ **9.** $-\frac{1}{2}e^{-x}(\cos x+\sin x)+C.$
10. $-x^2\cos x+2x\sin x+2\cos x+C.$ **11.** $y^2\sin y+2y\cos y-2\sin y+C.$ **12.** $-x^3\cos x+3x^2\sin x+6x\cos x-6\sin x+C.$

13. $\frac{x}{2}[\sin(\ln x) - \cos(\ln x)] + C$. **14.** $\frac{1}{2}(\ln|\tan\frac{1}{2}x| - \csc x \cot x) + C$.

15. (b) $-\frac{1}{5}\sin^4 x \cos x - \frac{4}{15}\sin^2 x \cos x - \frac{8}{15}\cos x + C$.

Miscellaneous Exercises 11.4, pages 404-405

1. $\frac{3}{2}$. **2.** Diverges. **3.** Diverges. **4.** $\frac{1}{2}$. **5.** 2. **6.** 0.
7. $\frac{10}{3}$. **8.** Diverges. **9.** 6. **10.** 1. **11.** 1. **12.** Diverges.
13. Diverges. **14.** π. **15.** Diverges. **16.** Diverges. **17.** Diverges.
18. $\pi/2$. **19.** a. **20.** $\pi a^2/4$. **21. (a)** $k > -1$; **(b)** $k < -1$.
22. $\frac{1}{2} + \ln 2$. **24. (a)** Diverges; **(b)** π cu. units. **25. (a)** $4\pi a^2$ sq. units; **(b)** $4\pi^2 a^3$ cu. units. **26.** ln 2. **27.** $\frac{1}{2}$. **28.** Diverges.

29. $\frac{1}{2a^2}\ln 2$. **30.** $\pi/2a$. **31.** $\frac{1}{a}\ln(1 + \sqrt{2})$.

Exercises 12.1 [A], page 412

1. (a) $(\sqrt{3}, 1)$; **(b)** $(\sqrt{2}/2, -\sqrt{2}/2)$; **(c)** $(7.7, -6.4)$, approx.; **(d)** $(0, 0)$. Many possible answers for polar coordinates.

8. Ex. 2: $x = a$. Ex. 4: $x^4 + 2x^2y^2 + y^4 - 4x^3 - 4xy^2 + 3x^2 - y^2 = 0$. Ex. 7: $y^2 = 20(x + 5)$.

9. (a) $r = 6\sin\theta$; **(b)** $r^2 = k^2 \csc 2\theta$; **(c)** $r^2 = a^2\cos 2\theta$.

10. (a) $(a, \pi/2)$, $(a, 3\pi/2)$; **(b)** $(1.6, \sin^{-1}\frac{3}{5})$ and the origin.

Exercises 12.1 [B], page 412

1. Many possible answers.

8. Ex. 2: $y = a$. Ex. 5: $(x^2 + y^2)^{3/2} = a^2(x^2 - y^2)$.

10. (a) $(a, \pi/6)$, $(a, 5\pi/6)$, $(a, 7\pi/6)$, $(a, 11\pi/6)$; **(b)** $(1, \pi/4)$, $(1, 5\pi/4)$, and the origin.

Miscellaneous Exercises 12.1, pages 412–413

13. (a) $y^2 = -4a(x - a)$; **(b)** $x\cos\alpha - y\sin\alpha = a$; **(c)** $x = \pm a\sqrt{2}$.
14. $r = -a + b\cos\theta$. **15. (a)** $(0, 0)$, $(\sqrt{2}/2, \pi/4)$.
16. (a) $(0, 0)$, $(a + a\sqrt{2}/2, 3\pi/4)$, $(a - a\sqrt{2}/2, 7\pi/4)$; **(b)** $\{(r, \theta): r = a\tan\theta\}$.

Exercises 12.2 [A], pages 414–415

1. $d = \sqrt{r_1^2 + r_2^2 - 2r_1r_2 \cos(\theta_1 - \theta_2)}$. 2. (a) $r = a \sec\theta$; (b) $r = a \csc\theta$.

3. (a) $r = 2a\cos\theta$; (b) $r = 2a\sin\theta$; (c) $r = a\sqrt{2}(\sin\theta + \cos\theta)$.

4. (a) $r = \dfrac{6}{1 - \cos\theta}$; (b) $r = \dfrac{10}{1 + \cos\theta}$; (c) $r = \dfrac{12}{1 + \sin\theta}$.

5. $|\alpha - \beta|$. 6. $(8/3, 0)$. 7. $(2, \pi)$, $(6, \pi)$. 10. $r = a\cos\theta$.

Exercises 12.2 [C], pages 415–416

1. $r = a\sin 2\theta$. 2. $r = a(1 + \cos\theta)$. 3. $r = 2a(\sec\theta - \cos\theta)$.
4. $r = a\cos\theta \pm b$. 5. $r^2 = 2a^2\cos 2\theta$. 6. $r = a\sec\theta \pm b$.
7. $r = a(\sec\theta \pm \tan\theta)$. 8. Part of $r = p\cos\theta$.
9. $r^4 - 2a^2r^2\cos 2\theta + a^4 = b^4$.
10. $r_1r_2\sin(\theta_1 - \theta_2) + r_2r\sin(\theta_2 - \theta) + rr_1\sin(\theta - \theta_1) = 0$.

Exercises 12.3 [A], page 419

2. (a) $\tan^{-1}(\pi/2)$, $\tan^{-1}\pi$; (b) $\tan^{-1}(-\frac{1}{3})$; (c) $\tan^{-1}\left(\cot\theta_1 - \dfrac{1}{e}\csc\theta_1\right)$.
5. $\tan^{-1}\frac{1}{2}$.
6. (a) $(0, 0)$, $(3a/2, 2\pi/3)$, $(3a/2, 4\pi/3)$; (b) $(a/2, \pi/3)$, $(2a, \pi)$, $(a/2, 5\pi/3)$.
7. $\dfrac{a}{k}(e^{2k\pi} - 1)\sqrt{k^2 + 1}$ units.

Exercises 12.3 [B], page 419

1. (a) $-\pi/6$; (b) $\pi/2$; (c) $-\pi/4$. 2. $\pi/3$. 4. $\tan^{-1} 2$.
5. $a(1 + \sqrt{3})$ units. 6. (a) $ak\sin\theta$; (b) $2ak\,|\sin\frac{1}{2}\theta|$.

Miscellaneous Exercises 12.3, page 420

1. $\dfrac{8a}{3}[(1 + \pi^2)^{3/2} - 1]$ units. 2. $3\pi a/2$ units. 3. (a) $\pi/4$; (b) infinitely many; (c) $\sqrt{2}$ units. 4. $8a$ units. 5. (a) $\tan^{-1} n\sqrt{3}$; (b) $\pi/2$.

Exercises 12.4 [A], page 422

1. πa^2 sq. units. 2. πa^2 sq. units. 3. $3\pi a^2/2$ sq. units.
4. $9\pi/2$ sq. units. 5. $2a^2$ sq. units. 6. $\frac{1}{8}\pi a^2$ sq. units.
7. $\frac{1}{2}a^2(1 + \sqrt{3})$ sq. units.

Exercises 12.4 [B], page 422

1. $3\pi a^2/2$ sq. units. **2.** a^2 sq. units. **3.** 11π sq. units.
4. $3\pi a^2/2$ sq. units. **5.** $\pi a^2/4n$ sq. units. **6.** $\frac{1}{4}$.

Miscellaneous Exercises 12.4, page 422

1. $\pi a^2/2$ sq. units. **2.** $8a^2/3$ sq. units. **3.** **(a)** $\frac{1}{4}$; **(b)** $\frac{1}{2}$.
4. $(\pi + 3\sqrt{3})$ sq. units. **5.** $\frac{4}{3}$ sq. units.

Exercises, page 435

2. (**a**) $(7, 2)$ (**b**) $(-11, 9)$ (**c**) 5
3. (**a**) $(3, 1)$ (**b**) $(2, -4)$
4. (**a**) $\mathbf{r}(t) = t(2, 1) + (-1, 2)$; $x = 2t - 1$, $y = t + 2$; $x - 2y + 5 = 0$
   (**b**) $\mathbf{r}(t) = t(4, 2) + (3, 4)$; $x = 4t + 3$, $y = 2t + 4$; $x - 2y + 5 = 0$
   (**c**) $\mathbf{r}(t) = t(-1, 3) + (2, 0)$; $x = -t + 2$, $y = 3t$; $3x + y - 6 = 0$
   (**d**) Same as (**c**).

Index

ABCDEFGHIJ-08543
PRINTED IN THE UNITED STATES OF AMERICA